McGRAW-HILL MATHEMATICS

DOUGLAS H. CLEMENTS

KENNETH W. JONES

LOIS GORDON MOSELEY

LINDA SCHULMAN

McGraw-Hill School Division

New York Farmington

PROGRAM AUTHORS

Dr. Douglas H. Clements

Kenneth W. Jones

Lois Gordon Moseley

Dr. Linda Schulman

CONTRIBUTING AUTHORS

Christine A. Fernsler

Dr. Liana Forest

Dr. Kathleen Kelly-Benjamin

Maria R. Marolda

Dr. Richard H. Moyer

Dr. Walter G. Secada

MULTICULTURAL AND EDUCATIONAL CONSULTANTS

Rim An

Sue Cantrell

Mordessa Corbin

Dr. Carlos Diaz

Carl Downing

Linda Ferreira

Judythe M. Hazel

Roger Larson

Josie Robles

Veronica Rogers

Telkia Rutherford

Sharon Searcy

Elizabeth Sinor

Michael Wallpe

Claudia Zaslavsky

COVER PHOTOGRAPHY Jade Albert for MMSD; t.m. John Eastcott and Yva Monatiuk/Animals/Animals; l. MMSD.

PHOTOGRAPHY CREDITS All photographs are by the Macmillan/McGraw-Hill School Division (MMSD) and Scott Harvey and Ken Karp for MMSD except as noted below.
Table of Contents iii: m. Tom Sanders/Stock Market; • iv: t. Lee Snyder/Photo Researchers; m. Rafael Macia/Photo Researchers; b. Henry Horenstein/Viesti Associates; • v: m. Ed Bock/The Stock Market; b. Mike Powell/Allsport; • vi: m. Superstock, Inc.; • vii: m. Peter Skinner/Photo Researchers; b. Susan McCartney/Photo Researchers; b. Photo Researchers; • viii: m. Coco McCoy/Rainbow; b. Superstock, Inc.; • ix: t. Tom & Pat Leeson/Photo Researchers; m. Archivo Cameraphoto Venezia/Art Resource, Inc.; b.r. Bob Daemmrich/The Image Works; • x: m. Richard Pranitzke/Photo Researchers; b. PhotoDisc; • xi: t. Cynthia Johnson/Gamma Liaison; **Chapter 1** 7: Johan Elbers/Sygma; b. Perkins/Beckett, Summer 1989, American Museum of Natural History; • 8: Hiro Komine; b. Dave Mager for MMSD; • 12: Yoichiro Miyazaki/FPG International; Tom Sanders/The Stock Market; • 13: Reuters/Corbis-Bettmann; • 18: t. AP/Wide World Photos; • 19: P & F Communications for MMSD; • 20: Pierre Perrin/Sygma; • 27: Nick Dolding/Tony Stone Images; • 30: t.r. Peter J. Bryant/Tony Stone Images; m.r. Jeff Greenberg/Photo Researchers, Inc.; m.l. Tim Davis/Photo Researchers, Inc.; b. Patrick Forestier/Sygma; • 36: t. Daniel J. Cox/Natural Exposures; m. Charles V. Angelo/Photo Researchers, Inc.; • 36-37: Tim Davis/Photo Researchers, Inc.; b.l. Michael Durham/Ellis Nature Photos; • 37: b. A.H. Rider/Photo Researchers, Inc.; t. Daniel J. Cox/Liaison International; **Chapter 2** 38: b. Henley & Savage/The Stock Market; • 38-39: Wes Thompson/The Stock Market; • 44: Eising/StockFood America; • 48: Corbis-Bettmann; • 49: Corbis-Bettmann; • 56: t.l. Sepp Seitz/Woodfin Camp and Associates; t.r. Zigy Kaluzny/Woodfin Camp and Associates; b.r. Timothy Eagen/Woodfin Camp and Associates; • 62: Dept. of the Treasury Bureau of Engraving; • 68: Jose Carrillo/Stock Boston; • 69: Graham Lawrence/Picture Perfect; • 72: Sepp Seitz/Woodfin Camp & Assoc.; • 78: t.m. Lee E. Battaglia/Photo Researchers; • 78: b.l. Ancient Art and Architecture Collection; • 79: t.l. Herbert W. Booth III/Liaison International; m. Ken Cavanagh/Photo Researchers; b.r. Lee F. Snyder/Photo Researchers; r.m. Russell D. Curtis/Photo Researchers; **Chapter 3** 80-81: Henry Horenstein/Viesti Associates; • 80: m. i. Myrleen Cate/Tony Stone Images; • 80–81: Mark Stephenson/Westlight; • 82: Lawrence Migdale; • 87: b. Michele Burgess/The Stock Market; • 89: Jason Bleibtreu/Sygma; • 93: Peter Fisher/The Stock Market; • 100: Photographer Photos Co./Gamma-Liaison; • 101: b.l. Tom McHugh/Photo Researchers; b.m. Tim Flach/Tony Stone Images; • 106: t. Susanah Druck; • 109: North Wind Picture Archive; • 112: t.l. George A. Dillon/Stock Boston, Inc.; t.r. Monkmeyer/Collins; • 113: Lawrence Migdale.• 119: Ned Matura/Liaison International; **Chapter 4** 122: Stephen Frisch/Stock Boston; • 129: t.r. Bob Daemmrich/The Image Works; • 132: t. Bob Daemmrich/Stock Boston, Inc.; m. Bob Daemmrich/Stock Boston, Inc.; • 134: t. Dave Tillotson/The Houston Chronicle; • 137: t. Archive Photos; b. Bridgeman/Art Resource, Inc.; • 144: b. Ed Bock/Stock Market; t.r. Jerry Koontz/The Picture Cube, Inc.; • 145: Laura Dwight; • 146: Daemmrich/Stock Boston; • 153: Eric Neurath/Stock Boston; • 158: LeDuc/Monkmeyer; • 160: t.m.r. Bob Krist/Stock Market; • 160: b.r. Sonlight Images for MMSD; • 161: John Terence Turner/FPG International; • 166: t.r. Tom McHugh/Photo Researchers; m.l., b.r. Don Mason/The Stock Market; • 167: t.l. Tom Brakefield/The Stock Market; m.r. Dan Gair/The Picture Cube, Inc.; t.m. J. Giannotti/The Picture Cube, Inc.; b.m. ZEFA-Cutouts/The Stock Market; b.l. Kennan Ward/The Stock Market; **Chapter 5** 168: b. Oz Charles; • 168-169: PhotoDisc; • 170: Erik S. Lesser/Silver Image; • 171: UPI/Corbis-Bettmann; • 172: Mike Powell/Allsport USA; • 181: Picture Network International, Ltd. • 188: Superstock, Inc.; • 189: William Taufic/The Stock Market; • 191: Roger Wood/(c) Corbis; • 192: Dennis Cox/ChinaStock; • 193: Steven Ferry; • 194: t. Dennis MacDonald/PhotoEdit; • 195: from Marketing; • 196: l. Chris Sorensen/The Stock Market; b.r. James Kozyra/Liaison International; • 197: t.r. Superstock, Inc.; • 202: l. Calvin Larsen/Photo Researchers; • 202: r. Liane Enkens/Stock Boston, Inc.; **Chapter 6** 204: b. Seth Resnick/Liaison International; • 204-205: John Matchett/Picture Perfect; • 206: D. Young-Wolff/PhotoEdit; • 213: John M. Roberts/Stock Market; • 214: Gary Braasch/Woodfin Camp & Assoc.; • 220: t. Nick Vedros, Vedros & Assoc./Tony Stone Images; b. Tony Freeman/Photo Edit; • 220: t.l. C.T. Tracy/FPG International; • 223: David Lissy/FPG International; • 226: t.r. courtesy Jeff Mantus; • 229: m.r. courtesy Megan McAtee; • 230: t.l. Eunice Harris/The Picture Cube; 231: b.r. Ellen B. Senisi/Photosynthesis; **Chapter 7** 240-241: Roger Steene/Ellis Nature Photography; • 240: i. Pedro Coll/The Stock Market; • 242: t.r. Courtesy of Kim Harrison; • 244: m. Peter Skinner/Photo Researchers; b. Fred McConnaughey/Photo Researchers; • 247: t. Jeremy Stafford-Deitsch/Ellis Nature Photography; Tom McHugh/Steinhart Aquarium/Photo Researchers; • 256: t. David Lissy/Folio; • 257: t.l. Charles V. Angelo/Photo Researchers, Inc.; t.m.l. PhotoDisc; b.r. Brent Petersen/The Stock Market; • 261: Lawrence Migdale/Stock Boston, Inc.; • 262: Bill Barley/Photri; • 263: m.r. Suzanne and Joseph Collins/Photo Researchers, Inc.; t.r. Kennan Ward/The Stock Market; b.l. David Young-Wolfe/PhotoEdit; • 263: b.r. Superstock, Inc.; • 265: b. Bob Gibbons/Photo Researchers; • 265: Mark Richards/PhotoEdit; • 268: b.l. Jon Feinaersh/The Stock Market; m.l. Tony Freeman/PhotoEdit; t.r. David McGlynn/FPG International; t.l. Tom McHugh/Steinhart Aquarium/Photo Researchers; r.m.b. Michael Gadomski/Photo Researchers; b.r. Tony Freeman/PhotoEdit; • 274: Stephen Dalton/Photo Researchers; **Chapter 8** 276: m. J. Mil Stein/The Image Works; • 278: t. Dowling/Gamma Liaison; • 280: t. Gary Wolinski/Stock Boston, Inc.; • 281: t. Joseff Neltu/Stock Boston, Inc.; • 290: t.r. Toyohiko Yamada/FPG International; • 293: b.r. Erich Lessing/Art Resource, Inc.; • 297: t. Chuck Pefley/Stock Boston, Inc.; • 298: b.r. PhotoDisc vol. 1; b.l. PhotoDisc vol. 24; t.l. PhotoDisc vol. 1; t.r. Stephanie Myers; • 300: t. Focus On Sports; • 308: t.r. Archive Photos; • 310: m.l. Bob Randall for MMSD; b.r. Coco McCoy/Rainbow; t.r. Bob Raskid/Monkmeyer; m. Superstock, Inc.; m. Phyllis Picordi/Stock Boston, Inc.; • 316: b.l. Photo Researchers; • 316: t.r. Michael Holford; • 317: b.l. Coco McCoy/Rainbow; **Chapter 9** 318: Art Wolfe/Tony Stone Images; • 320: t.r. Superstock, Inc.; • 323: t.r. Will & Deni McIntyre/Photo Researchers; • 326: Sinclair Stammers/Photo Researchers; t.r. Stephen Johnson/Tony Stone Images; 328: Joseph Stella/Art Resource, Inc.; 329: m. Tom & Pat Leeson/Photo Researchers; l. Hans Lutz/Okapia/Photo Researchers; • 332: t.m. Tom & Pat Leeson/Photo Researchers; l. Jim Levin for MMSD.• 334: m. Photri/The Stock Market; r. Rockefeller Center Photo; r. David Young-Wolfe/Photo Researchers; • 336: t.l. NASA; • 338: Cordon Art Museum B.V.; b. Robert Frerck/Tony Stone International; r. NYU/Peter Freed; Cordon Art Museum B.V.; • 340: r. Denver Art Museum Collection; • 344: Jim Harrison/Stock Boston, Inc.; • 345: m.l. Jim Corwin/Stock Boston, Inc.; m.r. Alfred Pasieka/Science Photo Library/Photo Researchers; l. Raymond Barnes/Tony Stone Images; • 346: Joel Gordon; • 351: Wolfgang Hoyt; • 355: b.l. Michal Heron/The Stock Market; • 361: Grenoble Musee/Art Resource, Inc.; • 362: b.r. European Photo/FPG International; t.l. Scala/Art Resource, Inc.; b.l. Archivo Cameraphoto Venezia/Art Resource, Inc.; t.r. Eric Corle/Stock Boston, Inc.; • 363: Coco McCoy/Rainbow; **Chapter 10** 373: Allen/Liaison International; • 379: b. William S. Nawrocki/Nawrocki Stock Photo, Inc.; • 380: t.r. Universtiy Museum Archives/University of PA; • 382: t.l. Mario Ruiz/The Stock Market; • 384: t.r. Brian Smith; • 400: b.l. Monkmeyer/Leduc; b.l. The Stock Market; • 407: Jacob Taposchauer/FPG International; **Chapter 11** 408: m.i. Bill Aron/Photo Edit; bkgnd. Aaron Rezny/The Stock Market; • 415: b.r. Richard Pranitzke/Photo Researchers; • 420: t.r. Collins/Monkmeyer; • 423: Giraudon/Art Resource; • 428: t. courtesy Andre Pierre Lincy; • 438: b.r. David Ball/The Picture Cube; t.r. Don Mason/The Stock Market; • 439: b.l. Constance Hansen/The Stock Market; b.l. Al Assid/The Stock Market; b.l. Esbin Anderson/The Image Works; • 440: b.r. Gabe Palmer/The Stock Market; b.r. Andy Levin/Photo Researchers; b.l. Elena Roonaid/PhotoEdit; m.r. Jose Paiaez/Stock Market; • 446: b.l. D. Young-Wolff/PhotoEdit; b.r. Eising/Stock Food America; David Young-Wolff/PhotoEdit; • 447: r., t.l. Photo Disc vol. 8; Chapter 12 • 448: m. Cynthia Johnson/Gamma Liaison; • 454: t.r. Thayer Syme/FPG International; • 456: Kevin Schafer/Tony Stone Images; • 462: Fredrik D. Bodin/Stock Boston, Inc.; • 464: t. Michelle Michaels; • 468: t.l. Tom & Dee Ann McCarthy/The Stock Market; • 470: m. Roger Ressmeyer/Corbis; t.r. NASA/Photo Researchers; • 474: b.r. Michael Newman/PhotoEdit; • 480: m.r., m.l. Bettmann/Corbis; • 481: Frank Rossotto/The Stock Market.

ILLUSTRATION CREDITS Bernard Adnet: 249, 250 • Ken Bowser: 205, 208, 219, 220, 224, 409, 414, 424, 433, 449, 452, 462, 479 • Hal Brooks: 2. 12. 17. 20. 25. 28 • Tom Cardamone: 7, 87, 137, 191, 202, 213, 265, 293, 297, 327, 343, 379, 415, 465 • Susan Johnston Carlson: 277, 278. 280, 285, 291, 292, 294, 297, 300, 302, 308 • Anthony Cericola: 252, 258, 264, 268, 319, 322, 336, 351 • Eldon Doty: 473 • Brian Dugan: 18, 98, 217, 270, 388, 389, 406, 407 • Annie Gusman: 26. 35. 150 160, 161, 166 • Henry Hill: 243-247. 269. 271, 274, 275 • WB Johnston: 254, 266, 267, 334, 348, 368-370. 398 • Stanford Kay: 90, 96, 118, 349, 450, 453, 459, 473-475 • Deborah Haley Melmon: 40, 47 • Marion Nixon: 220, 230, 231, 393 • Lori Osieki: 70, 71 • Michael Racz: 20, 21 • Margaret Robinson: 202 • Douglas Schneider: 354 • Stephen Schudlich: 52, 53, 55, 58, 72, 76 • Remy Simard: 84-86, 93, 94, 108, 112, 113, 117, 124, 136, 138, 165 • Matt Straub: 152, 248, 259, 273, 280, 291, 295, 302, 308, 309. 315. 380, 395 • Beata Szpura: 14, 15 • Gary Torrisi: 324-326 • Joe VanDerBos: 3. 33. 39, 42. 43. 63, 77, 214, 215, 233 • Nina Wallace: 104, 105, 114, 125, 126, 129, 130, 132, 151, 153, 156, 162, 164, 186, 187, 196, 201 • Susan Williams: 183, 196, 201, 372, 373, 375, 405 • Rose Zgodzinski: 209, 223, 233, 367, 379, 383

McGraw-Hill School Division

A Division of The ***McGraw-Hill*** *Companies*

McGraw-Hill School Division
1221 Avenue of the Americas
New York, New York 10020

Printed in the United States of America
ISBN 0-02-110319-4 / 4
5 6 7 8 9 043/071 04 03 02 01 00 99

Contents

1 Place Value and Number Sense

2 Money, Addition, and Subtraction

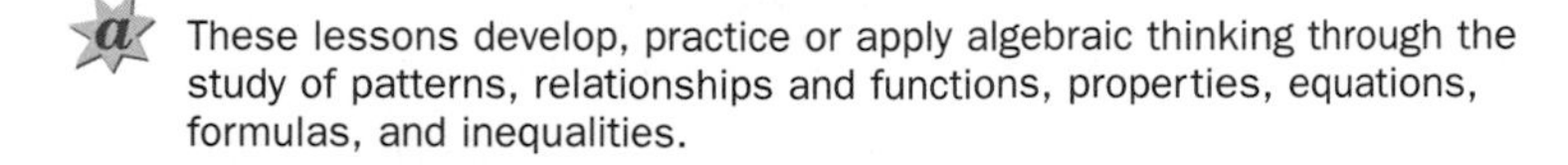

These lessons develop, practice or apply algebraic thinking through the study of patterns, relationships and functions, properties, equations, formulas, and inequalities.

3 Time, Data, and Graphs

4 Multiplication and Division Facts

 These lessons develop, practice or apply algebraic thinking through the study of patterns, relationships and functions, properties, equations, formulas, and inequalities.

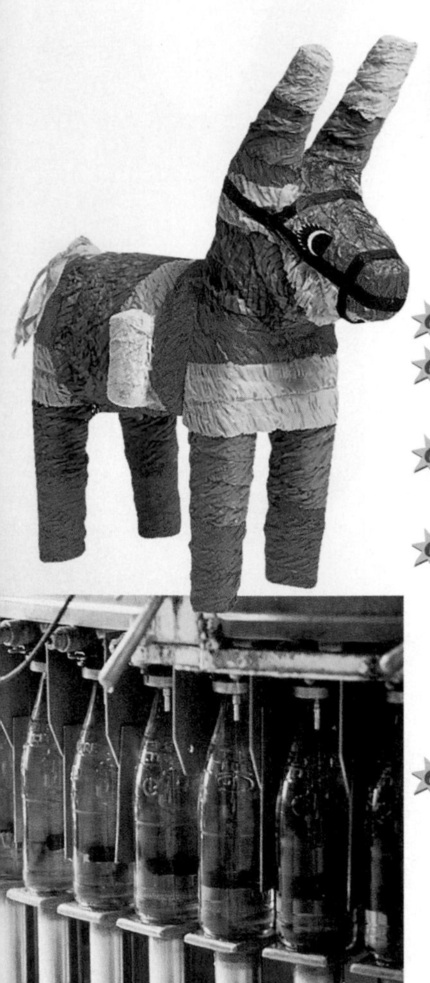

5 Multiply by 1-Digit Numbers

6 Multiply by 2-Digit Numbers

7 Measurement

These lessons develop, practice or apply algebraic thinking through the study of patterns, relationships and functions, properties, equations, formulas, and inequalities.

8 Divide by 1- and 2-Digit Numbers

9 Geometry

10 Fractions and Probability

 These lessons develop, practice or apply algebraic thinking through the study of patterns, relationships and functions, properties, equations, formulas, and inequalities.

11 Using Fractions

12 Decimals

These lessons develop, practice or apply algebraic thinking through the study of patterns, relationships and functions, properties, equations, formulas, and inequalities.

CHAPTER 1

PLACE VALUE AND NUMBER SENSE

THEME

Amazing Facts

People are always looking for ways to break world records. In this chapter, you will learn about some world records and other amazing facts. You may even create your own record!

What Do You Know 

Estimate the number of stars you think you can draw in 1 minute.

1 Work with a partner to test your estimate. Time each other to see how many stars each of you can really draw in 1 minute. Copy and complete the table.

Minute Draw

Item	Estimate	Exact Number

2 Write a statement that compares your estimate to the exact number of stars you drew.

3 Portfolio Josh's group drew 134 stars in 1 minute. Show how you can group 134 things so that they are easy to count. Draw a picture of the grouped stars or of place-value models.

Use Illustrations

Tess drew 25 shapes in one minute. What shapes do you think she drew? Illustrate them.

When you are reading, the illustrations can give you a better idea of what the author means. Sometimes a picture gives additional information.

1 What information does your illustration give that is not in the paragraph?

2 How could you illustrate the paragraph to show tens and ones?

Vocabulary

place value, p.4
digit, p.8
standard form, p.8
expanded form, p.8
period, p.8
is less than (<), p.12
is greater than (>), p.12
round, p.20
line plot, p.28
data, p.28

Numbers in Your World

In the newspaper you can find many ways that numbers are used in our world.

Work Together

Work in a group to read two or three pages in a newspaper. Record as many phrases with numbers in them as you can find.

Decide on different ways or categories you will use to organize your numbers.

Create a table like the one below. List your categories. Tally, then count the number of examples in each category. Include an example from the newspaper for each category.

▶ How did you choose the categories that you used?

Cultural Note

The first printed newspaper was a Chinese circular, *The Peking Gazette,* that was printed around A.D. 700.

Make Connections

Mike's group organized their numbers this way.

How We Organized the Numbers We Found

Category	Tally	Number of Examples	Example in Newspaper
To count	𝍸 𝍸 ||	12	343,000 bees
To show order	|||	3	The 23rd Olympic Games
To name	𝍸 𝍸 𝍸	15	WR 67 Derbyshire Fire & Rescue Service
To measure	𝍸 𝍸 𝍸 𝍸 |	21	80 pounds

- What categories did your group use? How are they different from those shown on page 2?

- Find examples from your data that can be placed in the categories Mike's group chose.

Check for Understanding

CHECK

Tell if the number is used to count, show order, name, or measure.

1. The World Trade Center in New York City has about 50,000 people working in it and about 90,000 visitors daily.

2. Bernard Lavery grew a cabbage that weighed 124 pounds.

3. The first American woman in space was Sally Ride.

Critical Thinking: Summarize **Explain your reasoning.**

4. Give examples of how numbers can answer questions such as *What? When? Where? How big?* and *How many?*

5. What other types of questions can be answered using numbers?

Practice

PRACTICE

Tell if the number is used to count, show order, name, or measure.

1. The largest single windows are 164 feet high.

2. About 350 million people in the world visit zoos each year.

3. The first woman to climb Mount Everest was Junko Tabei.

4. The survey shows that people prefer the Jaguar XJ7.

5. George Adrian picked 15,830 pounds of apples in 8 hours.

6. The New York City subway system has 469 stations.

7. Write four newspaper headlines. Use numbers to count, show order, name, and measure. Have other students categorize each number.

8. **Data Point** Predict the number that you read, say, or hear most often in a day. Use a tally table to record all the numbers you find in a day. Was your prediction correct?

Building to Thousands

You can use place-value models to show numbers in different ways.

> **You will need**
> - *ones, tens, hundreds, and thousands place-value models*

Work Together

Work with a partner. Use a table to record your results.

Number	Thousands	Hundreds	Tens	Ones
43 (ones only)				
43 (tens and ones)				

> **place value** The value of a digit depending on its place in a number.

Show 43 using:
- ▶ only ones models.
- ▶ tens and ones models.

Show 121 using:
- ▶ only tens and ones models.
- ▶ hundreds, tens, and ones models.

Show 1,310 using:
- ▶ only hundreds and tens models.
- ▶ thousands, hundreds, and tens models.

Talk It Over

- ▶ How would you show each of these numbers using the least number of models? the greatest number of models?

A 1,001

B 1,010

C 1,100

Make Connections

A place-value chart can help you understand numbers greater than 1,000.

You can think of 1,257 as:
1 thousand 2 hundreds 5 tens 7 ones,
or 12 hundreds 5 tens 7 ones,
or 1,257 ones.

thousands	hundreds	tens	ones
1	2	5	7

▶ How can you find 1 hundred more than 1,257? 1 hundred less? What are the new numbers? Explain your reasoning.

Check for Understanding

Write the number.

1 

2 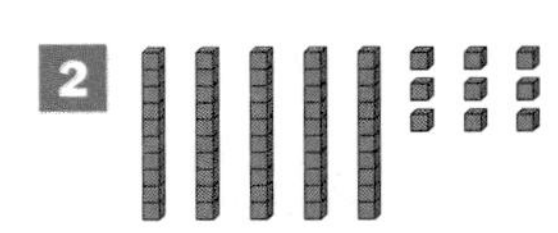

3 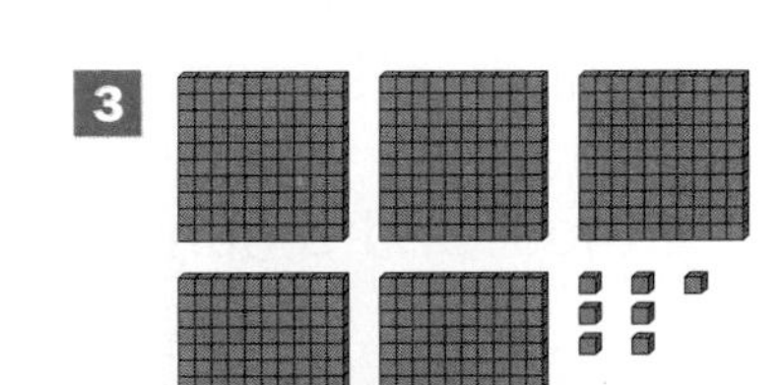

4 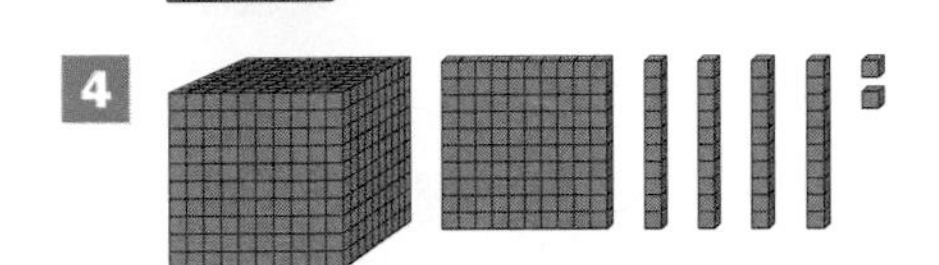

5 

6 9 hundreds 5 tens 8 ones

7 1 thousand 7 hundreds 9 tens

Complete.

8 1,082 = 1 thousand ■ tens 2 ones

9 ■ = 1 thousand 5 tens

10 958 = 9 ■ 5 tens 8 ones

11 131 = 1 hundred 3 tens ■ ones

12 1,900 = 1 ■ 9 ■

13 460 = 4 hundreds 6 ■

Critical Thinking: Generalize **Explain your reasoning.**

14 Write steps that explain to another student how you would find the following:

a. 10 more than 765 **b.** 100 less than 765 **c.** 1,100 more than 765

CHECK

Turn the page for Practice.

Practice

Write the number.

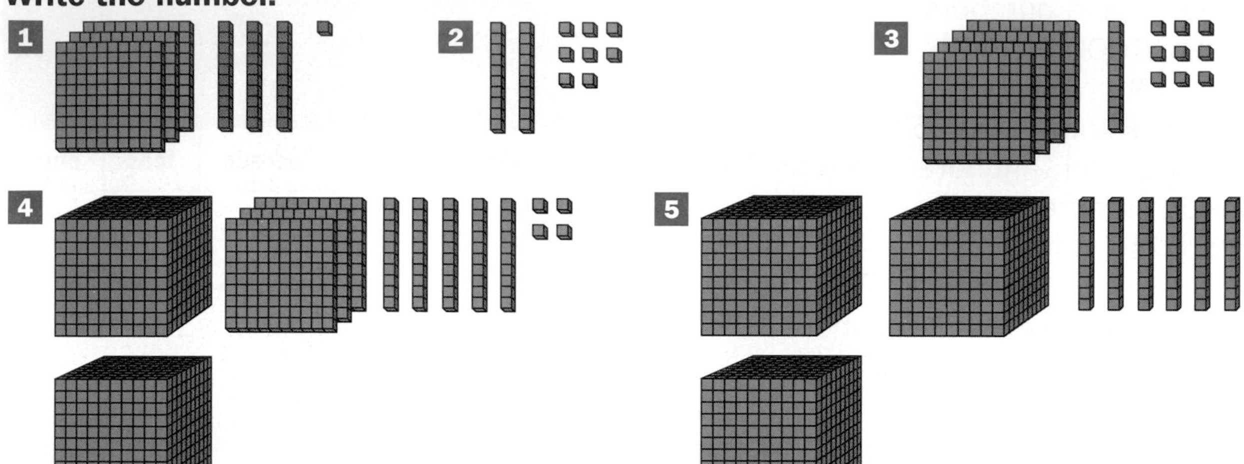

6. 1 thousand 4 hundreds 1 ten 2 ones
7. 8 hundreds 9 tens 2 ones
8. 1 hundred 9 tens
9. 3 hundreds 4 tens
10. 6 hundreds 5 ones
11. 1 thousand 3 hundreds
12. 1 thousand 9 tens
13. 1 thousand 5 ones
14. 87 tens 5 ones
15. 14 hundreds
16. 100 tens

Complete.

17. 385 = ■ hundreds 8 tens 5 ones
18. 1,090 = 1 thousand ■ tens
19. 990 = 9 hundreds 9 ■
20. 1,504 = 1 thousand 5 ■ 4 ones
21. ■ = 7 hundreds 3 ones
22. ■ = 1 thousand 7 ones
23. Find 10 more than 1,719.
24. Find 10 less than 821.
25. Find 100 more than 652.
26. Find 100 less than 1,050.
27. Find 1 more than 800.
28. Find 1 less than 1,000.
29. Find 1,000 more than 604.
30. Find 1,000 less than 1,229.

Make It Right

31. Luis wrote the number in standard form. Explain what the mistake is.

7 hundreds 8 ones = 78

MIXED APPLICATIONS
Problem Solving

32 Una tiled her bathroom floor. She used square mats of 100 tiles, strips of 10 tiles, and some single tiles. If she used 9 mats, 8 strips, and 7 single tiles, how many tiles did she use in all?

33 **Logical reasoning** I am a four-digit number. My hundreds digit is less than 1. You write my tens and ones digits as 67 ones. If you add my digits you get 14. What number am I?

34 Each story of the World Trade Center is about 10 feet (ft) high. About how many stories high was Philippe Petit? SEE INFOBIT.

35 **Data Point** Survey your classmates to find which season they like the most. Show your results in a tally. How many students like spring most?

INFOBIT
In 1974, Philippe Petit walked across a high wire between the towers of the World Trade Center. At 1,350 feet, this was the highest high-wire feat ever done.

Cultural Connection Inca Quipu

About 500 years ago, the Inca lived along the western coast of South America. They kept records using *quipus* (KEE-pooz) like the one shown.

Knots were tied in different positions on each cord to show hundreds, tens, and ones.

hundreds
tens
ones
26 143

SOUTH AMERICA
Inca Empire

Give the number that is shown on the cords.

1 tens, ones

2 tens, ones

3 hundreds, tens, ones

Thousands

LEARN

IN THE WORKPLACE

Sharon O'Connell is the founder of the Sadako/Paper Crane Project.

Do you know that over 435,500 origami paper cranes are hanging from a memorial at Peace Park in Hiroshima, Japan? In the United States, the Sadako/Paper Crane Project collects origami cranes to donate to the Peace Park.

A place-value chart shows the value of the **digits** in a number.

Each group of three digits is called a **period.**

Thousands Period			Ones Period		
Hundreds	Tens	Ones	Hundreds	Tens	Ones
4	3	5	5	0	0

You can read or write numbers in different ways.

Standard Form: 435,500 **Note:** Commas are used to separate periods.

Word Name: four hundred thirty-five thousand, five hundred

Expanded Form: 400,000 + 30,000 + 5,000 + 500

Check Out the Glossary
- digit
- period
- standard form
- expanded form

See page 544.

Talk It Over

- How do the places in the thousands period relate to each other? How are they similar to the places in the ones period? How do the places help you read the number?
- What is the value of each digit in 435,500? How are the values used in the expanded form?

Cultural Note

In Japan, it is customary to make a garland of origami paper cranes for a friend who is ill.

What if you have 629,509 folded-paper cranes. How many cranes will you have if you make 10 more? 100 more?

A place-value chart can help you see changes in the places of a number.

Thousands Period			Ones Period			
Hundreds	Tens	Ones	Hundreds	Tens	Ones	
6	2	9	5	0	9	
6	2	9	5	1	9	← **10 more cranes**
6	2	9	6	0	9	← **100 more cranes**

▶ How would you show 200 fewer cranes? 2,000 more cranes?

More Examples

A 240,000 can be written as two hundred forty thousand.

B 57,480 can be written as 50,000 + 7,000 + 400 + 80.

C 800,975 can be written as 800 thousand, 975.

D 2,000 less than 690,000 is 688,000.

E 10 more than 79,995 is 80,005.

Check for Understanding

Write the word name for the number. Then write the number in expanded form.

1 8,425 **2** 19,841 **3** 75,003 **4** 605,911

5 Find 300 less than 8,769. **6** Find 4,000 more than 21,695.

Critical Thinking: Analyze **Explain your reasoning.**

7 Aida says you can rename 35,725 as 3 + 5 + 7 + 2 + 5. Do you agree or disagree?

8 When you increase a number by 1,000, do you need to change only the thousands place? Show an example to support your reasoning.

CHECK

Turn the page for Practice.

PRACTICE

Practice

Write the number in standard form and in expanded form.

1

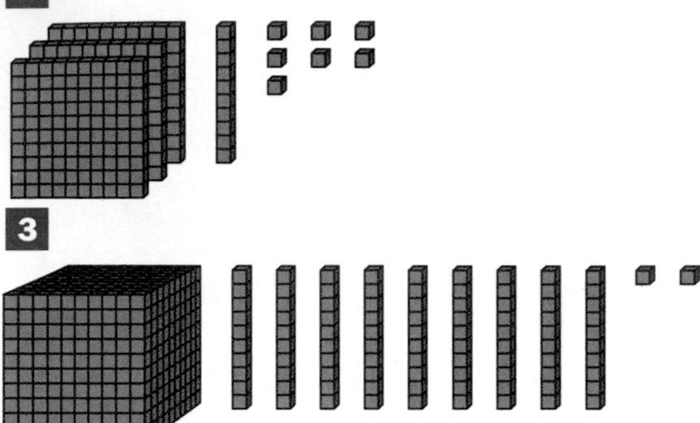

2

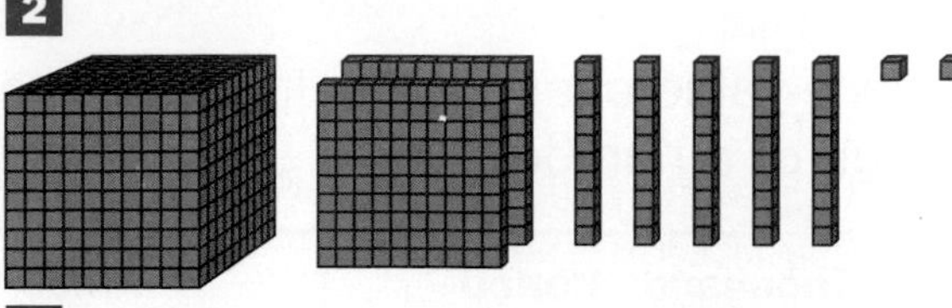

3

4

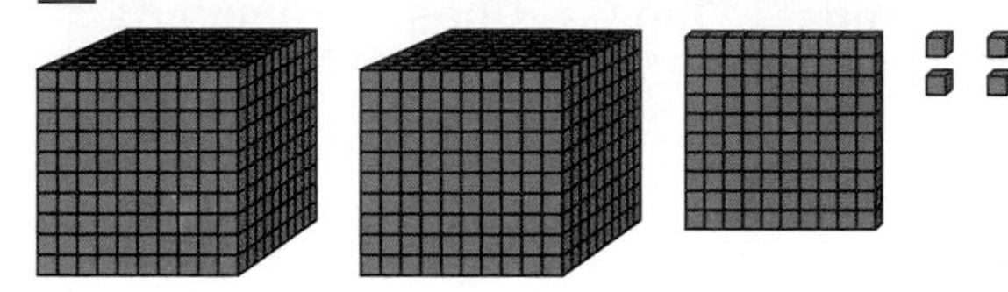

5 four hundred twenty

6 eight thousand, thirteen

7 twenty-three thousand

8 six hundred thousand

9 seven hundred nine thousand

10 four hundred thousand, five

Write the word name for the number. Then write the number in expanded form.

11 478

12 6,125

13 9,501

14 74,076

15 81,250

16 110,560

17 460,203

18 903,511

ALGEBRA Find the rule. Then complete the table.

19

Rule: ■					
1,009	7,968	50,900	81,913	100,050	193,200
9	6,968	49,900	■	■	■

Fill in the missing numbers.

20 3,039 = 3,000 + ■ + 9

21 80,720 = 80,000 + ■ + 20

22 73,296 = 70,000 + ■ + 200 + 90 + ■

23 456,750 = 400,000 + ■ + 6,000 + 700 + ■

24 Find 100 more than 6,985.

25 Find 3,000 less than 81,290.

26 Find 2,000 less than 8,903.

27 Find 300,000 more than 265,900.

Number Tic-Tac-Toe Game!

First, make one set of cards for each of the numbers 0 through 9. Make another set with the words *ones, tens, hundreds,* and *thousands.* Put each set of cards in a separate pile.

You will need
- *counters*
- *index cards*

thousands hundreds tens ones 1 2

Next, make a game sheet like the one shown. You may choose any 4-digit numbers to fill in the grid.

7,128	4,309	9,000
2,093	5,100	1,631
3,298	6,214	8,123

Play the Game

- ▶ Mix up each pile of cards. Decide the order in which you will play.
- ▶ When it is your turn, pick a card from each pile. Then place counters on any number on your game sheet that has the digit in the place shown.
- ▶ The first player to get three counters in a row, column, or diagonal wins the round. The winner gets a point.
- ▶ Create new game sheets. Continue playing until a player has 5 points.

How did you decide what numbers to place on your game sheet?

mixed review • test preparation

Write the number in standard form.

1 8 hundreds 2 tens

2 6 hundred thousands 7 thousands 8 tens

3 two thousand-seven

4 4 thousands 6 hundreds 3 tens 8 ones

Compare and Order Numbers

LEARN

Here are some amazing heights reached by helicopters and balloons! Compare them.

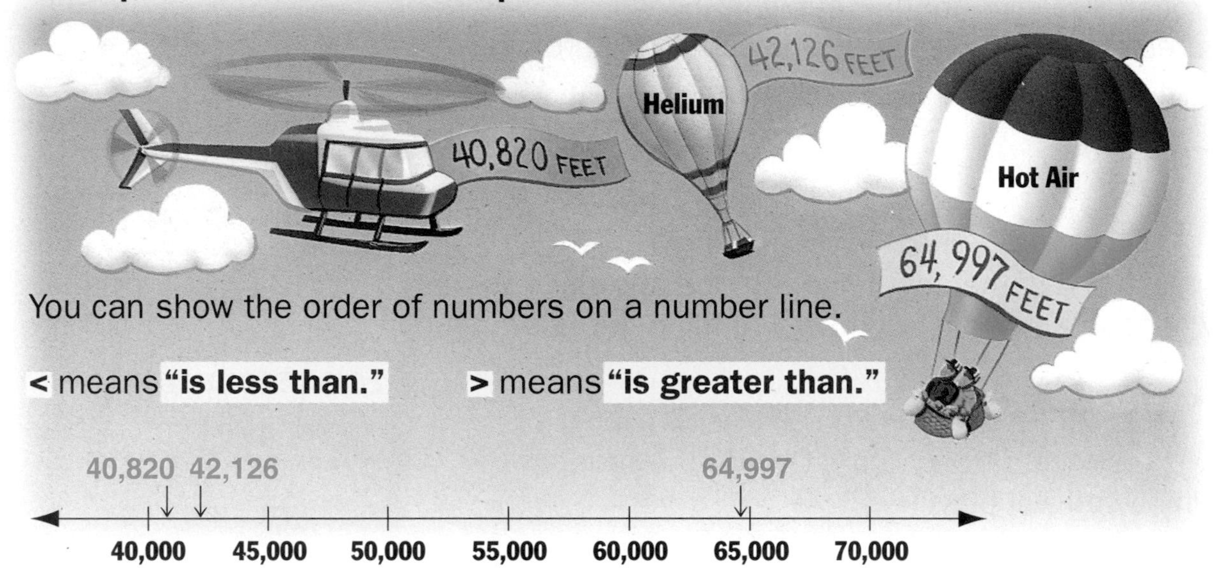

You can show the order of numbers on a number line.

< means **"is less than."** > means **"is greater than."**

You can also use place value to compare and order numbers.

Line up the ones. Start with the greatest place. Compare the digits.	**Compare the digits in the next place.**	**Write the numbers in order.**
40,820 64,997 42,126 **Think:** $6 > 4$, so 64,997 is the greatest number.	40,820 42,126 **Think:** $2 > 0$, so $42{,}126 > 40{,}820$.	From greatest to least: 64,997; 42,126; 40,820 From least to greatest: 40,820; 42,126; 64,997

The hot-air balloon rose the highest, then the helium balloon, then the helicopter.

Check Out the Glossary
is less than,
is greater than
See page 544.

CHECK

Check for Understanding

Order the numbers from greatest to least.

1 17,750; 12,000; 17,540 **2** 8,060; 3,594; 6,333 **3** 7,934; 7,958; 3,000

Critical Thinking: Generalize **Explain your reasoning.**

4 Journal How is writing numbers in order like writing words in alphabetical order?

Practice

Use the number line to compare. Write >, <, or =.

1 5,900 ● 4,080 **2** 6,400 ● 6,799 **3** 3,184 ● 3,108

Compare. Write >, <, or =.

4 878 ● 794 **5** 8,673 ● 8,654 **6** 1,516 ● 1,112

7 8,504 ● 8,515 **8** 10,198 ● 101,980 **9** 254,811 ● 250,811

Order the numbers from least to greatest.

10 639; 504; 648 **11** 8,799; 9,411; 9,059 **12** 5,770; 7,707; 5,077

13 6,402; 3,499; 3,480 **14** 11,450; 7,760; 8,046 **15** 4,518; 45,108; 4,718

MIXED APPLICATIONS
Problem Solving

16 Order the distances traveled from greatest to least. SEE INFOBIT.

17 The first supersonic flight, in a Bell XS-1 rocket plane, reached an altitude of 42,000 ft. The aircraft with the heaviest takeoff load, an Antonov An-225, reached an altitude of 40,715 ft. Which reached the greater altitude?

18 **Make a decision** You want to fly to Toronto on March 12. You can fly on AZ airline for $175, but you have to change planes and wait 1 hour. You can fly on DRZ airline for $240 nonstop, but you have to leave on March 13. What would you do?

INFOBIT
Amazing records: Rick Hansen wheeled his wheelchair 24,901 miles. Plennie Wingo walked 8,000 miles backward. Johann Hurlinger walked on his hands 870 miles.

mixed review • test preparation

1 548 **2** 2,000 + 300 + 10

3 732,907 **4** 40,000 + 1,000 + 900 + 20 + 5

Extra Practice, page 485

Read
Plan
Solve
Look Back

Make a Table

LEARN

Gene, Callie, and Claire read about different world records. Gene reads about the drums. Callie does not read about Scrabble. Which record does each person read about?

Mr. Hughes had seen over 20,000 movies!

Use the questions to help you solve the problem.

Read What do you know? What do you need to find?

Gene reads about the drums. Callie does not read about Scrabble.

Which record does each person read about?

Jeffery Carlo has 112 pieces in his drum set!

Phil Appleby scored 1,049 in a game of Scrabble!

Plan How can you solve the problem?

Choose a strategy to try: Make a table.

Use checks to show what each person read to show what they do not read.

Solve How can you carry out your plan?

	Movies	Drums	Scrabble
Gene	X	✔	X
Callie	✔	X	X
Claire	X	X	✔

Have you answered the question?

Yes. Gene reads about the drums, Callie reads about movies, and Claire reads about Scrabble.

Look Back Does your answer make sense?

Check. They read about different world records. The answer matches the information given.

CHECK

Check for Understanding

1 **What if** there is a fourth world record. Would you be able to tell which record each person reads about? Why or why not?

Critical Thinking: Analyze **Explain your reasoning.**

2 Solve the problem without using a table. Which method did you use? Which of the two methods do you prefer? Why?

MIXED APPLICATIONS

Problem Solving

1 Patrick has 18 CDs. He has twice as many country CDs as classical CDs. He has 3 more rock CDs than country CDs. How many of each type of CD does he have?

2 The world record amount of times that a needle has been threaded is 20,675 in 2 hours (h). Write a newspaper headline that rounds the number to the nearest thousand.

3 Viv Richards, Dan Steeles, and Russell Locke each have one world record. It is in either cricket, darts, or bowling. Viv Richards has the world record in cricket. Dan Steeles does not have the record in darts or cricket. Russell Locke does not have the record in bowling. What record does each person hold?

4 May surveyed students in her class to find out which sport they thought was the most exciting. Her survey showed that 5 named diving, 6 named downhill skiing, 9 named surfing, and 7 named mountain climbing. How many students thought water sports were the most exciting?

5 **Spatial reasoning** Show which shape you can *not* get by drawing a line through a square.

Use the table for problems 6–10.

6 Which dance had the greatest number of people? the least number of people?

7 Did the dancing dragon or the tap dancers have more people?

8 Which dance had about 120,000 people?

9 Order the dances in the table from greatest to least according to the number of people. Which dance is second?

10 Write newspaper headlines for each dance. Use estimates. Explain how you estimated.

World Records for Dancing

Type of Dance	Number of People Who Joined In
Electric slide line dance	30,000
Dancing dragon	1,019
Tap dancing	6,196
Conga line dancing	119,986

Extra Practice, page 486

Write the number in standard form and in expanded form.

1 seven hundred thirteen

2 thirty-six thousand, two hundred thirty-nine

3 forty thousand, seventy-five

4 two hundred forty thousand, seven hundred twenty-five

Write the number in word form and in expanded form.

5 413 6 7,075 7 42,680 8 240,418

Compare. Write >, <, or =.

9 927 ● 772

10 2,011 ● 10,031

11 8,902 ● 979

12 256,723 ● 28,602

13 6,400 ● 6,091

14 99,804 ● 100,984

Order the numbers from least to greatest.

15 73,420; 9,814; 9,836

16 4,480; 8,750; 8,040

Solve.

17 Lin, Rosa, and Lee like either classical music, rock music, or calypso music. None of them like the same type. Lin likes classical music. Lee does not like rock music. Rosa does not like either classical or calypso music. Which types of music do Lee and Rosa like?

18 Ann, Lynda, Paul, and Stephan either swim or play basketball, soccer, or baseball. None are on the same team. Lynda plays basketball. Paul does not swim or play soccer. Stephan does not play soccer or baseball. Which sports do Ann, Paul, and Stephan take part in?

19 By May 1994, the musical *The Fantasticks* had been performed 14,095 times; and the play *The Mousetrap,* 17,256 times. Which show had had more performances?

20 Journal Explain how you can use place value to order 7; 7,070; 707; and 70.

Determine If a Number Is Reasonable

What an amazing claim! Can a baby be 30 feet tall?

To check if 30 feet is reasonable, Rebecca and Ramón used other information they knew.

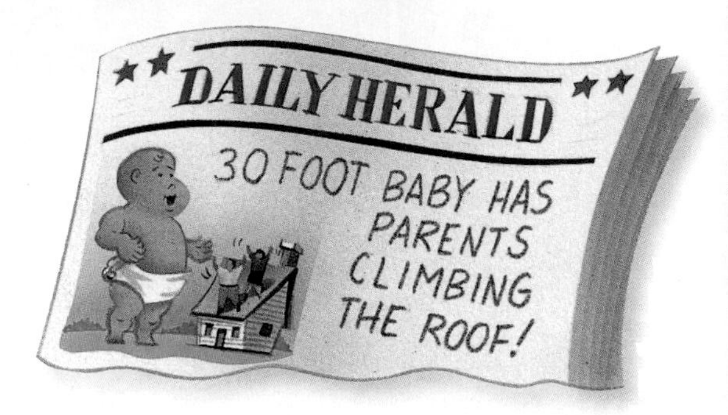

Rebecca used what she knows about her height.

I think there must be a mistake in the newspaper. I am ten years old, and I am less than 5 feet, which is much less than 30 feet.

Ramón used what he knows about the heights of members of his family.

Even my parents are not 30 feet! I think they must mean 30 inches instead of 30 feet. When my sister was born, she was 20 inches. So 30 inches seems reasonable.

Tell if the claim is reasonable. Explain your thinking.

1 **Claim:** Simone's mother weighs 5 pounds.
- **a.** Do you weigh more or less than a 5-pound bag of rice?
- **b.** Do you think you weigh more or less than an adult?

2 **Claim:** Hiko's cup holds 300 milliliters (mL) of water. He says that 289 milliliters of water will fill his bathtub.
- **a.** Compare 300 milliliters with 289 milliliters.
- **b.** Is the bathtub smaller or larger than the cup?

3 **Claim:** A newspaper article says, "It costs $8 million to clean up one of the beaches in our state!" and "The Governor will sign the state budget for $16 million."
- **a.** The state pays for services like schools, the police, firefighters, buses, trains, and parks. Do you expect that cleaning up a beach costs more than each of these services?
- **b.** Compare $8 million and $16 million. Do you think both statements can be correct? Explain your thinking.

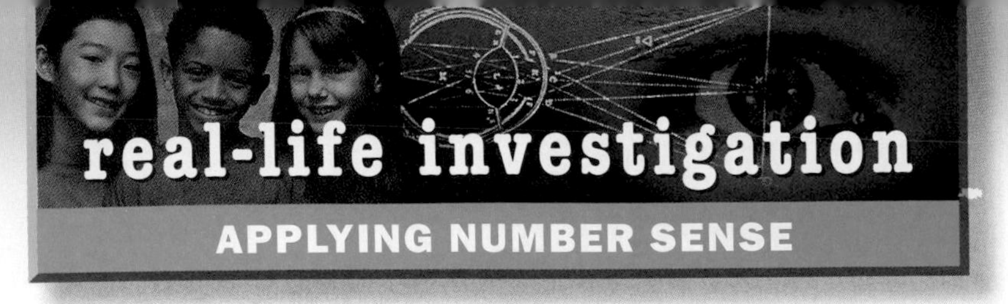

Your Own Book of Records

The Guinness Book of Records is a collection of amazing facts and feats performed around the world. In this activity, you will work in teams to create your own feats, make your own record-breaking attempts, and write a book of record-breaking results for your class.

Here are some ideas:

Largest selection of greeting cards

Tallest tower made by stacking number cards

Greatest number of up and down motions with a yo-yo!!

Longest time for hula-hooping

DECISION MAKING

Performing Feats

1 Choose feats you wish to measure, and record in your book.

2 Decide on the setup and the measurement tools needed to do each feat.

3 Decide on roles for each member of your team. Roles may include: a timer, a measurer of distances, a counter, and a judge who explains the rules clearly.

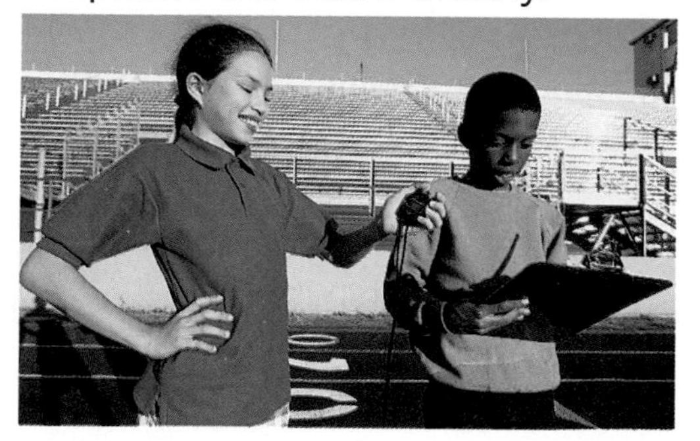

4 Compare measurements for each feat to find the best result—fastest, longest, or greatest.

Reporting Your Findings

5 

Prepare a book for your team that lists and explains each feat with the best result for that feat. Include the following:

- READING ARITHMETIC WRITING **Use Illustrations** Take photographs or make drawings of the feat.
- Choose one feat. Explain how to improve the performance of that feat.

6 Combine your book with the books of other teams to make a class book.

Revise your work.

- Are your measurements and comparisons correct?
- Is your book clear and organized?
- Did you proofread your work?

MORE TO INVESTIGATE

PREDICT how well you could perform a feat if you practiced for a while. Choose a feat. Perform the feat. Then compare the results to your prediction.

EXPLORE the different ways to measure feats for going to school, jumping rope, and creating a toothpick building.

FIND the world records for some interesting feats in *The Guinness Book of Records.*

Round Numbers

LEARN

When you want to give an idea of the size of a number, you can round.

If a newspaper rounded the temperature, what might the headline say?

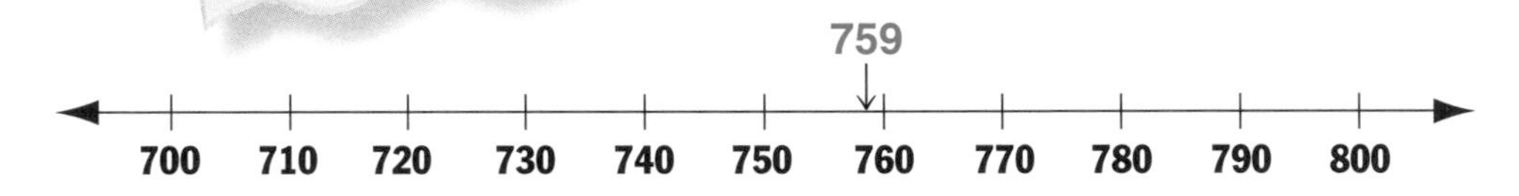

Round to the nearest ten degrees.

Think: 759 is nearer to 760 than to 750. Round 759° to 760°.

DAILY DISPATCH
HIGHEST RECORDED SEA TEMPERATURE WAS ABOUT 760°F!

Round to the nearest hundred degrees.

Think: 759 is nearer to 800 than to 700. Round 759° to 800°.

Talk It Over

- How would you round the number 737 to the nearest ten? to the nearest hundred? Explain your reasoning.
- What numbers when rounded to the nearest ten are 780? How do you know?

round Finding the nearest ten, hundred, thousand, and so on.

A British actor plans to travel around the equator for a television show. If a newspaper rounded the distance around the equator to the nearest thousand, what could the headline say?

Round 24,901 to the nearest thousand.

Step 1 **Look at the place to the right of the thousands place.**

24,901

Step 2 **If the digit is less than 5, round down. If the digit is 5 or greater, round up.**

$9 > 5$, so 24,901 rounds up to 25,000.

The headline might say, "It's just amazing! British actor to travel about 25,000 miles to get around the equator!"

More Examples

A Round to the nearest hundred.
4,256
↑
$5 = 5$, so 4,256 rounds up to 4,300.

B Round to the nearest thousand.
89,400
↑
$4 < 5$, so 89,400 rounds down to 89,000.

Check for Understanding

CHECK

1. Round 784 to the nearest ten.

2. Round 60,750 to the nearest hundred.

3. Round 100,642 to the nearest thousand.

Critical Thinking: Generalize **Explain your reasoning.**

4. When does rounding to the nearest hundred give the same answer as rounding to the nearest thousand? Give an example.

5. Journal Tell why you would or would not round the numbers.
   a. John has a fever of 102°F. What should he tell the doctor?
   b. The survey showed that 32,563 people watched the game. What will the television news report?

Turn the page for Practice.

Practice

Use the number line. Round to the greatest place.

1 728

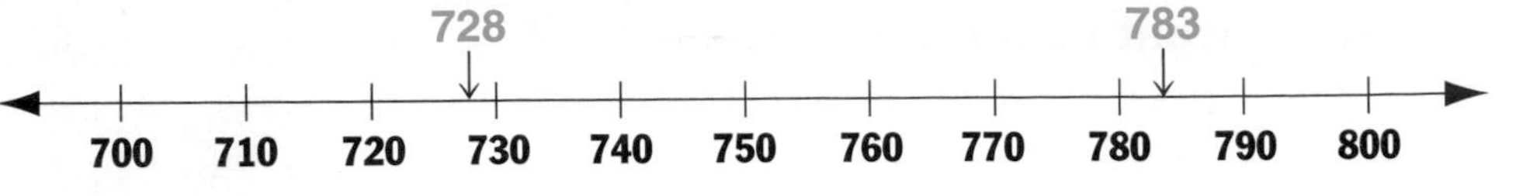

2 783

3 9,289

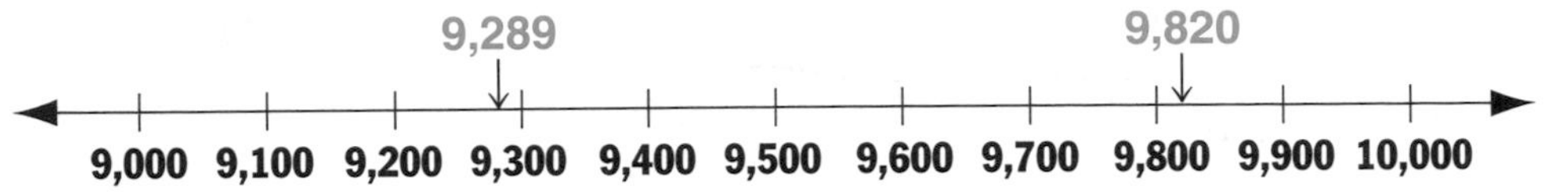

4 9,820

Round to the nearest ten.

5 451 **6** 23 **7** 9,095 **8** 389,995

Round to the nearest hundred.

9 4,754 **10** 1,506 **11** 99,984 **12** 508,046

Round to the nearest thousand.

13 8,911 **14** 2,099 **15** 76,548 **16** 109,612

ALGEBRA **Find the rule. Then complete the table.**

17

Rule: ■	
Input	Output
6,543	6,500
24,063	24,100
72,098	■

18

Rule: ■	
Input	Output
28,650	27,650
150,987	149,987
59,702	■

19

Rule: ■	
Input	Output
8,230	18,230
86,500	96,500
293,360	■

Match the number to a rounded number. Tell if the number is rounded to the nearest ten, hundred, or thousand.

20 6,280 **a.** 6,300

21 9,800 **b.** 410,000

22 409,950 **c.** 98,100

23 98,096 **d.** 10,000

Make It Right

24 Sue Ellen rounded 49,400 to the nearest thousand. Tell what the mistake is, then correct it.

49,400 rounds up to 50,400.

MIXED APPLICATIONS

Problem Solving

25 Steven Newman spent four years walking 22,500 miles for a world record. Which headline would you use? Why?

a. Man walks 200,000 miles
b. Record broken as man walks 20,000 miles
c. New 23,000-mile record to beat!

26 **Spatial reasoning** Show how to make this shape by combining the least number of triangles.

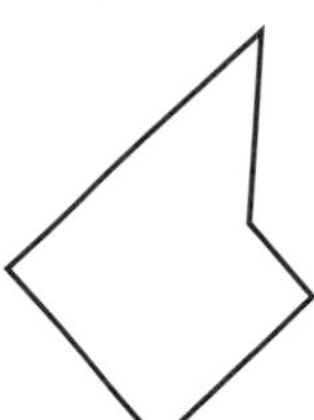

27 For which activity would you *not* use a rounded number? Explain why.

a. find distance for a bicycle trip
b. cut lengths to make a frame for a picture
c. tell how many people attend an event

28 **Make a decision** The longest paper chain had 400,000 links. The most kites flown in a single line was 11,284. The largest bubble made from dishwashing liquid and water was 50 feet long. Which record would you attempt to beat? Explain why you made your choice.

29 **Write a problem** in which you can find the answer by rounding a number to the nearest ten, hundred, or thousand. Solve it. Have others solve it.

30 What is the greatest number you can make using each of the digits 2, 3, 7, 5, and 1 only once? What is the least number you can make with the digits?

31 **Data Point** Use the Databank on page 532. Write a headline that uses rounded numbers to compare the average depth of the Indian Ocean with the average depth of another ocean.

32 The Arctic Ocean is the smallest ocean on Earth. Its average depth is 3,407 feet. Is this more or less than 1 mile? (Hint: A mile is 5,280 feet.)

mixed review • test preparation

Write the number in standard form and in expanded form.

1 eight hundred fifty-six

2 four thousand, eleven

3 sixty-three thousand, twenty-nine

4 seven thousand, two

5 eight hundred seven thousand, twenty-three

6 three thousand, nine hundred fourteen

Millions

LEARN

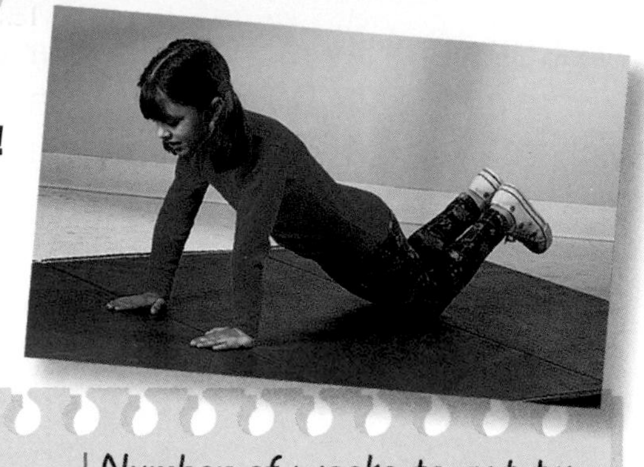

The world record for push-ups is held by Paddy Doyle. He did over 1,000,000 in one year!

Work Together

Work with a partner. Use a calculator to solve each problem.

If you did 1,000 push-ups each week, how many weeks would it take to get to:
- ▶ 10,000 push-ups?
- ▶ 100,000 push-ups?
- ▶ 1,000,000 push-ups?

If you did 10,000 push-ups each week, how many weeks would it take to get to:
- ▶ 100,000 push-ups?
- ▶ 1,000,000 push-ups?

If you did 100,000 push-ups each week, how many weeks would it take to get to:
- ▶ 1,000,000 push-ups?

Record your answers in a table.

	Number of weeks to get to:		
Push-ups each week	10,000 push-ups	100,000 push-ups	1,000,000 push-ups
1,000			

Talk It Over

- ▶ How many:
  - **a.** thousands are in 10,000? 100,000? 1,000,000?
  - **b.** ten thousands are in 100,000? 1,000,000?
  - **c.** hundred thousands are in 1,000,000?
- ▶ How many push-ups can you do in a week? How many weeks would you take to do 1,000,000 push-ups?
- ▶ How did you use the calculator to find your answers? Share your methods with classmates.

Make Connections

You can see the relationship of 1,000; 10,000; 100,000; and 1,000,000 in a place-value chart. Each place is ten times greater than the place to its right.

Millions	Thousands			Ones		
Ones	Hundreds	Tens	Ones	Hundreds	Tens	Ones
1	5	0	0	2	3	0

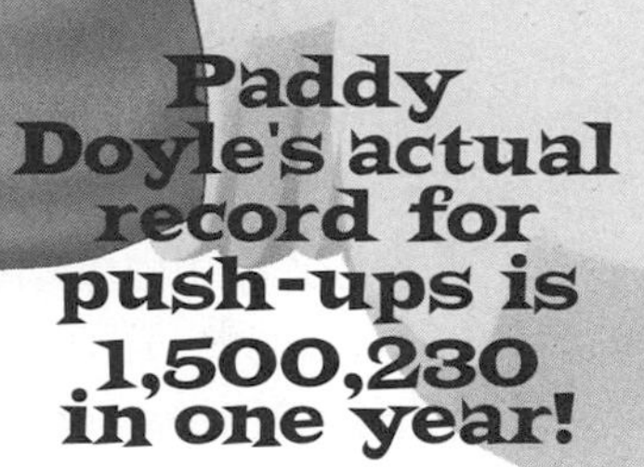

Paddy Doyle's actual record for push-ups is 1,500,230 in one year!

You can read or write the number in different ways.

Standard Form: 1,500,230

Word Name: one million, five hundred thousand, two hundred thirty
or
1 million, 500 thousand, 230

Expanded Form: 1,000,000 + 500,000 + 200 + 30

Check for Understanding

CHECK

Write the number in standard form.

1. three million, eighty-two thousand, ten
2. five million, seven hundred sixty-five thousand, one hundred six
3. seven million, three hundred nine thousand, five hundred thirty
4. eight million, four hundred forty thousand, two

Write the number in word form and in expanded form.

5. 1,748,235
6. 1,035,010
7. 6,526,085

Critical Thinking: Analyze **Explain your reasoning.**

8. What number is 1 greater than 6,999,999? In which places did the digits change?
9. Can 1,500,000 also be written as 15 hundred thousand?

Turn the page for Practice.

Practice

ALGEBRA: PATTERNS Complete the different names for the number.

1 2,000,000 = ■ thousands
■ ten thousands
■ hundred thousands

2 8,000,000 = ■ thousands
■ ten thousands
■ hundred thousands

Write the number in standard form and in expanded form.

3 one million, four hundred thirty-eight thousand

4 four million, eighty-seven thousand, three hundred seven

5 nine million, two hundred seventy thousand, ninety-one

6 six million, five hundred three thousand

Write the word name for the number.

7 875,623 **8** 2,300,000 **9** 908,607 **10** 600,004 **11** 8,398,012

12 1,004,600 **13** 7,072,234 **14** 5,020,100 **15** 3,205,000 **16** 4,000,050

17 What is 1,000,000 more than 896,000?

18 What is 1,000,000 less than 4,082,500?

19 What is 100,000 more than 6,980,400?

20 What is 10,000 less than 1,067,999?

21 Tell whether the amount is counted in the millions. Explain your reasoning.
a. the number of miles you walk to school
b. the number of people living in New York City
c. the number of dollars needed to run all the schools in your state

ALGEBRA Write *true* or *false*.

22 $A > D$

23 $B = C$

24 $B < A$

25 $C > D$

26 $B < D$

A = 1,438,095
B = 199,580
C = 1,000 less than 200,580
D = 1,000,000 more than 500,000

MIXED APPLICATIONS
Problem Solving

Pencil & Paper | Calculator | Mental Math

27 If you earned \$10,000 a year, how long would it take you to earn \$1,000,000? Explain your answer.

28 Use each digit only once to make the greatest number you can: 0, 2, 5, 7, 8, and 9.

29 About how long do you think it would take for 1 million dominoes to topple? Explain your reasoning. **SEE INFOBIT.**
a. about 15 minutes
b. about 5 hours
c. about 50 minutes
d. about 5 days

30 **Logical reasoning** A calculator display shows 9586425. . Later, it shows 9576425. . Tell how this change could have been made.

31 **Write a problem** using two or more numbers in the millions. Exchange problems with a classmate. Solve and discuss your answers.

INFOBIT
The greatest number of dominoes set up single-handedly was 320,236 by Klaus Friednich of Germany. He toppled 281,581 of the dominoes within 12 minutes 57.3 seconds.

more to explore

Counting to a Million

How long do you think it would take to count to 1,000,000?

To estimate the time it takes to count to 1,000, you can:
- measure the time it takes to count to 100.
- use your measurement to estimate the time it would take to count to 1,000.

Think: If it takes 2 minutes to count to 100, then it would take ten times as long, or 20 minutes, to count to 1,000.

Measure the time it takes you to count to 100. Then use a calculator to estimate the time it would take to count to:

a. 1,000. **b.** 10,000. **c.** 100,000. **d.** 1,000,000.

Problem Solvers at Work

Read
Plan
Solve
Look Back

Part 1 Interpret Data

LEARN

Fourth graders in the River School surveyed their classmates to compare the names to the amazing claim at the right.

Our data:
3, 3, 11, 2, 4, 12, 6, 5, 13, 7, 7, 11, 12, 4, 6, 5, 11,
2, 3, 3, 4, 6, 11, 6, 5, 4, 3, 6, 6, 4, 4, 12, 5, 11

AMAZING CLAIM
LONGEST FIRST NAME GIVEN TO BABY HAS 32 LETTERS

Erehwon, TX — They will have to attach another page to the birth certificate of the newest citizen of this hard-to-find small town. When asked about the unusual and almost unpronounceable name, the parents of the 7 pound 9 ounce baby girl said, "It's an old family tradition to name the first baby after their

The class used a **line plot** to show a picture of their **data.**

Number of Letters in First Name

		x		x							
	x	x		x					x		
	x	x	x	x					x		
	x	x	x	x					x	x	
x	x	x	x	x	x				x	x	
x	x	x	x	x	x				x	x	x
2	**3**	**4**	**5**	**6**	**7**	**8**	**9**	**10**	**11**	**12**	**13**

Work Together

Use the results of the survey to solve. Explain your methods.

1 How many fourth-grade students are there? How do you know?

2 Are there any clusters in the data? What does this tell you?

3 What statement can you make about the number of letters in the first names of the fourth-grade students?

4 **What if** there are four new students with first names that have 5, 5, 5, and 10 letters. How does the new data affect the statement you made?

5 READING ARITHMETIC WRITING **Use Illustrations** Compare the line plot and the list of data. Which one makes it easier to interpret the data? Why?

6 Use the survey to explain why having a first name with 32 letters is an amazing claim. If you had to predict the number of letters in a first name, what would your prediction be? Why?

line plot A graph that shows data using symbols that are lined up.

data Information.

Part 2 Write and Share Problems

Erica wrote this problem. Then she chose a paragraph and made a tally table to record the letters she counted.

Letter	Tally	Number of Times the Letter Appears
a	𝍸𝍸 |||	13
b	||||	4
c	𝍸	5
d	||||	4
e	𝍸𝍸𝍸𝍸	20

CHECK

Which letter appears most often: a, b, c, d, or e?

Erica Lester
Snowden Elementary School
Memphis, TN

7 Solve Erica's problem. Explain your reasoning.

8 Choose a paragraph of your own. Collect data about the five letters. Create a table like Erica's to show your data.

9 Solve the problem for your own data. Explain how the table helped you to solve the problem.

10 **Write a problem** of your own where you can use a table to record and organize data to solve the problem.

11 Trade problems. Solve at least three problems written by your classmates.

12 What was the most interesting problem that you solved? Why?

Turn the page for Practice Strategies.

PRACTICE

Menu

Choose five problems and solve them. Explain your methods.

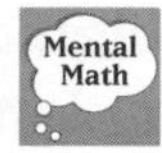

1 Katy has two books. The first book has 928 pages. The second book is 100 pages longer. How many pages does the second book have?

2 A tree has 2 limbs. Each limb has 3 branches. Each branch has 3 caterpillars on it. How many caterpillars are there?

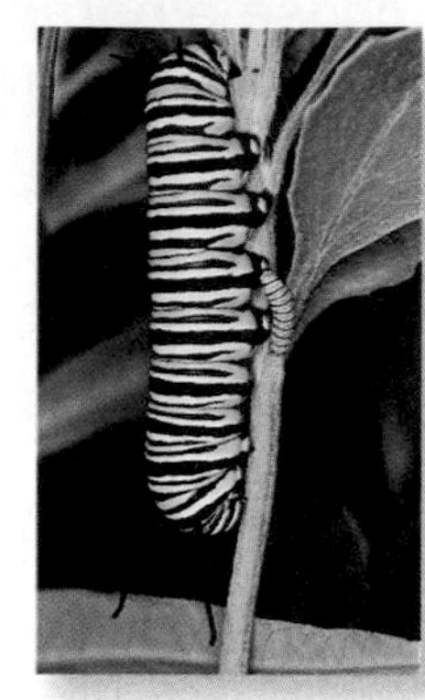

3 **Logical reasoning** I am a number with the digits 0, 3, 4, 5, and 9. If you round me to the nearest thousand or the nearest hundred, you get 44,000. The digit in my tens place is 5. What number am I?

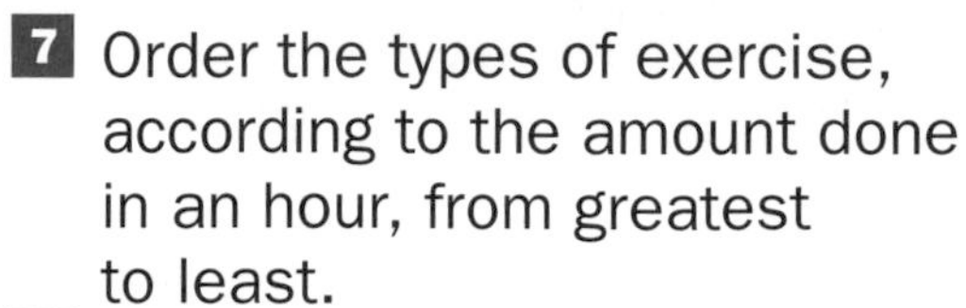

4 Moy, Tara, and Vincent live on different streets. Moy does not live on Main Street. Tara does not live on Glenwood Avenue or Village Place. Vincent lives on Village Place. Where do Moy and Tara live?

5 Which animal do most students have as a pet? fewest students?

6 Write a statement that describes the results of this survey.

arf!

Types of Pets Students Have

Cat	Dog	Fish	Snake	Bird
	x			
x	x			x
x	x	x		x
x	x	x	x	x

7 Order the types of exercise, according to the amount done in an hour, from greatest to least.

8 Which type of exercise was done about 3,600 times? Explain your answer.

Type of Exercise	World Record Number Done in an Hour
Squats	4,289
Squat thrusts	3,552
Parallel-bar dips	3,726
Burpees	1,840

Choose two problems and solve them. Explain your methods.

9 **a.** Which of these numbers are Flims? 1,476; 60,000; 412,070; 6,072; 7,307

b. Describe a Flim.

These numbers are Flims.	
1,450	2,000
9,678	4,000
7,396	9,996
3,044	2,700

These numbers are not Flims.	
675	750
1,735	6,087
13,401	700,000
82,500	1,975,728

10 Use the digits 0, 1, 2, 3, 4, 5, 6, 7, 8, and 9 one time each to replace these circles to make true statements. Compare your statements with other classmates.

a. 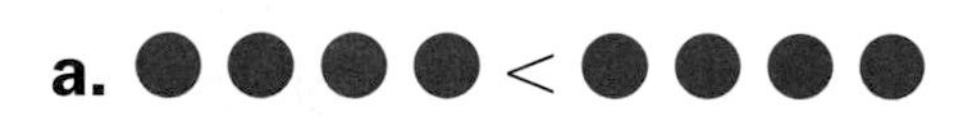

b. ● ● ● ● ● < ● ● ● ● ●

c. ● ● ● ● > ● ● ● ●

d. ● ● ● ● ● > ● ● ● ● ●

11 **Spatial reasoning** A design made from toothpicks is shown below.

a. Remove 4 toothpicks and leave only 4 squares that are all the same size.

b. Remove 5 toothpicks and leave only 3 squares that are all the same size.

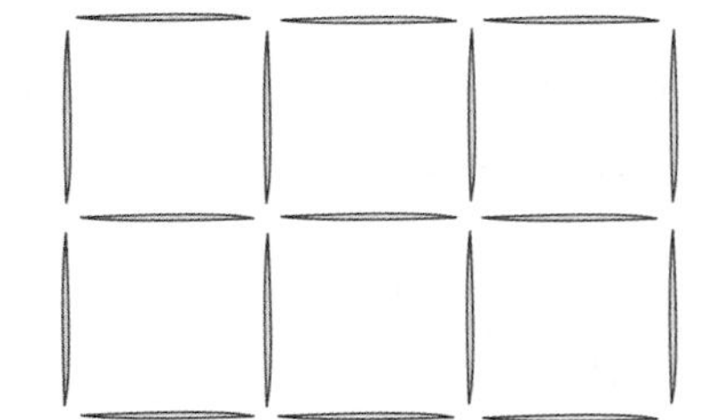

12 **At the Computer** When you stack cubes to form a column, you can see some faces but not others. For example, if you place one cube on a table, you can see five faces but not the face resting on the table.

Use a drawing program to build columns 1 cube high, 2 cubes high, and so on, up to 10 cubes high. Make a table like the one below. Then write about what you notice.

Number of Cubes High	1	2	3	■	■	■	■	■	■	■
Number of Faces You Can See	5	9	■	■	■	■	■	■	■	■

Language and Mathematics

Complete the sentence. Use a word in the chart. (pages 2–27)

1 4,000 + 600 + 7 is the ■ of the number 4,607.

2 A comma separates the millions and thousands ■ in the number 7,835,082.

3 84,600 > 82,900 means 84,600 ■ 82,900.

4 To round a number to the nearest thousand, you look at the digit in the ■ place.

Vocabulary
- periods
- hundreds
- expanded form
- millions
- is greater than
- is less than

Concepts and Skills

Write the number in standard form. (page 8)

5 six hundred forty-three

6 seven thousand, nine hundred

7 three thousand, two hundred seventy-eight

8 eight hundred sixty-one thousand, four hundred sixteen

9 six hundred four thousand, thirty-nine

10 six million, one hundred thousand, nine hundred two

Write the number in word form and in expanded form. (page 8)

11 9,460 **12** 2,807 **13** 8,074 **14** 50,072 **15** 652,913

Compare. Write >, <, or =. (page 12)

16 983 ● 9,831 **17** 356,700 ● 356,585 **18** 507,120 ● 510,000

Order the numbers from least to greatest. (page 12)

19 1,432; 2,232; 1,475

20 67,902; 66,400; 67,050

21 443,700; 88,500; 438,600

22 Round 9,742 to the nearest hundred.

23 Round 3,621 to the nearest thousand.

24 Round 865,398 to the nearest thousand.

25 Round 999,950 to the nearest thousand.

Think critically. (page 8)

26 Analyze. Explain what mistake was made. Then correct it.

34 thousand = 3,000 + 4,000

27 Generalize. Write *always, sometimes,* or *never.* Give examples to support your answer.

a. When you round a number to the nearest thousand, you increase the thousands digit by 1. Then, you change the digits that follow to zeros.

b. Choose any two whole numbers. The number with more digits is the greater number.

MIXED APPLICATIONS
Problem Solving

(pages 14, 28)

28 What do all these numbers have in common?

1,023,100	45,012	222,400
434,124	131,240	13,232

29 A student surveyed her classmates and made the line plot below. Write a statement that describes the survey.

How Many Animals Do You See Each Day?

	x	x						
	x	x						
	x	x						
x	x	x	x					
x	x	x	x	x	x			x
1	**2**	**3**	**4**	**5**	**6**	**7**	**8**	**9**

30 A chain of paper clips was made by 60 students in Singapore in 1992. The chain measured 18,087 feet in length. What was the length of the chain to the nearest thousand feet?

Use the information in the table for problems 31–33.

31 Which tunnel or bridge has the greatest length? the least length?

32 Which tunnel is longer: the Lincoln or the Mount Royal?

33 What is the difference between the length of the Golden Gate Bridge and the Verrazano-Narrows Bridge?

Tunnels or Bridges	Length
Fort McHenry Tunnel, MD	7,200 ft
Golden Gate Bridge, CA	4,200 ft
Lincoln Tunnel, NJ/NY	13,200 ft
Howrah Bridge, India	1,500 ft
Mount Royal Tunnel, Canada	16,900 ft
St. Gotthard Tunnel, Switzerland	53,800 ft
Verrazano-Narrows Bridge, NY	4,300 ft

Write the word name and expanded form for the number.

1. 2,506
2. 8,067
3. 30,470
4. 67,432

Write the number in standard form.

5. 5 hundreds 8 ones
6. 9 hundreds 2 tens 7 ones
7. 72 tens
8. 18 hundreds

Tell if the number is used to name, order, count, or measure.

9. Kelly is third in line.
10. The rope is 9 ft long.
11. Fifty people are here.
12. The MX6 airplane is in the hangar.
13. The concert will last seven days.

Compare. Write >, <, or =.

14. 375 ● 3,750
15. 7,612 ● 7,195
16. 4,365 ● 11,628

Order the numbers from least to greatest.

17. 5,826; 6,019; 5,819
18. 68,271; 6,083; 6,826
19. Round 8,304 to the nearest hundred.
20. Round 6,531 to the nearest thousand.
21. Round 59,827 to the nearest thousand.

Solve.

22. Use the record heights of 6 feet 11 inches, 112 feet, and 90 feet. The largest piggy bank was not 112 feet nor 90 feet high. The figures on Stone Mountain are not 112 feet high. What was the height of the highest recorded sea wave?
23. The actual mountain heights are 7,439 meters, 8,848 meters, and 6,194 meters. Everest is about 9,000 meters. Victory Peak is not about 7,000 meters. How high is McKinley? Order the mountains from greatest to least according to their heights.
24. The largest jar of jelly beans had 378,300 beans. The jigsaw puzzle with the most pieces had 204,484 parts. Were there more beans or jigsaw pieces?
25. Shakespeare's play *Hamlet* has 29,551 words in it. If a newspaper article rounded the number to the nearest thousand, what might a headline say about it?

What Did You Learn?

Write about the top three immigrant groups in the United States in 1993.

In 1993, there were 65,578 immigrants from China, 18,783 immigrants from Great Britain, 126,561 immigrants from Mexico, 63,457 immigrants from the Philippines, and 17,241 immigrants from Jamaica.

- Write a title that includes numbers.
- Organize the data in a table. How did you organize the numbers?
- Write two statements that compare the data about the top three immigrant groups.

A Good Answer

- organizes the data to allow easy comparisons
- includes a title with numbers that would make people want to read it
- gives accurate comparisons of the data

You may want to place your work in your portfolio.

What Do You Think

1. Can you read and write numbers through hundred thousands? If not, what do you find difficult to do?

2. List all the ways you might use to compare or order numbers.
   - Use a place-value chart.
   - Use place-value models.
   - Use a number line.
   - Other. Explain.

math science technology
CONNECTION

Endangered Species

The snow leopard lives in the mountains of Asia. It is killed for its spotted fur, which is made into fur coats. It is also killed by hunters and by people who think it will kill livestock. The snow leopard is an endangered species.

Reasons why species become endangered include:
a. natural causes such as volcanoes, floods, and disease.
b. loss of habitat (places to live).
c. killings by hunters, by species that are new to the region, or by cars and motorboats.
d. pesticides and poisons.

► What are some ways humans can help endangered species?

Use the Databank on page 533 for problems 1–5.

1 Which species of animals have populations that decreased? increased? How can you tell?

2 Use the populations in the 1990s. Write the animals in order from greatest to least. Which species of animals have populations less than 2,000?

At the Computer

3 Choose three animals from the Table of Endangered Animal Populations. Use graphing software to make a bar graph for each animal. Each bar graph should compare the animal population in the 1970s and 1990s. How does the bar graph show that the population increased? decreased?

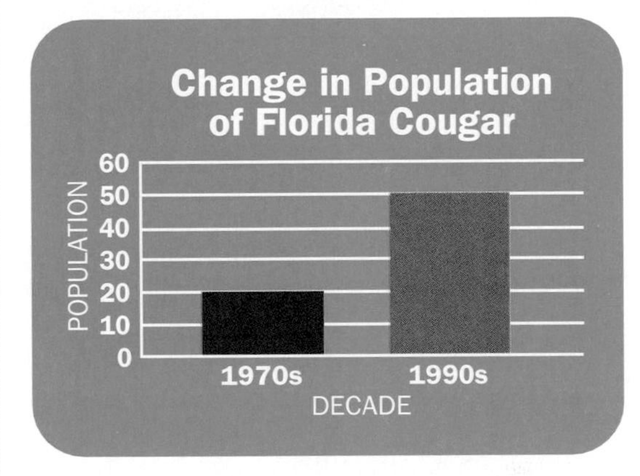

4 Write two statements about the animal populations you chose.

5 Use an on-line encyclopedia. Find new information about the population of the animals. Use the information in your graph. What do you notice?

CHAPTER 2

MONEY, ADDITION, AND SUBTRACTION

THEME

Smart Shopping

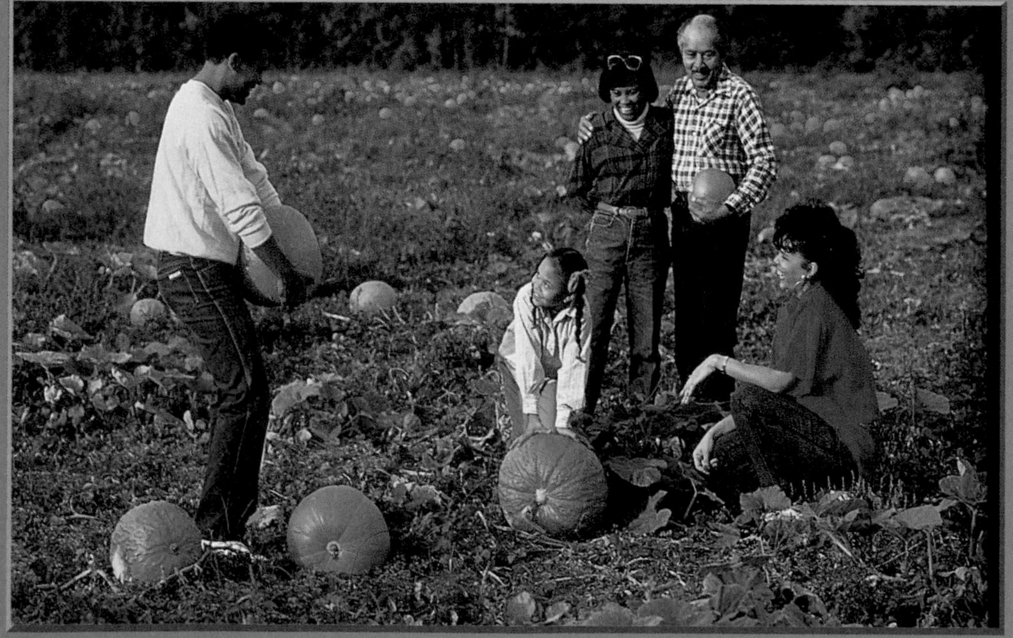

Smart shoppers always save money! In this chapter, you will decide if you will use coupons, buy at a sale, go to a street fair, or buy from a store.

What Do You Know 

Use the information in the price list for problems 1–3.

1. What is the cost of 2 pounds of apples and a 3-pound pumpkin?

2. **What if** you pay for 1 large mum, 1 small mum, and 1 small bag of peat moss with one $10 bill and two $5 bills. What will be the change? Give three ways to show the change.

3. Portfolio You have $50.00 to spend at the farm stand. Make a list of the items you can buy with your money. You can buy more than one of each item. Explain how you decided which items to buy.

Write a Paragraph

A paragraph is a group of sentences about one idea. The first sentence often states the idea. The other sentences tell more about the idea.

Suppose you went to a farm stand. Use the price list above to write a paragraph telling what you bought and how much you spent.

1. What is the topic of your paragraph?

2. What details did you include?

Vocabulary

estimate, p. 46
sum, p. 46
regroup, p. 48
Associative Property, p. 52
Commutative Property, p. 52
difference, p. 58
subtract, p. 58

Count Money and Make Change

LEARN

Go shopping at the flea market!

Work Together

Work with a partner. Choose an item you would like to buy. Pay for the item with one of the bills shown.

You will need
- *play money—bills and coins*

Have your partner use play money to find the change.

▶ Discuss your methods with another group. How do they compare?

Make Connections

Martina buys a puzzle with $5. She counts up to $5 to find the coins and bills she will get in change.

Think:

Start at $2.35. $2.40 → $2.50 → $2.75 → $3.00 → $4.00 → $5.00

To find the amount of change, Martina counts the bills and coins.

Think:

The amount of change is $2.65.

- ▶ What other coins could you use to count from $2.35 to $3.00?
- ▶ **What if** there are no quarters. How can you find the change to $5?

Check for Understanding

Find the amount of change that will be given.

1 Cost: $2.75
Amount given: $5.00

2 Cost: $3.98
Amount given: $5.00

3 Cost: $8.20
Amount given: $10.00

Critical Thinking: Analyze **Explain your reasoning.**

4 What method can you use to give change with the least number of coins and bills?

CHECK

Practice

PRACTICE

Write the money amount.

1 **2** **3**

Show the amount of money in two different ways.

4 67¢ **5** 98¢ **6** $3.42 **7** $12.54

Find the amount of change.

8 Cost: 47¢
Amount given: $1.00

9 Cost: $2.15
Amount given: $5.00

10 Cost: $6.42
Amount given: $10.00

Extra Practice, page 488

Compare, Order, and Round Money

LEARN

Suppose you have saved up for a skateboard. Which of these costs the least? the most?

Compare prices.
Show each amount using play money.

Step 1	Step 2	Step 3
Compare the three amounts	**Compare the other two amounts.**	**Write the amounts in order from least to greatest.**
\$17.99 \$15.99 \$17.25 **Think:** 5 < 7, so \$15.99 is the least.	\$17.99 \$17.25 **Think:** 2 < 9, so \$17.25 < \$17.99.	\$15.99, \$17.25, \$17.99

The Thunder costs the least. The Red Flash costs the most.

Which costs about $16.00?

Round each amount to the nearest dollar.

$17.99 → $18.00
$15.99 → $16.00
$17.25 → $17.00

The Thunder, at $15.99, costs about $16.00.

CHECK

Check for Understanding

Show each amount. Then write the amounts in order from least to greatest.

1 $8.43, $10.89, $5.75

2 $37.99, $73.15, $35.50

Round to the nearest dollar and to the nearest 10¢.

3 $4.35 **4** $9.80 **5** $16.55 **6** $99.90 **7** $231.85

Critical Thinking: Analyze **Explain your reasoning.**

8 Beth rounded $463.65 to $500.00. Sam rounded it to $460.00. Alicia rounded it to $463.70. Which one is correct?

Practice

Compare. Write >, <, or =.

1. $6.58 ● $4.30
2. $0.89 ● $1.65
3. 88¢ ● 49¢
4. $37.56 ● $9.98
5. $156.18 ● $159.14
6. $3,889.99 ● $3,890.99

Write in order from greatest to least.

7. $8.52, $8.25, $8.79
8. $27.15, $7.98, $14.35, $14.85
9. $38.10, $3.81, $38.81, $8.38
10. $47.50, $147.50, $4.75, $47.05

Round.

11. $7.42 to the nearest dollar
12. $0.98 to the nearest dollar
13. $7.12 to the nearest 10¢
14. $14.49 to the nearest 10¢
15. $6.50 to the nearest $10
16. $27.49 to the nearest $10

Tell what item or items can be bought with the amount.

17. $5.00
18. $10.00
19. $15.00
20. $20.00

MIXED APPLICATIONS
Problem Solving

21. At the street fair, Loni buys a book for $2.29 and hands the clerk a five-dollar bill. What amount should she receive in change?
22. Sam buys a T-shirt for $6.69 and gives the cashier $20.00. What is the least number of bills and coins he can receive?
23. If you have 5 quarters and 11 dimes, do you have enough to buy a ball for $2.00? Why or why not?
24. **Write a problem** about buying items shown above. Solve it and have others solve it.

mixed review • test preparation

Write the number in standard form.

1. 1,000 + 200 + 4
2. six hundred thousand
3. 80,000 + 4,000 + 900

Compare. Write >, <, or =.

4. 7,510 ● 7,502
5. 11,600 ● 11,690
6. 24,893 ● 106,000

Extra Practice, page 488

Addition Strategies

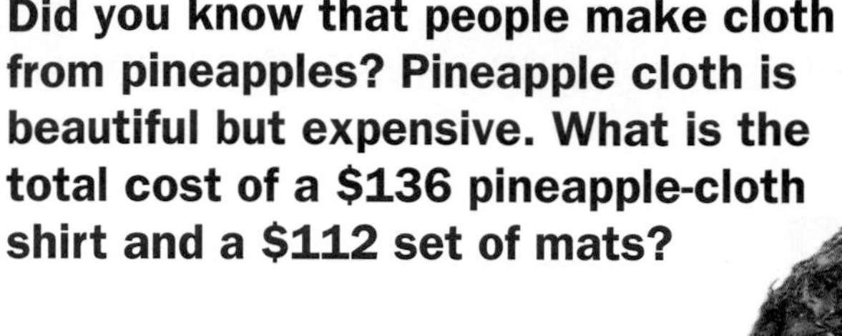

Did you know that people make cloth from pineapples? Pineapple cloth is beautiful but expensive. What is the total cost of a $136 pineapple-cloth shirt and a $112 set of mats?

Cultural Note

In the Philippines, pineapple leaves are used to make piña (pihn-YAH) cloth.

Add: $136 + $112

You can add left to right mentally.

$136 + 112 2	$136 + 112 24	$136 + 112 248
Think: $100 + $100 = $200	**Think:** $30 + $10 = $40	**Think:** $6 + $2 = $8

The total cost is $248.

What if a tourist buys a shirt for $149 and a curtain for $127. How much does the tourist pay?

Add: $149 + 127

You can also add mentally using this method.

$149
+ 127

Think: $149 + $100 = $249
$249 + $20 = $269
$269 + $7 = $276

The tourist pays $276.

CHECK

Check for Understanding

Add mentally. Explain your thinking.

1 283 + 414 **2** 276 + 121 **3** 295 + 131 **4** $183 + $716

Critical Thinking: Summarize

5 

Explain why you look for tens or hundreds when you add mentally. Give some examples.

PRACTICE

Practice

Add mentally.

1. 62 + 37
2. 53 + 24
3. 85 + 11
4. 73 + 22
5. 85 + 22
6. 75 + 24
7. 86 + 12
8. 129 + 160
9. 167 + 822
10. 246 + 731
11. $458 + $121
12. 248 + 550
13. 120 + 209
14. 498 + 101
15. 232 + 166
16. $749 + $130
17. 177 + 221
18. $354 + $215
19. 201 + 397
20. 586 + 313

ALGEBRA Complete. Use mental math.

21. 36 + ■ = 59
22. ■ + 246 = 546
23. ■ + $175 = $800
24. 391 + ■ = 692
25. 582 + ■ = 797
26. ■ + 672 = 776

MIXED APPLICATIONS
Problem Solving

27. Tomás counts the number of ties in stock. He finds 287 navy-blue ties and 112 forest-green ties. How many ties are there in all?

28. Mickey has $103 and wants to buy a jacket for $48.99 and 2 hats for $28.75 each. Does she have enough money to buy all the items? Explain.

29. Liza bought a skirt for $25.50. She got back a ten-dollar bill and a five-dollar bill in change. How much did she hand to the cashier?

30. In her purse Karen finds 4 one-dollar bills, 6 quarters, 7 dimes, 8 nickels, and 12 pennies. Find the total amount of these coins.

more to explore

Use Hundreds to Add

You can use this method to add mentally.

Add: 298 + 457
+2 ↓ ↓ −2
300 + 455 = 755

Think: Add a number to one addend to make a hundred. Subtract the same number from the other addend.

Add. Use mental math.

1. 396 + 175
2. $397 + $156
3. 295 + 698
4. 427 + 198
5. 139 + 203
6. 148 + 795
7. $437 + $399
8. $691 + $259

Extra Practice, page 488

Estimate Sums

LEARN

Store clerks sometimes survey customers. In this survey, about how many customers knew about the sale?

Estimate: 147 + 162

Round each number to find the **sum** mentally.

Think: Round to the nearest hundred.

147 + 162
↓ ↓
100 + 200 = 300

About 300 customers knew about the sale.

More Examples

A Estimate: \$0.28 + \$0.94

Think: Round to the nearest ten cents.
\$0.30 + \$0.90 = \$1.20

B Estimate: \$3.75 + \$5.14

Think: Round to the nearest dollar.
\$4 + \$5 = \$9

C Estimate: 3,654 + 225

Think: Round to the nearest hundred.
3,700 + 200 = 3,900

D Estimate: 4,954 + 1,425 + 3,567

Think: Round to the nearest thousand.
5,000 + 1,000 + 4,000 = 10,000

CHECK

Check for Understanding

Estimate the sum. Tell how you rounded.

1 \$0.68 + \$0.72

2 \$4.78 + \$2.32

3 \$41.72 + \$57.19

4 843 + 372

5 6,610 + 385

6 39 + 52 + 46

Critical Thinking: Generalize **Explain your reasoning.**

7 How will an estimated sum compare to an exact sum if you estimate by rounding up both? rounding down both? rounding one up and rounding down the other? Give examples.

8 Tell how you could round the numbers in ex. 3 and 5 another way to estimate the sums. What would the estimates be?

estimate To find an answer that is close to the exact answer.

sum The result of addition.

Practice

Estimate. Round to the nearest ten or ten cents.

1. 22 + 52
2. \$0.87 + \$0.72
3. 78 + 65
4. \$0.09 + \$0.78
5. \$0.78 + \$0.19
6. 499 + 66
7. \$3.62 + \$0.18
8. 43 + 34 + 25

Estimate. Round to the nearest hundred or dollar.

9. 413 + 841
10. \$8.32 + \$2.75
11. \$3.66 + \$0.88
12. 295 + 536
13. \$2.91 + \$7.23
14. 7,722 + 198
15. \$3.62 + \$3.76
16. 482 + 92 + 29

Estimate. Round to the nearest thousand or ten dollars.

17. \$52.43 + \$36.89
18. \$20.13 + \$32.45
19. 4,378 + 6,527
20. 5,643 + 989
21. \$87.50 + \$7.82
22. 6,740 + 271 + 867

MIXED APPLICATIONS
Problem Solving

Use the catalog for problems 23–26.

23. Sean orders one craft kit of each type from the catalog. About how much does his order cost?

24. Which two items can you buy together for less than \$50.00? Explain your answer.

25. Tot's Toys had 2,460 customers for its catalog last year. This year it has an additional 1,510 customers. About how many customers does it have this year?

26. **Write a problem** using the catalog page. Solve it. Have others solve it.

mixed review • test preparation

Write the number in standard form and in expanded form.

1. fifty-two thousand, ninety-four
2. two hundred thousand, one hundred

Write in order from least to greatest.

3. 467; 678; 464; 644
4. 98; 698; 812; 89
5. 4,048; 2,480; 8,420; 3,999

Extra Practice, page 489

Add Whole Numbers

LEARN

Selling Native American crafts shares beauty and keeps traditional arts alive. Each wampum bead takes about two hours to make. How much would a craftsperson earn by selling a wampum bracelet for $239 and a necklace for $185?

Cultural Note

The Iroquois and Chippewa used to trade wampum belts as a promise to keep treaties and to show friendship. Today, the belts are worn during some celebrations.

Add: $239 + $185

Estimate the sum.
Think: $200 + $200 = $400

Find the exact sum.

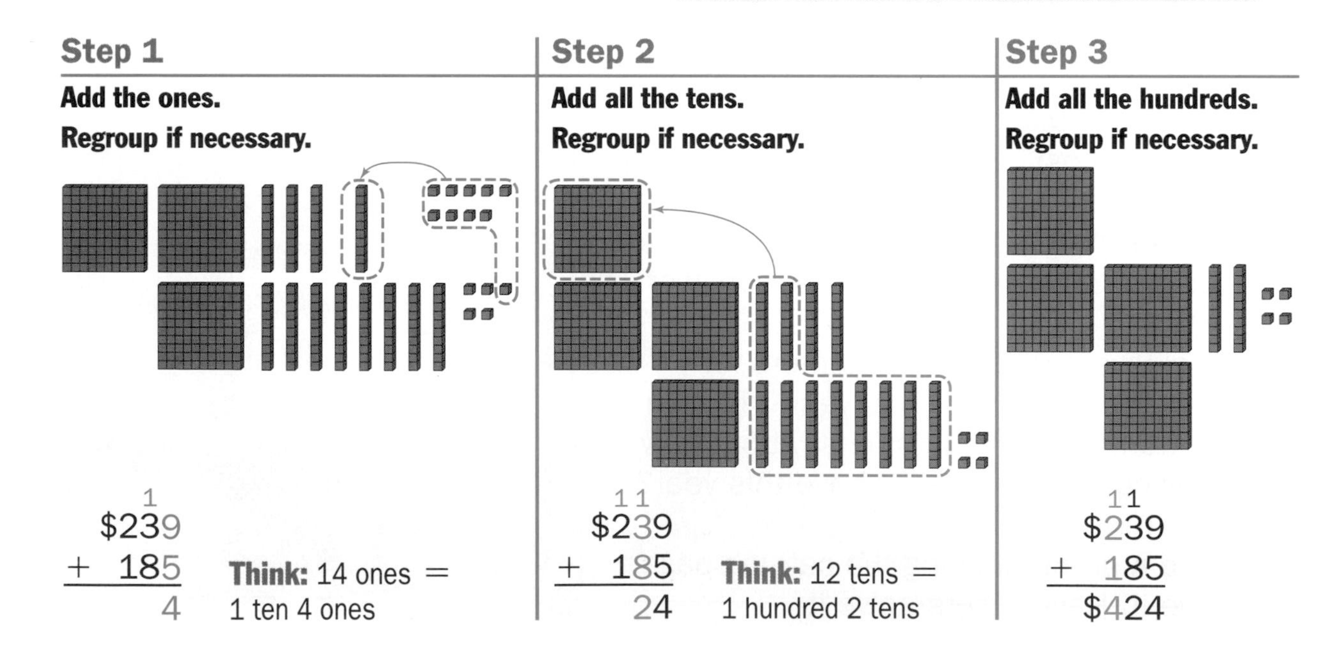

Step 1

Add the ones.
Regroup if necessary.

$$\begin{array}{r} ^{1} \\ \$239 \\ +\ 185 \\ \hline 4 \end{array}$$

Think: 14 ones = 1 ten 4 ones

Step 2

Add all the tens.
Regroup if necessary.

$$\begin{array}{r} ^{1\,1} \\ \$239 \\ +\ 185 \\ \hline 24 \end{array}$$

Think: 12 tens = 1 hundred 2 tens

Step 3

Add all the hundreds.
Regroup if necessary.

$$\begin{array}{r} ^{1\,1} \\ \$239 \\ +\ 185 \\ \hline \$424 \end{array}$$

The craftsperson would earn $424.

Talk It Over

- How can you tell that the answer is reasonable?
- Explain what is regrouped when adding 239 and 185.
- Add 348 and 562. Explain your method.

One design for a wampum belt uses 6,543 beads. The second design uses 5,762 beads. How many beads are needed to make both designs?

Add: 6,543 + 5,762

Estimate the sum. **Think:** 7,000 + 6,000 = 13,000

You can use paper and pencil to find the exact sum.

Step 1	Step 2	Step 3	Step 4
Add the ones. Regroup if necessary.	**Add all the tens. Regroup if necessary.**	**Add all the hundreds. Regroup if necessary.**	**Add all the thousands.**
$\begin{array}{r} 6{,}543 \\ +5{,}762 \\ \hline 5 \end{array}$	$\begin{array}{r} ^{1} \\ 6{,}543 \\ +5{,}762 \\ \hline 05 \end{array}$	$\begin{array}{r} ^{1\,1} \\ 6{,}543 \\ +5{,}762 \\ \hline 305 \end{array}$	$\begin{array}{r} ^{1\,1} \\ 6{,}543 \\ +5{,}762 \\ \hline 12{,}305 \end{array}$
	Think: 10 tens = 1 hundred 0 tens	**Think:** 13 hundreds = 1 thousand 3 hundreds	

You can also use a calculator. 6,543 + 5,762 = 12305.

To make both designs, 12,305 beads are needed.

More Examples

A
$$\begin{array}{r} ^{1\,1} \\ 823 \\ +578 \\ \hline 1{,}401 \end{array}$$

B
$$\begin{array}{r} ^{1} \\ 6{,}456 \\ +\quad 339 \\ \hline 6{,}795 \end{array}$$

C Calculator 5,743 + 198,400 = 204143.

Check for Understanding

Add. Estimate to check that your answer is reasonable.

1 $\begin{array}{r} 468 \\ +593 \\ \hline \end{array}$ **2** $\begin{array}{r} 863 \\ +\ 42 \\ \hline \end{array}$ **3** $\begin{array}{r} \$25.80 \\ +\ 17.90 \\ \hline \end{array}$ **4** $\begin{array}{r} 6{,}515 \\ +7{,}943 \\ \hline \end{array}$ **5** $\begin{array}{r} 229{,}650 \\ +\ 94{,}500 \\ \hline \end{array}$

Critical Thinking: Analyze **Show examples.**

6 Suppose you add two 4-digit numbers. What is the greatest number of digits that can be in the sum? the least number?

7 When adding two numbers, when would you use mental math? pencil and paper? a calculator?

CHECK

Turn the page for Practice.

PRACTICE

Practice

Add. Remember to estimate.

1 67 + 31

2 45 + 96

3 87 + 37

4 246 + 923

5 $2.50 + 7.75

6 9,125 + 8,993

7 2,988 + 699

8 12,548 + 7,647

9 $1,560 + 789

10 37,118 + 89,594

11 257 + 86

12 7,129 + 973

13 4,089 + 1,925

14 27,517 + 35,650

15 8,918 + 4,546

16 29,657 + 3,635

17 $45,851 + $675

18 895,420 + 175,950

Complete Mr. Valencia's checkbook balances.

	DATE	DESCRIPTION	AMOUNT	BALANCE
				$1,435.00
	11/7	Deposit: check from sale of computer	$376.00	$1,811.00
19	11/16	Deposit: paycheck	$590.00	■
20	11/18	Deposit: check from Aunt Ina	$135.00	■
21	11/24	Deposit: school loan	$475.83	■

Find only those sums greater than 5,000.

22 4,890 + 741

23 965 + 998

24 2,498 + 2,277

25 5,681 + 8,743

26 4,038 + 975

27 9,983 + 9,906

28 3,132 + 1,485

29 3,602 + 1,870

30 4,610 + 1,111

31 2,163 + 2,577

Write two addends for the sum. Each addend must be greater than 500.

32 1,460 **33** 2,500 **34** 9,654 **35** 18,785 **36** 234,112

Make It Right

37 André lined up the numbers and added. Tell what the mistake is, then correct it. Explain how estimating could help him.

75,896 + 9,420
Sum: 170,096

MIXED APPLICATIONS

Problem Solving

38 Frederick's Music Store is offering a special sale on a piano: Pay $900 now. Pay $3,400 in one year. What is the total cost of the piano?

39 Abe sold 465 pieces of pottery on Saturday and 637 pieces on Sunday. About how many pieces did he sell in all?

40 **Make a decision** A photography store is offering a special. If you have $25.00, which photo or photos would you choose? Explain.

41 **Logical reasoning** The sum of two numbers is 1,000. One addend is 50 more than the other addend. What are the two numbers?

42 **Write a problem** similar to the number puzzle in problem 41. Ask some classmates to solve it.

Cultural Connection

The Chinese Abacus

The Chinese abacus, or *suan pan* (swahn pan), can be used to add numbers.

The abacus uses place value. The rod at the right stands for the ones place. The next rod stands for tens, and so on. An upper bead placed next to the crossbar shows 5. A lower bead placed next to the crossbar shows 1.

Here is how you can add 5,637 and 1,826.

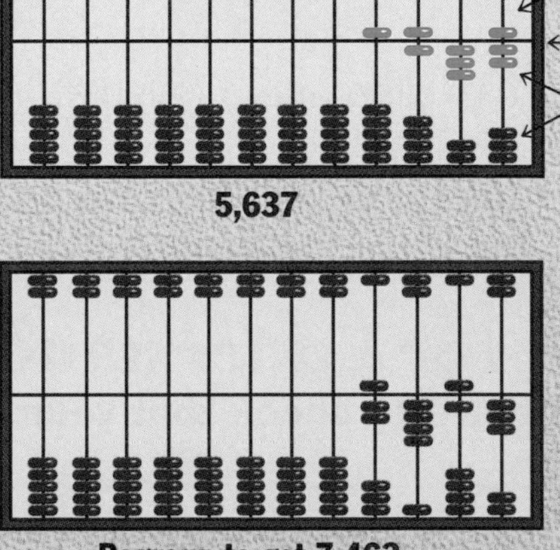

5,637

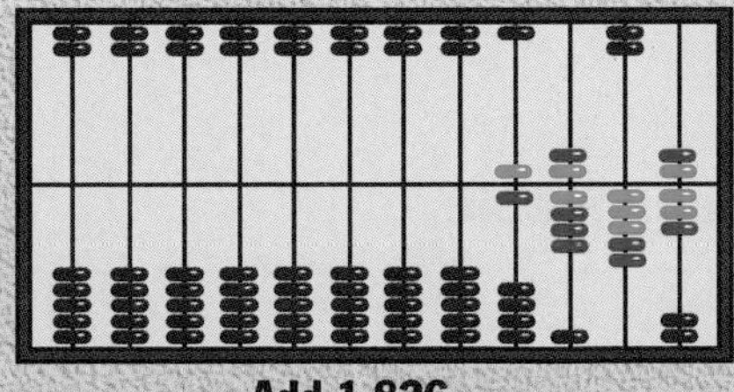

Add 1,826.

Regroup to get 7,463.

Use the abacus or drawings of the abacus to do ex. 11–13 on page 50.

Three or More Addends

LEARN

Does your family rent movies? How long would it take to watch these three movies?

Commutative Property
You can change the order of two addends without changing the sum.

Associative Property
You can change the grouping of three or more addends without changing the sum.

Add: 95 + 122 + 175

Estimate the sum. **Think:** 100 + 100 + 200 = 400

You can use the **Commutative** and **Associative** properties to help you find the exact sum.

Step 1	Step 2	Step 3
Add the ones. Regroup if necessary.	**Add all the tens. Regroup if necessary.**	**Add all the hundreds.**
$\begin{array}{r} {}^{1} \\ 95 \\ 122 \\ +175 \\ \hline 2 \end{array}$ **Think:** 5 + 5 = 10; 10 + 2 = 12	$\begin{array}{r} {}^{11} \\ 95 \\ 122 \\ +175 \\ \hline 92 \end{array}$ **Think:** 9 + 1 = 10; 10 + 7 + 2 = 19	$\begin{array}{r} {}^{11} \\ 95 \\ 122 \\ +175 \\ \hline 392 \end{array}$

You can also use a calculator. 95 + 122 + 175 = 392.

It would take them 392 minutes to watch all the movies.

CHECK

Check for Understanding

Add. Estimate to check that your answer is reasonable.

1 48 + 65 + 72 **2** 4,290 + 740 + 518 **3** 158 + 74 + 92 + 301

Critical Thinking: Analyze **Explain your reasoning.**

4 Why is it important to line up the ones digits when adding? Give an example using a 2-, a 3-, and a 4-digit number.

Practice

Add. Remember to estimate.

1 $\begin{array}{r} 47 \\ 54 \\ +\ 36 \\ \hline \end{array}$

2 $\begin{array}{r} 457 \\ 35 \\ +\ 153 \\ \hline \end{array}$

3 $\begin{array}{r} 1,268 \\ 793 \\ +\ \ 204 \\ \hline \end{array}$

4 $\begin{array}{r} \$24.69 \\ 73.45 \\ +\ \ 86.21 \\ \hline \end{array}$

5 $\begin{array}{r} 1,709 \\ 2,340 \\ 569 \\ +\ \ 753 \\ \hline \end{array}$

6 87 + 629 + 4,125

7 54 + 89 + 351 + 822

8 134 + 487 + 5,206 + 978

9 70 + 2,400 + 900

10 $12.52 + $7.45 + $11.89

11 10,500 + 2,750 + 3,478

MIXED APPLICATIONS
Problem Solving

Use the store ad for problems 12–13.

12 **What if** you can buy a CD player in July if you pay $7 each month until January and then pay $100.95. What would be the total cost? Is this a better price than buying the CD player from JC's Store?

13 **Make a decision** If you had $500, what would you buy to get the radio/cassette free? Explain.

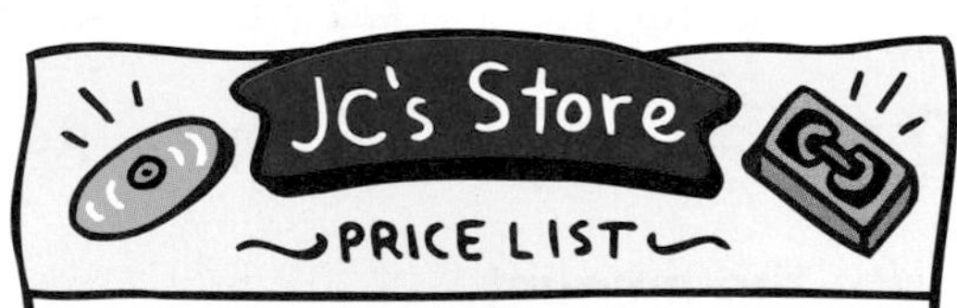

VHS Recorder: $199.95
Each video: $17.99

Rechargeable CD Player: $149.95

AM/FM/CD/Cassette Recorder: $99.95
Each CD: $11.99 Each cassette: $7.99

Zoom Supreme Design Camera: $399.95

Pay $30 now, pay rest next January
NO INTEREST!
Spend $500, get Radio/Cassette FREE!

more to explore

Other Addition Strategies

You can use other addition properties or strategies to help you add.

Add: 396 + 0 + 23

$\begin{array}{r} 396 \\ +\ \ 23 \\ \hline 419 \end{array}$

Think: 23 + 0 = 23
If you add zero to a number, you get the number.

Add: 143 + 313 + 242

$\begin{array}{r} 143 \\ 313 \\ +\ 242 \\ \hline 698 \end{array}$

Think:
Look for doubles.

Add. Explain your method.

1 582 + 344 + 142

2 2,639 + 0 + 632

3 1,056 + 356

midchapter review

Add mentally.

1 30 + 40 **2** 700 + 400 **3** 4,000 + 8,000 **4** 12,000 + 5,000

Estimate the sum.

5 64 + 19 **6** 395 + 654 **7** \$6.25 + \$5.54 **8** 4,186 + 765

9 212 + 689 **10** 552 + 74 **11** \$33.78 + \$65.00 **12** 5,425 + 14,650

Add. Use mental math when you can.

13 $\begin{array}{r} 75 \\ +18 \\ \hline \end{array}$ **14** $\begin{array}{r} 6,225 \\ +5,859 \\ \hline \end{array}$ **15** $\begin{array}{r} \$4.69 \\ +\ 8.05 \\ \hline \end{array}$ **16** $\begin{array}{r} 15,308 \\ +\ 6,475 \\ \hline \end{array}$ **17** $\begin{array}{r} 659 \\ +\ 74 \\ \hline \end{array}$

18 27 + 43 **19** 40 + 30 + 57 **20** 500 + 700 + 315

21 586 + 719 **22** 749 + 89 **23** \$712 + \$459 + \$1,570 + \$98

Write the amount.

24

25

Find the amount of change.

26 Price of memo pad: \$0.89
Cash given: \$5

27 Price of earrings: \$7.35
Cash given: \$10

Write in order from least to greatest.

28 \$62.05, \$6.50, \$26.50, \$62.50

29 \$19.62, \$119.15, \$1.67, \$16.92

Solve. Use any method.

30 The Renoir School sponsored an art show. There were 145 ceramic pieces, 274 paintings, and 78 pieces of jewelry. About how many pieces of art were there?

31 On Friday 650 people attended the show. On Saturday 893 people attended the show. On which day did more people attend?

32 Skye bought two pieces of students' work for \$8.50 and \$12.25. How much did he spend?

33 Journal Explain three different ways you can estimate 239 + 1,468 + 2,529.

developing number sense

MATH CONNECTION

Use Front Digits to Estimate

Dawn wants to buy a pair of skis that costs $115. She has $95 in her bank account. On her birthday, she gets $38. Does she have enough money?

Use the front digit of each number to estimate the answer.

Estimate: $95 + 38$

$\downarrow \quad \downarrow$

$90 + 30 = 120$ Compare: $\$120 > \$115$

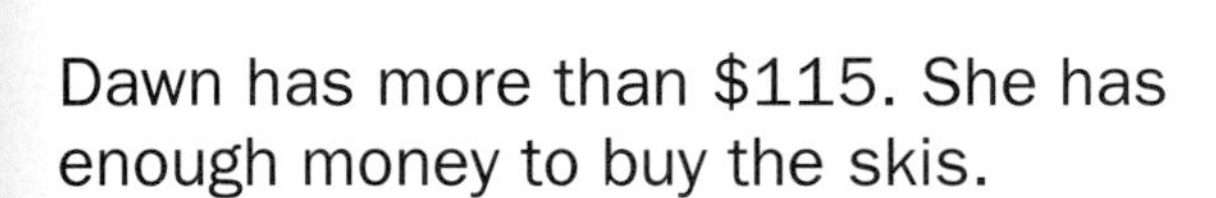

Dawn has more than $115. She has enough money to buy the skis.

Solve. Use front digits to estimate. Explain your reasoning.

1 Here are the results of a survey on how often customers shopped in Vesta's each month.
- More than 5 times: 247
- 3 to 5 times: 378
- Less than 3 times: 615

Were more or fewer than 1,000 people surveyed?

2 Vesta's will give away a free computer to the Sheerin School if the school can collect $5,000 worth of receipts. In November the school collected $2,650 in receipts. In December the school collected $3,600. Will the school get the free computer?

3 Ms. Clark has $200. She wants to buy a bicycle for $99, a fishing rod for $59, and a fishing reel for $75. Does she have enough money?

4 Dawn earns $22 dog-sitting and $16 recycling bottles and cans. Does she have enough money to buy a ski jacket for $29.99?

5 **Make a decision** Obando's local store sells the book *Long Journey* for $7.95. The book sells at Vesta's for $6.09. Obando can walk to the local store, but he must pay $2.25 for the bus fare to and from Vesta's. What should he do?

6 Freddy is collecting baseball cards. His album holds 200 cards. Last week he bought 64 cards at Vesta's. For his birthday he got 74 cards, and his neighbor gave him 92 cards. Can his album hold all these cards?

real-life investigation

APPLYING ADDITION AND SUBTRACTION

How We Spend Our Money

Americans spend their money in many different ways. In this activity, you will work with a group to decide what you would buy if you went on a $400 shopping trip.

Ways That Americans Spend Their Money

1 Health
2 Food
3 Rent
4 Items for the home
5 Transportation
6 Personal Expenses
7 Recreation
8 Clothing
9 Charities
10 School fees
11 Personal

What things would you buy?

Where would you buy them?

What prices would you pay?

DECISION MAKING

Spending Money

1 Shop for items your group would like to purchase by:

- ▶ visiting stores, markets, and other places where you can shop.
- ▶ collecting catalogs, newspapers, and store fliers.
- ▶ collecting advertisements, sales brochures, and coupons.

Calculate the prices you would need to pay for the items.

2 Decide on at least ten items you will buy with $400. List them in order of importance.

3 Describe where the items were bought. Include any coupons or advertisements you used.

4 Use a table to keep track of your spending. Record each item you buy, the amount already spent, and the balance or amount that is left.

Reporting Your Findings

5 Portfolio Prepare a report that describes what you buy. Include the following:

- ▶ Show your record of each item, the total amount spent, and the balance.

READING ARITHMETIC WRITING

Write a Paragraph Tell what items you bought and how much each one cost. Include the total you spent and how much you have left. Explain how you decided what items to buy and where to buy them.

6 Compare your report with the reports of other groups.

Revise your work.

- ▶ Are your calculations correct?
- ▶ Is your report clear and organized?
- ▶ Did you proofread your work?

MORE TO INVESTIGATE

PREDICT the greatest number of items you could buy from your original list with $250. Then check your prediction.

EXPLORE the items you buy in a week and how much money you actually spend.

FIND out why certain items are always on sale each year at the same time.

Subtraction Strategies

LEARN

Kits are popular with 8- to 12-year-olds. An explorer kit is $93 in the catalog. What is the sale price for the kit?

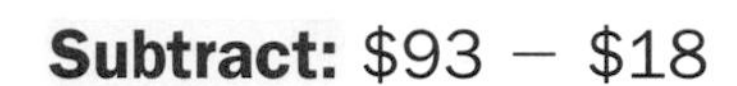

Subtract: $93 − $18

Look for ways to change the numbers so you can find the **difference** mentally.

Think: Add 2 to each number.

$$\begin{array}{l} \$93 - \$18 \\ +2 \downarrow \qquad \downarrow +2 \\ \$95 - \$20 = \$75 \end{array}$$

Think: Subtract 3 from each number.

$$\begin{array}{l} \$93 - \$18 \\ -3 \downarrow \qquad \downarrow -3 \\ \$90 - \$15 = \$75 \end{array}$$

The sale price of the kit is $75.

What if the sale is $22 off any purchase over $80. What would be the sale price of the kit?

Subtract: $93 − $22

You can also subtract mentally using this method.

$$\begin{array}{r} \$93 \\ -\ 22 \\ \hline \end{array}$$

Think: $93 − $20 = $73
$73 − $2 = $71

The sale price of the kit would be $71.

Check Out the Glossary
difference
subtract
See page 544.

CHECK

Check for Understanding

Subtract mentally. Explain your thinking.

1 28 − 18 **2** 85¢ − 52¢ **3** $276 − $99 **4** 485 − 260

Critical Thinking: Analyze **Explain your reasoning.**

5 Does subtraction have a Commutative Property like addition? Use place-value models to support your answer.

Practice

Subtract mentally.

1 26 − 8 **2** 31 − 7 **3** 29 − 19 **4** 48 − 29

5 74 − 18 **6** 96 − 35 **7** 86¢ − 19¢ **8** 153 − 28

9 293 − 76 **10** $7.77 − $1.98 **11** 645 − 497 **12** $8.04 − $6.79

13 592 − 103 **14** $0.47 − $0.14 **15** $2.64 − $0.99 **16** 435 − 197

ALGEBRA: PATTERNS Find each difference. What pattern do you notice?

17 **a.** 23 − 0 **b.** 59 − 0 **c.** 187 − 0 **d.** 365 − 0

18 **a.** 48 − 48 **b.** 546 − 546 **c.** 866 − 866 **d.** 738 − 738

MIXED APPLICATIONS
Problem Solving

Use the catalog page for problems 19–22.

19 Karen has $9.54. If she buys a Birds Around Us kit, how much money will she have left?

20 Sal has $25. Can he afford the tangram puzzles, electricity kit, and delivery?

21 Is it cheaper to buy the Electricity Exploration kit and the Build Your Own Radio kit or the Birds Around Us kit and two tangram puzzles? Explain.

22 **Write a problem** using the catalog. Solve it and have others solve it.

EXPLORATION COMPANY *Sale!*

Elecricity Exploration	$15.49
Build Your Own Radio	$17.99
Make Your Own Camera	$14.85
Plants Around Us	$8.69
Birds Around Us	$8.99
Tangram Puzzles	$9.87
Gift Box	$2.25

Take off an additional $5. for purchases over $25.

DELIVERY

$50.00 and less	$3.95
$50.01 - 100.00	$6.95
$100.01 and more	$9.95

mixed review • test preparation

Round to the nearest hundred and the nearest thousand.

1 721 **2** 854 **3** 2,319 **4** 3,975 **5** 25,542

Add mentally.

6 83 + 47 **7** 48 + 95 **8** 239 + 403 **9** 698 + 257

Extra Practice, page 490

Estimate Differences

LEARN

In-line skates are great, but they are expensive. Suppose you have saved $42 already. About how much more money do you need to save?

Estimate: $98 − $42

Round each number so you can find the difference mentally.

Think: Round to the nearest ten.

$$
\begin{array}{ccc}
\$98 & - & \$42 \\
\downarrow & & \downarrow \\
\$100 & - & \$40 = \$60
\end{array}
$$

You need to save about $60 more.

More Examples

A Estimate: 4,378 − 1,542

Think: Round to the nearest thousand.
4,000 − 2,000 = 2,000

B Estimate: 564 − 36

Think: Round to the nearest ten.
560 − 40 = 520

C Estimate: $19.75 − $6.52

Think: Round to the nearest dollar.
$20 − $7 = $13

D Estimate: $2.14 − $0.43

Think: Round to the nearest ten cents.
$2.10 − $0.40 = $1.70

CHECK

Check for Understanding

Estimate the difference. Tell how you rounded.

1 81 − 66 **2** $0.89 − $0.26 **3** 462 − 239 **4** $3.65 − $1.47

5 7,459 − 3,390 **6** $38.65 − $11.47 **7** 787 − 69 **8** $34.78 − $2.75

Critical Thinking: Analyze **Explain your reasoning.**

9 Tell how you could round the numbers in ex. 7 and 8 another way to estimate the differences. What would the estimates be?

Practice

PRACTICE

Estimate. Round to the nearest ten or ten cents.

1. 85 − 37
2. $0.69 − $0.24
3. 93 − 45
4. 82 − 26
5. $0.37 − $0.11
6. 73 − 33
7. 872 − 27
8. $2.58 − $0.38

Estimate. Round to the nearest hundred or dollar.

9. 418 − 149
10. $7.98 − $6.65
11. 645 − 265
12. 504 − 294
13. 774 − 286
14. 6,803 − 550
15. $27.62 − $5.28
16. 5,295 − 204

Estimate. Round to the nearest thousand or ten dollars.

17. 9,683 − 1,078
18. 2,063 − 1,066
19. $35.45 − $18.02
20. 6,993 − 6,063
21. $60.64 − $2.64
22. 9,255 − 754

MIXED APPLICATIONS

Problem Solving

23. This year 357 students registered for the In-line Skating program. Last year 289 registered. About how many more registered this year?

24. **ALGEBRA** Find three pairs of numbers that make each a true sentence.
a. △ − ○ = 50 **b.** ◇ + □ = 70

25. The sale price of a helmet is $35.99. You save $18.97 off the regular price. How much is the regular price for the helmet?

26. **Data Point** Ask 20 students what type of skating they like most. Record your data in a tally table. How many more students prefer ice skating to in-line skating?

more to explore

Using Front Digits

You can use the front digits to estimate an answer.

Estimate: 756 − 439 → 700 − 400 = 300

Think: 7 hundreds − 4 hundreds = 3 hundreds, or 300

Estimate the difference using front digits.

1. 35 − 24
2. 865 − 565
3. 3,712 − 1,630
4. 7,250 − 3,650
5. $0.92 − $0.54
6. $7.89 − $2.25
7. $9.43 − $6.25
8. $75.98 − $48.89

Extra Practice, page 490

Subtract Whole Numbers

LEARN

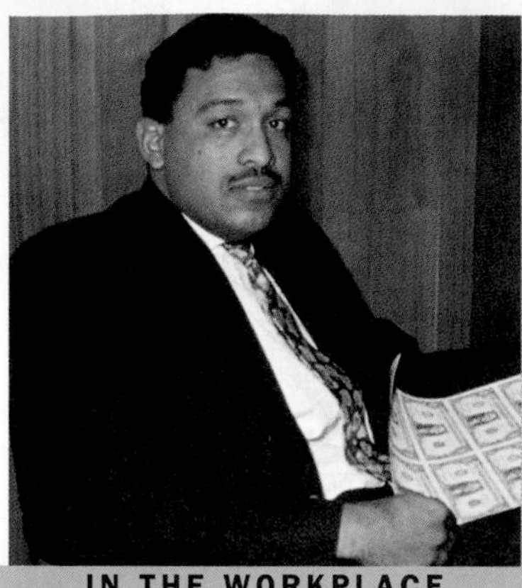

IN THE WORKPLACE

Larry Felix, Chief of the Office of External Relations, U.S. Dept. of the Treasury

The Department of the Treasury puts together money that has been burned, buried, or torn. Suppose a person brings $725 of burned money to the Treasury and only $458 can be restored. How much is not saved?

Subtract: $725 − $458

Estimate the difference.

Think: $700 − $500 = $200

Find the exact difference.

Step 1

Regroup if necessary.
Subtract the ones.

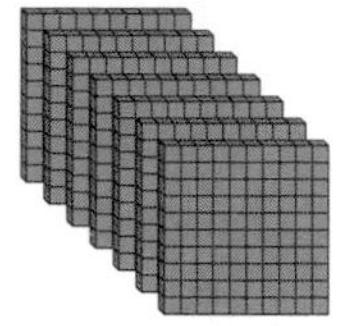

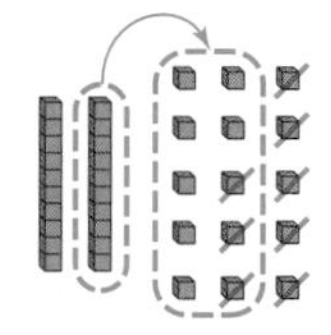

$$\begin{array}{r} \scriptstyle 1\ 15 \\ \$7\,2\,\not{5} \\ -\ \ 4\,5\,8 \\ \hline 7 \end{array}$$

Think: 2 tens 5 ones = 1 ten 15 ones

Step 2

Regroup if necessary.
Subtract the tens.

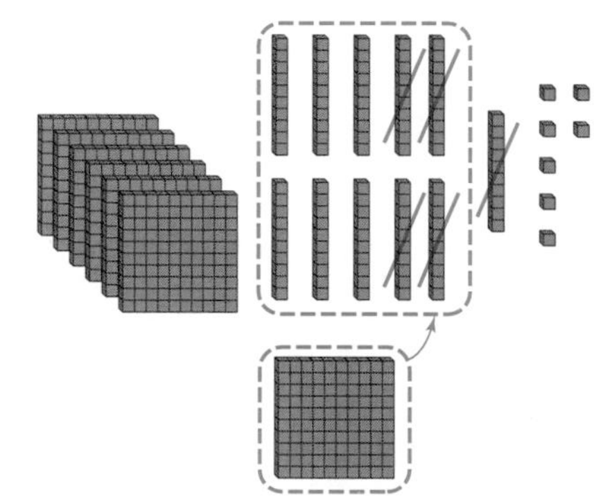

$$\begin{array}{r} \scriptstyle 11 \\ \scriptstyle 6\ \not{1}\ 15 \\ \$\not{7}\,\not{2}\,\not{5} \\ -\ \ 4\,5\,8 \\ \hline 6\,7 \end{array}$$

Think: 7 hundreds 1 ten = 6 hundreds 11 tens

Step 3

Regroup if necessary.
Subtract the hundreds.

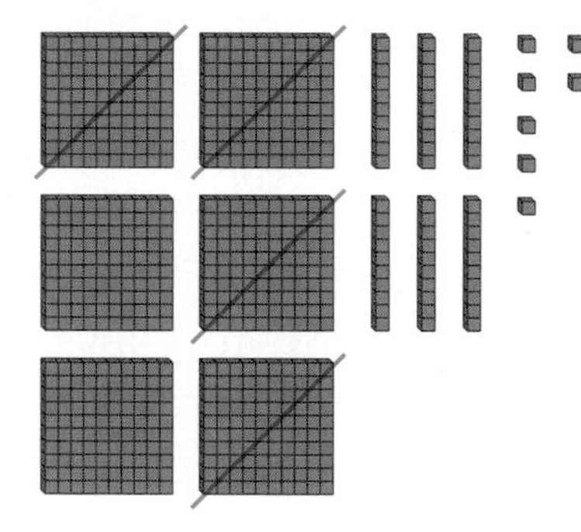

$$\begin{array}{r} \scriptstyle 11 \\ \scriptstyle 6\ \not{1}\ 15 \\ \$\not{7}\,\not{2}\,\not{5} \\ -\ \ 4\,5\,8 \\ \hline \$2\,6\,7 \end{array}$$

Check by adding.
$267 + $458 = $725

$267 is not saved.

Talk It Over

▶ How can you tell that the answer is reasonable?

▶ Subtract 283 from 564. Explain your method.

Suppose you buried $2,139 in your backyard. If insects destroyed some and only $1,351 was pieced together, how much money was destroyed?

Subtract: $2,139 − $1,351

Estimate the difference. **Think:** $2,000 − $1,000 = $1,000

You can use paper and pencil to find the exact difference.

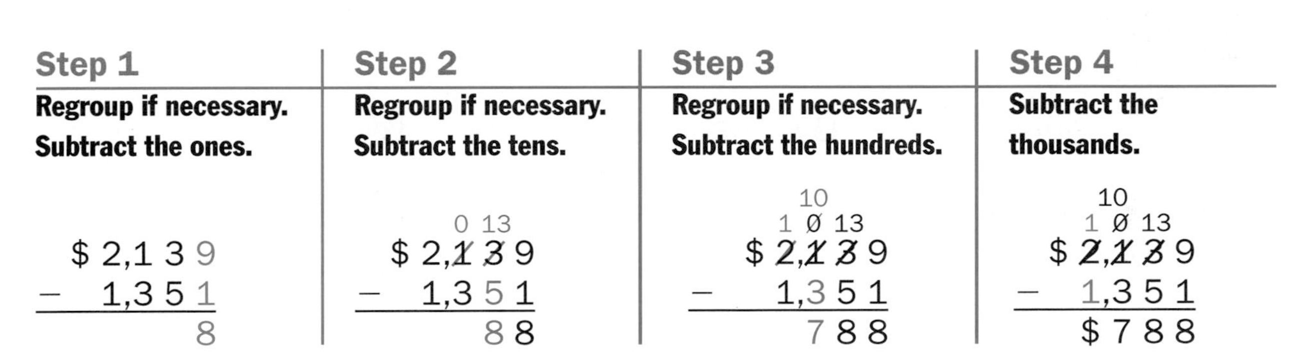

Step 1	Step 2	Step 3	Step 4
Regroup if necessary. Subtract the ones.	**Regroup if necessary. Subtract the tens.**	**Regroup if necessary. Subtract the hundreds.**	**Subtract the thousands.**
$2,139 − 1,351 8	(0 13) $2,1̸3̸9 − 1,351 88	(10) (1 0̸ 13) $2̸,1̸3̸9 − 1,351 788	(10) (1 0̸ 13) $2̸,1̸3̸9 − 1,351 $788

Check: 788 + 1,351 = 2,139

You can also use a calculator. 2,139 − 1,351 = 788.

The insects destroyed $788.

More Examples

A
(11)
(3 1̸ 15)
4̸,2̸5̸9
− 786
3,473

B
(14)
(7 4̸ 12)
$8̸5̸.2̸0
− 56.50
$28.70

C
(15)
(6 5̸ 12)
183,7̸6̸2̸
− 71,574
112,188

D  256,322 − 89,750 = 166572.

Check for Understanding

Subtract. Estimate to check that your answer is reasonable.

1 468 − 399

2 987 − 65

3 $37.25 − 18.57

4 8,458 − 6,973

5 328,475 − 90,580

Critical Thinking: Analyze **Show examples.**

6  To subtract two numbers, when would you use mental math? pencil and paper? a calculator?

CHECK

Turn the page for Practice.

PRACTICE

Practice

Subtract. Remember to estimate.

1 $\begin{array}{r} 66 \\ -\ 23 \\ \hline \end{array}$ **2** $\begin{array}{r} 91 \\ -\ 35 \\ \hline \end{array}$ **3** $\begin{array}{r} 227 \\ -\ 34 \\ \hline \end{array}$ **4** $\begin{array}{r} 768 \\ -\ 544 \\ \hline \end{array}$ **5** $\begin{array}{r} \$5.16 \\ -\ 3.08 \\ \hline \end{array}$

6 $\begin{array}{r} \$8.51 \\ -\ 2.72 \\ \hline \end{array}$ **7** $\begin{array}{r} 6{,}120 \\ -\ 3{,}600 \\ \hline \end{array}$ **8** $\begin{array}{r} 5{,}475 \\ -\ 884 \\ \hline \end{array}$ **9** $\begin{array}{r} 8{,}419 \\ -\ 5{,}206 \\ \hline \end{array}$ **10** $\begin{array}{r} \$32.65 \\ -\ 18.47 \\ \hline \end{array}$

11 $\begin{array}{r} 9{,}412 \\ -\ 4{,}063 \\ \hline \end{array}$ **12** $\begin{array}{r} 3{,}777 \\ -\ 899 \\ \hline \end{array}$ **13** $\begin{array}{r} 33{,}454 \\ -\ 26{,}847 \\ \hline \end{array}$ **14** $\begin{array}{r} \$2{,}163 \\ -\ 755 \\ \hline \end{array}$ **15** $\begin{array}{r} 45{,}223 \\ -\ 6{,}094 \\ \hline \end{array}$

16 \$585 − \$286 **17** 1,569 − 872 **18** 4,113 − 1,099

19 8,291 − 7,408 **20** \$75.79 − \$16.89 **21** 8,546 − 7,858

22 64,245 − 30,757 **23** 118,491 − 73,508 **24** 216,843 − 44,579

Use the sale sign to find the items that have:

25 the greatest difference in price.

26 the least difference in price.

27 a difference of about \$5.

28 a difference of \$4.04.

ALGEBRA Find the missing number.

29 79 + ■ = 90 **30** 84 − ■ = 54 **31** ■ − 65 = 200

32 947 − 540 = ■ **33** ■ + 183 = 544 **34** 33 + 79 + ■ = 212

Write the number as the difference of two numbers.

35 800 **36** 575 **37** 1,600 **38** 1,845 **39** 75,211

Make It Right

40 Here is how Ruth found 3,948 − 1,486. Explain what the mistake is, then correct it.

$\begin{array}{r} \overset{14}{3{,}948} \\ -\ 1{,}486 \\ \hline 2{,}562 \end{array}$

As Close As You Can Game!

First, make 30 cards—three cards for each of the digits 0 through 9.

Next, choose a 3-digit number, a 4-digit number, and a 5-digit number as target numbers.

You will need
- *index cards*

Player 1

Target Number	Number from cards	Score
364	362	2

0 9 1

Target Numbers

364 5,712 10,624

Play the Game

Play in a group of three or four students.

- ▶ Mix up the cards. Give six cards to each player.
- ▶ Take turns. Call out one target number. Players then use their cards to make a number as close as possible to the target number.
- ▶ To score, find the difference between the target number and the number that was made. Try to get the fewest points that you can.
- ▶ Record the information on score sheets. Play two more rounds. Use a different target number for each round. The player with the fewest points at the end of three rounds is the winner.

What strategy did you use to make the numbers with your cards?

mixed review • test preparation

1 28 + 19 **2** 67 + 39 **3** 544 + 183 **4** 947 + 540

Write the number in words.

5 350 **6** 2,306 **7** 38,009 **8** 90,014

Extra Practice, page 490

Choose the Operation

LEARN

Read **Bored on a trip? You can buy an electronic version of Wheel of Fortune for $25.50 or a travel version that is not electronic for $18.75. How much more is the electronic version?**

Plan To solve the problem, you need to decide which operation to use.

To find how much more the electronic version costs, find the difference.

Solve Subtract: $25.50 − $18.75 = $6.75

The electronic version costs $6.75 more than the travel version.

Look Back How can you check that your answer is correct?

CHECK

Check for Understanding

1 Karen buys two travel games that cost $8.99 and $5.30. What operation would you use to find how much she pays? Why would you use that operation?

Critical Thinking: Analyze **Explain your reasoning.**

Which operation would you use to find the answer?

2 You know the number of pieces in a construction set. How many pieces are in five sets?

3 How much less is the cost of a pint of craft paint than the cost of a quart of craft paint?

4 How much heavier is the electronic Wheel of Fortune than the travel version?

MIXED APPLICATIONS

Problem Solving

PRACTICE

1 Karen estimated that playing Clue would take 35 minutes. The actual time was 26 minutes. Was the actual time longer or shorter than her estimate? How much longer or shorter?

2 Rhoda is walking in a walkathon. She walks 10 kilometers, rests, walks 8 kilometers, rests, walks 6 kilometers, and rests. After how many kilometers do you think she will rest again? Why?

3 **Make a decision** Suppose you go on a class trip to a Native American museum. You have $7.00 for food and souvenirs. Postcards cost $0.65 each, a small drum costs $4.50, a bracelet costs $2.29, a necklace costs $5.75, fry bread costs $1.59, corn bread costs $1.80, and a fruit juice costs $0.80. How much would you spend?

4 **Spatial reasoning** Show four different ways to get from point *A* to point *B* by traveling along the lines.

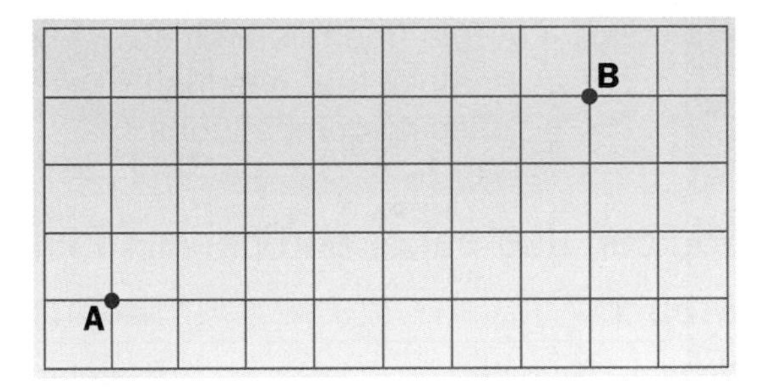

5 **Data Point** Use the Databank on page 533. Douglas is hiking up Mount Sunflower, Kansas. He stops for lunch at about 2,000 feet. Abdul is hiking up Mount Rogers, Virginia. He stops for lunch at about 3,000 feet. Who has farther to go to the top? About how much farther does he have to go?

6 **Write a problem** about buying items in a hardware store. The problem needs to be solved by using addition. Then use the same numbers to write a subtraction problem. Ask others to solve your problems and to explain why they chose the operation needed to solve each problem.

Use the line plot for problems 7–9.

7 How many more students preferred the zoo to City Hall? the Science Museum to City Hall?

8 How many students voted for the zoo or the Science Museum?

9 **What if** the class had also gone to Sea World. How do you think the line plot would change?

Favorite Class Trips

Science Museum	City Hall	Zoo
x x x x x x x	x x x x x x	x x x x x x x x x x x

Subtract Across Zero

LEARN

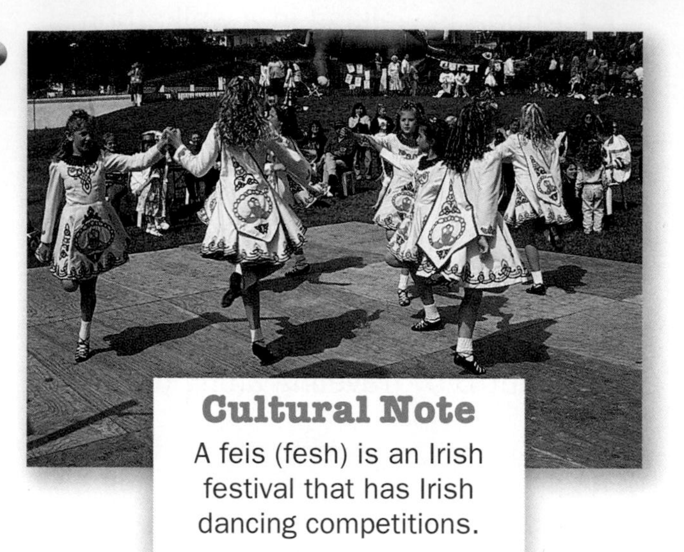

Cultural Note

A feis (fesh) is an Irish festival that has Irish dancing competitions.

It is fun to go to cultural dance competitions. There were 300 dancers at a feis. Of them, 124 won at least one prize. How many did not win prizes?

Subtract: 300 − 124

Estimate the difference.

Think: 300 − 100 = 200

You can use paper and pencil to find the exact difference.

Step 1

No ones. No tens.
Regroup the hundreds.

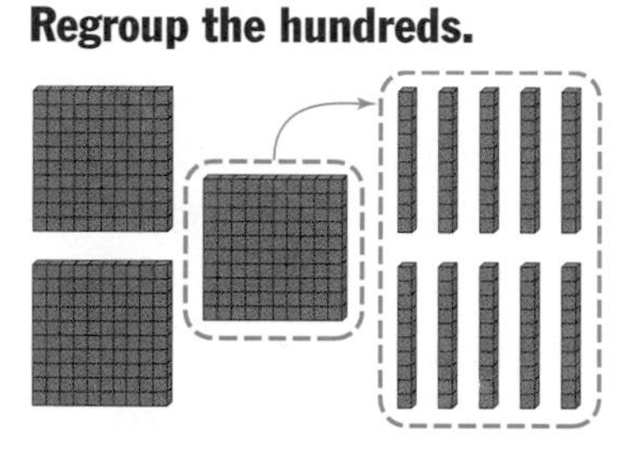

$$\begin{array}{r} \overset{2}{\cancel{3}}\,\overset{10}{\cancel{0}}\,0 \\ -\,1\,2\,4 \\ \hline \end{array}$$

Think: 3 hundreds = 2 hundreds 10 tens

Step 2

Regroup the tens.

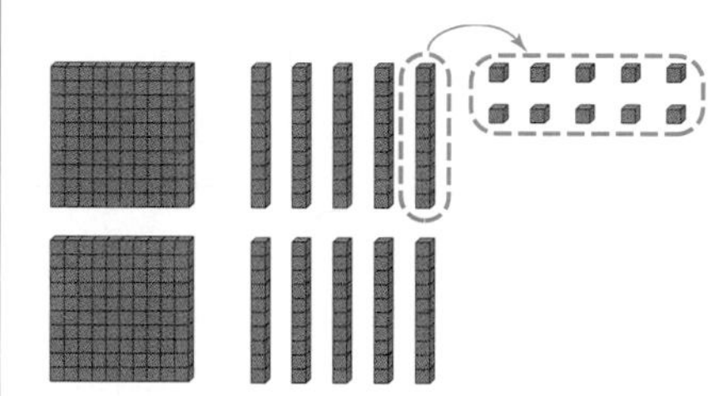

$$\begin{array}{r} \overset{2}{\cancel{3}}\,\overset{\overset{9}{\cancel{10}}}{\cancel{0}}\,\overset{10}{0} \\ -\,1\,2\,4 \\ \hline \end{array}$$

Think: 10 tens = 9 tens 10 ones

Step 3

Subtract.

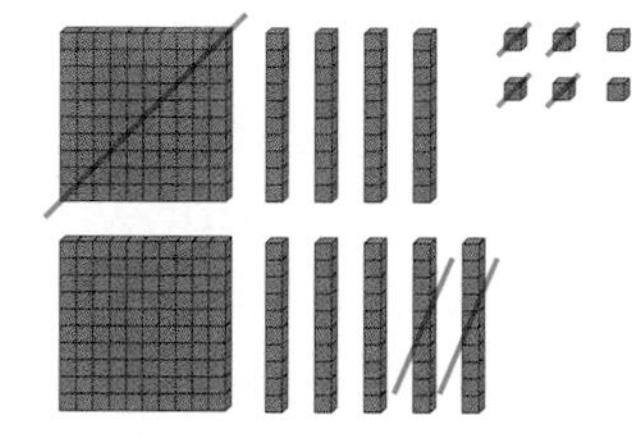

$$\begin{array}{r} \overset{2}{\cancel{3}}\,\overset{\overset{9}{\cancel{10}}}{\cancel{0}}\,\overset{10}{0} \\ -\,1\,2\,4 \\ \hline 1\,7\,6 \end{array}$$

176 dancers did not win prizes.

CHECK

Check for Understanding

Subtract. Estimate to check that your answer is reasonable.

1 802 − 421

2 2,502 − 726

3 $70.08 − 42.90

4 700 − 219

5 45,000 − 9,411

Critical Thinking: Analyze **Explain your reasoning.**

6 Give two ways to complete the regrouping:
600 = 5 hundreds ■ tens ■ ones.

PRACTICE

Practice

Subtract. Remember to estimate.

1 890 − 457

2 901 − 55

3 400 − 232

4 2,003 − 1,150

5 4,700 − 925

6 $27.00 − 12.60

7 $90.70 − 48.05

8 7,024 − 949

9 8,000 − 2,450

10 9,000 − 6,079

11 105 − 97 **12** 10,805 − 9,724 **13** $80.60 − $12.85 **14** $90.00 − $28.48

MIXED APPLICATIONS
Problem Solving

15 About how much higher is the Carrantuohill than the Wicklow Mountains? SEE INFOBIT.

16 Brenda drove 25 miles to the O'Reillys' house. Then she drove another 7 miles to the football game. How far did she drive?

17 **Make a decision** You want to buy potatoes to make Irish stew. A bag of 25 potatoes costs $1.67. You can get a free bag of 15 potatoes if you buy a bag of spinach for $2.38. What would you do? Why?

INFOBIT
The highest point in Ireland is Carrantuohill, which is 1,041 meters high. The Wicklow Mountains, also in Ireland, reach their highest point at 926 meters.

18 **What if** you drive 2 miles for gas and then 31 miles to the feis. What is the best estimate of the total distance?
a. 20 miles **b.** 30 miles **c.** 40 miles

19 Tickets to the feis for Mr. O'Reilly and his 9-year-old daughter, Orla, cost $9.00. Orla's ticket cost $3.25. How much did Mr. O'Reilly's ticket cost?

mixed review • test preparation

1 56 + 12 **2** 88 + 64 **3** 349 + 283 **4** 847 + 671

Write the number in expanded form.

5 803 **6** 4,098 **7** 51,800 **8** 1,702,400 **9** 930,032

Extra Practice, page 491

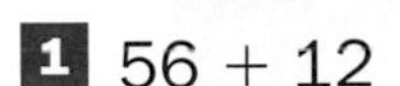

Problem Solvers at Work

Read
Plan
Solve
Look Back

LEARN

Part 1 Find Needed or Extra Information

Ramón eats breakfast at the Morningside Diner.

Work Together

Tell what information is needed to solve the problem. Tell what information is not needed. Then, solve the problem.

1 Which is less expensive, ordering the Hearty Meal or ordering the items separately?

2 **What if** you dislike the fruit cup. Should you still order the Late Riser? Tell why.

3 You want to order eggs Benedict and juice for breakfast. How much does that cost?

4 **Make a prediction** Predict what item would be chosen most often if your class orders breakfast from the Morningside Diner. How can you check your prediction?

Part 2 Write and Share Problems

Sabrina wrote the problem on the right.

5 What information is needed to solve Sabrina's problem? What information is extra?

6 Rewrite Sabrina's problem without the extra information. Then, solve it.

7 Replace the question in Sabrina's problem with a different question. Find the answer.

8 Write your own problem that includes extra information.

9 Trade problems. Solve at least three problems written by your classmates.

10 READING ARITHMETIC WRITING **Write a Paragraph** Tell about the most interesting problem that you solved. Explain why you thought it was interesting.

STUDENT TO STUDENT

CHECK

How much will it cost if you buy 2 waffles and 2 muffins?

Muffins	$1.18	Crackers	$0.50
Pancakes	$2.79	Toast	$1.25
Waffles	$2.29	Eggs (2)	$1.65
Juice	$1.37	Bacon	$2.19
Milk	$1.55	Cereal	$1.20

Sabrina Benitez
Shaughnessy School
Lowell, MA

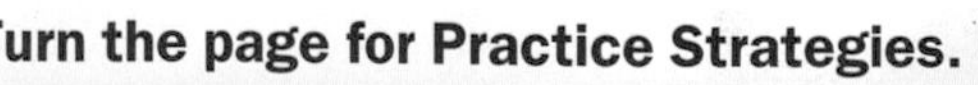
Turn the page for Practice Strategies.
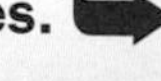

PRACTICE

Menu

Choose five problems and solve them. Explain your methods.

1 The theater sold 357 tickets for the 2 P.M. show and 825 tickets for the 7 P.M. show. About how many tickets were sold for the two shows?

2 Dolores bought from a catalog a sweater that cost $39.98. She paid $44.83 including delivery and tax. How much was the delivery and tax?

3 Bobby bought a notebook for $2.79 and a binder for $3.65, including tax. How much did he pay for the two items?

4 At a "Buy 2 for the Price of 1" sale, Noah bought 2 boxes of cereal for $4.39 and 2 bottles of fruit punch for $1.59. How much did Noah save?

5 Ellis bought a secondhand bicycle for $27. A similar bicycle sells new for $185. How much money did he save by buying the secondhand bicycle?

6 Petra and Axel buy food for a party. Drinks cost $7.69, cookies, chips, and crackers cost $15.54, fruit costs $14.35, and cheese costs $11.89. Will $50 cover their costs?

7 Allison and Maddie buy 1,000 glass and clay beads. There are 400 more glass beads than clay beads. How many glass beads are there?

8 Write the 1994 populations in order from greatest to least. Alabama: 4,218,792; Arizona: 4,075,052; California: 31,430,697; Delaware: 706,351; Montana: 856,047

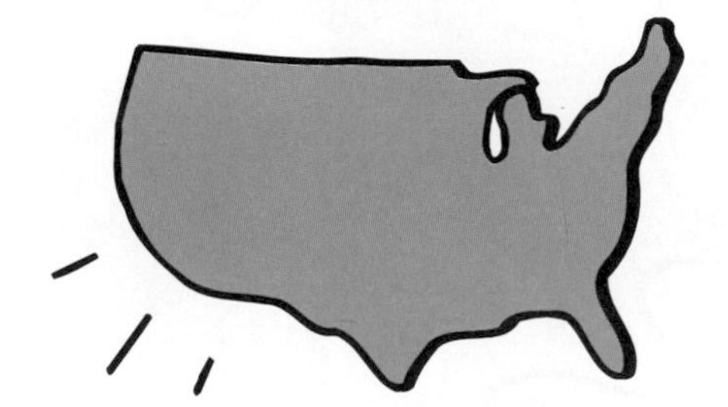

Choose two problems and solve them. Explain your methods.

9 **ALGEBRA** These two scales are balanced.

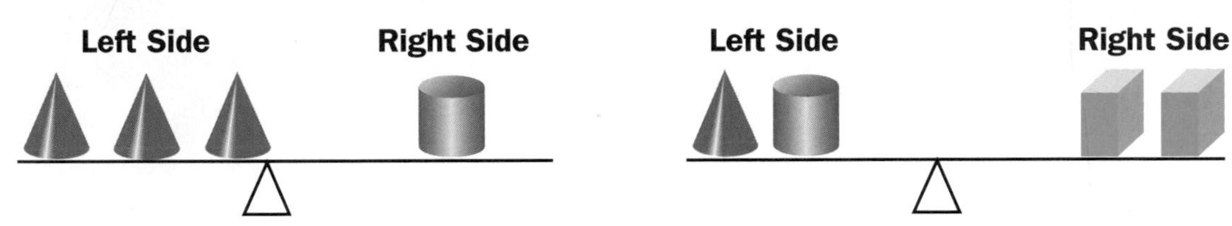

Suppose these shapes were put on a balance scale. Would the left and right sides balance? If not, tell what you can do to balance them.

a. **Left Side**  **Right Side**

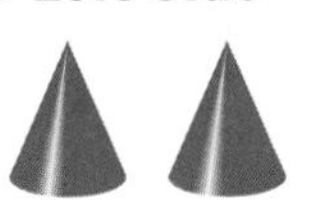 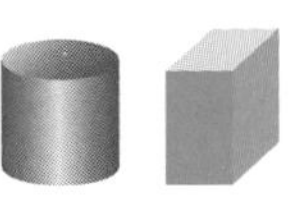

b. **Left Side** **Right Side**

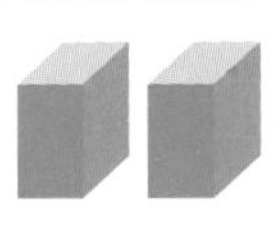 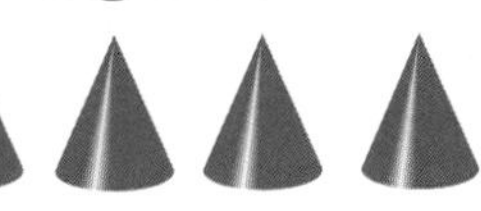

c. **Left Side**  **Right Side**

d. **Left Side** **Right Side**

 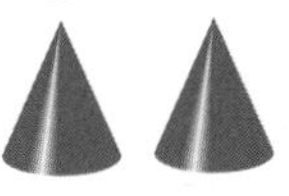

10 Even numbers are numbers like 2, 4, 6, 8, 10, 12, and 14. Odd numbers are numbers like 1, 3, 5, 7, 9, 11, and 13.

a. Tell if the sum will be even or odd. Give examples.

- ▶ sums of two odd numbers
- ▶ sums of two even numbers
- ▶ sums of one even and one odd number

b. Tell if the sum is correct or incorrect without adding.

$12 + 4 = 16$ $\quad$ $27 + 36 = 64$

$154 + 356 = 501$ $\quad$ $201 + 198 = 399$

11 **At the Computer** Create a spreadsheet so that you can record your income (any money that you get), your expenses (any money that you spend), and your balance (any money that you have left) for one week. Predict what your balance at the end of the week will be before finding it. Write about what you notice.

Extra Practice, page 491

Language and Mathematics

Complete the sentence. Use a word in the chart. (pages 40–69)

1 The ▪ of 100 and 40 is 60.

2 You can make change by ▪.

3 An example of the ▪ Property is
$(3 + 2) + 5 = 3 + (2 + 5)$.

4 You ▪ to change 45 to 3 tens 15 ones.

5 The ▪ Property lets you change 63 + 78 to 78 + 63.

Vocabulary
addends
sum
difference
regroup
Commutative
Associative
counting up

Concepts and Skills

Write the money amount. (page 40)

6

7

Find the amount of change. (page 40)

8 Price: $2.45
Amount given: $5

9 Price: $6.73
Amount given: $10

10 Price: $12.59
Amount given: $20

Compare. Write >, <, or =. (page 42)

11 $14.78 ● $14.53 **12** $25.87 ● $36.14 **13** $6,562 ● $6,852

Add or subtract mentally. (pages 44, 58)

14 40 + 37 **15** 12 + 30 + 42 **16** 45¢ − 6¢ **17** 325 − 298

Estimate the sum or difference. (pages 46, 60)

18 93 + 85 **19** 938 − 654 **20** $8.56 + $7.89 **21** 7,150 − 975

22 154 + 712

23 1,248 − 879

24 5,625 − 3,569

25 $84.65 + 39.15

26 65,125 − 48,279

Add or subtract. Use mental math when you can. (pages 44, 58)

27 $\begin{array}{r} 58 \\ +29 \\ \hline \end{array}$ **28** $\begin{array}{r} \$7.85 \\ +\ 3.17 \\ \hline \end{array}$ **29** $\begin{array}{r} 4{,}180 \\ +5{,}932 \\ \hline \end{array}$ **30** $\begin{array}{r} 7{,}405 \\ 401 \\ +\ 889 \\ \hline \end{array}$ **31** $\begin{array}{r} 37{,}956 \\ 9{,}586 \\ +\ 740 \\ \hline \end{array}$

32 $\begin{array}{r} 63 \\ -19 \\ \hline \end{array}$ **33** $\begin{array}{r} 504 \\ -\ 78 \\ \hline \end{array}$ **34** $\begin{array}{r} \$72.35 \\ -\ 18.29 \\ \hline \end{array}$ **35** $\begin{array}{r} 8{,}047 \\ -3{,}918 \\ \hline \end{array}$ **36** $\begin{array}{r} 26{,}045 \\ -\ 8{,}550 \\ \hline \end{array}$

37 39 + 51 **38** 6,610 + 2,590 **39** 75 + 420 **40** \$843 + \$372

41 38 − 19 **42** 750 − 297 **43** \$68.37 − \$39.55 **44** 6,000 − 4,514

Think critically. (page 48)

45 Analyze. Tell what the mistake is, then correct it.

$\begin{array}{r} 4{,}972 \\ +\ 5{,}885 \\ \hline 9{,}757 \end{array}$

46 Generalize. Write *always, sometimes,* or *never.* Give examples to support your answer.

a. The sum of two 3-digit numbers is a 3-digit number.

b. The difference of two 4-digit numbers is a 5-digit number.

c. The sum of two addends is less than one of the addends.

d. The difference of two numbers is greater than one of the numbers.

MIXED APPLICATIONS
Problem Solving

(pages 66, 70)

Use the price list for problems 47–50.

The Party Store

hats (8)..........\$1.89
bugles (8).......\$1.59
plates (8).......\$2.39
cups (8)..........\$1.59
napkins (8)..........\$0.79
balloons (8).........\$0.59
party favors (8)...\$2.75
Twister..............\$3.47

Party Packages.......**\$8.99**
(consists of 8 plates, 8 cups, 8 napkins, 8 balloons, 8 party favors, Twister game)
(prices include tax)

47 Carla has \$10. Does she have enough to buy 8 hats, 8 plates, 8 napkins, 8 cups, and 8 balloons?

48 Jamal hands the cashier a five-dollar bill to pay for party favors and balloons. What is his change?

49 Karen buys 8 plates, 8 cups, and 8 napkins. She gets \$6.03 back in change. How much did she give the cashier?

50 Is the Party Package a good deal? Do you save any money by buying the package instead of buying each item separately? If so, how much?

Write the money amount.

1

2

Find the amount of change.

3 Price of notebook: $3.29
Amount given: $5

4 Price of tape: $7.25
Amount given: $10

5 Price of shirt: $16.58
Amount given: $20

6 Price of CD: $14.95
Amount given: $20

Compare. Write >, <, or =.

7 $17.27 ● $17.24

8 $32.56 ● $29.17

9 $4,052 ● $4,115

Estimate the sum or difference.

10 258 + 316

11 2,681 − 992

12 $75.60 + 47.10

13 56,720 − 43,951

Add or subtract. Use mental math when you can.

14 73 + 17

15 275 + 125

16 $56 − $10

17 418 − 397

18 $5.28 + 6.19

19 7,216 + 504 + 372

20 $92.14 − 68.26

21 $26,000 − 18,543

Solve. Use the prices for problems 22–24.

22 Karen bought a T-shirt and shorts. She gave the clerk $30. How much change did she get?

23 Tell if there is needed or extra information. Sue bought 3 sweatshirts and 2 pants. How much did she spend?

24 Bob got $20 for his birthday. Does he have enough to get 3 caps? Explain your reasoning.

25 The Petrified Forest contains 93,533 acres. Yosemite has 761,236 acres. Which is larger?

What Did You Learn?

Three fourth-grade classes collected money for a children's charity.

Collections for One Week	
Mr. Potter's class	\$15.45
Ms. Hale's class	\$20.75
Ms. Angelini's class	\$12.00

Bob estimates that the total amount collected for the week is about \$50.00.

Anita's estimate is \$48.00.

- ▶ Explain how Bob and Anita might have made their estimates.
- ▶ Decide which is a better estimate. Explain your reasoning.

A Good Answer

- explains how each estimate is made using estimation strategies
- explains logically why one estimate is better by comparing the estimates to the actual total

You may want to place your work in your portfolio.

What Do You Think

1. Can you write, compare, and estimate money amounts? If not, what do you do when you find it difficult to finish a problem?

2. What type of estimation are you comfortable doing?
   - Rounding to the greatest place
   - Rounding to the nearest ten
   - Rounding to the nearest hundred
   - Rounding to the nearest thousand
   - Other. Explain.

math science technology CONNECTION

Cultural Note

About 600 B.C. in Lydia, an ancient country that is now part of Turkey, the first "coins" were made. The government of Lydia shaped electrum, a mixture of gold and silver, into bean-shaped lumps of the same weight and stamped them with special symbols. The word *coin* is from a word that means "stamp."

THE MANY FORM$ OF MONEY

Scientists know of about 100 different elements that are found on Earth. Of these, about 80 of them are metals. Metals are easy to form into different shapes and are shiny when polished, so they can easily be used to make coins.

Thousands of years ago, people used natural objects such as feathers, shells, and salt bars, and metals such as iron, copper, gold, and silver as money.

For centuries, coins were made of metals equal to the value of the coin. In the past, coins made of gold, silver, and copper were most common in the United States. Gold was last used for coins in 1933, and silver was removed from most coins in 1965. The five-cent coin (nickel) is still made with some nickel metal in it. All other coins are made from mixtures of copper and zinc.

Feather money was used by the Pacific Islanders of Santa Cruz. Tiny red feathers, glued together, were tied to fiber coils up to 32 ft long.

Today, people use different types of money to buy things. Sometimes, money can be exchanged electronically, using a computer.

- ▶ Describe the different types of metals that you have seen.
- ▶ Where do you think we get metals from?

Some Notes and Coins

In addition to coins, countries have paper money called notes. The note in the United States is the dollar.

You can make a dollar four ways using only one type of commonly used coin—4 quarters, 10 dimes, 20 nickels, and 100 pennies.

1 Suppose you have 7 coins in your pocket that equal $1.00. What could they be?

2 What is the least number of coins you can use to make 87¢?

3 Use the Databank on page 534. Look at the money of Botswana. Show three ways to make 47 thebes.

At the Computer

4 What would it be like to have a world without money? What advantages and disadvantages do you think there might be? Use a word processing program to write about your ideas.

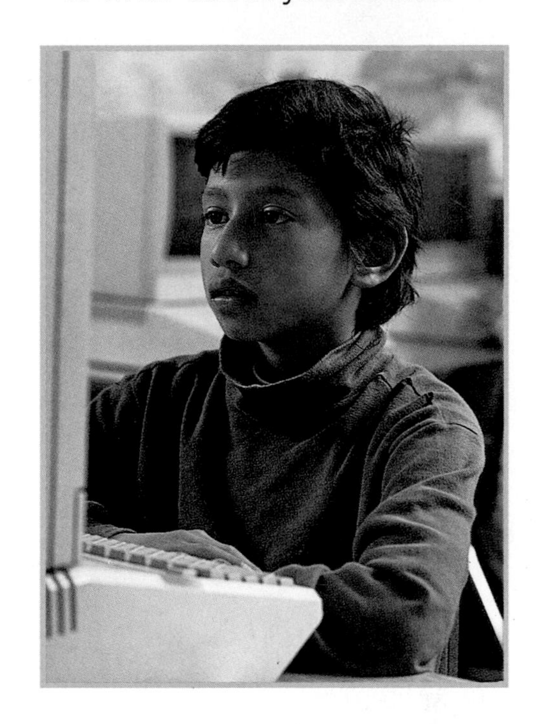

CHAPTER 3

TIME, DATA, AND GRAPHS

THEME

All About Us

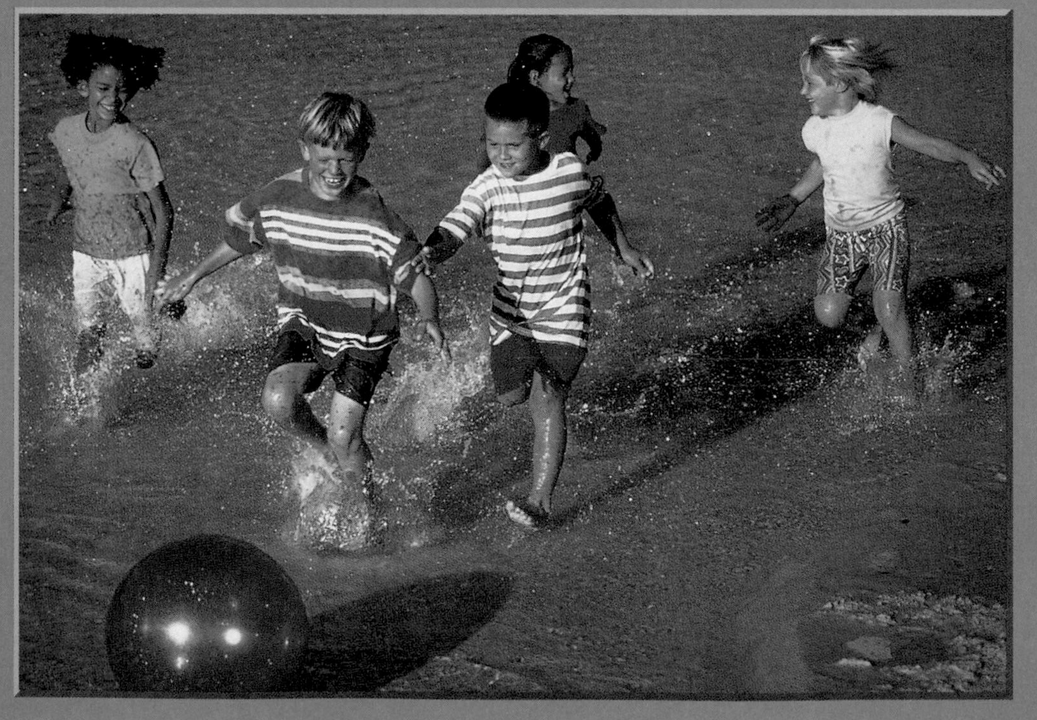

What do you think fourth graders are like? In this chapter, you will collect information about yourself and others. The information will show interesting facts about you!

What Do You Know

What are some of the activities you do during summer or winter vacations? On the right are the results of a survey of fourth graders.

Summer or Winter Activities

Activities	Number of Students
Ballet	𝍸
Bowling	𝍸 𝍸 ‖
Ice skating	𝍸 𝍸 𝍸
In-line skating	𝍸 𝍸 𝍸 𝍸
Skiing	𝍸 ‖
Swimming	𝍸 𝍸 𝍸 𝍸 𝍸

1 What activity is most popular? How do you know?

2 Make three statements based on the data.

3 Plan an afternoon of activities. Make a schedule that shows how you spend the time. Include activities that take less than a half hour and those that take more than an hour.

Write: Compare/Contrast **When you read, you often find out what something is like by comparing it to something else. Sometimes the writer gives you clues or you have to look for the likenesses and differences.**

The table above shows things that fourth grade students like to do during summer or winter vacations. Choose two activities to compare and contrast.

1 How are the two activities alike? How are they different?

Vocabulary

A.M., p. 84
P.M., p. 84
elapsed time, p. 84
ordinal numbers, p. 85
range, p. 90
median, p. 90
mode, p. 90
pictograph, p. 94
key, p. 94
bar graph, p. 100
ordered pair, p. 104
line graph, p. 106

Time

LEARN

Lena, a fourth-grade student, gets up every weekday morning at 7:00 A.M. to get ready for school.

Below are the times for her morning activities.

She gets dressed.

Read: seven-fifteen *or* fifteen minutes after seven *or* a quarter after seven

Write: 7:15

She eats her breakfast.

Read: seven-thirty *or* thirty minutes after seven *or* half past seven

Write: 7:30

She leaves for school.

7:45

Read: seven forty-five *or* forty-five minutes after seven *or* a quarter to eight

Write: 7:45

► Lena's brother gets up 15 minutes (min) earlier than she does. At what time does he get up? How do you know?

CHECK

Check for Understanding

Write the time in two different ways.

1 

2 

3 

Critical Thinking: Generalize

4 Name three activities that you enjoy doing that may take this amount of time.
a. about one minute **b.** about one hour (h) **c.** about one day (d)

Practice

Write the time in two different ways.

1 

2 

3 

4 

PRACTICE

Choose the most reasonable unit of time (seconds, minutes, hours, days, weeks).

5 It takes Lena about 15 ■ to walk her dog.

6 A fourth grader goes to school about 180 ■ in a year (y).

Which estimate of time is more reasonable?

7 see a movie
120 seconds (s) or 120 minutes

8 learn to play a guitar
24 hours or 24 weeks (wk)

9 brush your teeth
2 minutes or 2 days

10 drive 600 miles
10 hours or 10 weeks

MIXED APPLICATIONS

Problem Solving

11 Sammy buys two notebooks that cost $1.60 each. He gives the cashier a $5 bill. How much change does he get?

12 Kathy goes to a karate class at 4:15 every Tuesday after school. The class lasts an hour. What time does Kathy finish her karate class?

13 **Logical reasoning** Karl is putting numbers into three groups: Numbers Greater than 25, Even Numbers, and Numbers with 2 in the Tens Place. Which number fits in all three groups?
a. 29 **b.** 22 **c.** 26

14 **Data Point** Make a class pictograph that shows the time students leave for school. Use clock faces with students' names and place them beside a time. What does the graph tell you about your class?

mixed review • test preparation

1 $52 + $5

2 $300 − $200

3 $88 − $59

4 $127 − $95

5 $185 + $705

6 $97 − $9

Extra Practice, page 492

LEARN

Elapsed Time

You and your friends begin watching a video of last year's Community Dance Expo at 11:30 A.M. The video runs for 2 hours 30 minutes. Will the video be over by the time soccer practice begins at 2:30 P.M.?

You can count on to find the time you will finish the video.

First count on the hours.

Think: 2 hours (2 h)
11:30 A.M. ⟩ 1 hour
12:30 P.M. ⟩ 1 hour
1:30 P.M.

Then count on the minutes.

Think: 30 minutes (30 min)
1:30 P.M. ⟩ 30 minutes
2:00 P.M.

The video will be over at 2:00 P.M., before soccer practice begins at 2:30 P.M.

Your dance partner for this year's Community Dance contest arrives at 4:30 P.M. and leaves at 7:45 P.M. How long do you practice together?

You can count on to find the **elapsed time.**

First count on the hours.

Think: From 4:30 P.M. to 7:30 P.M. is 3 hours (3 h).

Then count on the minutes.

Think: From 7:30 P.M. to 7:45 P.M. is 15 minutes (15 min).

You practice for 3 hours 15 minutes.

1st Place

Talk It Over

- Why do we need A.M. and P.M. when referring to a time?
- **What if** the video lasted 1 hour 50 minutes. At what time would you have to begin the video to be finished by 2:00 P.M.? Explain your answer.

Check Out the Glossary
A.M.
P.M.
elapsed time
ordinal numbers
See page 544.

You can also use a calendar to find elapsed time.

Rehearsals for this year's Dance Expo will begin on the Monday after Thanksgiving. They will run every day for 12 days. On what day and date will the last rehearsal be?

Step 1
Use the calendar. Find the Monday after Thanksgiving.

Note: Use **ordinal numbers** when you talk about dates.

The Monday after Thanksgiving is November 30th.

Step 2
Use November 30th as the first day and count on 12 days.

Rehearsals will start on November 30th and end on Friday, December 11th.

The last rehearsal will be on December 11th.

November

Sun	Mon	Tu	Wed	Th	Fri	Sat
1	2	3 Election Day	4	5	6	7
8	9	10	11 Veterans Day	12	13	14
15	16	17	18	19	20	21
22	23	24	25	26 Thanks-giving	27	28
29	30					

December

Sun	Mon	Tu	Wed	Th	Fri	Sat
		1	2	3 Visit Senior Home	4	5 4th grade play
6	7	8	9	10	11	12
13	14	15	16	17	18	19
20	21	22 Winter Begins	23	24	25	26
27	28	29	30	31		

Check for Understanding

Complete the table.

1	Start time: 10:20 P.M.	Elapsed time: 2 h 30 min	End time: ■
2	Start time: 9:30 A.M.	Elapsed time: ■	End time: 3:50 P.M.
3	Start time: ■	Elapsed time: 5 h 45 min	End time: 1:00 P.M.
4	Start date: September 1	Elapsed time: 38 days	End date: ■
5	Start date: January 12	Elapsed time: ■	End date: February 2

Critical Thinking: Analyze **Explain your reasoning.**

6 The school art show lasts 3 weeks and ends on December 31st. When does it start?

CHECK

Turn the page for Practice.

PRACTICE

Practice

Tell what time it will be:

1 in 20 min.

2 in 3 h 30 min.

3 in 5 h 5 min.

4 2 h 50 min after 7:30 A.M.

5 3 h 30 min after 9:15 A.M.

Tell what time it was:

6 a half hour ago.

7 1 h 10 min ago.

8 2 h 15 min ago.

9 2 h 30 min before 10:15 A.M.

10 5 h 20 min before 4:20 P.M.

Use the calendars for February and March for problems 11–14.

11 What is the date of the third Tuesday in February?

12 On what day of the week is Groundhog Day?

13 If the Presidents' show runs from Presidents' Day through the end of the month, for how many days does the show run?

14 If you leave on Valentine's Day and return 16 days later, on what day and date do you return?

February

Sun	Mon	Tu	Wed	Th	Fri	Sat
1	2 Groundhog Day	3	4	5	6	7
8	9	10	11	12	13	14 Valentine's Day
15	16 President's Day	17	18	19	20	21
22	23	24	25	26	27	28

March

Sun	Mon	Tu	Wed	Th	Fri	Sat
1	2	3	4	5	6	7

Complete the table.

15	Start time: 8:15 A.M.	Elapsed time: 3 h 50 min	End time: ■
16	Start time: 5:50 A.M.	Elapsed time: ■	End time: 10:40 A.M.
17	Start date: May 10	Elapsed time: 45 days	End date: ■

MIXED APPLICATIONS
Problem Solving

18 Joe left on February 20th and returned on March 6th. Wes left on February 18th and returned on March 20th. Whose trip was longer? How much longer?

19 **Make a decision** You have $7.20. What would you buy if a fruit salad cost $2.75, a tuna sandwich cost $2.90, a chicken salad cost $3.50, and a cheese melt cost $2.50?

20 How many days have passed since the start of the school year? How many days are left?

21 **Write a problem** about elapsed time. Use a calendar. Have a classmate solve your problem.

22 **Data Point** Make a line plot for your class like the one shown. Place one check mark above all the ways you have traveled. Can you tell how many students are in your class from the graph? Why or why not?

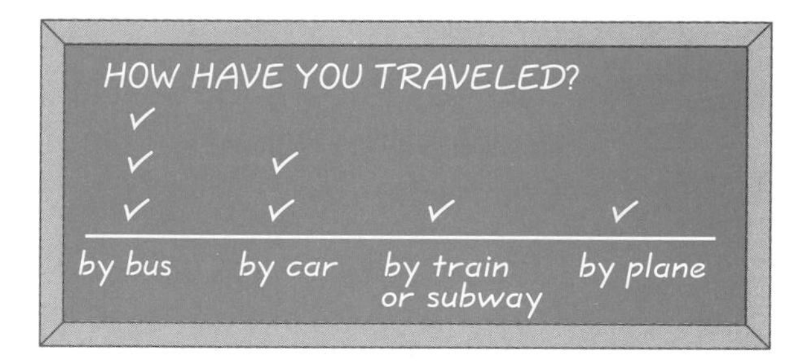

Cultural Connection Swahili African Time

Many people in the East African countries of Kenya (KEN-yuh) and Uganda (yoo-GAN-duh) start counting the hours of the day at sunrise. Their 1:00 is the same as 7:00 A.M. standard time.

	Breakfast	Lunch	Dinner
Standard Time	7:00	12:00	6:00
Swahili Time	1:00	6:00	12:00

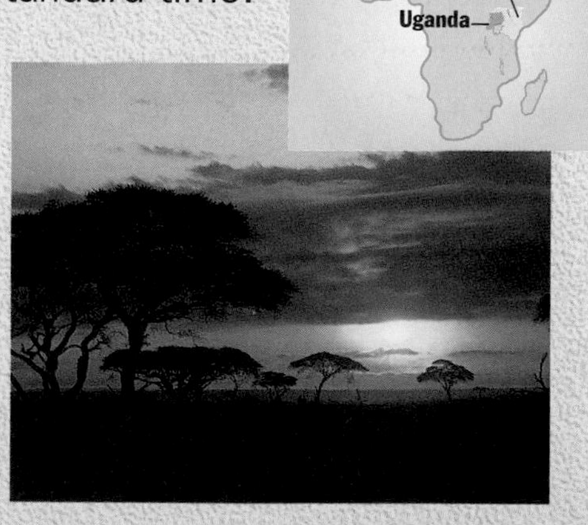

Use the chart to answer the question.

1 How can you change Swahili time to standard time?

2 At what time would you get to school in Swahili time?

Extra Practice, page 492

Problem-Solving Strategy

Read
Plan
Solve
Look Back

Work Backward

LEARN

Read **Tyler and her mother want to go by bus to a 5:30 P.M. swimming class. This class is at the Midwood Community Center. It takes them about 25 minutes to walk to the Hillsvale bus stop. Which bus should they take? When should they leave home?**

BUS SCHEDULE

LOCATION	P.M.	P.M.	P.M.	P.M.	P.M.
Bus Port	2:15	3:00	4:12	4:30	5:00
Hillsvale	2:25	3:10	4:20	4:45	5:15
Junction North	2:35	3:20	4:35	4:50	5:30
Maple	2:40	3:25	4:42	4:59	5:40
Ridgewood	2:45	3:30	4:50	5:05	6:01
Tenpal	2:50	3:35	4:55	5:10	6:09
Midwood Community Center	3:02	3:47	5:05	5:15	6:20
Corroll	3:12	3:57	5:19	5:30	6:35
Astor	3:17	4:02	5:30	5:41	6:47

Plan You can work backward to solve this problem. Use the bus schedule.

Solve Start by finding the last bus that can arrive at the Midwood Community Center before 5:30 P.M.

Think: A bus arrives at the Community Center at 5:15 P.M. It leaves Hillsvale at 4:45 P.M.

Then find the time Tyler and her mother need to leave home.

Think:

End time	4:45 P.M.
Elapsed time	25 minutes
Start time	4:20 P.M.

They should take the 4:45 P.M. bus and leave home by 4:20 P.M.

Look Back How can you work forward to check your answers?

CHECK

Check for Understanding

1 **What if** Tyler and her mother want to arrive at the Center about a half hour before the class. Which bus should they take? When should they leave home? Explain your method.

Critical Thinking: Summarize

2 Give examples of problems where you can work backward to find:
a. the amount of change. **b.** the time that a concert ends.

MIXED APPLICATIONS

Problem Solving

PRACTICE

1 Mei wants to finish her exercises by 7:10 P.M. She usually takes 30 minutes to exercise. What is the latest time that she can start?

2 Lee has $5.00 to start. He uses 7 quarters at the arcade. Then he buys a bottle of juice for 89¢. How much money does he have left?

3 Kyle likes to create patterns using his calculator. He chooses a number, adds 8, subtracts 4, adds 2, then subtracts 1. His result is 27. What number did he choose?

4 Karlene has $57.30 in her savings account at the end of the month. During the month she took out $25 and put in $14.98. How much money did she have at the beginning of the month?

Use the schedule for problems 5–7.

5 How much longer is the swimming class than the dance class?

6 Which class is the longest?

7 Choose two activities from the schedule that you would like to do. How long is each activity? How much time would you have between the activities?

Midwood Community Center

Schedule

Activity	Begin	End
Gymnastics	10:00 A.M.	11:15 A.M.
Soccer	11:35 A.M.	12:45 P.M.
Dance	1:00 P.M.	2:25 P.M.
Swimming	3:05 P.M.	4:40 P.M.

8 **Spatial reasoning** Three views of the same cube are shown. What shape is on the opposite face from the circle?

9 What was the difference between the times Mihir Sen took to swim the Panama Canal and the Palk Strait? SEE INFOBIT.

10 **Write a problem** that you can work backward to solve. Use the bus schedule on page 88 or use a bus schedule of your own.

INFOBIT
In 1966, Mihir Sen of India swam the Panama Canal in 4 hours. He also swam the Palk Strait in 34 hours 15 minutes.

Range, Median, and Mode

Did you know that the more you read, the better you get at it? This table shows the amount of time five friends spend reading. There are many ways to describe the data.

How Much Time Do We Spend Reading

Friend	Numbers of Hours
Alex	2
Sue	1
Roberto	3
Katie	7
Sharella	2

Work Together

Work with a partner. Use inch graph paper to make a strip for the number of hours each friend reads.

You will need
- *inch graph paper*
- *scissors*

Arrange the strips in order from least to greatest.

SUE   ROBERTO

- ▶ Compare the shortest strip with the longest strip. The difference between the greatest number and the least number in a set of data is called the **range.**
- ▶ Look at the strip in the middle. The number in the middle of a set of data that is in order is called the **median.**
- ▶ Find any strips that are the same length. The number that occurs most often in a set of data is called the **mode.**

Check Out the Glossary
range
median
mode
See page 544.

Survey seven of your classmates about the time they spend reading books each week. Display your data in a table. Then find the median, mode, and range of your data.

Talk It Over

- ▶ What are the range, median, and mode of the data above? What do they tell you about time spent reading?
- ▶ What are the range, median, and mode of your data? What do they tell you about how much your friends read?

Make Connections

Tim surveyed 19 students. He showed the data on a line plot and wrote questions he wanted to answer.

You can find the range, median, and mode of the data from the line plot. Each X on the line plot stands for a student's response.

Range Subtract the least number on the scale from the greatest number. $7 - 1 = 6$. The range is 6 books.

Median Cross off an X from each end of the line plot. Repeat this step until only one X is left. Then read the median from the scale below the X. The median is 3 books.

Mode Look at the tallest column of Xs. Then read the mode from the scale. The mode is 2 books.

How many books did you read last month?

Data

6 5 1 2 3 1 2 5 7 2
6 2 5 3 3 1 2 1 3

Number Of Books Read By Students Last Month

1	2	3	4	5	6	7
xxxx	xxxxx	xxxx		xxx	xx	x

1. Did more students read 3 books than any other number of books?
2. What is the greatest difference in the number of books read?
3. Did at least half the students read 3 or more books?

Check for Understanding

1 Find the range, median, and mode from the line plot on the right.

Estimate of the Number of Pages in the Last Book Read

80	90	100	110
xxxx	xxx	xx	xx

2 **What if** you surveyed one more student who read a book that was 90 pages long. Would the range, median, or mode change? Why or why not?

Critical Thinking: Analyze **Explain your reasoning.**

3 Which of Tim's questions can you answer with the range? the median? the mode?

4 Journal Which is easiest to read from the line plot, the range, median, or mode? Which is hardest? Why?

CHECK

Turn the page for Practice.

PRACTICE

Practice

Find the range, median, and mode for the set of data.

1 Heights of students (in inches):
56 40 36 60 40

2 Prices of cassette players:
$60 $80 $80 $38 $72

3 Numbers of letters in names:
13 10 5 27 8 16 11

4 Ages of members of Carla's family:
35 70 2 43 9 7 68

5 Numbers of students in a class:
21 29 31 28 25 36 25

6 Scores of math quiz:
65 90 100 80 75 80 90

Use the data in ex. 1–6 to tell whether the sentence is *true* or *false*. Then explain your answer.

7 More students are 60 inches tall than any other height.

8 At least half the students got 80 or more on the quiz.

9 The longest name is only 2 letters longer than the shortest name.

10 The members of Carla's family have many different ages.

Use the data at the right for ex. 11–13.

11 What is the range, median, and mode for the data?

12 Write three questions about the data that can be answered using the range, median, or mode.

13 Write two statements about the data. Exchange your statements with others. Explain why their statements are true or false.

Number of minutes spent studying yesterday by seven students:

35 25 60 40 25 30 25

Make It Right

14 Kara found the median score. Explain what mistake was made, then give the correct answer.

My scores:
80 50 90 40 100
My median score is 90.

MIXED APPLICATIONS

Problem Solving

Game	1	2	3	4	5
Ada	60	60	60	60	60
Jan	40	40	60	80	90

Use the data at the right for problems 15–16.

15 How can the score 60 describe Ada's and Jan's scores? Use the words *range, median,* or *mode* in your answer.

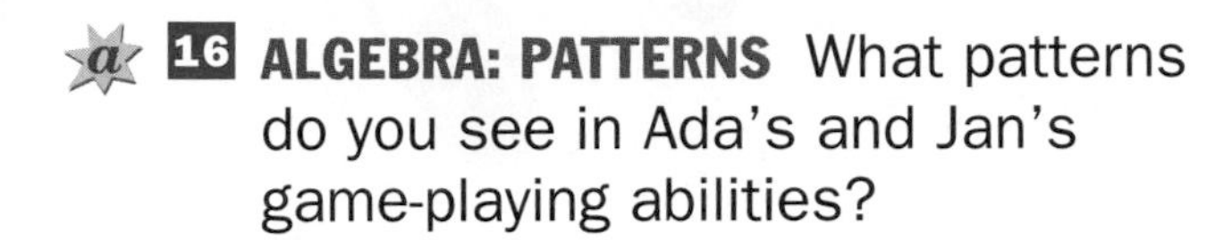

16 **ALGEBRA: PATTERNS** What patterns do you see in Ada's and Jan's game-playing abilities?

17 Seth's game scores are 40, 60, and 80. Sarah's scores are 20, 90, and 90. Which student has the greater total score? By how much?

18 The least expensive game in the arcade costs \$0.25 to play. The range in the costs of the games is \$1.50. How much does the most expensive game cost to play?

19 Rewrite the problem, leaving out any extra information.

Brandon is meeting his friends at the mall at 2:45 P.M. He leaves home at 2:10 P.M. and gets to the mall at 2:59 P.M. Is he early or late?

20 **Data Point** Use your data about your classmates' reading time. Write two questions that can be answered and two that cannot. Ask others to answer your questions or explain why they cannot.

more to explore

Where's the Median?

Six students line up according to height. No student has the median height.

45 in. 45 in. 46 in. 48 in. 49 in. 52 in.

To find the median, think about the height halfway between 46 inches and 48 inches. So the median height is 47 inches.

Find the median height, in inches (in.), of the group.

1 40 in., 42 in., 44 in., 45 in.

2 37 in., 44 in., 46 in., 48 in., 49 in.

3 48 in., 48 in., 53 in., 55 in., 56 in., 59 in.

4 39 in., 40 in., 47 in., 53 in., 56 in., 56 in.

LEARN

Pictographs

Does everyone you know like the same kinds of music? The frequency table shows the results of a fourth-grade survey. A pictograph of the data has been started.

Work Together

Work with a partner. Copy and complete the pictograph. Then answer the questions.

What is your favorite type of music?		
Type of music	**Tally**	**Frequency**
Pop	~~||||~~ ~~||||~~ ||	12
Rock	~~||||~~ ~~||||~~ ~~||||~~ ~~||||~~ ||||	24
Country	~~||||~~ |||	8
R&B	~~||||~~ ~~||||~~ ~~||||~~ ~~||||~~	20
Rap	||||	4
Other	~~||||~~ |||	8

What is your favorite type of music?	
Pop	웃 ㅓ
Rock	웃 웃 웃
Country	
R&B	
Rap	
Other	
Key: 웃 stands for 8 votes.	
ㅓ stands for 4 votes.	

► How did you decide what symbols to use to show the number of students who chose country music? R&B?

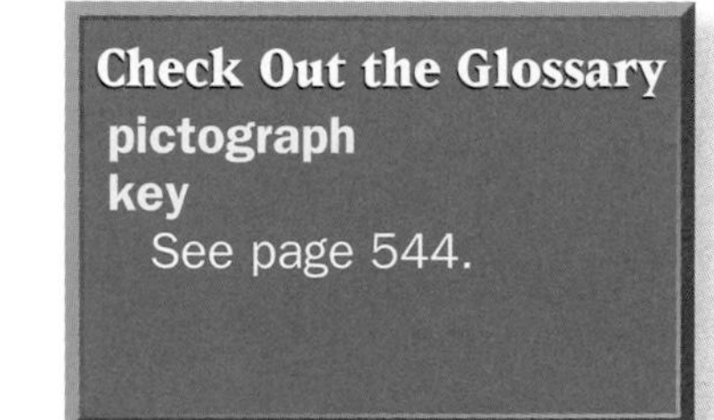

Check Out the Glossary
pictograph
key
See page 544.

► What type of music is the favorite most of the time? How can you tell by looking only at the pictograph?

► **What if** 4 students who chose pop music and 4 students who chose R&B decide they like rock music better. How will the pictograph change?

Make Connections

To make a pictograph:

- ▶ list the items (type of music).
- ▶ choose a symbol to represent the responses (number of votes).
- ▶ use a key to tell the number each symbol stands for.
- ▶ draw symbols to show the responses for each item.
- ▶ write a title.

CHECK

Check for Understanding

1 **What if** you use [symbol] to stand for 4 votes. How will this change the pictograph? Which key do you prefer? Why?

Critical Thinking: Compare and Contrast

2 **Write: Compare/Contrast** Look at the table and the pictograph on page 94. Which was easier to use to compare the data?

PRACTICE

Practice

Use the table and the pictograph for problems 1–4.

Favorite sport of students		
Sport	**Tally**	**Total**
Baseball	~~IIII~~ ~~IIII~~ ~~IIII~~	15
Basketball	~~IIII~~ ~~IIII~~ ~~IIII~~ ~~IIII~~ ~~IIII~~ ~~IIII~~	
Tennis	~~IIII~~ IIII	9
Soccer	~~IIII~~ ~~IIII~~ ~~IIII~~ ~~IIII~~ I	

Favorite sport of students	
Baseball	
Basketball	
Tennis	
Soccer	
Key: stands for 6 students. stands for 3 students.	

1 Copy and complete the table and the pictograph. Which sport do most students say is their favorite?

2 How many more students said baseball, rather than tennis, was their favorite sport?

3 How many students were surveyed? Explain how you know.

4 **Data Point** Choose a topic for a "favorite" survey. Survey your classmates or other students in your school. Make a pictograph to show the results.

midchapter review

Write the time in two different ways.

1 

2 

3 

4 

Complete the table.

	Activity	Start Time	Elapsed Time	End Time
5	Trip to museum	7:30 A.M.	■	12:15 P.M.
6	Baseball game	3:45 P.M.	3 hours 20 minutes	■
7	Piano class	■	50 minutes	5:20 P.M.
8	Trip to Peru	■	20 days	May 16
9	Summer camp	June 18	■	July 2

Find the range, median, and mode for the set of data.

10 Minidisk player prices:
$89 $101 $95 $95 $101

11 Number of hours spent studying:
1 3 1 4 4 3 3 3 1

12 Number of plays seen:
12 9 3 2 2 7 3

13 Number of languages spoken:
1 1 2 1 2 2 1 3 1

14 Number of best friends:
3 4 2 2 5

15 Number of states visited:
5 1 3 7 2 8 0

Use the pictogaph for problems 16–19.

16 Which football team do most people say is their favorite?

17 How many people say the Buffalo Bills is their favorite team?

18 How many people were surveyed?

19 Write a paragraph that compares the data shown in the pictograph.

20 Journal Describe how you choose a symbol for a pictograph and what the symbol stands for.

Favorite Football Team

Miami Dolphins (3 full footballs)
Dallas Cowboys (5 full footballs, 1 half football)
San Francisco 49ers (4 full footballs)
Buffalo Bills (4 full footballs, 1 half football)

KEY: (football) stands for 6 people.
(half football) stands for 3 people.

Graphs and Tables

A computer can help you create tables and graphs that are linked to each other. Create a table, then click on the graph tool to display the data in a bar graph.

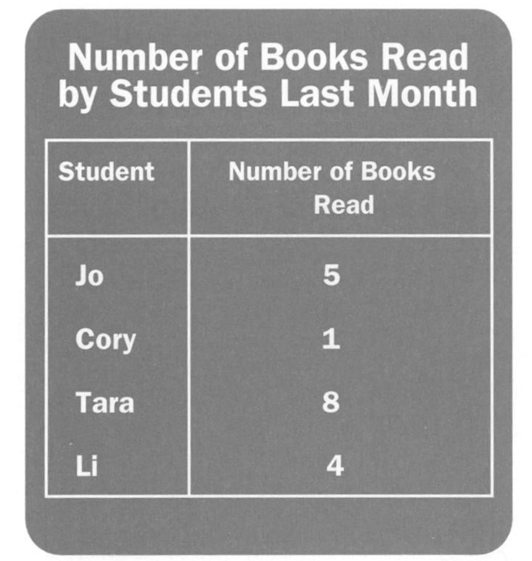

Number of Books Read by Students Last Month

Student	Number of Books Read
Jo	5
Cory	1
Tara	8
Li	4

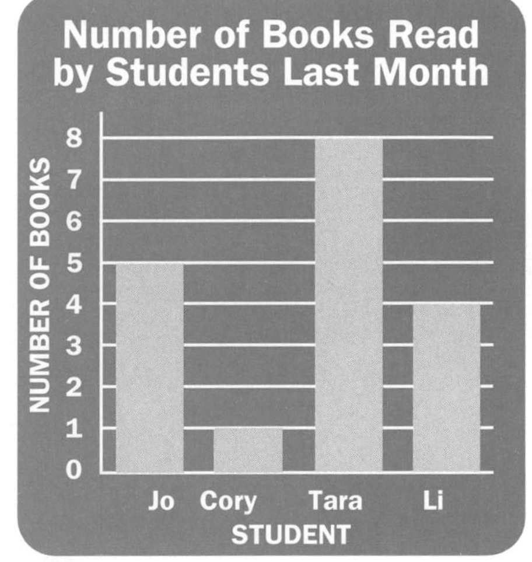

- Change the table so that the number of books Li read is 5. How does this affect the graph? Explain.

- Lower the bar above Tara to 5. How does this affect the table? Explain.

1 Show the data below in the table and graph. Include your own data also.
Lena read 2 books last month, Maria read 5 books, and Carl read 3 books.

2 Predict how the graph would change if you read 5 more books and Tara read 2 fewer books.

Critical Thinking: Generalize

Explain your reasoning.

3 Explain why it is useful to make tables and graphs with a computer.

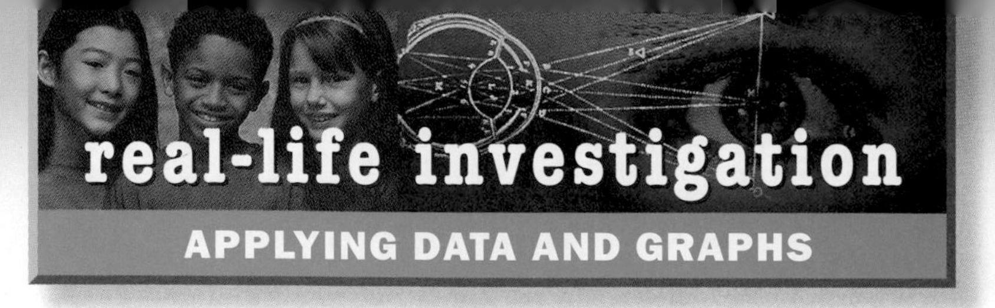

Measure by Measure

How does a doctor know whether you are big or small for your age? Doctors use standard height and size tables. You can create your own tables for fourth graders and report on what you find.

DECISION MAKING

Measuring Yourselves

1 Form groups of six or more students. Work in pairs within your group to find the following measurements in centimeters. Copy and complete the table.

How We Measure Up
Height:
Head Size:
Arm Span:
Foot Size:
Length of Giant Step:

2 Decide how you will use each student's measurements to make tables of group measurements. Display all the group tables for everyone to see.

3 Use your group's data for each measurement to predict the range, mode, and median for your entire class. Explain your thinking.

Reporting Your Findings

4 Portfolio Prepare a group report on one of the measurements you investigated. Include the following:

- A record of that measurement for everyone in your class. Organize the data into a table.
- Make a graph to show the class data for the measurement.
- Write statements about what the graph tells you about your class. Use the words *range*, *median*, and *mode* whenever you can.

5 Explain how your group's predictions compare to the entire class data.

Revise your work.

- Did you measure each person in exactly the same way?
- Do your statements make sense?
- Is your report clear?

MORE TO INVESTIGATE

PREDICT what the range, mode, and median measurements might be for a fifth-grade class.

EXPLORE how a measurement other than the one in your report compares with others in your class.

FIND how feet, arms, and hands were used as units of measurement in the past.

Bar Graphs

Fourth graders care a lot about people, animals, and the environment.
In this survey, students were asked to choose their favorite wild animals.

Use a graph to compare the data quickly and make it easy to read.

Cultural Note

Two pandas were given to the National Zoo in Washington, D.C., as gifts from the Chinese government

Work Together

Work with a partner. Copy and complete the **bar graph**. Then answer the questions.

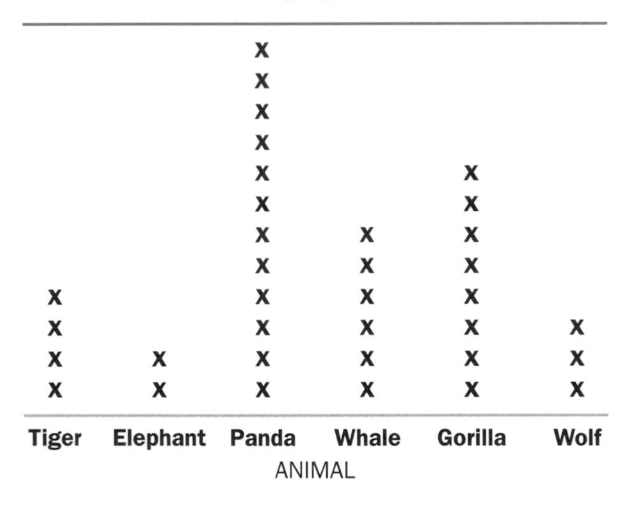

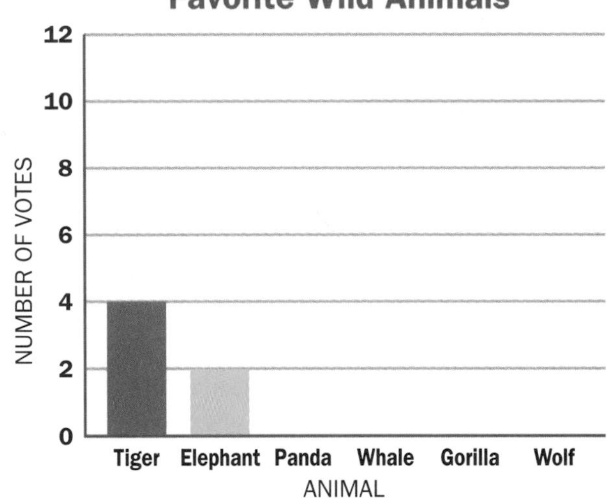

Talk It Over

bar graph A graph that displays data using bars of different lengths.

- ▶ How did you decide what the heights of the bars for the panda, whale, gorilla, and wolf should be?
- ▶ Both graphs help you compare the data quickly. Which graph makes it easier to make comparisons? Which makes it easier to read exact numbers? Tell why.
- ▶ Your classmate makes the following statements. Do you agree with her? Why or why not?
  a. More students like whales than gorillas.
  b. Only 2 students chose elephants.
  c. Most students like pandas, gorillas, and whales.

Make Connections

To make a bar graph:

- ▶ list the items (animals) along the bottom of the graph.
- ▶ write a scale with numbers along the left side of the graph, starting with 0.
- ▶ draw a bar to match the numbers in the data collected.
- ▶ write a title.

How do the scale numbers in the bar graph increase? What other scale numbers could you have used?

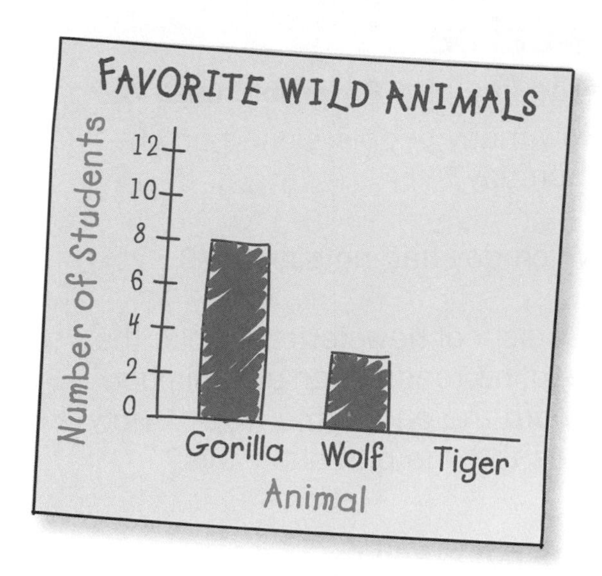

Check for Understanding

Use the bar graph for problems 1–4.

1 Which animal had the greatest increase in number? How do you know?

2 What was the increase in the number of acouchis (a-KOO-cheez)? the number of vicuñas (vī-KOO-nuhz)?

3 What is the difference between the largest and smallest increases of endangered animals? How do you know? How can you find this from the bar graph?

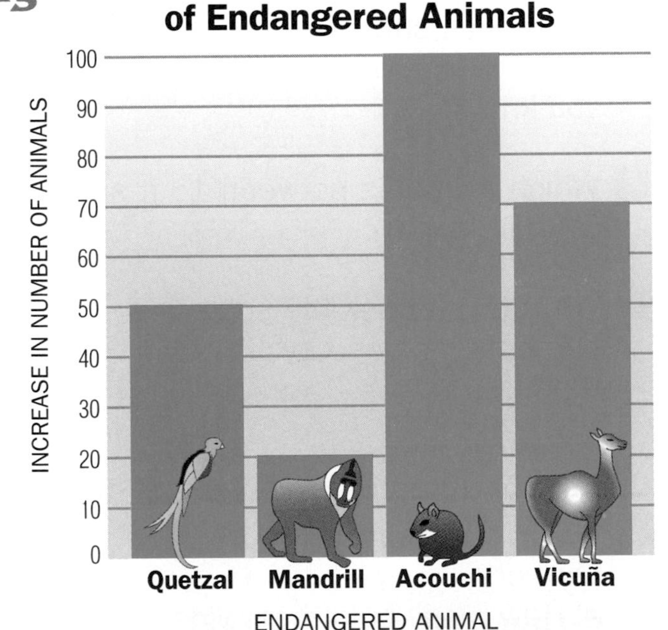

Critical Thinking: Generalize

Explain your reasoning.

4 When is a bar graph more useful than a line plot? than a pictograph?

CHECK

PRACTICE

Practice

Use the line plot for problems 1–3.

1 How many people signed up on Thursday?

2 Which day had no sign-ups?

3 In a school newsletter article, the headline read: "Sign-ups climbed toward the end of the week!" How does the line plot show this?

The Comeback of Whales

Mon.	Tues.	Wed.	Thurs.	Fri.
				x
				x
			x	x
			x	x
x			x	x
x	x		x	x
x	x		x	x

PEOPLE WHO SIGNED UP EACH DAY

Use the bar graph for problems 4–7. Explain how you found your answer.

4 How long is the dugong? How tall is the panda?

5 How much shorter is the python than the whale? Explain how you know.

6 Which animal is between 15 ft and 30 ft in length?

7 Why do you think the scale numbers increase by 10 rather than by 1?

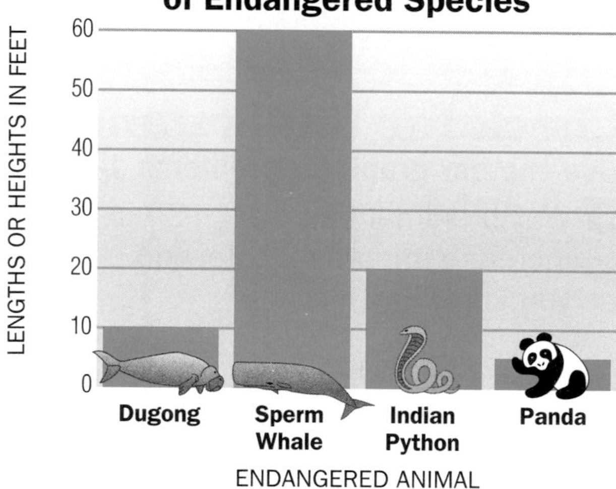

Use the bar graph for problems 8–9.

8 Which question can you *not* answer?

a. How many polar bear videos were borrowed?

b. Can you borrow an alligator video today?

c. Are there more than 3 videos on endangered animals?

9 **What if** the library had a video on gorillas and it was borrowed 10 times. Copy the graph and add the new data.

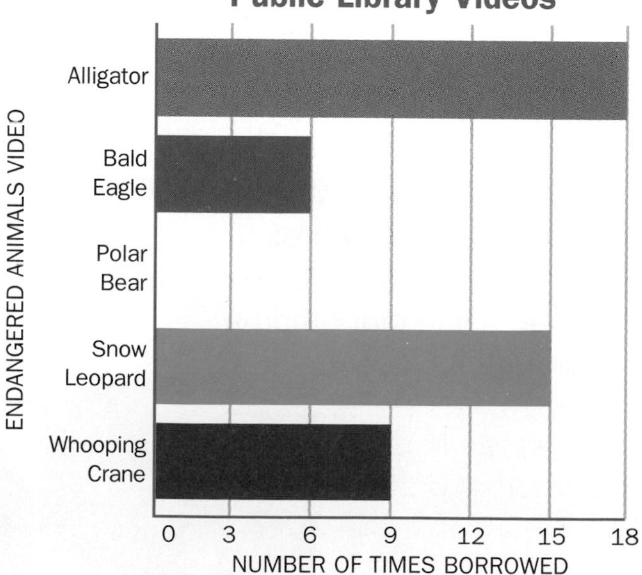

MIXED APPLICATIONS
Problem Solving

Use the table for problems 10–13.

Types of Endangered Species	
Species	**Number of Species**
Mammals	338
Birds	243
Fish	113
Reptiles	112
Others	148

10 **Make a decision** If you wanted to make a bar graph using the data in the table, what scale would you use? Explain.

a. 5 **b.** 10
c. 100 **d.** 200

11 What is the total number of species that are endangered?

12 **What if** the number of endangered mammals decreased by 50. How would a bar graph of the data change?

13 Which species has about three times the number of endangered animals as reptiles? about twice the number?

14 **Write a problem** using the data from any of the graphs on pages 100–102. Solve it. Ask others to solve it.

15 **Data Point** Survey your classmates to find out what pets they have. Make a bar graph to show the results.

more to explore

Stem-and-Leaf Plots

A stem-and-leaf plot is another way to organize data.

Data:	40	37	27	37	29
	35	42	42	32	37

1 The stems are the tens digits of the data. What do the leaves stand for?

2 What is the largest class size? the smallest?

3 How many fourth-grade classes are there?

4 Make a stem-and-leaf plot of students in fourth-grade classes in the Union School District: 36 18 32 36 25 28 25 32 20

Thorndike School District
Students in Fourth-Grade Classes

Stem	Leaf				
2	7	9			
3	2	5	7	7	7
4	0	2	2		

Key: 2 | 7 stands for 27 students.

Extra Practice, page 494

Ordered Pairs

LEARN

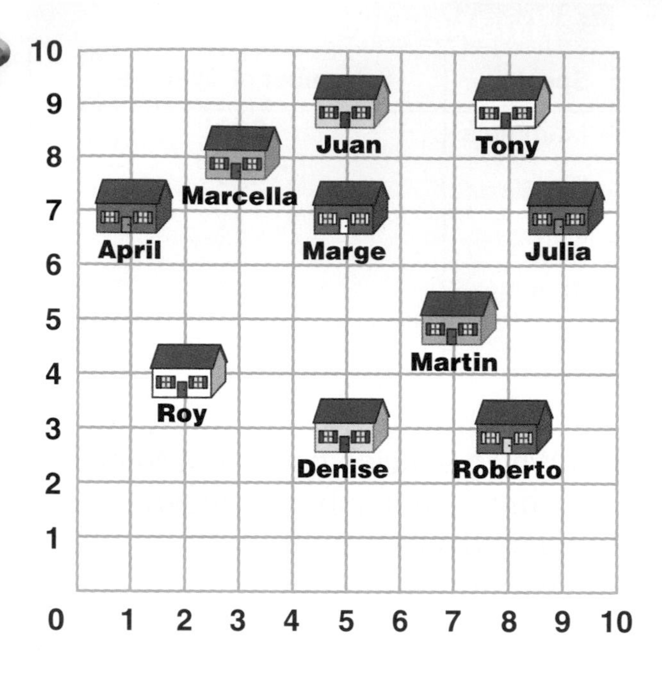

Have you ever made a map to show where you live or where you need to go? This grid map uses both horizontal and vertical scale numbers to describe the locations of places on the map.

To locate a point on the grid, start at zero, go right, then go up.

To tell where Marge lives, Tanya starts at zero. She counts 5 blocks to the right and then 7 blocks up. The 5 and the 7 are an **ordered pair.** They are written as (5, 7).

- How would you find (3, 8)? Explain.
- Is this location the same as (8, 3)? Explain.
- Why is the order of the numbers in a pair important?

ordered pair A pair of numbers that gives the location of a point on a grid.

CHECK

Check for Understanding

1 Give the ordered pairs for where Denise, Juan, and Julia live.

2 Tell who lives at (2, 4), (7, 5), and (8, 9).

3 **What if** Justin lives a block away from Roy. Where could you place his house on the map?

Critical Thinking: Analyze **Explain your reasoning.**

4 What is similar and different about finding the locations (5, 3) and (5, 9)? (1, 7) and (9, 7)?

5 What can you say about the locations of any group of ordered pairs that have the same first number? the same second number?

Practice

Use the map to find the ordered pair for the location.

1 Pirate's Bluff

2 Parrot Peak

3 Ship Bay

4 Treasure Trove

5 Rainbow Gold

6 Morgan's Patch

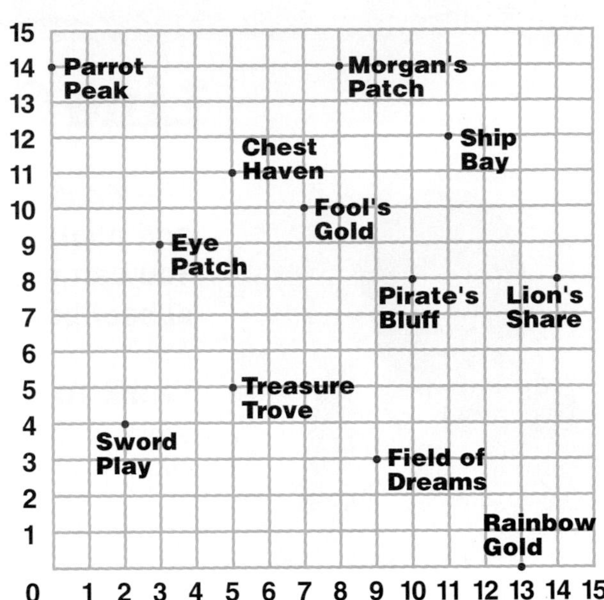

Tell what is found at the location on the map.

7 (5, 11) **8** (14, 8) **9** (2, 4) **10** (9, 3) **11** (3, 9) **12** (7, 10)

MIXED APPLICATIONS
Problem Solving

13 **Logical reasoning** Kettle Climb is directly above Rainbow Gold and to the right of Treasure Trove. What is the ordered pair that describes its location?

14 Tanya leaves Parrot Peak at 10:30 A.M. She arrives at Rainbow Gold at 1:45 P.M. How long did she take to get to Rainbow Gold?

15 **Data Point** Use a grid to create a neighborhood of your own or draw a map of your neighborhood. Give coordinates for places on your map. Have others find the locations.

16 **Spatial reasoning** You have two squares of the same size. You cut each square to get two equal triangles. What different shapes can you create with the four triangles?

mixed review • test preparation

Estimate the sum or difference.

1 67 + 43

2 498 − 261

3 \$4.93 + \$2.38

4 78,967 − 21,638

Find the range, median, and mode of the game scores.

5 24, 59, 38, 26, 93, 26, 47

6 75, 80, 82, 76, 95, 95, 95, 60, 70

Extra Practice, page 494

LEARN

Line Graphs

IN THE WORKPLACE

Susannah Druck is the program coordinator of *EcoSphere*, a conservation magazine produced by children.

The *EcoSphere* children research, interview, draw, and photograph for each issue of the magazine. You can show the number of new *EcoSphere* subscribers in a line graph.

Work Together

Work with a partner. Copy and complete the line graph. Then answer the questions.

line graph A graph that uses lines to show changes in data.

Month	Number of New Subscribers
1	24
2	30
3	30
4	34
5	33
6	32
7	32
8	28

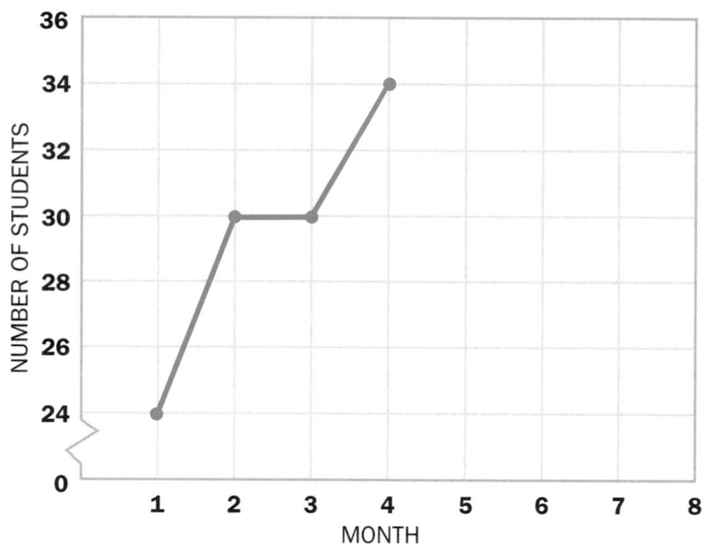

Talk It Over

- ▶ How did you show the number of new subscribers for months 5, 6, 7, and 8?
- ▶ Which month had the fewest new subscribers? the most?
- ▶ Which months had the same number of new subscribers?

Make Connections

To make a line graph:

- ▶ choose scale numbers and label the graph. You can use a broken graph to make a scale shorter.
- ▶ plot points that show the data and connect them with a line.
- ▶ write a title.

How do you choose a scale for the graph? What do you have to think about?

A line graph helps you to see changes that happen over time.

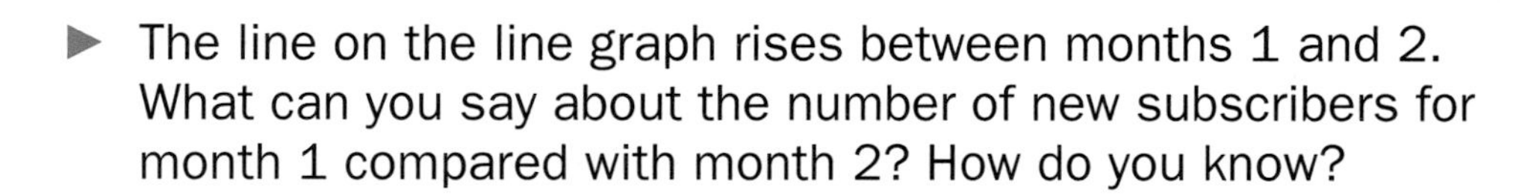

- ▶ The line on the line graph rises between months 1 and 2. What can you say about the number of new subscribers for month 1 compared with month 2? How do you know?
- ▶ What does it mean when the line goes down? when the line is flat? Give examples.
- ▶ What does this graph tell you about the number of new subscribers for the first eight months? How do you know?

Check for Understanding

Use the line graph for problems 1–4.

1 In which month is the depth of the river the greatest? the least?

2 Between which two months does the depth of the river change the most?

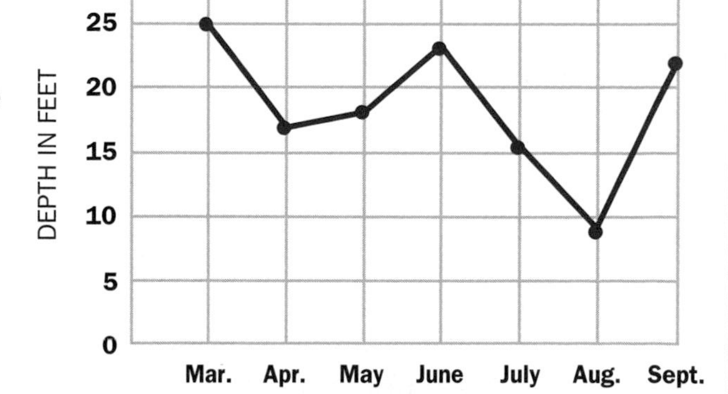

Critical Thinking: Analyze

Explain your reasoning.

3 What statements can you make based on the data in the graph?

4 Journal Why do you think a line graph is a good way to show data?

CHECK

Turn the page for Practice.

Practice

Use the line graph for problems 1–6.

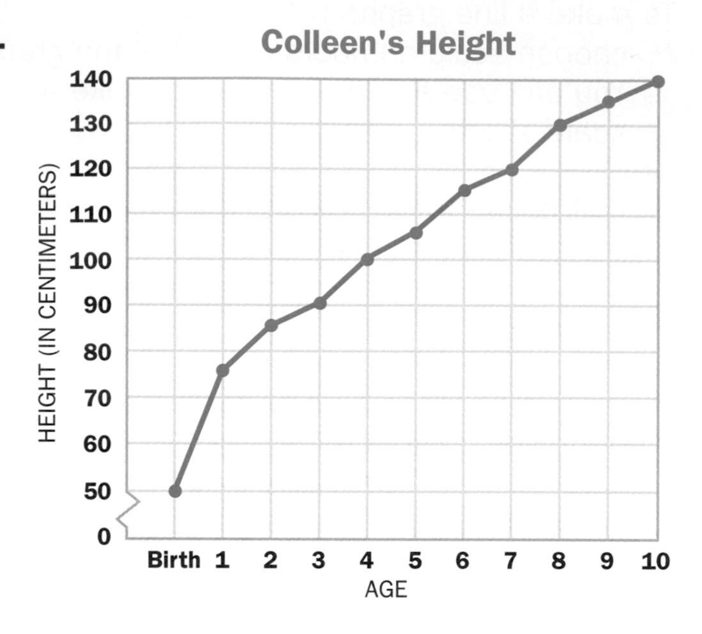

1 How tall was Colleen when she was 4?

2 Between which two ages did Colleen grow the most?

3 How much did Colleen grow between the ages of 5 and 6?

4 How old was Colleen when she was 120 cm tall?

5 What was Colleen's height at age 1? How do you know?

6 What does this graph tell you about how Colleen grew?

Use the table for problems 7–11.

7 Make a line graph using the data in the table.

8 Tell how you chose the scale you used.

9 What was the difference between the highest and lowest temperatures? What does this tell you about the school day?

10 Write a short paragraph describing what the weather was like on that school day.

Temperature During Schooltime

Time	Temperature (in degrees Celsius, °C)
A.M.	
8	4
9	8
10	10
11	12
Noon 12	12
P.M.	
1	12
2	16
3	20

Make It Right

11 Tyler wrote this statement. Explain what mistake was made, then correct it.

If the temperature at 4:00 P.M. is 20° C, the line in the graph will fall because the line rises only if the temperature increases.

MIXED APPLICATIONS

Problem Solving

12 **Make a decision** Match the graph to a title. Explain your decision.

a. The Numbers of Hats We Owned Last Year
b. Numbers of Hours Spent Swimming Each Month Last Year
c. My Puppy's Weight Each Month Last Year

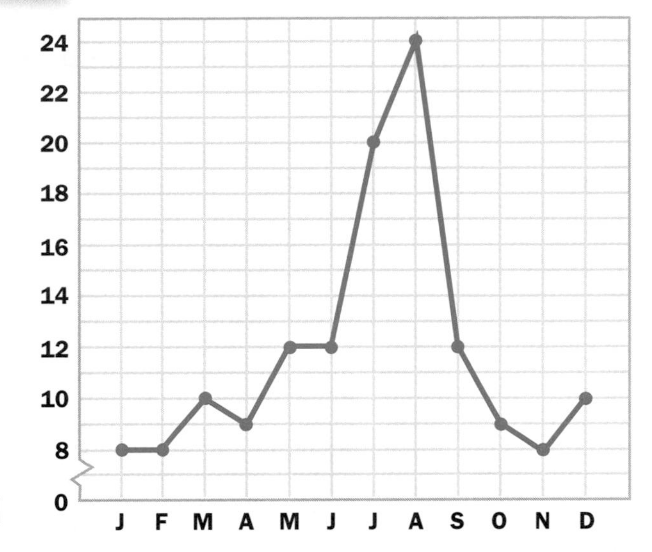

13 **Spatial reasoning** If you drew lines to connect the ordered pairs (1, 1), (4, 1), and (2, 4), what shape would you get?

14 **Data Point** Use the table on page 535 in the Databank to make a line graph. Write a sentence about what the graph tells you.

more to explore

Time Lines

Benjamin Banneker was a well-known African American mathematician, astronomer, and writer. This time line shows some of the important events in his life.

Life of Benjamin Banneker

Born 1731

Built wooden clock 1751

Predicted eclipse 1789

First published almanac 1792

1730 1750 1770 1790 1810

Began survey for Washington, D.C. 1791

Died 1806

1 How is a time line like a number line? How is it different?

2 How many years are between each mark on the time line?

3 How old was Banneker when he first published an almanac?

4 Make a time line that shows some important events in your own life.

Read
Plan
Solve
Look Back

LEARN

Part 1 Interpret Data

Mikel gathered data on membership in the after-school clubs in his school.

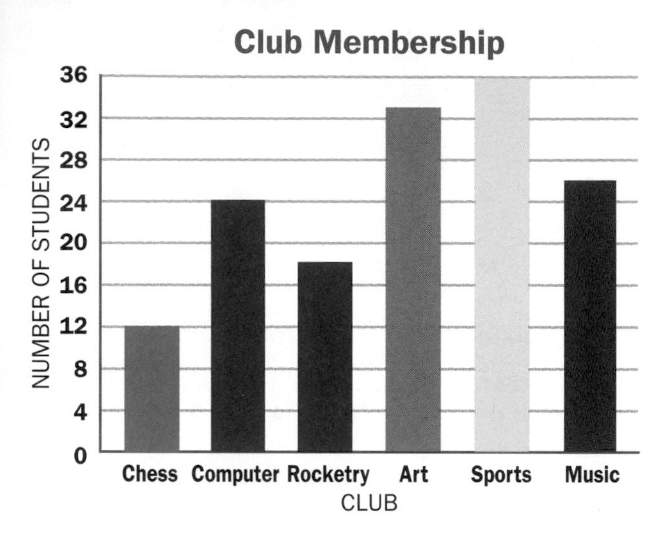

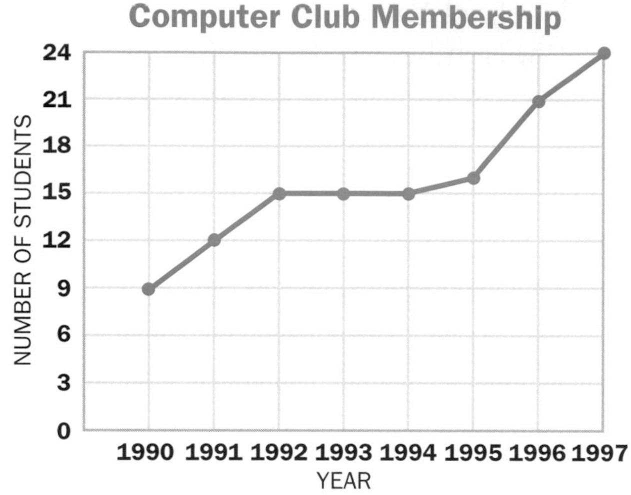

Work Together

Solve. Tell which graph you used.

1. How many more students belong to the Sports Club than belong to the Chess Club?

2. When was the greatest increase in membership in the Computer Club? How much was the increase?

3. If 12 members of the Rocketry Club also belong to the Sports Club, how many students are in both clubs in all?

4. **What if** membership in the Computer Club continues to increase each year by the same amount as it increased between 1996 and 1997. In what year would there be at least 40 members?

5. READING ARITHMETIC WRITING **Write: Compare/Contrast** Compare the data on each graph. Write a paragraph to compare and contrast the data shown.

Part 2 Write and Share Problems

Chris Wellman used the data in the line graph to write a problem.

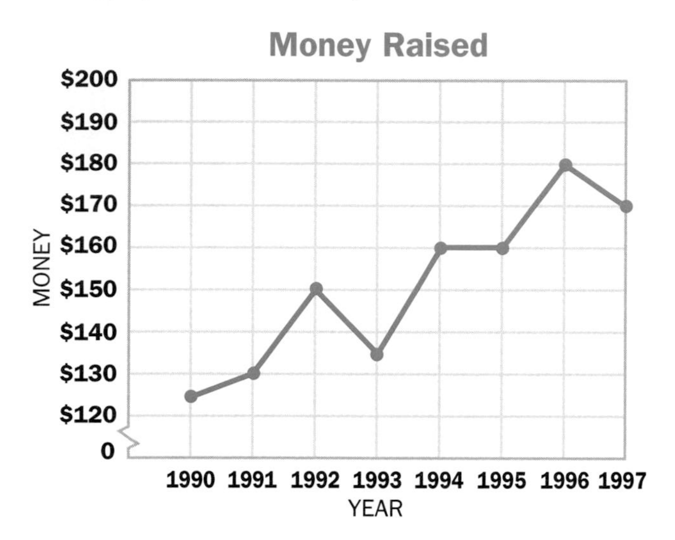

Each year, the PTA raises money for science equipment. Which years did they raise the same amount of money?

CHECK

Chris Wellman
Mandarin Oaks Elementary School
Jacksonville, FL

6 Explain how to solve Chris's problem using only words.

7 **Write a problem** of your own that uses information from the graph. Explain how your problem and Chris's problem are alike and different.

8 Trade problems. Solve at least three problems written by your classmates.

9 What was the most interesting problem that you solved? Why?

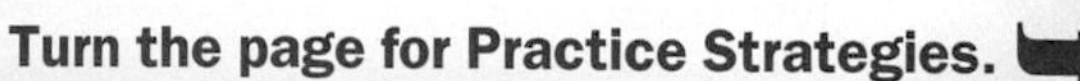
Turn the page for Practice Strategies.

Choose five problems and solve them. Explain your methods.

1 Desmond caught the train at 10:34 A.M. He rode on the train for 1 hour 40 minutes. At what time did Desmond get off the train?

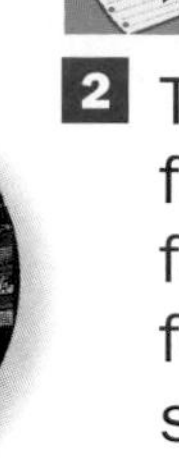

2 Tara buys a pen for $2.75, a notebook for $3.65, and a ruler for $0.98. Does she spend over $8.00? How do you know?

3 Kai collected 48 cans and 59 bottles. Bette collected 36 cans and 58 bottles. Who collected more cans and bottles? How many more?

4 Hallie spends $9.53 on school supplies. She gives the clerk a $10 bill. What is the least number of coins she can receive as change?

5 Joel's game scores are 1,243; 2,765; 3,142; 2,541; and 3,689. Pam's game scores are 2,165; 1,978; 3,047; 3,156; and 3,002. Who has the higher median score? By how much?

6 In Rio de Janeiro, Brazil, a crowd of 199,854 attended a World Cup soccer game in 1950. In 1990, about 180,000 attended a rock concert there. Which event had the greater attendance? By about how many more people?

7 Rory bought a coat for $160. He paid for it with 1 hundred-dollar bill and 6 ten-dollar bills. What other combinations of one-, ten-, and hundred-dollar bills could he have used?

8 **ALGEBRA: PATTERNS** Rhonda collected $15 for a charity in the first month, $20 in the second month, and $25 in the third month. If she continues with this pattern, how much should she expect to collect in the fifth month? Explain.

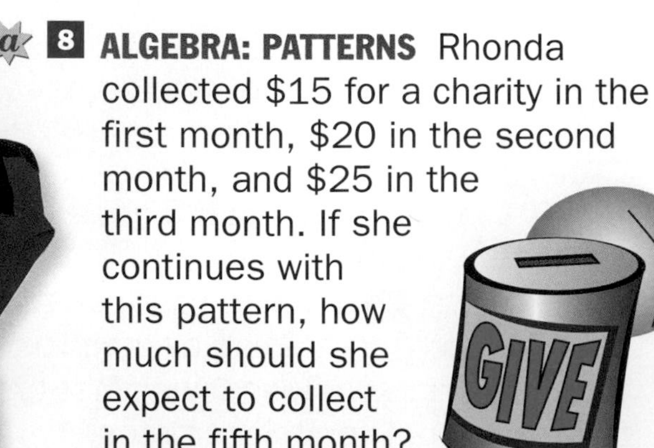

Choose two problems and solve them. Explain your methods.

9 In the Chinese calendar, a year is named for one of 12 animals. After 12 years, the calendar starts at the first animal again. At the right is a chart of two 12-year cycles.

a. Which years between 1900 and 1999 are named after the monkey?

b. For which animal is the year 2000 named?

c. Karen was 14 years old in 1996. She was born in the year of the Dog. Tell how old she will be when it is the year of the Dog again. What year is that?

Rat	1864	1876
Ox	1865	1877
Tiger	1866	1878
Rabbit	1867	1879
Dragon	1868	1880
Snake	1869	1881
Horse	1870	1882
Goat	1871	1883
Monkey	1872	1884
Rooster	1873	1885
Dog	1874	1886
Pig	1875	1887

10 Plan the next school week for your class. Include the following in the schedule:

- at least 4 hours of mathematics classes
- 3 hours at most of reading and spelling every day
- at least two science classes
- two gym classes

Your schedule must have at least lunch and four classes on any day. No two classes can be scheduled for the same time in a day.

11 Suppose you have 26¢. Can you spend exactly this amount in the school store? How? Is there more than one way to spend exactly 26¢? If so, tell how.

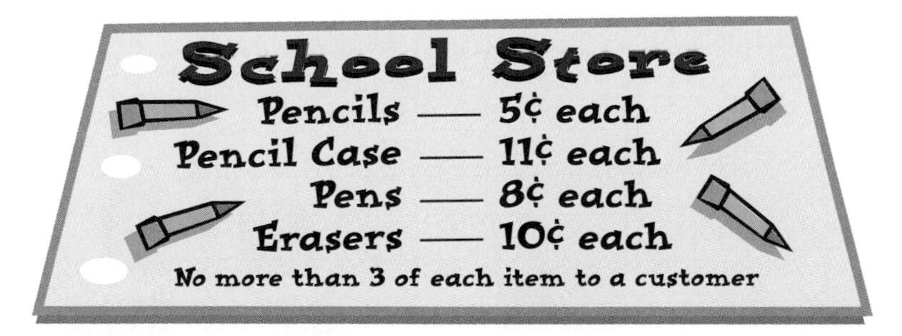

12 **At the Computer** Choose a question to ask your classmates about some feature or topic that interests you. Ask ten classmates your question. Then use a graphing program to display their responses. Write a paragraph describing what the data tells you about your classmates.

Extra Practice, page 495

Language and Mathematics

Complete the sentence. Use a word in the chart. (pages 82–109)

1 An ■ describes the location of a point on a grid.

2 You can use a ■ to show changes in data over time.

3 The ■ is the middle number in an ordered set of data.

4 Five minutes after 12 noon is 12:05 ■.

Vocabulary
- median
- A.M.
- P.M.
- line plot
- ordered pair
- line graph

Concepts and Skills

Complete. (page 84)

5 Start time: 2:15 P.M.
Elapsed time: 3 h 50 min
End time: ■

6 Start time: ■
Elapsed time: 4 h
End time: 1:30 P.M.

7 Start date: April 22
Elapsed time: ■
End date: May 14

Find the range, median, and mode for the set of data. Tell how each number describes the data. (page 90)

8 Math scores
325 352 368 375 412

9 Height of seven students (in inches):
49 50 55 61 50 53 54

Use the map for ex. 10–15.

Find the ordered pair for the location. (page 104)

10 Town Hall

11 Museum

12 School

Tell what is found at the location. (page 104)

13 (4, 2)

14 (5, 8)

15 (8, 5)

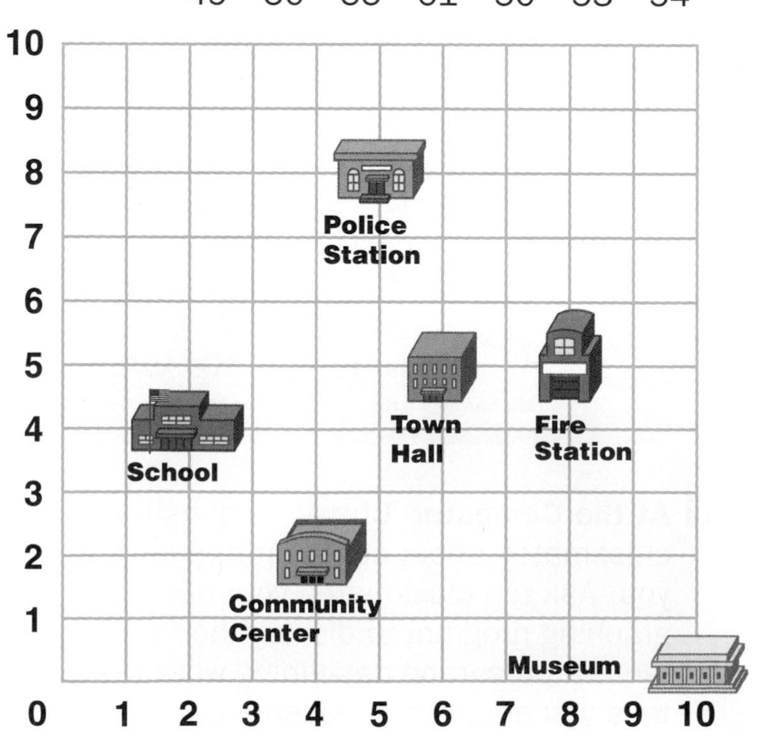

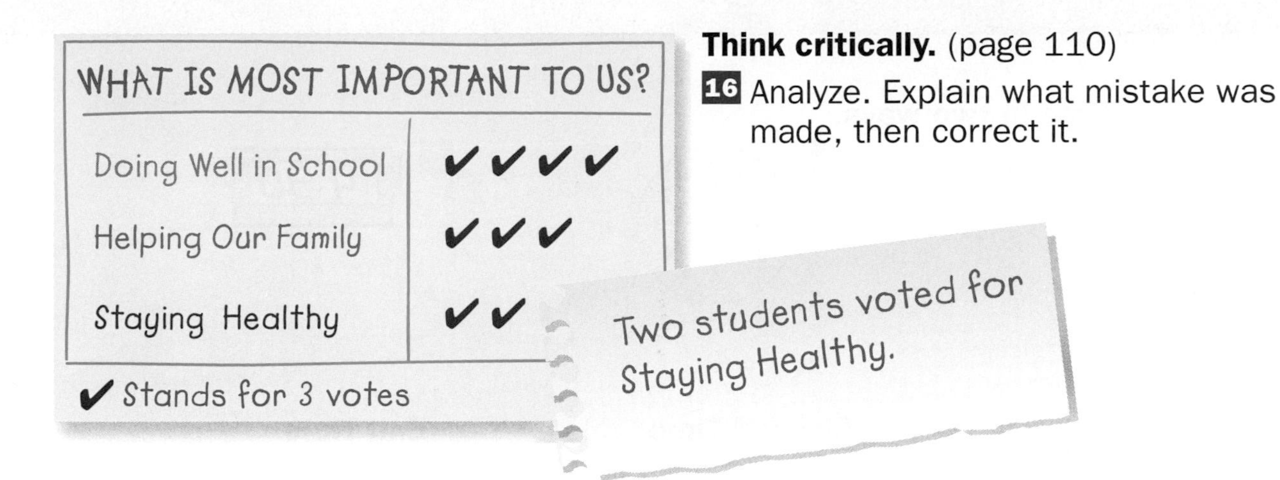

Think critically. (page 110)

16 Analyze. Explain what mistake was made, then correct it.

MIXED APPLICATIONS
Problem Solving

(pages 88, 110)

Use the line graph for problems 17–20.

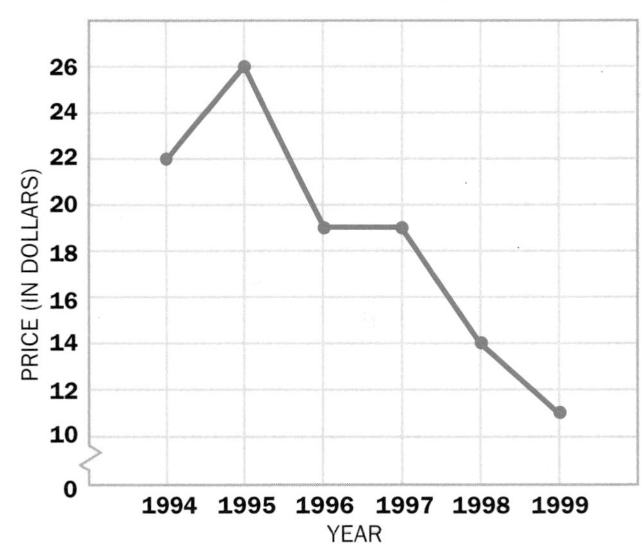

17 Which question can you *not* answer? Tell why.

a. Was there any change in price of Super Digidisc from 1994 to 1999?

b. In which year were the most number of Super Digidiscs sold?

c. In which year was the price of Super Digidisc the greatest? the least?

18 Describe what happened to the price between 1994 and 1999.

19 Between which two years was the greatest change in price? What was the change? Was it an increase or decrease?

20 In 1998 Doreen bought a Digidisc. She used a coupon good for $2.50 off the price. How much did she pay?

Write each time in two ways.

1 

2 

3 

Complete.

Start Time	Elapsed Time	End Time
4 1:15 P.M.	2 hours 55 minutes	■
5 9:55 A.M.	■	11:15 A.M.
6 ■	30 days	June 13
7 September 13	29 days	■

Find the range, median, and mode for the set of data.

8 25, 15, 15

9 2, 3, 9, 3, 9

10 4, 4, 4, 4, 0

11 10, 15, 10, 18, 2

12 11, 6, 3, 3, 1

13 2, 49, 37, 19, 1, 11, 8

Use the bar graph for ex. 14–16.

14 How many students were surveyed?

15 Which pet did students like least?

16 How many more students chose the rat than chose the fish?

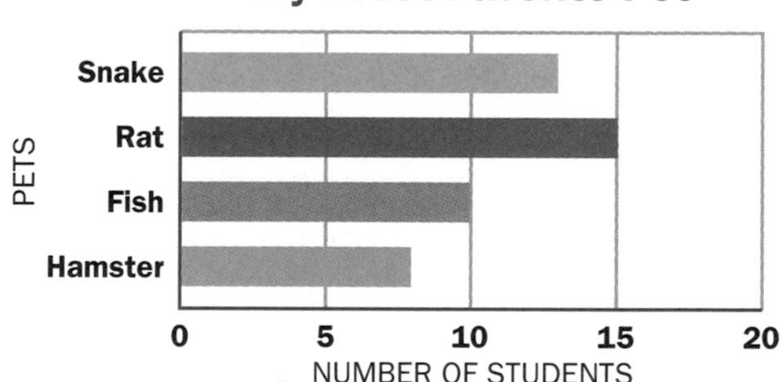

Solve.

17 Sean wants to be at a party at 4 P.M. He takes 3 h to mow the lawn and 40 min to walk to the party. What is the latest time he can start mowing the lawn?

18 If each ● stands for 4 balls in a pictograph, how many ● would you draw to show 38 balls?

19 Vern entered a number in his calculator. He added 23 and then subtracted 10 to get 28. What number did he start with?

20 Rita bought a camera originally priced at $139.65. Because it was on sale, she saved $13.96. How much did she pay for the camera?

What Did You Learn?

Help Seymour write a report to his teacher about how he spends his time after school. Include the following:

- A graph to display the data. Tell why you chose that type of graph.
- At least three statements about the data and the graph that use the terms *range, median,* and *mode* when appropriate.
- A description of how you spend your time after school. Tell how you would collect and display your data.

Seymour's Data

After-School Activity	Time Spent (in minutes)
Homework	55
Eating	45
Reading	30
Television	60
Playing	120

You will need
- *graph paper*

A Good Answer
- includes an appropriate graph that is labeled correctly and shows the data accurately
- clearly describes the data using the terms *range, median,* or *mode* correctly

You may want to place your work in your portfolio.

What Do You Think

1. Are you able to read and make graphs successfully? Why or why not?

2. What do you need to think about when you make a graph?
   - What kind of graph is best for the data?
   - Is a symbol or key needed?
   - What scale will you use?
   - Other. Explain.

math science technology

CONNECTION

Keeping the Beat

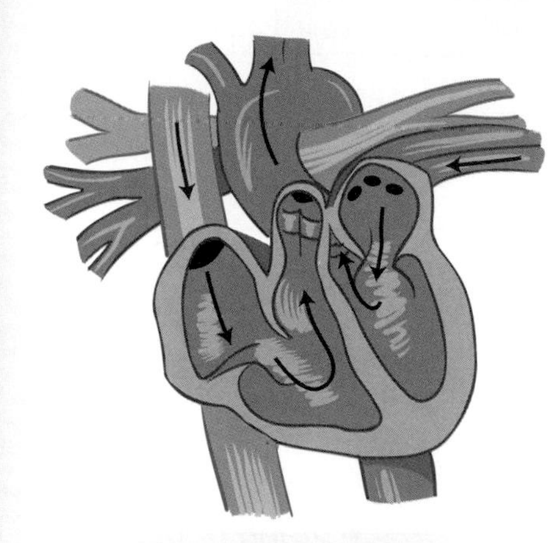

Flow of blood through the heart

Aerobic exercise helps keep your muscles and heart in shape. Your heart pumps blood throughout your body. Each beat is a pumping action.

Your pulse matches each beat of your heart. By counting your pulse, you can discover how fast your heart beats. You can also see how exercise and rest affect your heart rate.

- How does your breathing change when you exercise? your temperature? What other changes do you notice?

2 ways to find your pulse (a) neck (b) wrist

Measure Your Heart Rate

Work with a partner to record your heart rates at rest and exercising.

Get Ready
- What exercises will you do?
- How will you keep time?
- How will you record your heart rates?

Rest for 2 minutes, then take your pulse for 10 seconds. After you take your pulse, do the following steps. Take your pulse for 10 seconds after each step.

Step 1 Exercise for 1 minute.
Step 2 Rest for 1 minute.
Step 3 Rest for another minute.

Repeat Steps 1, 2, and 3 two more times. Remember to take your pulse after each step.

When you are done, rest for a minute one final time. Then take your pulse for 10 seconds.

Use the data you collected for problems 1–2.

1 Make a line graph that shows your heart rate during the 10 minutes of exercise and rest.

2 Write about what your graph shows.

At the Computer

3 Heart rates are usually given in beats per minute. Use a spreadsheet to change the heart rates for 10 seconds to heart rates for each minute. Then graph the data. (Hint: 60 seconds = 1 minute)

4 How is this graph similar to the one you made earlier? How is it different?

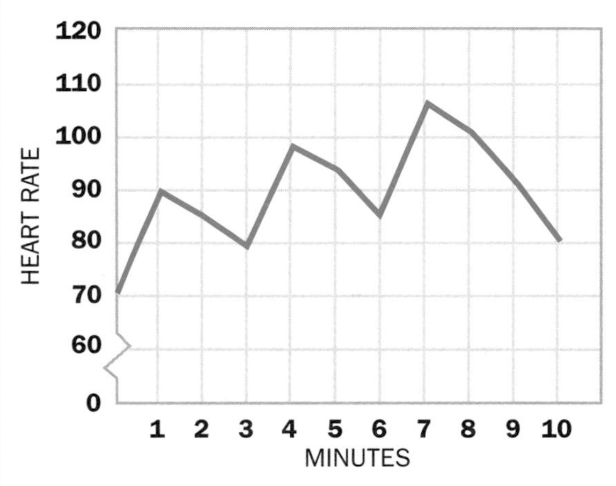

Cultural Note

The stethoscope is used to listen to the heartbeat. It was invented in 1816 by René T. H. Laënnec, a French doctor.

TEST PREPARATION

Choose the letter of the best answer.

1 Which names the same number as 2,037?

A 200 + 30 + 7
B 2,000 + 30 + 7
C 2,000 + 300 + 7
D 20,000 + 30 + 7

2 Which statement describes the graph on the number line below?

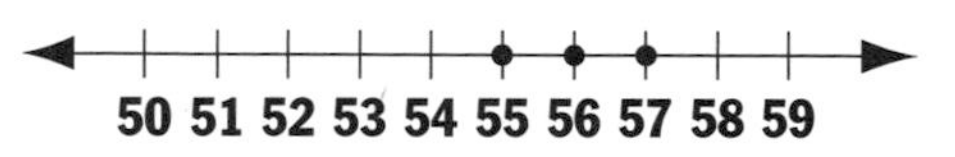

F All whole numbers greater than 55 and less than 57
G All whole numbers greater than 54 and less than 58
H All whole numbers greater than 54
J All whole numbers less than 58

3 Pablo leaves home for school at 8:25 A.M. He arrives at school at 9:05 A.M. How long does it take Pablo to get to school from his home?

A 25 min
B 30 min
C 40 min
D 35 min

4 What is Δ?

$$\begin{array}{r} 3\Delta 4 \\ -\ 287 \\ \hline 17 \end{array}$$

F 9
G 5
H 3
J 0

5

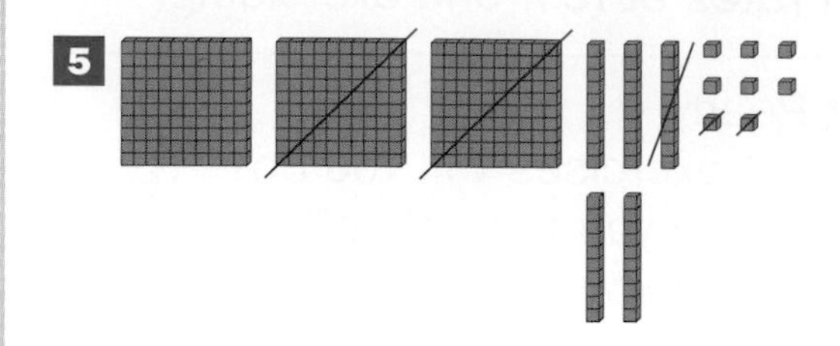

Which subtraction is shown?

A 212 – 146
B 358 – 146
C 358 – 212
D not here

6 Ian collects 179 cans. Sal collects 131 cans. Which is the best estimate of how many cans they both collect?

F 30
G 100
H 200
J 300

7 Marsha has an hour to use her computer before bedtime. She plays a game for 10 minutes. Then, she writes a math report for 25 minutes. How much computer time does she have left?

A 15 min
B 35 min
C 25 min
D 30 min

8 If each * stands for 2 dogs, how can you show 12 dogs?

F * *
G * * *
H * * * * * *
J * * * * * * * * * * * *

9 Paul has $0.77. Which group of coins can he *not* have?

A 7 dimes, 1 nickel, 2 pennies
B 6 dimes, 4 nickels, 2 pennies
C 1 quarter, 5 dimes, 2 pennies
D 3 quarters, 2 pennies

10 Which number is 1 thousand more than 565,400?

F 665,400
G 575,400
H 566,400
J not given

11 Complete the sentence.
96 + ■ = 163

A 259
B 76
C 67
D 29

12 Lea has $15 at the end of the week. During the week she took out $5 and put in $3. Which method could you use to find how much money she had at the beginning of the week?

F Add $3 to $15 then subtract $5
G Subtract $5 from $15 then add $3
H Subtract $3 from $15 then add $5
J Add $5 and $3 to $15

13 Which set of data has a mode of 54?

A 50, 89, 104, 54, 76
B 49, 51, 55, 56, 54
C 23, 54, 67, 54, 58
D 48, 60, 53, 55, 54

14 What is 75,514 rounded to the nearest thousand?

F 80,000
G 76,000
H 75,000
J 70,000

Use the bar graph for problems 15–16.

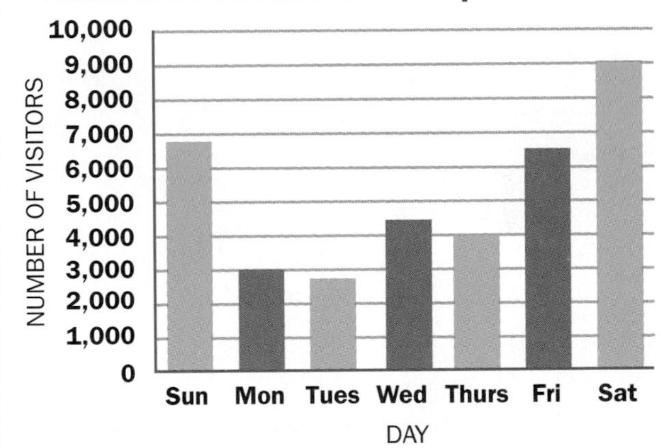

15 Which statement describes the bar graph?

A The Sports Center is more popular than the mall.
B More visitors go to the center on Saturday than on any other day.
C Most children enjoy the center.
D Discount tickets are sold on Wednesdays.

16 How many more visitors were there on Friday than on Monday?

F 3,500
G 3,000
H 6,500
J 9,500

TEST PREPARATION

CHAPTER 4

MULTIPLICATION AND DIVISION FACTS

THEME

Sports

Different sports are played around the world. In this chapter, you will learn how you can use math to help solve sports problems and to understand sports rules.

What Do You Know?

Michael collects sports caps and puts them on his wall.

1 How are the caps arranged? How many caps are there? Explain how you found the answer.

2 **What if** the caps were arranged in two equal rows. How many caps would be in each row? Explain.

3 Portfolio Think about arranging 24 caps in equal rows on a wall. Draw as many different ways as you can to arrange them.

Sequence of Events

Four runners begin a race at the same time. Tyler is the winner. Cora gets second prize, and Ana gets third. Ryan plans to do better next time. In what sequence did these runners finish?

When you read, it helps to understand the sequence, or order, of the events. Sometimes authors give you sequence words to help you understand. Sometimes they don't.

1 What sequence word would you choose for each runner?

2 How could you use a table or diagram to show race results?

Vocabulary

array, p. 125
factor, p. 125
product, p. 125
skip-count, p. 128
multiple, p. 128
Commutative Property, p. 140
Associative Property, p. 140
dividend, p. 146
divisor, p. 146
quotient, p. 146
fact family, p. 154
remainder, p. 157
divisible by, p. 157

Meaning of Multiplication

LEARN

Some day, you may have your own business. Suppose your company custom packages brightly colored street-hockey pucks. You offer these five packages.

STREET HOCKEY PUCK

Have It Your Way

1 Pack 2 Pack 3 Pack 4 Pack 5 Pack

Work Together

Work with a partner to fill orders for hockey pucks.

You will need
- *0–9 spinner*
- *counters*
- *graph paper*

Spin the spinner to get the number of 1-packs you need to fill. Your partner may use counters or graph paper to find the total number of pucks needed to fill the order.

Take turns spinning to fill two more orders for 1-packs. Then work together to fill three orders for 2-packs, 3-packs, 4-packs, and 5-packs.

Record your work in a table like the one shown.

Number of Packs Ordered	Number in Each Pack	Total
3	1	3
0	1	0

Talk It Over

- ▶ What method did you use to find the totals? Explain.
- ▶ **What if** you have an order for three 5-packs and another order for five 3-packs. What do you notice about the totals?
- ▶ Find the total for six 4-packs. Predict the total for four 6-packs. Explain your reasoning. Then check your prediction.

Make Connections

You can make an array to show 9 groups of 4.

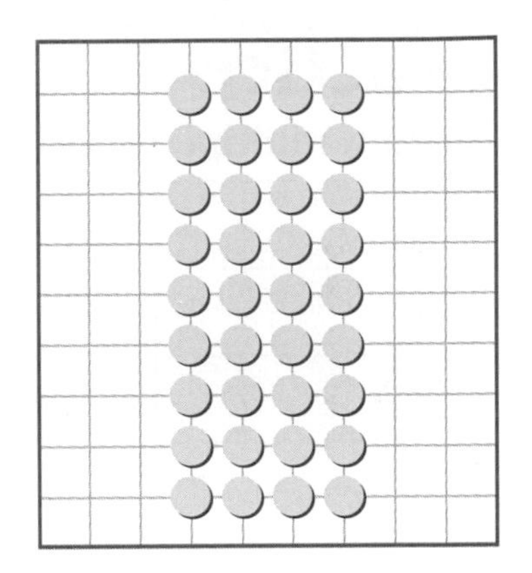

Number of Packs Ordered	Number in Each Pack	Total
9	4	36

You can also write a multiplication sentence.

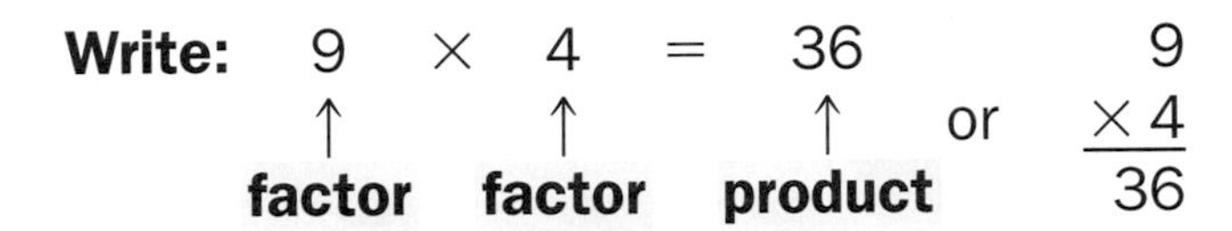

Write: $9 \times 4 = 36$ or $\begin{array}{r} 9 \\ \times 4 \\ \hline 36 \end{array}$

9 → **factor**, 4 → **factor**, 36 → **product**

Read: 9 times 4 is equal to 36.

▶ Write number sentences that have 0 as a factor and 1 as a factor. Use the patterns you notice to write a definition of the Zero Property and the Property of One.

array Objects displayed in rows and columns.

factor Numbers that are multiplied to give a product.

product The result of multiplication.

CHECK

Check for Understanding

Write a multiplication sentence for the picture.

1

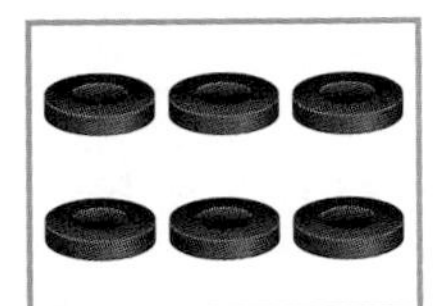

2

Multiply. You may use counters or graph paper.

3 5×4 **4** 2×7 **5** 3×2 **6** 5×6 **7** 8×3

8 6×6 **9** 9×4 **10** 9×8 **11** 8×0 **12** 5×5

Critical Thinking: Analyze

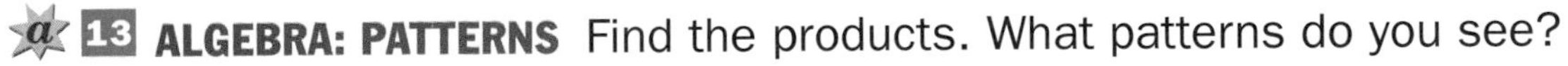

13 **ALGEBRA: PATTERNS** Find the products. What patterns do you see?

a. 6×1 8×1 1×7 1×5
b. 4×0 0×9 0×6 7×0
c. 3×9 9×3 8×7 7×8

14 Nicole finds 5 groups of 7 by adding 7 + 7 + 7 + 7 + 7. Can she use multiplication to find the total? Explain. Show the multiplication sentence if you can.

Turn the page for Practice.

Practice

Write a multiplication sentence for the picture.

1 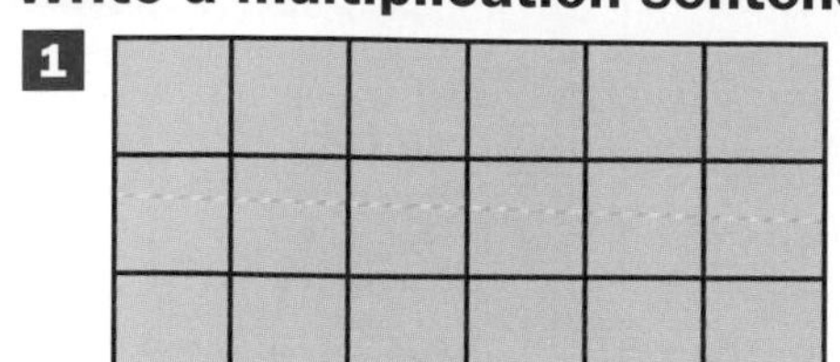

2 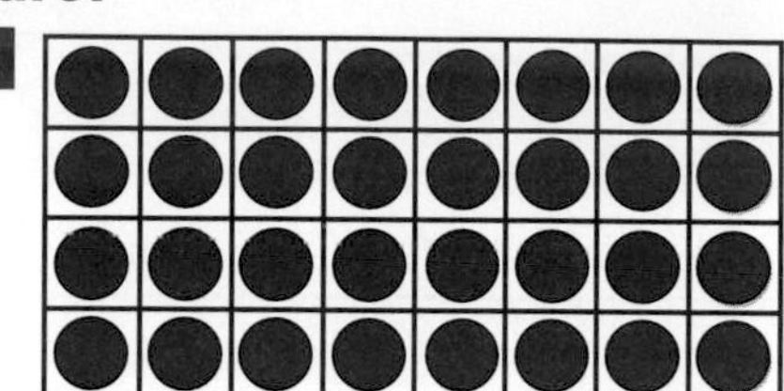

3 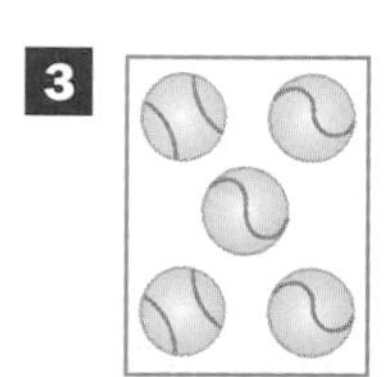 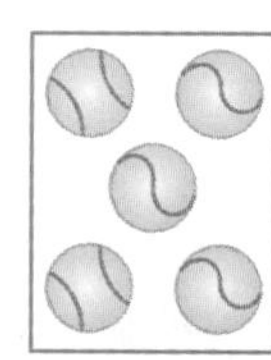 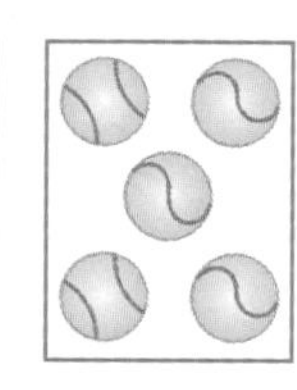 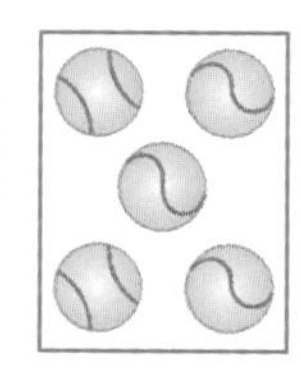

4 

Make or draw an array. Find the product

5 2×7 **6** 3×4 **7** 6×5 **8** 7×3 **9** 4×6

Multiply using any method.

10 9×5 **11** 3×8 **12** 6×5 **13** 8×4 **14** 7×7

15 4×1 **16** 9×6 **17** 5×8 **18** 0×8 **19** 1×9

20 3×3 **21** 5×3 **22** 8×6 **23** 7×5 **24** 6×2

25 $\begin{array}{r} 8 \\ \times 9 \\ \hline \end{array}$ **26** $\begin{array}{r} 2 \\ \times 5 \\ \hline \end{array}$ **27** $\begin{array}{r} 9 \\ \times 9 \\ \hline \end{array}$ **28** $\begin{array}{r} 7 \\ \times 0 \\ \hline \end{array}$ **29** $\begin{array}{r} 5 \\ \times 8 \\ \hline \end{array}$ **30** $\begin{array}{r} 6 \\ \times 7 \\ \hline \end{array}$

Write the letter of the factors with the same total.

31 $7 + 7 + 7 + 7 + 7 + 7$ **a.** 6×2

32 $4 + 4 + 4 + 4 + 4 + 4 + 4$ **b.** 7×9

33 $2 + 2 + 2 + 2 + 2 + 2 + 2 + 2$ **c.** 7×4

34 4×3 **d.** 6×7

35 9×7 **e.** 8×2

MIXED APPLICATIONS

Problem Solving

36 At a tennis competition, each of 6 schools brings a team of 6 tennis players. How many tennis players attend altogether?

37 In a classroom, there are 5 tables with 5 students at each table. How many students are in the classroom?

38 Skye is buying paddles for the table-tennis team. Each of the 7 players will get 3 paddles. How many paddles will Skye buy?

39 Fifty-six students play soccer. Thirty-nine students play basketball. Which sport has more players? How many more?

40 Complete this problem using your own numbers.
■ students worked on school projects. They worked in ■ groups of ■ students.

41 Hanna lines up the hockey sticks in the supply closet after practice. She puts them in rows of 6. She makes 7 rows. How many hockey sticks are there?

42 Rule: The first two numbers are even. The third number is odd. Which does *not* follow the rule?
a. 2, 200, 101
b. 10, 330, 207
c. 6, 63, 65

43 Ezra's school has 3 fourth-grade classes. One class has 22 students, another has 28 students, and a third has 31 students. How many students are in the fourth grade?

mixed review • test preparation

1 $24 + 16$ **2** $52 - 49$ **3** $57 + 109$ **4** $1{,}280 - 1{,}053$

5 $45 + 86$

6 $\$5.35 - 2.75$

7 $728 + 93$

8 $375 + 468$

9 $1{,}205 - 896$

10 $2{,}637 - 598$

11 $\$63.45 + 79.39$

12 $\$5{,}108 - 965$

13 $5{,}253 - 4{,}557$

14 $72{,}005 + 9{,}997$

Extra Practice, page 496

2 Through 5 as Factors

LEARN

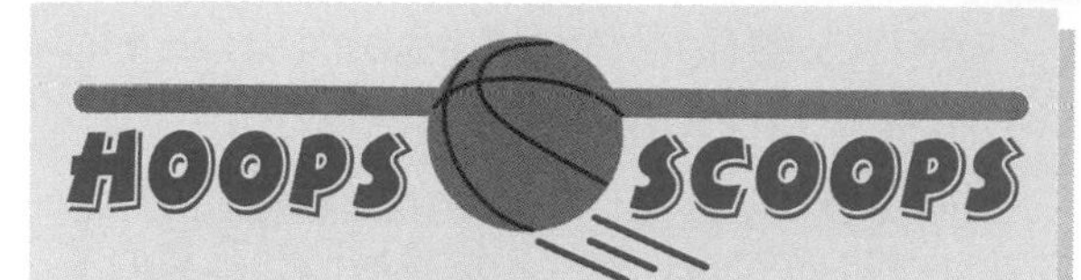

"There were four teams of five players in the Fourth Grade Basketball tournament. The only points made in the final minute were by three players on the winning team!"

In the final minute of the semifinal game, your team made its third 2-point shot to win the game. How many points did your team get in the final minute? How many fourth graders played in the tournament?

Skip-counting is a strategy that can help you find products.

Cultural Note

Basketball was invented in 1891 in Springfield, Massachusetts, by Dr. James Naismith. Today, the game is played all over the world.

To find the total number of points, you can skip-count by twos 3 times.

Think: Each of 3 players scores 2 points.

$3 \times 2 = 6$

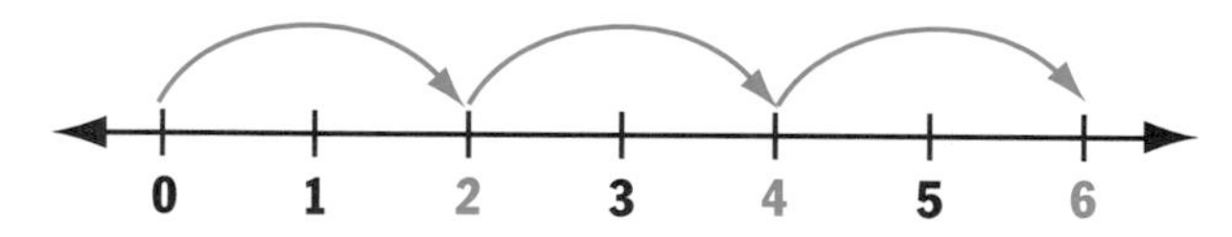

2, 4, and 6 are **multiples** of 2. Your team got 6 points.

skip-count Counting by twos, threes, and so on.

multiple The product of a number and any whole number.

To find the total number of players, you can skip-count by fives 4 times.

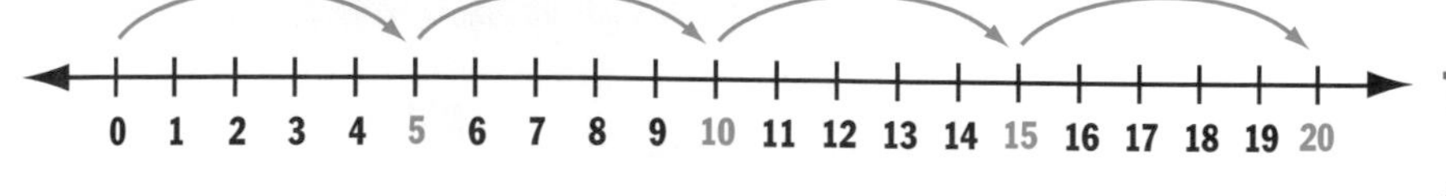

Think: 4 teams of 5 players

$4 \times 5 = 20$

5, 10, 15, and 20 are multiples of 5.

There were 20 fourth graders in the tournament.

Talk It Over

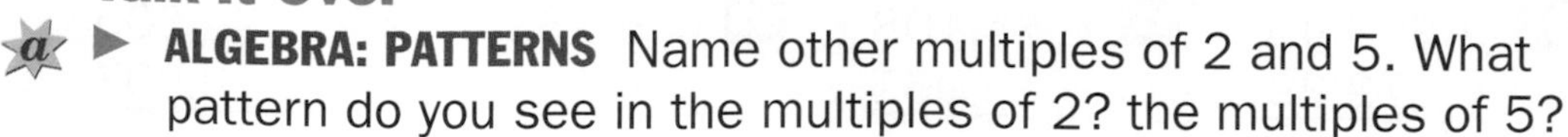

► **ALGEBRA: PATTERNS** Name other multiples of 2 and 5. What pattern do you see in the multiples of 2? the multiples of 5?

Here are some strategies to help you find products when 3 and 4 are factors.

Multiply: 3×6

You can find a known fact then add on.

Think: 2 groups of 6 plus 6 more

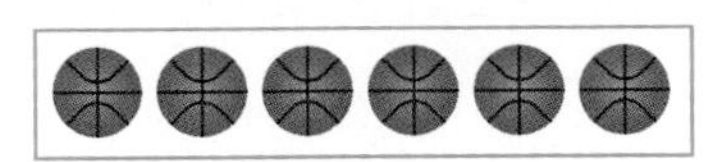

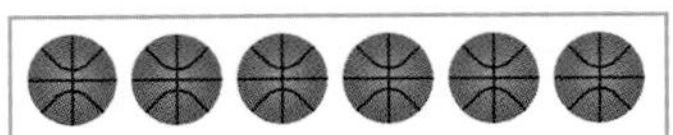

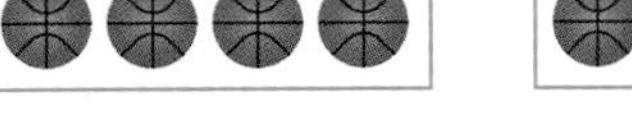

$$\underbrace{2 \times 6}_{12} + 6 = 18$$

So $3 \times 6 = 18$.

Multiply: 4×7

You can double a known fact.

Think: 2 groups of 7 plus 2 groups of 7

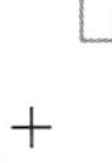

$$\underbrace{2 \times 7}_{14} + \underbrace{2 \times 7}_{14} = 28$$

So $4 \times 7 = 28$.

Check for Understanding

Multiply.

1. 3×5
2. 7×5
3. 6×5
4. 8×5
5. 5×3
6. 6×3
7. 4×2
8. 9×2
9. 7×2
10. 8×2
11. 3×3
12. 4×4
13. 4×9
14. 3×7
15. 5×5

Critical Thinking: Analyze **Explain your reasoning.**

16. **ALGEBRA: PATTERNS** What is 2×3? 3×2? What is 4×5? 5×4? What pattern do you see?

17. If you know 3×7, what other fact do you know?

CHECK

Turn the page for Practice.

PRACTICE

Practice

Write a multiplication sentence for the picture.

1

2
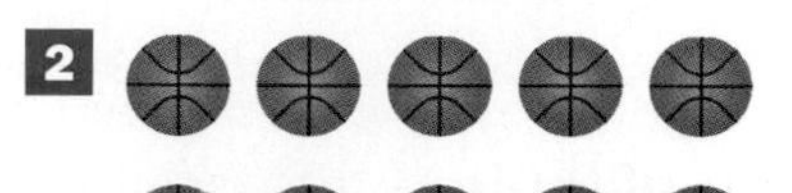
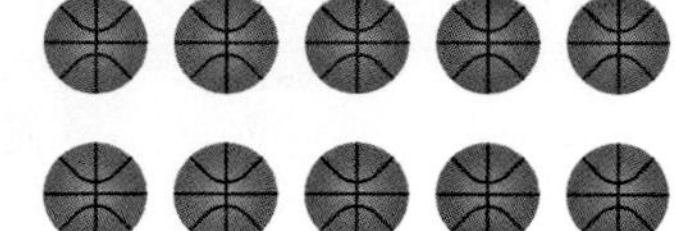

3
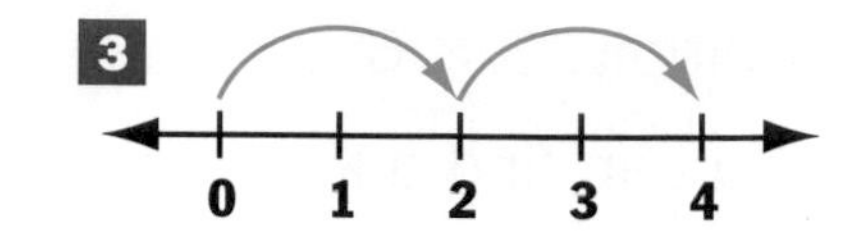

4
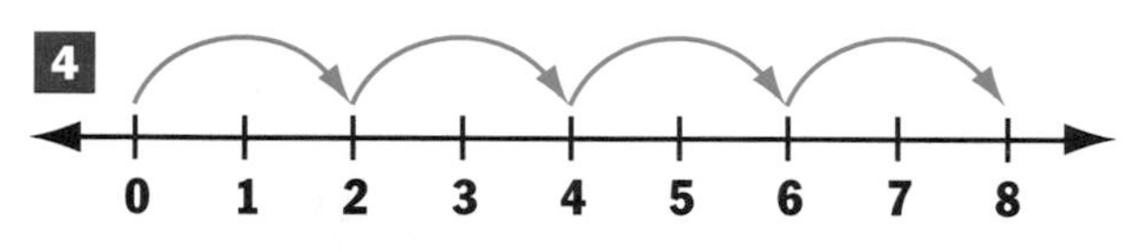

5
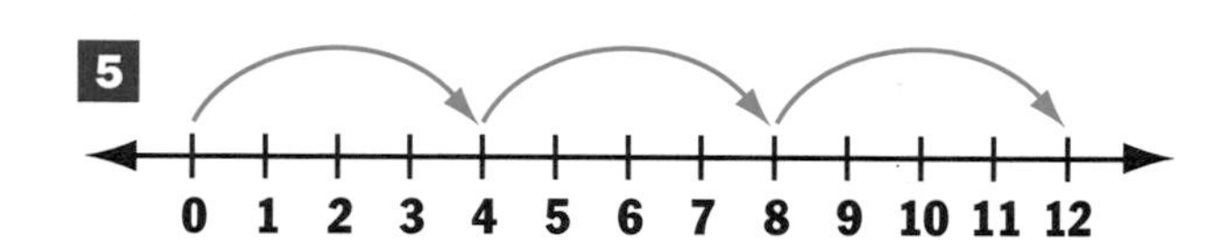

Draw an array. Find the product.

6 9×2 7 1×5 8 2×5 9 6×3 10 7×4

Multiply using any method.

11 7×2 12 5×3 13 2×6 14 5×9 15 5×8

16 4×7 17 3×8 18 3×9 19 4×2 20 4×8

21 2×8 22 3×7 23 4×6 24 2×5 25 5×7

26 $\begin{array}{r} 4 \\ \times 8 \\ \hline \end{array}$ 27 $\begin{array}{r} 5 \\ \times 4 \\ \hline \end{array}$ 28 $\begin{array}{r} 9 \\ \times 2 \\ \hline \end{array}$ 29 $\begin{array}{r} 7 \\ \times 4 \\ \hline \end{array}$ 30 $\begin{array}{r} 6 \\ \times 5 \\ \hline \end{array}$ 31 $\begin{array}{r} 9 \\ \times 5 \\ \hline \end{array}$

ALGEBRA **Find the rule. Then complete the table.**

32
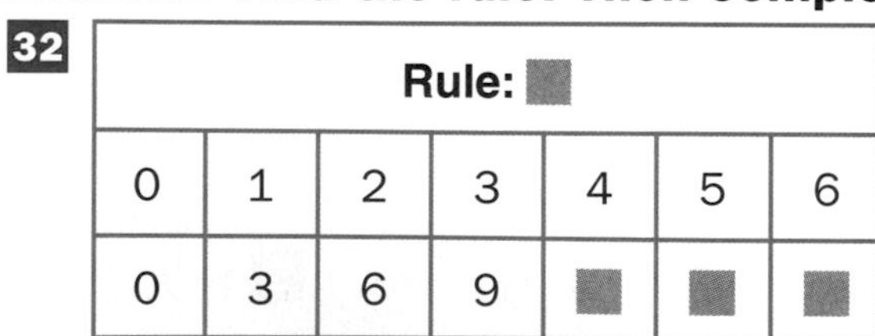

Rule: ■						
0	1	2	3	4	5	6
0	3	6	9	■	■	■

33

Rule: ■						
0	1	2	3	4	5	6
0	■	8	■	16	■	■

Tell if the number is a multiple of both 2 and 5.

34 12 35 15 36 20 37 25 38 40

Make It Right

39 Here is how Linda found 3×8. Tell what the mistake is, then give the correct answer.

$$3 \times 8 = \underbrace{2 \text{ groups of } 8}_{16} + \underbrace{1 \text{ group of } 3}_{3} = 19$$

MIXED APPLICATIONS

Problem Solving

National Basketball Association Scoring Champion						
Season	Player	Games	2-point Field Goals	3-point Field Goals	Free Throws	Total Number of Points
1986-1987	Michael Jordan	82	1,086	12	833	3,041
1987-1988	Michael Jordan	82	1,062	7	723	2,868
1988-1989	Michael Jordan	81	939	27	674	2,633
1989-1990	Michael Jordan	82	942	92	593	2,753
1990-1991	Michael Jordan	82	961	29	571	2,580
1991-1992	Michael Jordan	80	916	27	491	2,404
1992-1993	Michael Jordan	78	911	81	476	2,541
1995-1996	Michael Jordan	82	805	111	548	2,491

Use the table for problems 40–43.

40 In the 1987–1988 season, how many points did Michael Jordan score with 3-point field goals?

41 Which is worth more points: five 3-point field goals or seven 2-point field goals? Explain.

42 Did Michael Jordan score more than 17,000 points from the 1986–1987 season through the 1995–1996 season? Explain how you used estimation to find the answer.

43 The scoring leader for the 1994–1995 season was Shaquille O'Neal with 2,315 points. Did Michael Jordan score more points in the 1986–1987 season? If he did, how many more points?

more to explore

Counting Nickels

You can use nickels to find products with 5 as a factor.

To find 6×5, think of 6 nickels.

Think: 6 nickels is 30¢. $6 \times 5 = 30$

Tell how to find the product using nickels.

1 2×5 **2** 3×5 **3** 5×5 **4** 7×5 **5** 8×5

6 and 8 as Factors

LEARN

You can use strategies to find products when 6 and 8 are factors. You can double a known fact.

Multiply: 6×4 **Think:** 3 groups of 4 plus 3 groups of 4

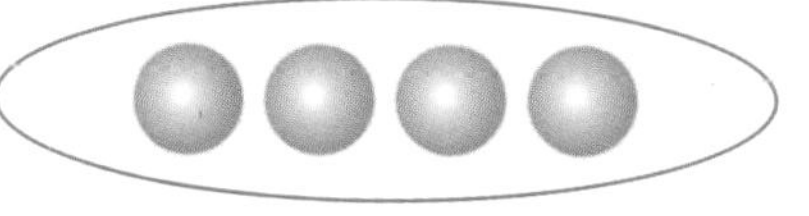
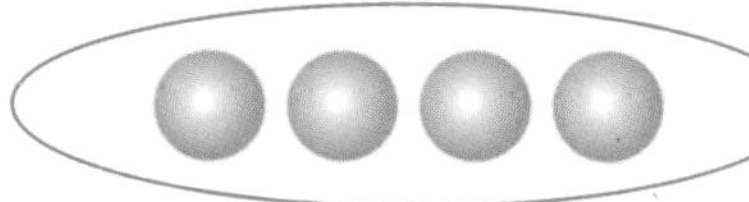
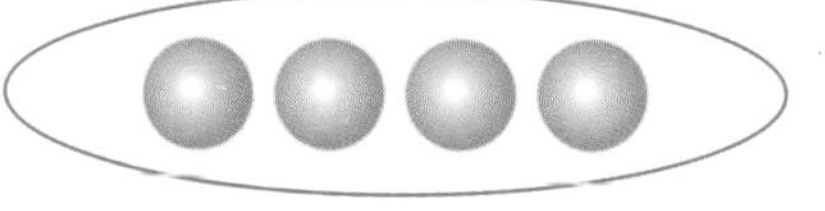
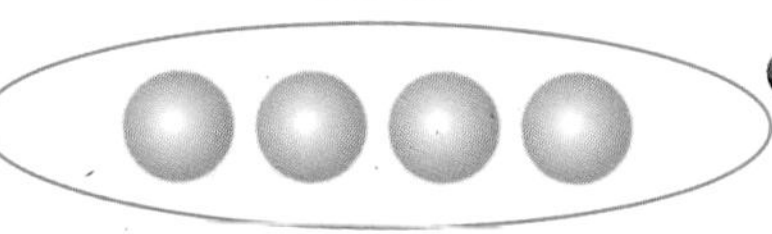
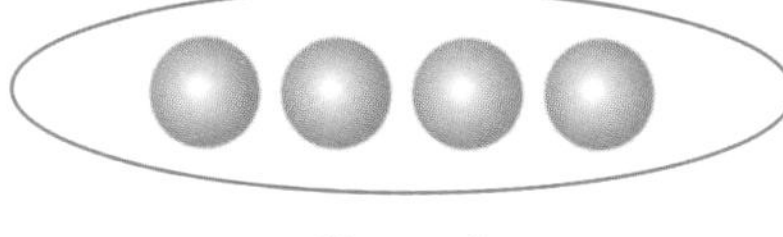

$$\underbrace{3 \times 4}_{12} + \underbrace{3 \times 4}_{12} = 24$$

So $6 \times 4 = 24$.

Multiply: 8×3 **Think:** 4 groups of 3 plus 4 groups of 3

$$\underbrace{4 \times 3}_{12} + \underbrace{4 \times 3}_{12} = 24$$

So $8 \times 3 = 24$.

CHECK

Check for Understanding

Multiply. Show your methods.

1 6×3 **2** 8×8 **3** 6×6 **4** 7×6 **5** 3×8

Critical Thinking: Analyze

6 Find two ways to double a known fact to solve $4 \times 8 = \blacksquare$.

Practice

Multiply using any method.

1. 6×2
2. 8×4
3. 6×5
4. 6×7
5. 7×8
6. 8×1
7. 6×0
8. 8×7
9. 2×8
10. 9×6
11. 8×0
12. 3×6
13. 6×6
14. 5×8
15. 6×8
16. 1×6
17. 2×8
18. 8×6
19. 4×6
20. 8×8
21. 9×8

ALGEBRA Complete the multiplication sentence.

22. $8 \times ■ = 56$
23. $■ \times 5 = 30$
24. $6 \times ■ = 0$
25. $■ \times 8 = 32$
26. $8 \times 9 = (4 \times 9) + (■ \times 9)$
27. $8 \times 7 = (4 \times 7) + (4 \times ■)$
28. If $\triangle = 7$, then how much is $\triangle\triangle\triangle$?
29. If $\square = 5$, then how much is $\square\square\square\square\square\square$?

MIXED APPLICATIONS
Problem Solving

30. In the fourth-grade hockey game, there were three periods, each 10 minutes long. How long was the playing time?
31. In the 1995–1996 season, the Philadelphia Flyers won 45 games, lost 24 games, and tied 13. How many games did they play in all?
32. **Logical reasoning** "I am a 2-digit number between 20 and 30. I am a multiple of 3 and 4. My ones digit is two more than my tens digit. What number am I?"
33. The Howe family went to a hockey game. They bought 3 adult tickets at \$7 each and 6 children's tickets at \$3 each. How much did they spend in all?

mixed review • test preparation

Find the range, median, and mode of the data.

1. 26, 23, 26, 18, 21
2. 85, 80, 60, 90, 75
3. 98, 100, 92, 80, 100
4. 143, 387, 288, 143, 288, 648, 322

Tell the time that is 15 minutes before the time shown.

5. 11:15 A.M.
6. 6:30 P.M.
7. 8:08 A.M.
8. 12 noon

Extra Practice, page 496

7 and 9 as Factors

Seven members of the Cypress Academy gymnastics club made it to the 1996 Women's Junior National Team. If each member competed in 5 events at the Junior Olympics, how many times did the club compete?

Multiply: 7×5

The Junior National Team
Cypress Academy, Houston, TX

You can use different strategies to find the product.

You can add to a known fact.

Think: 6 groups of 5 + 5 more

$$\underbrace{6 \times 5}_{30} + 5 = 35$$

You can also break apart a factor to get known facts.

Think: 5 groups of 5 + 2 groups of 5

$$\underbrace{5 \times 5}_{25} + \underbrace{2 \times 5}_{10} = 35$$

The club competed 35 times.

What if 9 members from Cypress Academy had made the team. If each member competed in 5 events, how many times would the club have competed?

Multiply: 9×5

To find products with 9 as a factor, you can first find products with 10 as a factor and then subtract.

Think: 10 groups of 5 − 1 group of 5

$$\underbrace{10 \times 5}_{50} - \underbrace{1 \times 5}_{5} = 45$$

Talk It Over

▶ In what ways can you break apart 7 to find 7×9? How can you use 10 as a factor to find the product? Show your methods.

ALGEBRA: PATTERNS You can also use a pattern to find 9×5. Look at these products with 9 as a factor.

- The tens digit of the product is always 1 less than the second factor.

$2 - 1$	$3 - 1$	$4 - 1$	$5 - 1$
↓	↓	↓	↓
$9 \times 2 = 18$	$9 \times 3 = 27$	$9 \times 4 = 36$	$9 \times 5 = 45$
$1 + 8$	$2 + 7$	$3 + 6$	$4 + 5$

- The sum of the digits in the product is always 9.

The club would have competed 45 times.

More Examples

A 7×8

Think: 5 groups of 8 + 2 groups of 8
$40 + 16 = 56$

B 9×8

Think: 10 groups of 8 − 1 group of 8
$80 - 8 = 72$

C 9×7

Think: $7 - 1 = 6$
$6 + ■ = 9$
$6 + 3 = 9$
$9 \times 7 = 63$

D 9×6

Think: $6 - 1 = 5$
$5 + ■ = 9$
$5 + 4 = 9$
$9 \times 6 = 54$

Check for Understanding

CHECK

Multiply using any method.

1 3×7	**2** 6×9	**3** 7×9	**4** 4×7	**5** 3×9
6 7×1	**7** 9×9	**8** 6×7	**9** 9×1	**10** 2×7
11 7×3	**12** 4×9	**13** 2×9	**14** 7×0	**15** 5×7

Critical Thinking: Analyze **Explain your reasoning.**

16 How can you find 7×4 if you know 6×4?

17 Is 9 a factor of each product below?
a. 54 **b.** 48 **c.** 81 **d.** 72 **e.** 91

18  

How can you break apart 9 to find 9×8?

Turn the page for Practice.

PRACTICE

Practice

Multiply mentally.

1. 7×2
2. 9×3
3. 9×0
4. 9×5
5. 7×6
6. 1×9
7. 4×4
8. 7×8
9. 8×9
10. 9×9
11. 7×7
12. 1×7
13. 6×6
14. 9×8
15. 7×8
16. 6×5
17. 7×9
18. 3×7
19. 5×3
20. 9×2
21. 7×6
22. 7×7
23. 1×5
24. 4×6
25. 6×3
26. 5×0
27. 8×5

Show at least two ways to get the product.

28. 9
29. 12
30. 18
31. 24
32. 72
33. 0

Complete the order form.

	Title of Book	Price	Number Ordered	Total
34	*Women in Gymnastics*	$9	7	■
35	*How to Tumble*	$5	1	■
36	*Getting to the Olympics*	$7	2	■
37	*All About the Parallel Bars*	$3	8	■

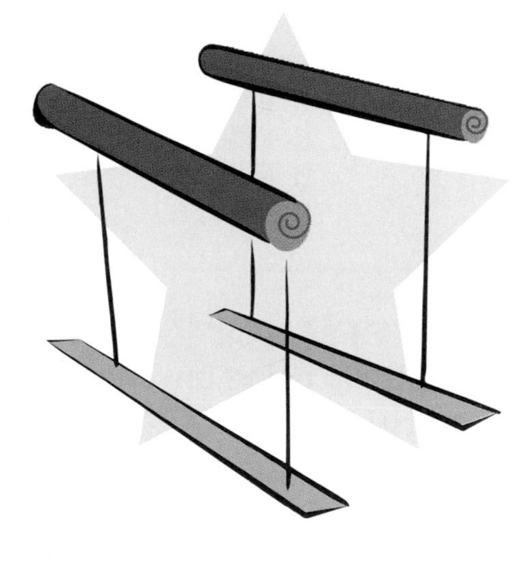

ALGEBRA Complete the sentence.

38. $7 \times ■ = 28$
39. $■ \times 6 = 54$
40. $5 \times ■ = 7 \times 5$
41. $3 \times ■ = 6 \times 4$
42. $11 \times ■ = 11$
43. $23 \times 14 = ■ \times 23$
44. $6 \times 6 =$ ■ groups of 6 + ■ groups of 6
45. $9 \times 4 =$ ■ groups of 4 + ■ groups of 4
46. $8 \times 7 =$ ■ groups of 7 + ■ groups of 7
47. $7 \times 4 =$ ■ groups of 7 + ■ groups of 7

MIXED APPLICATIONS
Problem Solving

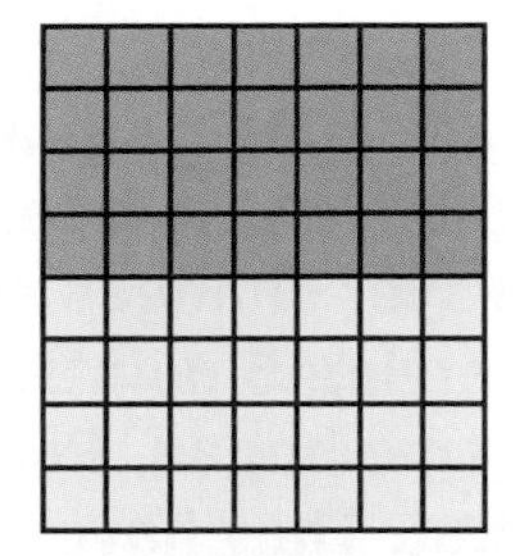

48 Tsung uses the area model at the right to find 8 × 7. He says that multiplication using the area model is the same as grouping counters. Explain how this is so.

49 **Spatial reasoning** What shape can you not get by combining four equal-sized triangles?
a. circle **b.** square **c.** triangle

50 There are four events in the Olympic women's gymnastics competition. If a gymnast scores a 9 in three events and a 7 in one event, what is her total number of points?

51 The Olympics is held every 4 years. How many team gold medals in a row did the USSR win in the Olympics? SEE INFOBIT.

INFOBIT
The USSR Olympic gymnastics team won team gold medals in every Olympics from 1952 until 1980.

Cultural Connection Egyptian Math Puzzle

The Ahmes Papyrus was written by a scribe named Ahmes. He recorded discoveries that Egyptians made about 4,000 years ago. Here is a part of a mathematical puzzle he wrote.

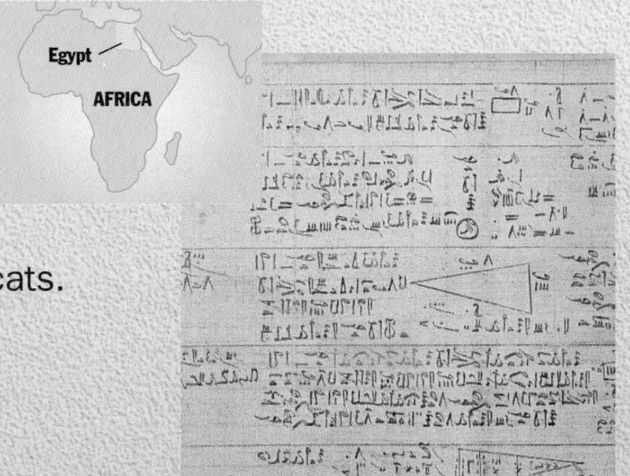

In each of seven houses, there are seven cats.
Each cat eats seven mice.
Each mouse eats seven ears of grain.
Each ear holds seven measures of grain.

Use a calculator to find the number of:

1 cats. **2** mice eaten. **3** ears of grain eaten. **4** measures of grain eaten.

Read
Plan
Solve
Look Back

Find a Pattern

LEARN

Read **Amy and Pedro are playing a game of table tennis. Each player changes serve after serving 5 points. When the game began, Pedro served first. They have played for 17 points so far. Who will serve the next ball?**

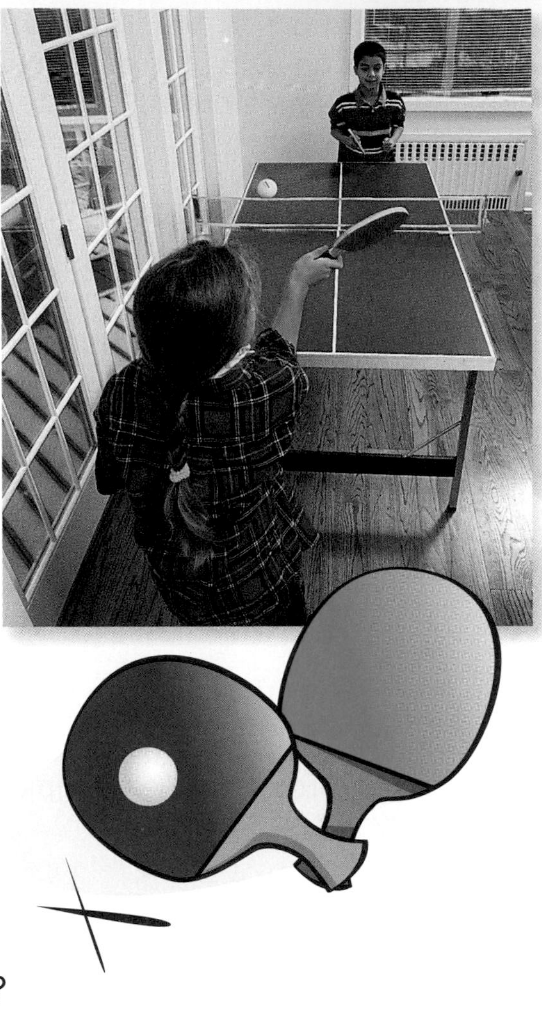

Plan To solve the problem, you can find a pattern.

Solve Find the serving pattern.
Pedro serves points 1 to 5.
Amy serves points 6 to 10.
Pedro serves points 11 to 15.
Amy serves points 16 to 20.
Pedro serves points 21 to 25.

Think: 17 points have been served.

The next ball is the 18th point.
Amy serves the 18th point.

Look Back Has the question been answered?

CHECK

Check for Understanding

1 **What if** the score was 5 to 18 in favor of Amy. What is the next point? Who will serve the ball? Explain your reasoning.

Critical Thinking: Analyze **Explain your reasoning.**

2 In American football, a touchdown is worth 6 points. A placekick over the crossbar following a touchdown is worth 1 point. If a team scores 26 points in a game, what combination of touchdowns and placekicks did they make?

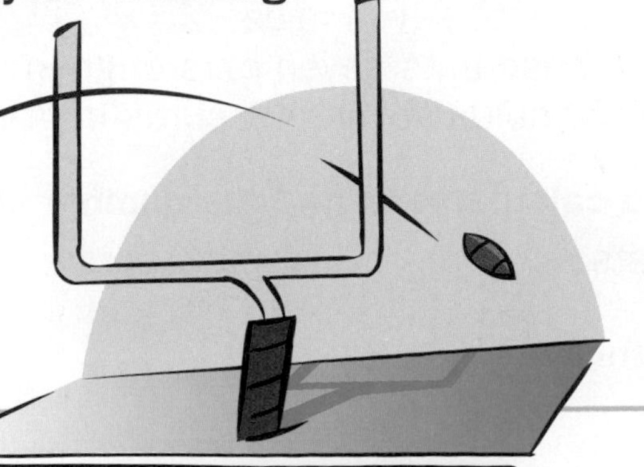

MIXED APPLICATIONS

Problem Solving

1 A magic square has rows, columns, and diagonals that add to the same number. The total is called the magic number. Complete the magic square below.

7	■	3
■	9	■
■	■	11

2 **Spatial reasoning** A football field is lined with chalk every 5 yards. How many chalk lines are there from goal line to goal line?

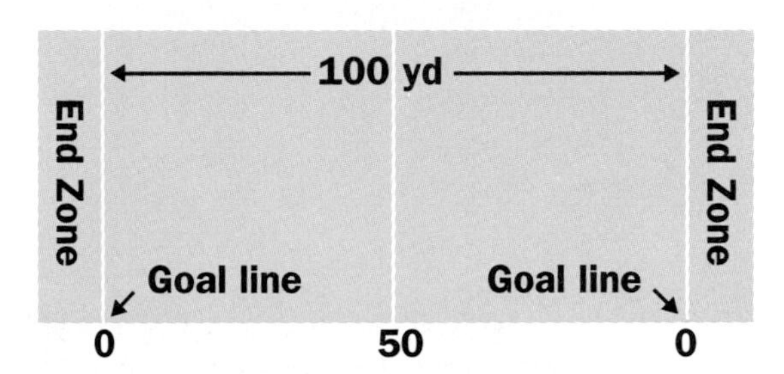

3 There are five Olympic speed-skating events for men: 500 m; 1,000 m; 1,500 m; 5,000 m; and 10,000 m. If a skater competes in all five events, what is his total distance?

4 In the 1996 NCAA Women's Basketball Tournament championship game, Tennessee scored 5 more points than Connecticut in overtime. Their combined scores were 171. How many points did each team score?

5 Keith Gledhill won a table-tennis championship in 1926. He won another championship in 1987. How many years had passed?

6 There are 16 cans in the 8th row, 14 in the 7th row, and 12 in the 6th row. If the pattern continues, how many cans will be in the 1st row?

Use the graph for problems 7–10.

7 Which ball weighs the most? the least?

8 Which of the balls shown weigh more than 4 ounces?

9 **What if** soccer were listed in the graph. A soccer ball weighs about 16 ounces. How many balls would you use to show the weight of a soccer ball?

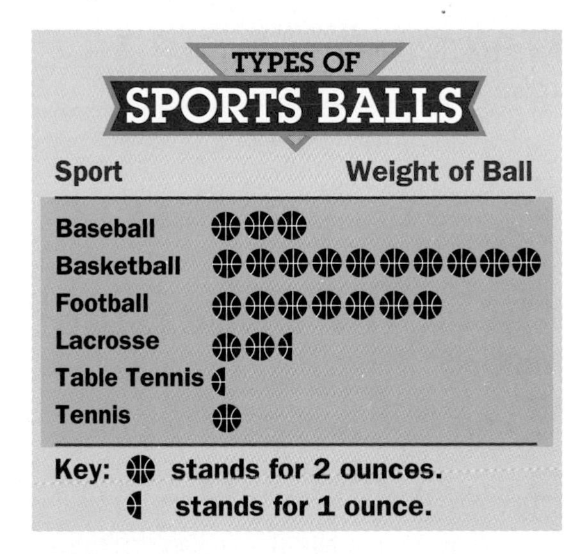

10 **Write a problem** using the data in the pictograph. Solve it and have others solve it.

11 **Data Point** Find how many different types of sports balls you have at home. Create a pictograph.

LEARN

Three Factors

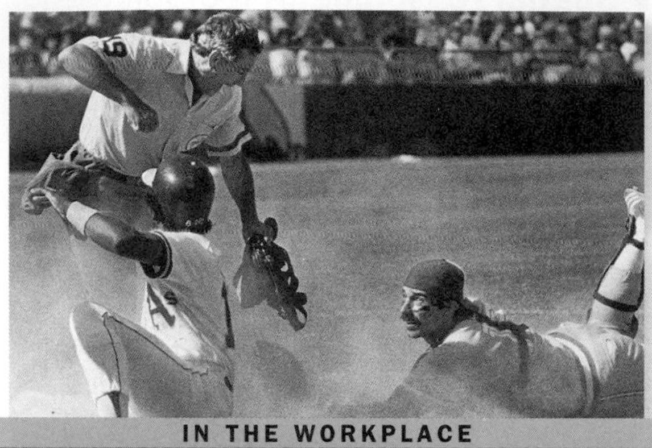

IN THE WORKPLACE

Rich Garcia, major league umpire, American League

Have you ever watched a baseball game that goes on and on? Rich Garcia rules whether a batter is out. If there are 9 complete innings and 3 outs per inning for each of the two teams, it takes a long time to see all those outs. How many outs are there?

You can multiply $2 \times (9 \times 3)$ to find the number of outs. Using properties may keep the factors easy to multiply.

$2 \times (9 \times 3) = 2 \times (3 \times 9)$ ⟵ **Use the Commutative Property.**

$= (2 \times 3) \times 9$ ⟵ **Use the Associative Property.**

$= 6 \times 9$ ⟵ **Multiply.**

$= 54$ There are 54 outs.

Commutative Property
You can change the order of two factors without changing the product. $9 \times 3 = 3 \times 9$

Associative Property
You can change the grouping of three or more factors without changing the product. $2 \times (3 \times 9) = (2 \times 3) \times 9$

More Examples

A $(8 \times 2) \times 5 = 8 \times (2 \times 5)$
$= 8 \times 10$
$= 80$

B $3 \times (2 \times 7) = (3 \times 2) \times 7$
$= 6 \times 7$
$= 42$

CHECK

Check for Understanding

Multiply.

1 $(3 \times 6) \times 2$ **2** $3 \times (5 \times 3)$ **3** $(8 \times 4) \times 2$ **4** $2 \times (7 \times 4)$

5 $5 \times (9 \times 2)$ **6** $(2 \times 7) \times 3$ **7** $(5 \times 7) \times 2$ **8** $2 \times (8 \times 4)$

Critical Thinking: Analyze **Explain your reasoning.**

9 How are the Commutative and Associative Properties for multiplication similar to those for addition?

Practice

Multiply.

1. (5 × 2) × 2
2. (9 × 3) × 2
3. 8 × (7 × 1)
4. (6 × 3) × 3
5. (2 × 6) × 4
6. 2 × (5 × 4)
7. (3 × 5) × 2
8. (2 × 9) × 3
9. 3 × (4 × 3)
10. (8 × 3) × 2
11. 9 × (0 × 8)
12. (7 × 1) × 7

ALGEBRA Complete the multiplication sentence.

13. (7 × 4) × 1 = ■
14. 6 × (5 × 0) = ■
15. 8 × 2 = ■ × 8
16. 6 × 9 = 6 × (■ × 3)
17. (2 × 3) × 4 = (2 × 4) × ■

MIXED APPLICATIONS
Problem Solving

18. Tori runs 2 miles each day on 5 days of each week. About how many miles does she run during February?

19. **Logical reasoning** Find as many ways as you can to show the number 90 using three numbers and at least two operations.

20. **Make a decision** Look at the dartboard at the right. What sections should you aim for to get a score of 50 using the fewest number of darts? Why?

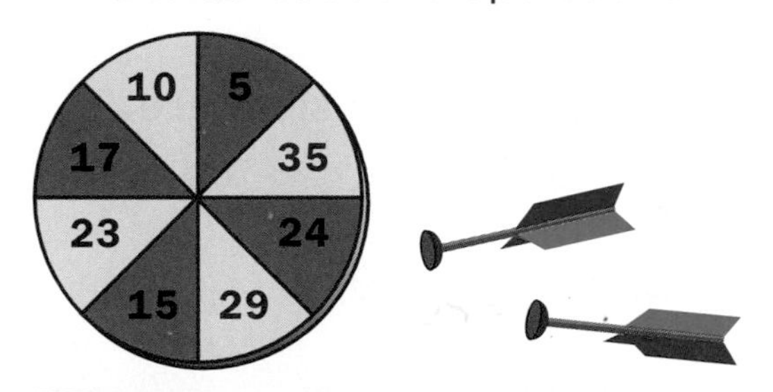

21. Estimate which of these products will be between 50 and 100. Explain your reasoning.
    a. 15 × 3 **b.** 25 × 10
    c. 9 × 8 **d.** 16 × 4

22. **Data Point** Survey your classmates to find the different types of exercises they do. Use a bar graph to show your data. Write a sentence about your graph.

mixed review • test preparation

Write in standard form.

1. 5,000 + 200 + 90 + 3
2. 70,000 + 800 + 50
3. 300,000 + 40 + 2
4. six thousand, eleven
5. eighty thousand, four hundred four

Compare. Write >, <, or =.

6. 8,742 ■ 8,800
7. 20,306 ■ 20,360
8. 5 million ■ 500,000

Extra Practice, page 497

Write a multiplication sentence for the picture.

1 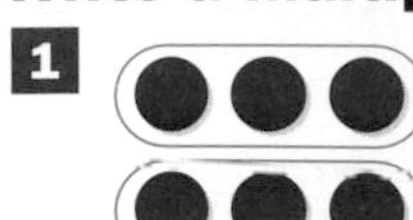 

2   

3 

Multiply. Draw an array or use counters if you wish.

4 6×1 **5** 1×7 **6** 0×4 **7** 8×0 **8** 5×3

9 0×1 **10** 2×3 **11** 5×6 **12** 2×8 **13** 8×8

14 3×5 **15** 4×3 **16** 6×5 **17** 3×8 **18** 6×7

19 4×4 **20** 9×3 **21** 9×8 **22** 7×7 **23** 5×9

24 $3 \times 5 \times 3$ **25** $8 \times 5 \times 2$ **26** $2 \times 9 \times 3$

27 $4 \times 6 \times 2$ **28** $2 \times 7 \times 4$ **29** $8 \times 3 \times 3$

ALGEBRA Complete the sentence.

30 $4 \times 8 \times 0 = ■$ **31** $7 \times 6 \times 1 = ■$ **32** $4 \times 9 = 9 \times ■$

33 $3 \times ■ \times 9 = 9 \times 9$ **34** $6 \times 4 = ■ \times 3 \times 4$

ALGEBRA Find the rule. Then complete the table.

35

Rule: ■					
0	1	2	3	4	5
0	■	16	■	32	■

36

Rule: ■					
0	1	2	3	4	5
0	7	■	21	■	■

Solve. Use any method.

37 At the basketball game, 7 rows of 8 seats were set aside for students. How many seats were set aside?

38 The band marches across the football field in 9 rows. Each row has 6 students. How many students are in the band?

39 Sybil bought a box of golf balls. The box had 2 layers of balls. Each layer had 5 rows of 4 balls. How many balls were in the box?

40 

Explain two different ways you can find 9×3.

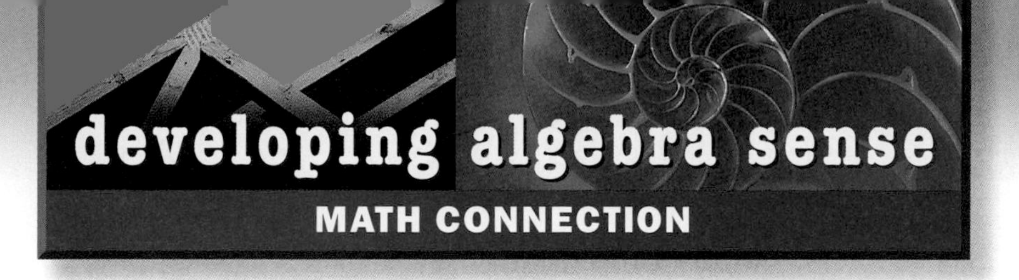

Patterns and Properties

Complete the multiplication table below. The number 20 is placed in Row 5 and in Column 4 because 20 is the product of 5 and 4. What is the product of 9 and 6?

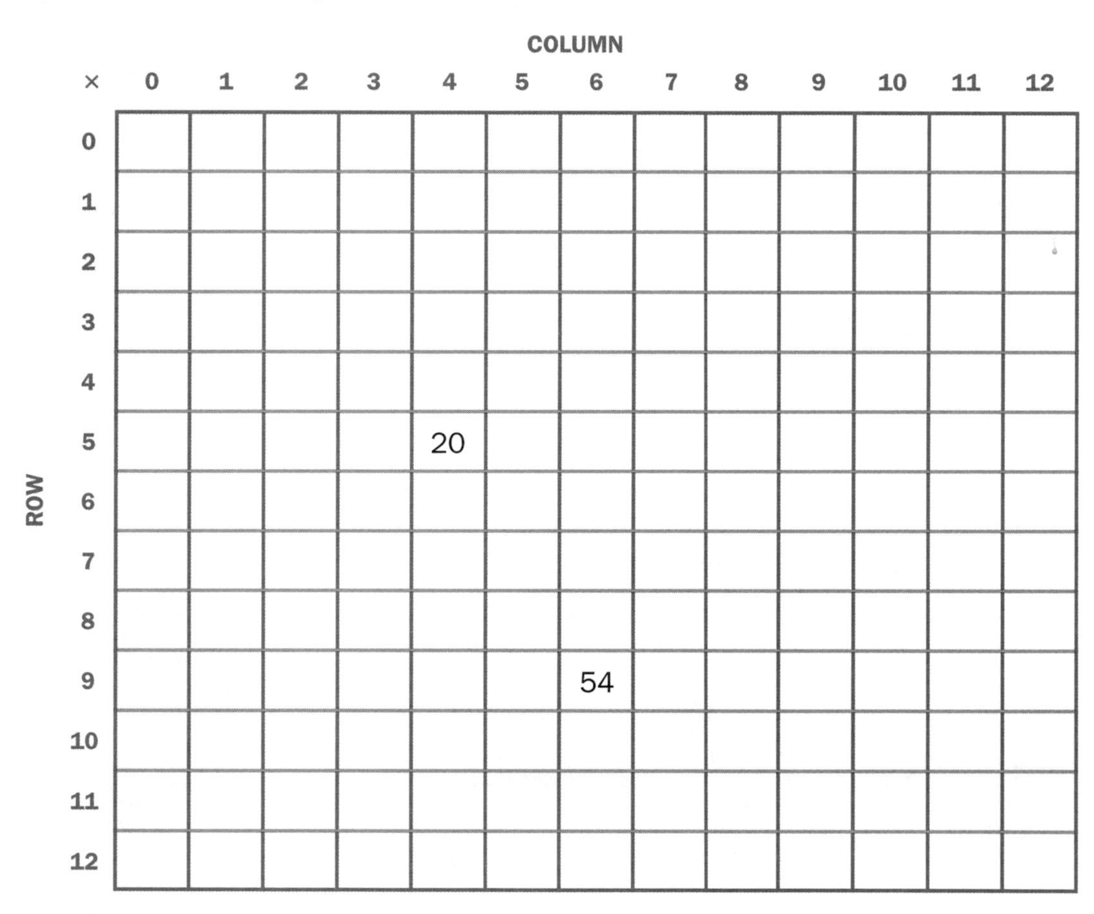

COLUMN

ROW ×	0	1	2	3	4	5	6	7	8	9	10	11	12
0													
1													
2													
3													
4													
5					20								
6													
7													
8													
9							54						
10													
11													
12													

ALGEBRA: PATTERNS Solve. Use the multiplication table.

1 What pattern do you see in Row 1? in Column 1?

2 What property can you use to complete Row 0 and Column 0?

3 Look at the products in Row 2. Compare these products with the products in Row 4. What pattern do you see?

4 If you complete Columns 0–3, which rows can you complete? Why?

5 Square numbers can be shown as squares. Where are these numbers in the table? What can you say about their factors?

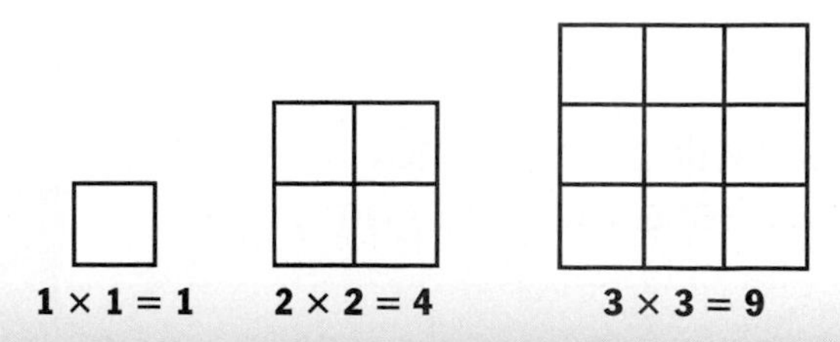

Changing the Rules

Rules in sports are very important. They tell how many people should play and what equipment they should use. They also tell how to play the sport and who the winner should be. In this activity, you will work in teams to create a sport of your own. That way you make all the rules!

- ▶ What do you need to think about when you create your sport?
- ▶ Will you create an entirely new sport or change one that you already know?

NEW SPORT

CALANDRA!

- ▶ How many players will there be? What roles will they play? What will they use or wear?
- ▶ When will players score points? How will they score points? What events should get a higher score?
- ▶ How will players know that they have won the game?

DECISION MAKING

Creating Your Sport

1 Work with a small group. Think of different ideas for sports.

2 Decide what your sport will be. Choose one that your classmates will enjoy playing at school. Make up a catchy name for your sport.

3 READING ARITHMETIC WRITING **Sequence of Events** Write the rules of your sport using sequence words. Make sure that your sport is safe.

4 Play your sport with the members of your group. Revise the sport or the rules if you need to make them simple and easy to understand.

Reporting Your Findings

5 Portfolio Prepare a poster that explains your sport. Include the following:

- ▶ Explain the rules and show drawings or photographs that help others understand how to play.
- ▶ Show some possible ways to get between 40 and 50 points. Use multiplication when you find the totals.
- ▶ Explain how you would group your entire class to play.

6 Compare your poster with those of other groups.

Revise your work.

- ▶ Does your poster include all the information that players need?
- ▶ Are your rules and scoring clear and easy to understand?
- ▶ Did you proofread your work?

MORE TO INVESTIGATE

PREDICT all the different ways to win the game by getting a score you choose.

EXPLORE ways to change your sport so that younger children can play. What do you need to think about to do this step?

FIND out who invented your favorite sport and what changes have been made to the sport since it was invented.

Meaning of Division

LEARN

Do you ever run relay races in gym class? With 4 runners on each team, how many teams can you make with 24 runners?

Work Together

Work with a partner. Use any method you want to model the problem. Record your work in a table like the one shown.

Total Number	Number in Each Team	Number of Teams
24	4	

You will need
- *counters*
- *graph paper*

How many teams of 3, 4, 5, and 6 can you make from your class?

Talk It Over

▶ Explain your method for finding the number of teams.

Make Connections

You used a table to record your work. You can also write a division sentence to record how you found the number of teams.

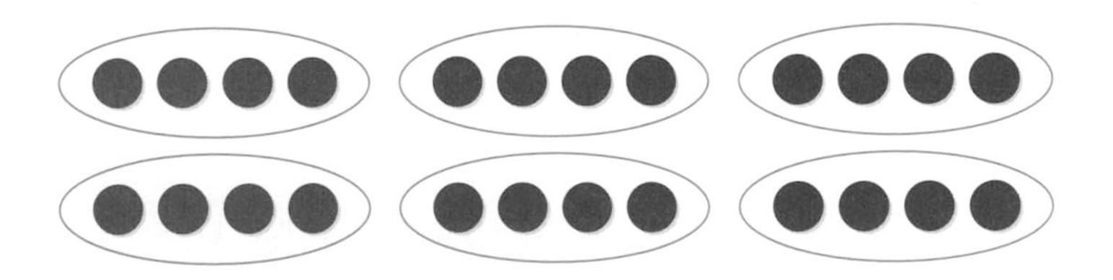

dividend A number to be divided.

divisor A number by which the dividend is divided.

quotient The result of division.

Total Number	Number in Each Team	Number of Teams
24	4	6

Write: 24 (Total, ↑ **dividend**) ÷ 4 (Number in each group, ↑ **divisor**) = 6 (Number of groups, ↑ **quotient**) or $4\overline{)24}$ with quotient 6

Read: 24 divided by 4 is equal to 6.

You can also divide to find the number in each group.

	Total		Number of groups		Number in each group
Write:	24	÷	6	=	4
	↑		↑		↑
	dividend		**divisor**		**quotient**

▶ Compare the multiplication sentence $6 \times 4 = 24$ to the division sentences above. Why do you think they are called related sentences?

▶ **What if** you divide 32 students into 4 teams. How many students will be on each team? Is the quotient the number of groups or the number in each group?

Check for Understanding

Complete the division sentence. Tell whether you divided to find: *the number of groups* or *the number in each group*.

1

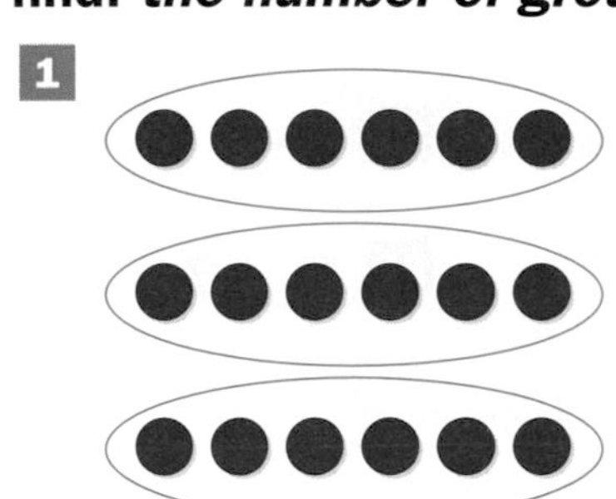

$18 \div 6 = ■$

2

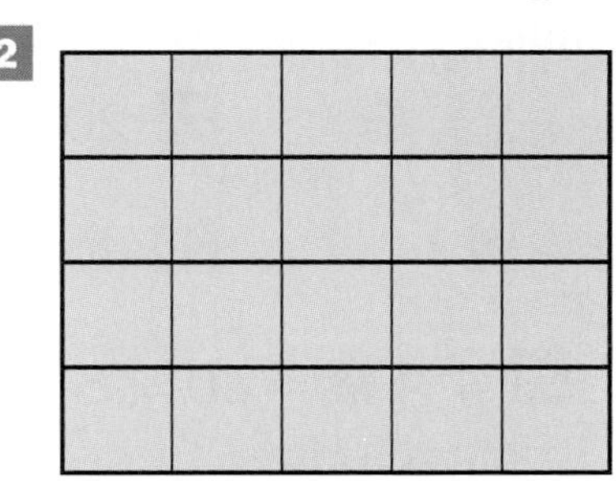

$20 \div 4 = ■$

3

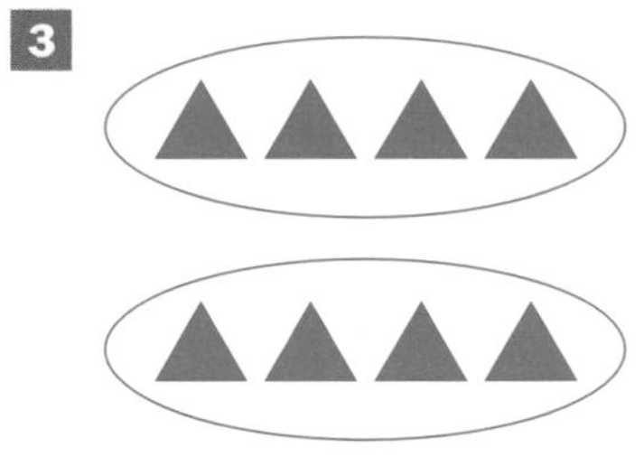

$8 \div 4 = ■$

Divide. Draw a picture to show the division.

4 $9 \div 3$ **5** $12 \div 6$ **6** $18 \div 9$ **7** $25 \div 5$ **8** $28 \div 4$

9 $12 \div 3$ **10** $21 \div 7$ **11** $18 \div 3$ **12** $32 \div 8$ **13** $48 \div 6$

Critical Thinking: Analyze

14 What does *division* mean? Use the words *separate* and *group* in your explanation.

15 Allie found $24 \div 6$ using the method below. Explain why her method works.

24 − 6 = 18 *18 − 6 = 12* *12 − 6 = 6* *6 − 6 = 0*

I subtracted 6 four times, so 24 ÷ 6 = 4.

CHECK

Turn the page for Practice.

Practice

Complete the division sentence. Tell whether you divided to find *the number of groups* or *the number in each group*.

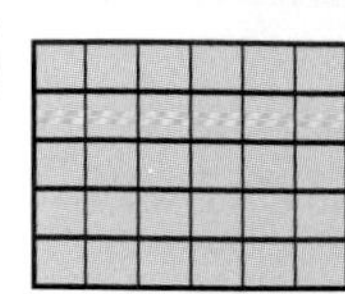

$30 \div 6 = ■$

2

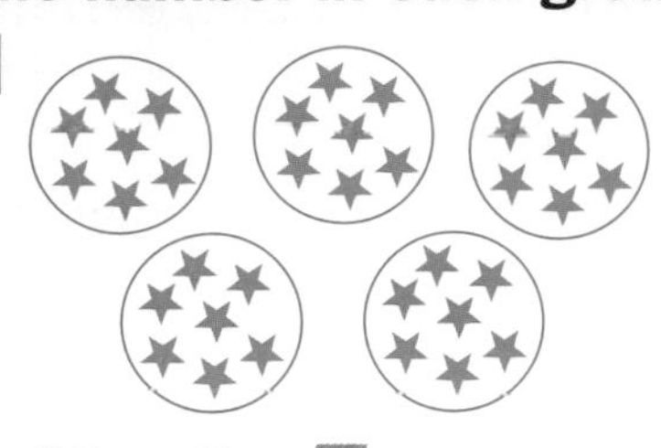

$35 \div 7 = ■$

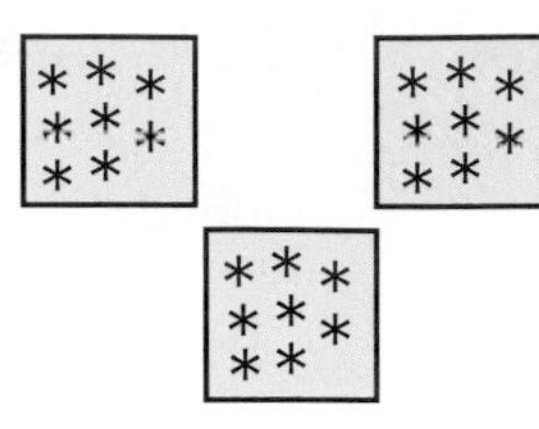

$24 \div 3 = ■$

Draw an array. Find the quotient.

4 $12 \div 4$ **5** $15 \div 3$ **6** $21 \div 7$ **7** $28 \div 4$ **8** $18 \div 2$

Divide using any method.

9 $12 \div 6$ **10** $30 \div 5$ **11** $24 \div 3$ **12** $27 \div 9$ **13** $32 \div 8$

14 $2\overline{)16}$ **15** $5\overline{)35}$ **16** $6\overline{)30}$ **17** $4\overline{)32}$ **18** $7\overline{)28}$

ALGEBRA **Find the missing number.**

19 $8 \div 4 = ■$ **20** $16 \div 2 = ■$ **21** $20 \div 5 = ■$ **22** $36 \div 9 = ■$

23 $■ \div 6 = 6$ **24** $42 \div ■ = 7$ **25** $40 \div ■ = 8$ **26** $■ \div 7 = 8$

27 $8 \div 2 = ■ \times 4$ **28** $4 \times ■ = 40 \div 5$ **29** $24 \div ■ = 21 \div 7$

Make It Right

30 Here is how Calvin found $12 \div 3$. Explain what the mistake is, then give the correct answer.

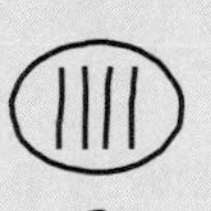

1 2 3

$12 \div 3 = 3$

MIXED APPLICATIONS

Problem Solving

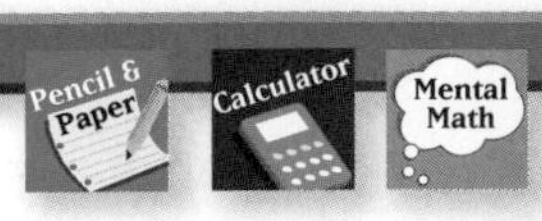

31 Your gym class has 35 students. For each basketball game you need two teams of 5 players each. Can your class play 4 basketball games on different courts at the same time? Why or why not?

32 Your 6-player hockey team is playing another team in the semi-finals. Two other 6-player teams are also in the semi-finals. How many players are in the semi-finals?

Dividing the Bases Game!

You will need
- *number cube*
- *two-color counters*

Play the Game

Play with a partner.

▶ Start at the batter's box. Roll the number cube to get a divisor.

▶ If 14 can be equally divided by the divisor, move one space on the game board.

▶ Take turns. Roll a number cube at each base. See who gets to home plate first.

mixed review • test preparation

Tell the time.

1 2 h 30 min after 3:15 P.M.

2 5 h 15 min after 9 A.M.

3 4 h 30 min before 5:30 P.M.

4 3 h 45 min before 1:00 P.M.

Round to the nearest hundred.

5 374

6 2,536

7 5,027

8 12,360

Extra Practice, page 498

2 Through 5 as Divisors

LEARN

Whitewater rafting is exciting! A group of 20 people is rafting. If each raft has 5 people, how many rafts are there?

Divide: $20 \div 5$

To find the number of rafts, you can skip-count backward on a number line.

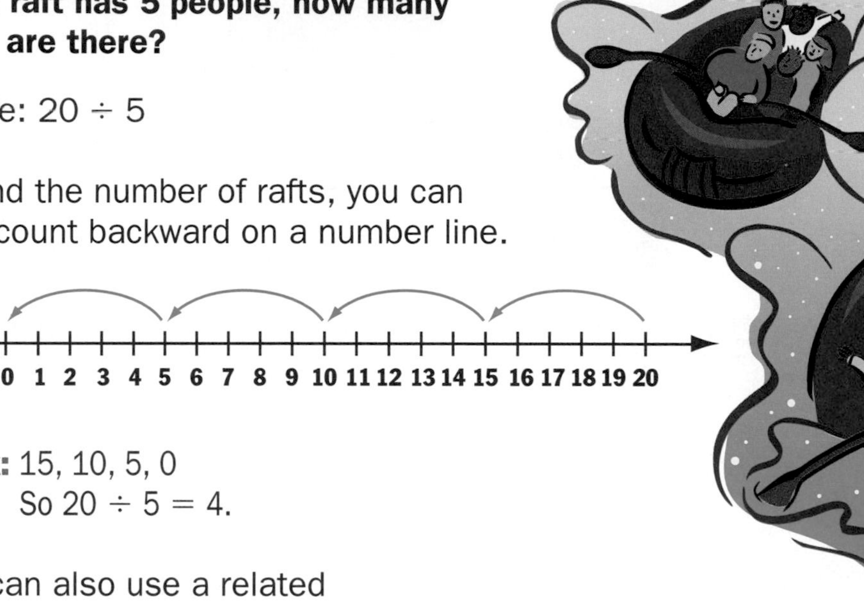

Think: 15, 10, 5, 0
So $20 \div 5 = 4$.

You can also use a related multiplication sentence to find the quotient.

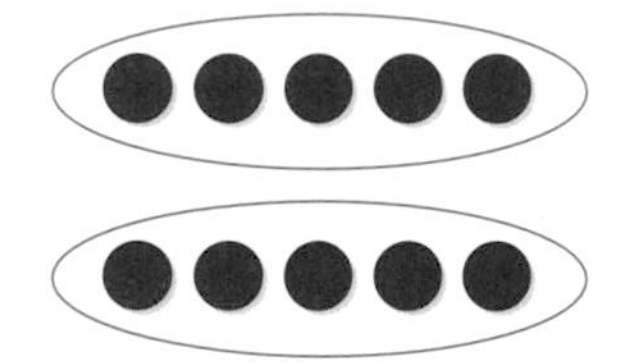

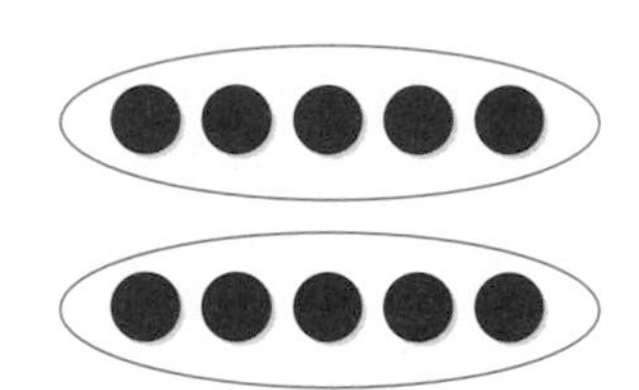

Think: What number times 5 is 20?
$\blacksquare \times 5 = 20$
$4 \times 5 = 20$
So $20 \div 5 = 4$.

There are 4 rafts.

CHECK

Check for Understanding

Divide.

1. $10 \div 2$
2. $6 \div 3$
3. $12 \div 4$
4. $15 \div 5$
5. $9 \div 3$
6. $16 \div 4$
7. $21 \div 3$
8. $8 \div 2$
9. $4 \div 4$
10. $5 \div 1$

Critical Thinking: Generalize **Explain your reasoning.**

11. Journal Addition and subtraction are called *inverse* operations because one operation undoes the other. Are multiplication and division inverse operations? Give an example.

Practice

Write a division sentence for each picture.

1 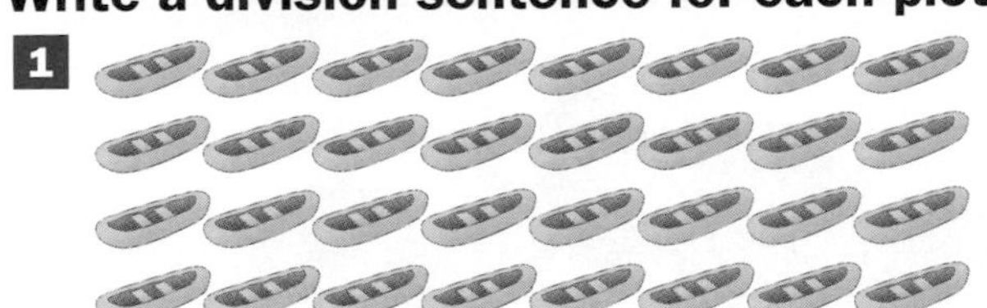

2 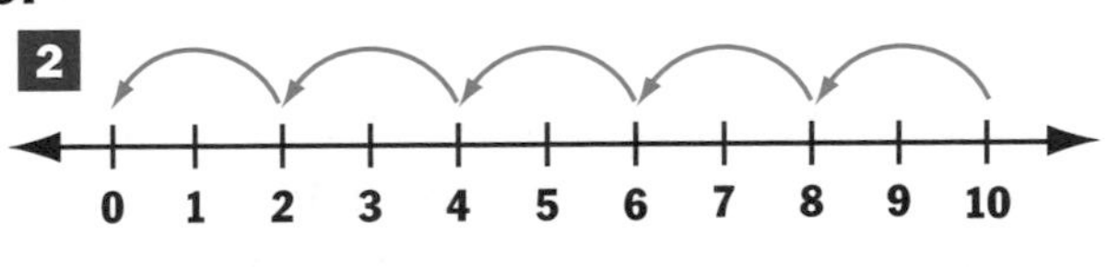

Divide.

3 $25 \div 5$ **4** $8 \div 4$ **5** $0 \div 3$ **6** $14 \div 2$ **7** $36 \div 4$

8 $4\overline{)24}$ **9** $3\overline{)12}$ **10** $5\overline{)35}$ **11** $4\overline{)28}$ **12** $5\overline{)5}$

13 $5\overline{)30}$ **14** $6\overline{)18}$ **15** $2\overline{)16}$ **16** $7\overline{)21}$ **17** $4\overline{)36}$

MIXED APPLICATIONS
Problem Solving

18 **What if** there are 4 rafts and 28 people. How many people can ride each raft if each raft has the same number of people?

19 A life jacket at the Outdoor Store costs $23.00. Shani has a store coupon for $5.45. How much will she pay for the jacket?

20 **Write a problem** where you divide to make groups of the same size. Solve your problem and have others solve it.

21 **Data Point** Use the Databank on page 536. Which tree is higher: the coconut palm or the western hemlock? By how many feet?

more to explore

Using a Calculator to Skip-Count

Multiply: 5×9

Enter: [0] [+] [5]

Press [=] nine times. Display: 45.

Starting at 0 and counting by 5s nine times gives 45. So $5 \times 9 = 45$.

Divide: $36 \div 4$

Enter: [3] [6] [−] [4]

Press [=] until you reach 0. Count the number of times.

Starting at 36 and counting back by 4s nine times gives 0. So $36 \div 4 = 9$.

Multiply or divide. Use your calculator to skip-count.

1 3×8 **2** $27 \div 3$ **3** $30 \div 5$ **4** 9×6 **5** $16 \div 2$

6 Through 9 as Divisors

LEARN

A fourth-grade class wants to play the Coconut Shell Relay. There are 21 students. They need to form teams of 7 students. How many teams can they make?

Divide: $21 \div 7$

Cultural Note

The Coconut Shell Relay is played by the Aborigines, a group of people native to Australia.

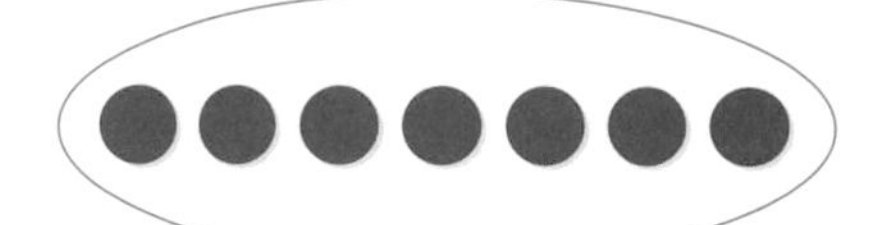

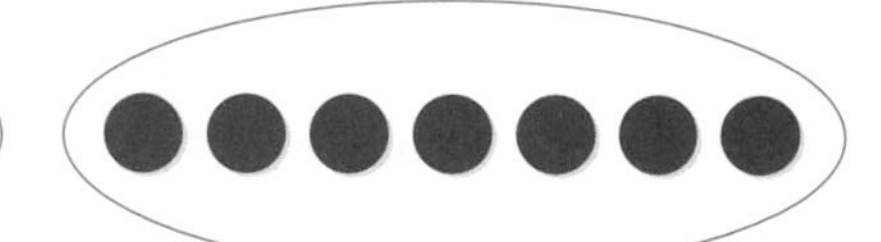

You can use a multiplication fact to find the quotient.

Think: What number times 7 is 21?

$\blacksquare \times 7 = 21$

$3 \times 7 = 21$

So $21 \div 7 = 3$.

You can also use a related division fact to find the quotient.

Think: 21 divided by what number is 7?

$21 \div \blacksquare = 7$

$21 \div 3 = 7$

So $21 \div 7 = 3$.

The class can make 3 teams.

CHECK

Check for Understanding

Divide.

1. $14 \div 7$
2. $24 \div 8$
3. $12 \div 6$
4. $18 \div 6$
5. $27 \div 9$
6. $40 \div 8$
7. $21 \div 7$
8. $36 \div 6$
9. $72 \div 9$
10. $8 \div 8$

Critical Thinking: Summarize

11. Explain the method you would use to find each quotient.
    a. $24 \div 6$ **b.** $56 \div 7$ **c.** $16 \div 8$ **d.** $81 \div 9$

Practice

Write a division sentence for the picture.

1

2 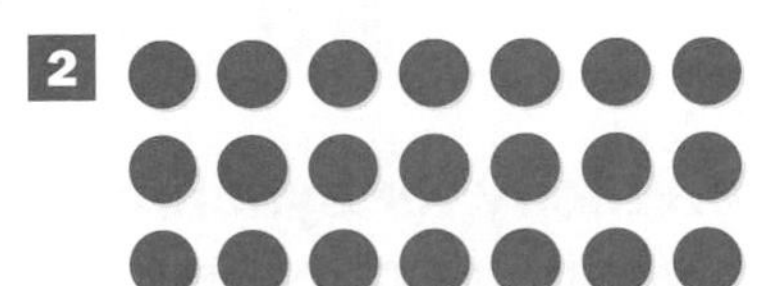

Divide.

3 $18 \div 9$ **4** $28 \div 7$ **5** $7 \div 7$ **6** $32 \div 8$ **7** $45 \div 9$

8 $54 \div 6$ **9** $56 \div 8$ **10** $49 \div 7$ **11** $0 \div 9$ **12** $48 \div 6$

13 $6\overline{)30}$ **14** $9\overline{)36}$ **15** $8\overline{)48}$ **16** $7\overline{)42}$ **17** $7\overline{)63}$

MIXED APPLICATIONS
Problem Solving

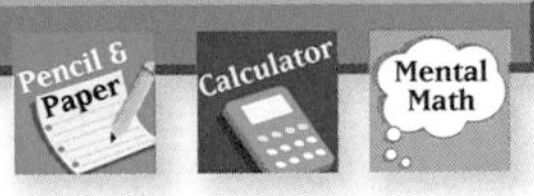

18 In a game of Coconut Shell Relay, there are 5 teams of 7 players. How many players are there in all?

19 **Make a decision** Your class has 24 students. How would you arrange the teams to play the Coconut Shell Relay?

20 How many tickets can be bought with $64 if each ticket costs $8?

21 About how many coconuts grow on a tree? **SEE INFOBIT.**

INFOBIT
About ten clusters of coconuts may be on a tree. There are 10 to 20 coconuts in a cluster.

mixed review • test preparation

Use the pictograph to answer questions 1 and 2.

1 If 21 life jackets were rented on Monday, what is the key?

Number of Jackets Rented	
Monday	
Tuesday	
Wednesday	

2 How many jackets were rented on Tuesday? on Wednesday?

Multiply.

3 9×4 **4** 6×8 **5** 4×5 **6** 7×7 **7** 3×9

Extra Practice, page 498

Fact Families

LEARN

At the medal ceremony for synchronized swimming, 24 swimmers crowded the platform for the awarding of the 3 team medals. How many swimmers were on each team?

You can use a **fact family** to find the number of swimmers on each team.

fact family Related facts using the same numbers.

$3 \times 8 = 24$
$8 \times 3 = 24$
$24 \div 8 = 3$
$24 \div 3 = 8$

There were 3 teams with a total of 24 swimmers.
There were 8 swimmers on each team.

More Examples

A $2 \times 4 = 8$
$4 \times 2 = 8$
$8 \div 2 = 4$
$8 \div 4 = 2$

B $9 \times 1 = 9$
$1 \times 9 = 9$
$9 \div 1 = 9$
$9 \div 9 = 1$

C $3 \times 3 = 9$
$9 \div 3 = 3$

CHECK

Check for Understanding

Find the fact family for the numbers.

1 16, 8, 2 **2** 4, 5, 20 **3** 6, 1, 6 **4** 4, 9, 36 **5** 8, 8, 64

Critical Thinking: Analyze **Explain your reasoning.**

6 Can you find $6 \div 0$ using a fact family? Why or why not?

7 What can you say about any number divided by zero?

8 **ALGEBRA: PATTERNS** Divide. What patterns do you see?

a. $0 \div 3$ $0 \div 4$ $0 \div 6$ $0 \div 9$
b. $3 \div 3$ $7 \div 7$ $8 \div 8$ $1 \div 1$
c. $4 \div 1$ $5 \div 1$ $7 \div 1$ $9 \div 1$

Practice

Find the fact family for the numbers.

1 3, 6, 18 **2** 4, 7, 28 **3** 5, 25, 5 **4** 72, 8, 9

5 9, 45, 5 **6** 14, 7, 2 **7** 63, 9, 7 **8** 1, 8, 8

9 4, 8, 32 **10** 7, 49, 7 **11** 7, 56, 8 **12** 9, 81, 9

ALGEBRA Find the missing number.

13 $■ \div 9 = 4$ **14** $48 \div 6 = ■$ **15** $6 \times ■ = 54$ **16** $64 \div ■ = 8$

17 $8 \times ■ = 16$ **18** $42 \div ■ = 7$ **19** $■ \div 8 = 5$ **20** $■ \div 9 = 6$

MIXED APPLICATIONS
Problem Solving

One scoring system in the Olympics gives 10 points for each gold, 5 points for each silver, and 4 points for each bronze.

21 What is the total score for Cuba?

22 Which nation has the greatest score? the least score?

23 Suppose a nation has a score of 100. What are possible medals its athletes could have won?

24 **Write a problem** using information from the table. Trade problems with other students. Solve.

Medals by Country
1996 Summer Olympics

Nation	Gold	Silver	Bronze
Australia	9	9	23
China	16	22	12
Cuba	9	8	8
France	15	7	15
Germany	20	18	27
Hungary	13	10	12
Italy	33	21	28
Korea	7	15	5
Russian Federation	26	21	16
Ukraine	9	2	12
United States	44	32	25

mixed review • test preparation

Subtract.

1 \$5 − \$1.98 **2** \$2 − \$1.15 **3** \$10 − \$5.75 **4** \$5.02 − \$1.27

Show the change in two different ways.

5 86¢ **6** 29¢ **7** \$1.94 **8** \$7.33 **9** \$15.75

Extra Practice, page 499

Remainders

LEARN

A class is going to a soccer game. Your job is to decide how many cars are needed to take the students to the game.

Cultural Note

Some historians trace soccer to the Japanese game of *kemari* which was played as early as 600 B.C.

You will need
- *counters*
- *graph paper*

Work Together

Work with a partner.

Take turns. Your partner chooses a 2-digit number for the total number of students. You choose a 1-digit number for the number in each group.

Use any method you want to find the number of equal groups and the number left over.

KEEP IN MIND
- ▶ Check your work to make sure that your answer is correct.
- ▶ Be ready to explain your methods to the class.

Record your results in a table.

Total Number of Students	Number in Each Group	Number of Equal Groups	Number Left Over	Number of Cars
23	5			

▶ How did you decide how many cars are needed?

Make Connections

Sometimes when you divide, the numbers do not come out evenly. They have a **remainder.**

remainder the number left over when dividing.

$$\begin{array}{r} 4 \text{ R3} \\ 5\overline{)23} \\ -20 \\ \hline 3 \end{array}$$

Think: $5 \times 4 = 20$ $\quad 20 < 23$

$5 \times 5 = 25$ $\quad 25 > 23$

So $23 \div 5 = 4$ R3.

- When there is no remainder, we say that the dividend is **divisible by** the divisor. What numbers in your table are divisible by 2? by 3? by 6?
- If a number is divisible by 2 and by 3, is it also divisible by 6?
- What happens when the dividend is not divisible by the divisor?

Check Out the Glossary
divisible by
See page 544.

Check for Understanding

CHECK

Divide.

1. $9 \div 2$
2. $11 \div 3$
3. $15 \div 4$
4. $13 \div 5$
5. $25 \div 4$
6. $7\overline{)37}$
7. $2\overline{)17}$
8. $5\overline{)47}$
9. $9\overline{)26}$
10. $3\overline{)29}$

Critical Thinking: Analyze **Explain your reasoning.**

11. Compare the remainder and divisor in several division problems. Is the remainder always equal to, greater than, or less than the divisor?

Practice

PRACTICE

Divide.

1. $22 \div 5$
2. $13 \div 3$
3. $25 \div 7$
4. $36 \div 9$
5. $38 \div 6$
6. $44 \div 8$
7. $50 \div 7$
8. $45 \div 6$
9. $55 \div 9$
10. $56 \div 8$
11. $3\overline{)19}$
12. $6\overline{)35}$
13. $7\overline{)28}$
14. $8\overline{)60}$
15. $5\overline{)34}$
16. $9\overline{)87}$
17. $4\overline{)18}$
18. $2\overline{)19}$
19. $5\overline{)34}$
20. $3\overline{)22}$
21. $8\overline{)77}$
22. $9\overline{)83}$
23. $5\overline{)41}$
24. $3\overline{)26}$
25. $7\overline{)48}$

Extra Practice, page 499

Read
Plan
Solve
Look Back

Part 1 Choose the Operation

LEARN

The town of Woodside sponsored a 10-km run for marathon runners. The top six runners are listed below.

Name of Runner	Starting Time	Ending Time
Fouz Alaoyi	11:00:00	11:31:48
Pedro Arcy	11:05:00	11:35:46
Stacy Flynn	11:00:00	11:40:45
Paul Malloy	11:07:00	11:37:59
Trevor Reed	11:08:00	11:49:28
Lena Wicks	11:05:00	11:42:16

Note: 11:31:48 stands for 11 hours 31 minutes 48 seconds.

Work Together

Solve. Tell which operation you used.

1. Which runner ran the race in the shortest time? the longest time?

2. **What if** Jane ran 2 km in 10 min. If she continues at about this pace, how long will it take her to run 10 km? About how far can she run in 30 min?

3. Tim stops to drink water every 3 km. In a 24-km run, how often will he stop to drink water before finishing the race?

4. **ALGEBRA: PATTERNS** To train for a race, Russell runs 30 min each day of the first week, 45 min each day of the second week, 60 min each day of the third week, and 75 min each day of the fourth week. If he continues this pattern, how long will he run each day of the fifth week? the sixth week?

5. READING ARITHMETIC WRITING **Sequence of Events** Write a paragraph for the Woodside newspaper giving the results of the race. Describe the race in the order of the events that happened.

Part 2 Write and Share Problems

Matthew used information about a runner to write the problem.

Bob runs 5 miles every day for 6 weeks. He needs a new pair of running shoes every 6 months. A pair of his running shoes costs $45.

6. Solve Matthew's problem. What operation did you use to solve the problem?
7. Change Matthew's problem so that you need information to solve it.
8. Explain what type of information is needed and how you might be able to collect the information.
9. **Write a problem** of your own about running. You should be able to solve the problem using multiplication or division.
10. Trade problems. Solve at least three problems written by your classmates.
11. What was the most interesting problem that you solved? Why?

If Bob runs 5 miles every day for 6 weeks, how many miles will Bob run altogether?

Matthew Niemi
Latson Road Elementary School
Howell, MI

CHECK

Cultural Note

The marathon is named in honor of the famous Greek messenger from the plains of Pheidippides (fī-DIP-uh-deez), who ran from Marathon to Athens.

Turn the page for Practice Strategies.

PRACTICE

Menu

Choose five problems and solve them. Explain your methods.

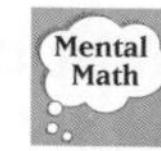

1 A sports drink comes in 8-ounce bottles. How many ounces are in 4 bottles?

2 Charlie's school day starts at 8:15 A.M. and ends at 2:45 P.M. How long is his school day?

3 Ace in-line skates have 4 wheels on each skate. The stockroom has 75 wheels. How many pairs of in-line skates can the company make? How many wheels will be left over?

4 Students use the gym for basketball every night from 5:00 P.M. to 8:00 P.M. How many hours each week is the gym used for basketball?

5 Use only the following keys on a calculator: 2 3 4 + − × ÷ =.

Show five different ways to get 24.

6 The tennis teacher wants to buy at least 25 tennis balls. The store sells tennis balls in containers of 3 balls for $4. How many containers does she need to buy? What is the total cost?

7 How many times does the digit 2 appear on the March calendar? Explain your answer.

8 **Write a problem** about baseball that can be solved by finding 9×12. Solve it. Have others solve it. Did you use mental math, pencil and paper, or a calculator? Why?

Choose two problems and solve them. Explain your methods.

9 You want to have a relay race with 30 students. Each team has the same number of students. There are at least 2 students on a team. How many different ways can you form teams? Name them.

10 Each letter stands for a digit. Find the numbers that match this addition problem.

$$\begin{array}{r} \text{ANT} \\ +\ \text{ANT} \\ \hline \text{TEEN} \end{array}$$

11 **ALGEBRA: PATTERNS** On your calculator, find products of two odd numbers, products of two even numbers, and products of one even and one odd number. Record each multiplication sentence and look for a pattern in the three cases.

a. What is the product of two odd numbers—odd or even?
b. What is the product of two even numbers—odd or even?
c. What is the product of an even number and an odd number?

Use patterns to predict if the product is even or odd.

d. 7×8 **e.** 9×7 **f.** 15×7 **g.** 18×14 **h.** 57×16

Use patterns to help you tell whether the product is correct or incorrect.

i. $6 \times 8 = 47$ **j.** $9 \times 7 = 63$ **k.** $15 \times 12 = 185$ **l.** $21 \times 11 = 232$

12 **At the Computer** Create a spreadsheet that can record the points scored in a basketball game. Watch a game and record the number of 2-point shots, 3-point shots, and free throws that are made for each team. Use the spreadsheet to calculate the total number of points for each team and compare your total to the total shown at the end of the game.

Basketball Scores

Player	Number of 2-point Shots	Number of 3-point Shots	Number of Free Throws	Total Points

Extra Practice, page 499

Language and Mathematics

Complete the sentence. Use a word in the chart. (pages 124–157)

Vocabulary
factor
multiple
quotient
remainder
Commutative
Associative

1 The ■ of 15 divided by 5 is 3.

2 When a number does not divide another number evenly, there will be a ■.

3 Since $7 \times 9 = 63$, 63 is a ■ of 7.

4 The number sentence $4 \times 8 = 8 \times 4$ shows the ■ Property.

Concepts and Skills

Write a multiplication sentence for the picture. (page 124)

5         

6  

7     

Draw an array. Find the product or quotient. (page 124, 146)

8 2×7 **9** 8×3 **10** $36 \div 9$ **11** $8 \div 1$ **12** 6×7

Find the product or quotient. (page 124)

13 8×6 **14** 9×5 **15** 4×4 **16** 6×0 **17** $7 \div 7$

18 $35 \div 5$ **19** $28 \div 4$ **20** $10 \div 3$ **21** $25 \div 9$ **22** $27 \div 3$

ALGEBRA Complete the sentence. (pages 124, 146, 154)

23 $6 \times$ ■ $= 24$ **24** ■ $\times 7 = 49$ **25** $8 \times 9 =$ ■

26 ■ $\div 4 = 8$ **27** $50 \div 8 =$ ■ R2 **28** $40 \div$ ■ $= 6$ R4

ALGEBRA Find the rule. Then complete the table. (page 124)

29

Rule: ■					
0	1	2	3	4	5
0	■	6	9	■	■

30

Rule: ■					
1	2	3	4	5	6
■	10	15	20	■	■

Think critically. (page 156)

31 Analyze. Explain what mistake was made, then correct it.

$39 \div 6 = 5R9$

32 Generalize. How are the operations related to each other? Give examples to show your answer. (page 158)

a. addition and subtraction
b. addition and multiplication
c. multiplication and division
d. division and subtraction

MIXED APPLICATIONS
Problem Solving

Pencil & Paper · Calculator · Mental Math

(pages 138, 158)

33 Eric plans to bike 75 miles. If he bikes 8 miles each hour, will he travel this distance in less than 9 hours? How do you know?

a **34** **ALGEBRA: PATTERNS** A group of 18 people is waiting for tables at a diner. If each table seats 6 people, how many tables do they need?

35 In 1924, the Olympic 100-meter freestyle was won in under a minute. It took 52 more years for it to be won in under 50 seconds. In what year did this happen?

36 To prepare for a bicycle tour, Jerry wants to buy a helmet for $34.89, a pair of gloves for $16.29, and a jacket for $45.99. If he has $100, does he have enough money?

Use the table for problems 37–40.

37 Who biked the longest distance? the shortest distance?

38 What is the difference between the longest and shortest distances?

39 Who took the most time to complete the bike tour?

40 About how many days did it take for the Slaughters to complete their bike tour?

Bike Tours

Who?	When?	Distance (miles)
Jay Aldous and Matt DeWaal	Apr. 2 to July 16, 1987	14,290
Tal Burt	June 1 to Aug. 17, 1992	13,523
John Hathaway	Nov. 10, 1974, to Oct. 6, 1976	50,600
Ronald and Sandra Slaughter	Dec. 30, 1989, to July 28, 1991	18,077

chapter test

Write a multiplication sentence for the picture.

1

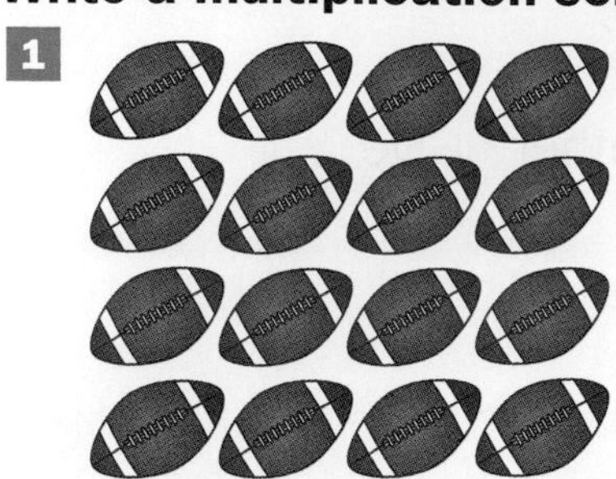

2

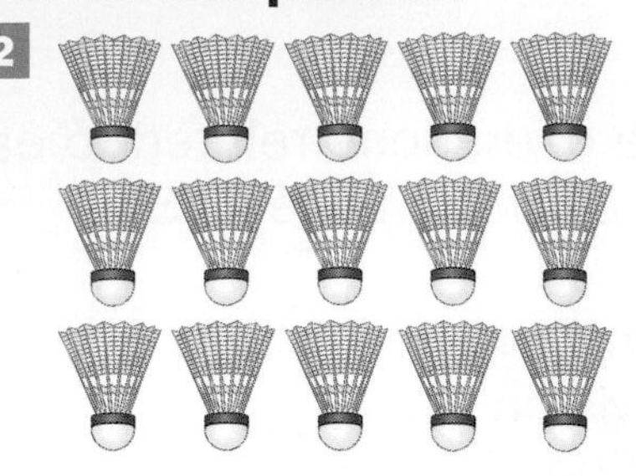

3

Draw an array. Find the product or quotient.

4 3×6

5 9×5

6 $25 \div 5$

7 $54 \div 6$

Find the product or quotient.

8 $36 \div 4$

9 $30 \div 6$

10 8×6

11 7×7

Complete the sentence.

12 $8 \times$ ■ $= 8$

13 ■ $\times 2 = 0$

14 $4 \times 7 = 7 \times$ ■

15 $(3 \times 5) \times 2 = 3 \times ($■ $\times 2)$

Write the fact family.

16 2, 3, 6

17 6, 7, 42

18 8, 8, 64

19 4, 9, 36

20 4, 4, 16

21 5, 6, 30

Solve.

22 Mario bakes 28 cookies. He has 7 friends over. How many cookies does each friend get if he shares the cookies equally among them?

23 Kelsey bought 3 tickets for the Olympics. Two cost $52.50 each, and the other one cost $38. How much did he spend?

24 Isabelle has a scarf. Each foot of her scarf has the following colored stripes on it: red, black, black, white. If there are 8 black stripes on the entire scarf, how long is the scarf?

25 Totsi rides her bike 1 mile on Monday, 1 mile on Tuesday, 2 miles on Wednesday, 3 miles on Thursday, and 5 miles on Friday. If she continues this pattern, how far will she ride on Saturday and Sunday?

What Did You Learn?

Mr. Jefferson works for a sporting goods store. He spilled some ink on this order form and cannot read some of the numbers.

- ► There can be from 1 through 9 items in each box. Name as many ways as you can think of to complete the bottom two rows of the form.

- ► For each row, explain your strategy for selecting numbers that might be under the ink spots.

ORDER FORM

Item	Number of Items in a Box	Number of Boxes Ordered	Total Number of Items Ordered
Hockey Pucks	6	5	30
Baseballs		4	
Table Tennis Paddles			24

INK

A Good Answer

- shows different ways to complete the order form correctly
- clearly describes the reasons for choosing possible missing numbers

You may want to place your work in your portfolio.

What Do You Think 

1. Can you find answers to multiplication and division facts easily? Why or why not?

2. List some methods you might use to multiply and some you might use to divide.

3. What would you do if you could not decide whether to multiply or divide to solve a problem?
   - Use counters to model the problem.
   - Draw a picture.
   - Other. Explain.

math science technology

CONNECTION

Nature's Protection

Cultural Note

Throughout history, humans have protected themselves using armor made from soft materials such as feathers and quilting, or hard materials such as bone, leather, and metal. Armor of crocodile skin was made in Egypt, helmets of porcupine fish skin were worn in the Pacific Islands, and covering made from short rods of wood were used by Native American tribes from the northwest coast of America.

Animals have different ways of protecting themselves against other animals that hunt them. The animals that hunt are called *predators.* The animals that are hunted are called *prey.*

The chameleon changes its color so that it blends in with the things around it. This makes it difficult for predators to find it.

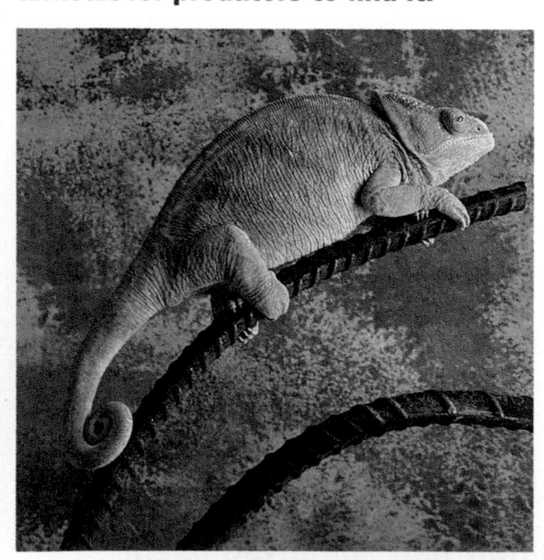

The squid discharges a brownish-black substance called ink. The ink clouds the water and dulls the sense of smell in fishes that pursue the squid by scent.

The porcupine uses its quills to defend itself against predators. It turns its back and stiffens its quills, rushing backward toward the enemy.

► How do the animals you know protect themselves from predators?

Comparing Speeds

Some animals' protection is their speed or agility. The table below shows the greatest speeds of some animals.

Animal	Speed (in miles per hour)
Chicken	9
Human	25
Grizzly bear	30
Coyote	43
Cape hunting dog	45
Lion	50
Gazelle	50
Cheetah	70

1 Which do you think hunt gazelles for food, coyotes or cheetahs? Explain.

2 Which animals are more than three times faster than a chicken?

3 **What if** you were told that a predator of the chicken is more than five times faster. If the predator is not the gazelle, what animal could it be?

At the Computer

4 Research to find other animals to add to the table. Use a graphing program to make a bar graph of the running speeds of the different animals.

5 Prepare a report that compares the speeds of the animals. Tell which animals are two times, three times, or four times faster than a chicken.

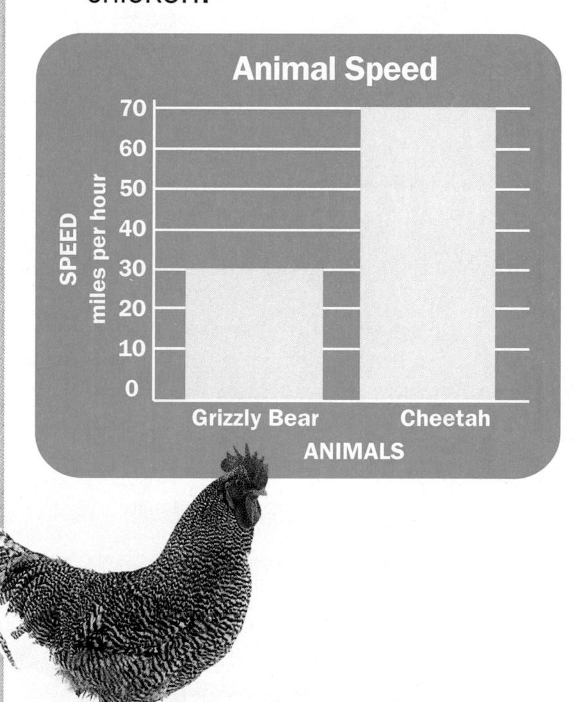

CHAPTER

5 MULTIPLY BY 1-DIGIT NUMBERS

THEME

Factories

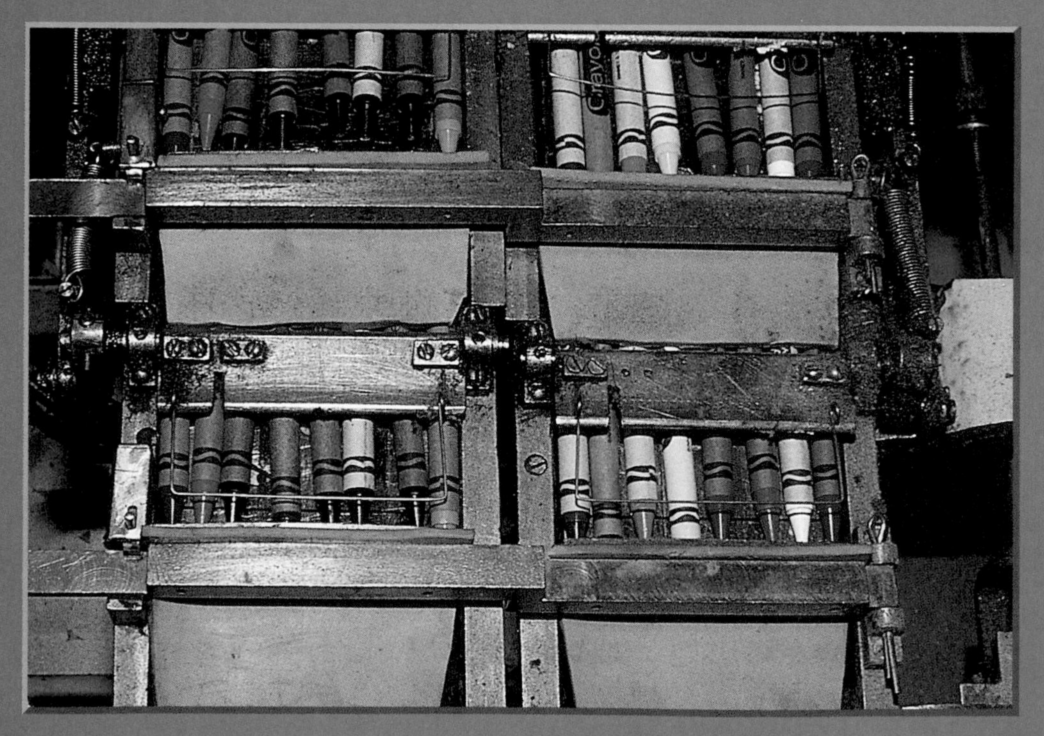

To run a factory, you need teamwork. In this chapter, you will work in teams and use multiplication as you learn about factory products such as crayons, computers, and piñatas.

What Do You Know?

1 How many crayons are there altogether in 3 boxes of 8 crayons?

2 If each person in your class got a box of 8 crayons, how many crayons would there be altogether? Explain your method.

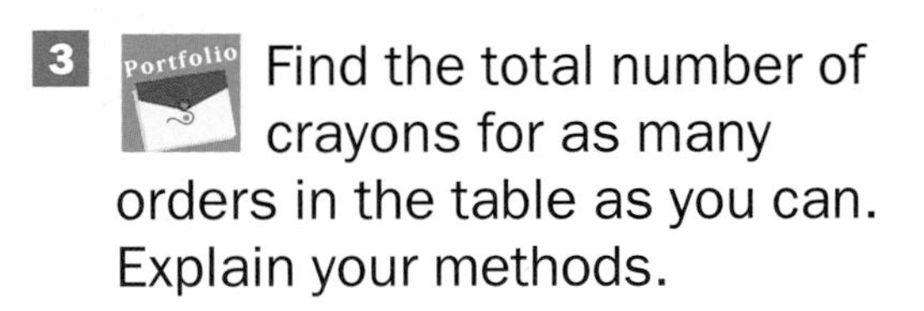

3 Find the total number of crayons for as many orders in the table as you can. Explain your methods.

Number of Crayons in Each Box	Number of Boxes Ordered	Total Number of Crayons
8	10	
16	4	
32	6	

Draw Conclusions

A carton arrives from the crayon factory for your class. The carton holds 30 boxes of crayons.

To draw a conclusion, you use information that the writer provides. You also use what you already know.

1 What did you already know about crayons?

2 What conclusion can you draw about the box?

Vocabulary

estimate, p. 172
product, p. 172
factor, p. 172
round, p. 172
regroup, p. 178

Multiplication Patterns

LEARN

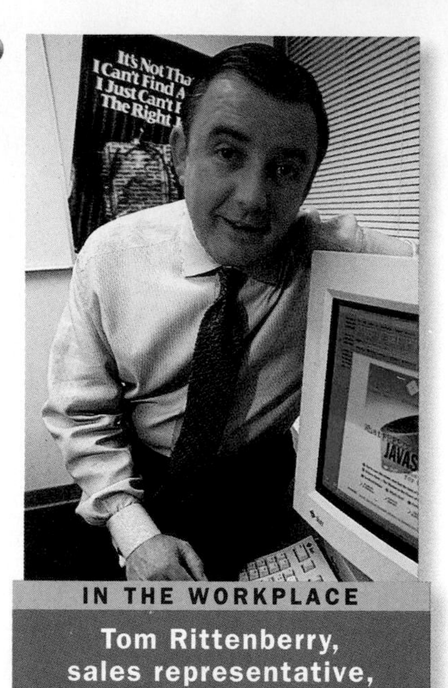

IN THE WORKPLACE
Tom Rittenberry, sales representative, Sun Microsystems, Knoxville, TN

Sales representatives sell products to people or companies. A Java Internet device costs about $1,000, and an Ultra Sparc desktop computer costs about $3,000. About how much do 6 of each product cost?

ALGEBRA: PATTERNS You can use patterns to find products mentally. What patterns do you see below?

$6 \times 1 = 6$
$6 \times 10 = 60$
$6 \times 100 = 600$
$6 \times 1{,}000 = 6{,}000$

Six Internet devices cost about $6,000.

$6 \times 3 = 18$
$6 \times 30 = 180$
$6 \times 300 = 1{,}800$
$6 \times 3{,}000 = 18{,}000$

Six desktop computers cost about $18,000.

More Examples

A
$8 \times 1 = 8$
$8 \times 10 = 80$
$8 \times 100 = 800$
$8 \times 1{,}000 = 8{,}000$

B
$9 \times 7 = 63$
$90 \times 7 = 630$
$900 \times 7 = 6{,}300$
$9{,}000 \times 7 = 63{,}000$

C
$5 \times 4 = 20$
$5 \times 40 = 200$
$5 \times 400 = 2{,}000$
$5 \times 4{,}000 = 20{,}000$

CHECK

Check for Understanding

Multiply mentally.

1. 7×100
2. $1{,}000 \times 3$
3. 7×200
4. 8×50
5. 4×80
6. $8{,}000 \times 3$
7. 5×20
8. 400×9
9. 3×700
10. $6 \times 1{,}000$
11. $5{,}000 \times 8$
12. 90×9

Critical Thinking: Analyze **Explain your reasoning.**

13. **ALGEBRA: PATTERNS** How can you use patterns to multiply mentally?

14. Why does the product of 5 and 600 have one more zero than the product of 4 and 600?

Practice

Multiply mentally.

1. 4 × 100
2. 100 × 5
3. 2 × 1,000
4. 1,000 × 4
5. 3 × 80
6. 4 × 7,000
7. 40 × 4
8. 5 × 200
9. 6 × 900
10. 1,000 × 5
11. 60 × 6
12. 5 × 6,000
13. 4 × 800
14. 30 × 4
15. 7 × 1,000
16. 8,000 × 5
17. 8 × 90
18. 900 × 9
19. 8 × 4,000
20. 6,000 × 6

ALGEBRA Find the missing number.

21. 7 × ■ = 350
22. 3 × ■ = 900
23. 4 × ■ = 8,000
24. ■ × 3 = 2,100

MIXED APPLICATIONS
Problem Solving

25. HAL, Inc., makes computer chips and packs 300 in each box. How many chips are in 9 boxes?

26. How many more additions in a second could the microcomputer do in 1987 than the ENIAC could in 1946?
SEE INFOBIT.

27. **Data Point** Use the Databank on page 537. Can you buy an Orange X610 laptop computer, a Z14 external modem, and a 3.8Y CD-ROM drive if you do not want to spend more than $4,500? Explain.

INFOBIT
In 1946, the ENIAC, the earliest digital machine, could do 5,000 additions in one second. By 1987 a microcomputer could do 400,000 additions in one second.

more to explore

Exponents

ALGEBRA You can write powers of 10 using exponents.

$10 \times 10 \times 10 = 1{,}000 = 10^3$ Read: 10 to the third power

Think: 10 is used as a factor 3 times. The exponent is 3.

Write the power of 10 using an exponent.

1. 10 × 10 × 10 × 10
2. 10 × 10 × 10 × 10 × 10
3. 10 × 10

LEARN

Estimate Products

The Acme Bicycle Factory makes 725 racing bicycles each day. Will Acme be able to complete this order in time?

Date	December 5, 1997
Item	Racing Bicycle
Number of items	4,000
Due Date	December 10, 1997
Ship To	Smith Bicycle

From December 5 to December 10 is 6 days.

You can **estimate** to solve this problem.

To estimate a **product, round** the greater **factor** so you can multiply mentally.

Estimate: 6×725
Think: $6 \times 700 = 4,200$

Compare this number with the number of items ordered.

Check Out the Glossary
For vocabulary words
See page 544.

Since 4,200 is greater than 4,000, you can predict that the order will be completed in time.

More Examples

A 3×45 **Think:** $3 \times 50 = 150$

B $8 \times \$4,612$ **Think:** $8 \times \$5,000 = \$40,000$

CHECK

Check for Understanding

Estimate the product. Tell how you rounded.

1 4×19 **2** 3×251 **3** $9 \times \$667$ **4** $7 \times 2,408$

5 79×6 **6** 332×5 **7** $\$589 \times 8$ **8** $3,641 \times 2$ **9** $\$4,398 \times 3$

Critical Thinking: Generalize **Explain your reasoning.**

10 How does knowing how to multiply multiples of 10, 100, and 1,000 help you estimate?

Practice

Estimate the product.

1 6 × 25 **2** 5 × $153 **3** 3 × 775 **4** 6 × 981

5 $\begin{array}{r} 5{,}029 \\ \times \quad 7 \\ \hline \end{array}$ **6** $\begin{array}{r} 3{,}642 \\ \times \quad 8 \\ \hline \end{array}$ **7** $\begin{array}{r} \$4{,}882 \\ \times \quad 4 \\ \hline \end{array}$ **8** $\begin{array}{r} 7{,}730 \\ \times \quad 5 \\ \hline \end{array}$ **9** $\begin{array}{r} \$3{,}498 \\ \times \quad 9 \\ \hline \end{array}$

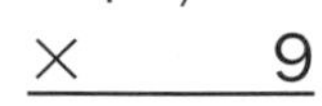

ALGEBRA Estimate. Write > or <.

10 4 × 284 ● 645 **11** 6 × 309 ● 2,400 **12** 21,000 ● 3 × 7,857

MIXED APPLICATIONS

Problem Solving

13 **Spatial reasoning** The tallest unicycle ever made is about 101 feet high. It is about as tall as a:
a. child **b.** chair **c.** building

14 The bicycle race attracted about 1,890 people each day. The race lasted 5 days. About how many people watched the race?

15 **Logical reasoning** Katie takes 20 minutes to sweep the yard. She says that if her brother helps her, they should take 40 minutes. Do you agree or disagree? Why?

16 **Make a decision** Wang's Bicycle Shop has $6,000. They can buy folding bicycles for $1,788 each or deluxe models for $2,100 each. What should they do?

Estimate the cost.

17 7 pairs of gloves

18 8 helmets

19 4 jackets

20 5 speedometers and 2 helmets

Bicycle Price List

Bicycle Supplies	Price
Helmet	$36
Gloves	$14
Speedometer	$25
Jacket	$109

mixed review • test preparation

Estimate the sum or difference.

1 51 + 67 **2** 859 − 601 **3** $6.03 + $1.81 **4** 57,937 − 11,836

Write the expanded form.

5 29 **6** 304 **7** 3,445 **8** 60,085 **9** 560,100

Extra Practice, page 500

Use Models to Multiply

LEARN

You can use what you know about place value and facts to multiply greater numbers.

You will need
- *graph paper*

Work Together

Work in a group to find 7×26.

On graph paper draw a rectangle that is 7 squares high and 26 squares long. The number of squares in the rectangle is the product of 7 x 26.

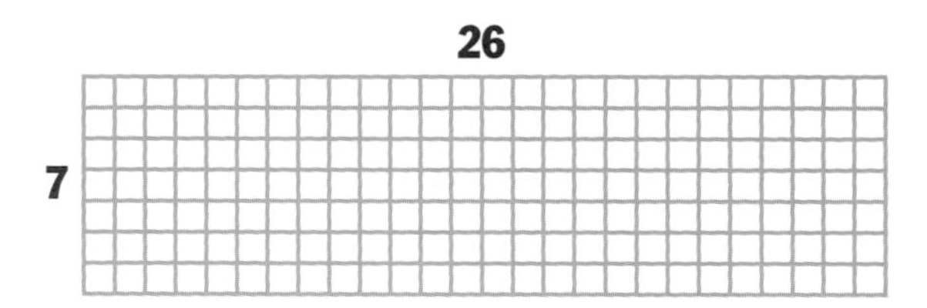

KEEP IN MIND
- Think creatively and logically.
- Be prepared to present one of your methods to the class.

Use your rectangle to find 7×26 in at least three different ways.

Talk It Over

- How many squares are there inside the rectangle? How did you find the number of squares?
- How could you use estimation to know if your answer is reasonable?

Make Connections

Here is how two students found 7 × 26.

Wayne

7

20 7x20

6 7x6

$$\begin{array}{r} 26 \\ \times\ 7 \\ \hline 140 \leftarrow 7 \times 20 \\ +\ 42 \leftarrow 7 \times 6 \\ \hline 182 \end{array}$$

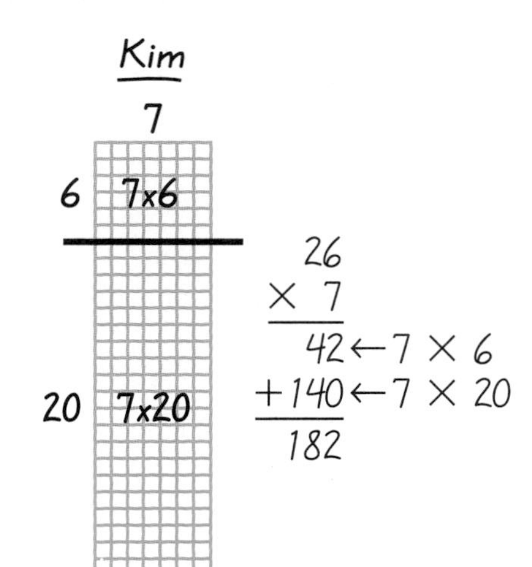

- ▶ How are Wayne's and Kim's methods the same? How are they different?
- ▶ How did they use place value, multiplication, and addition?

CHECK

Check for Understanding

Find the total number of squares in the rectangle without counting all the squares.

1 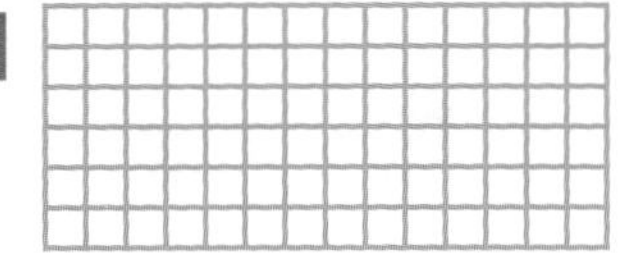

2 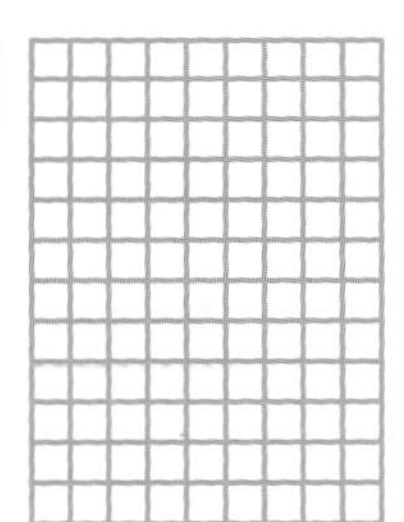

3 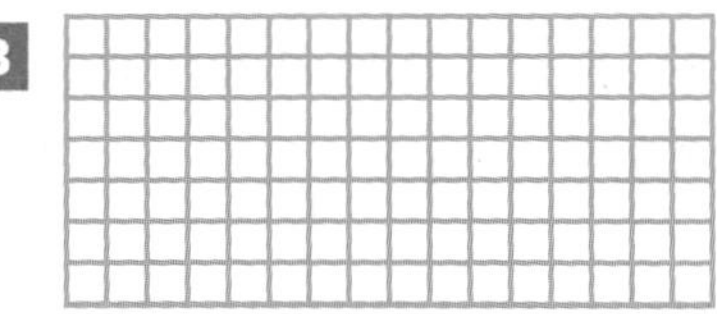

Find the product. You may use graph paper if you wish.

4 23×3

5 16×4

6 43×7

7 38×4

8 35×5

9 4 × 37

10 6 × 25

11 7 × 42

12 9 × 19

Critical Thinking: Analyze

13 Tell how you could use tens and ones models to find 4 × 34. Draw a picture to show your model.

Turn the page for Practice.

Practice

Find the total number of squares in the rectangle without counting.

1

2 

3 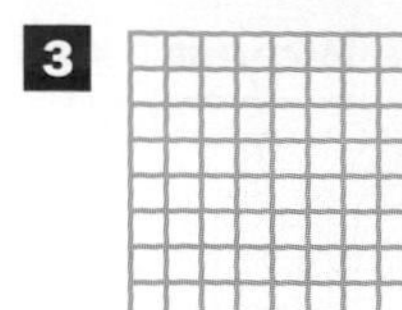

4 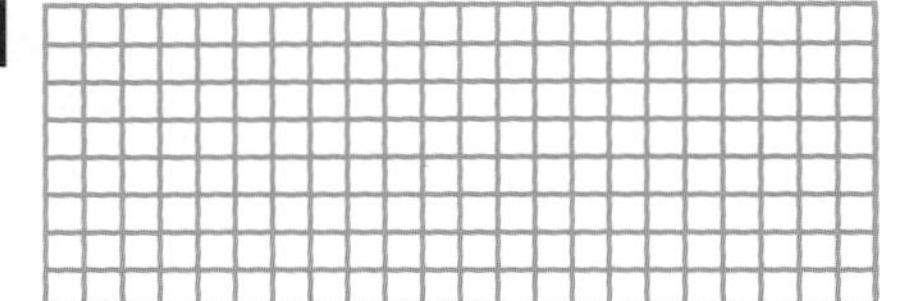

5 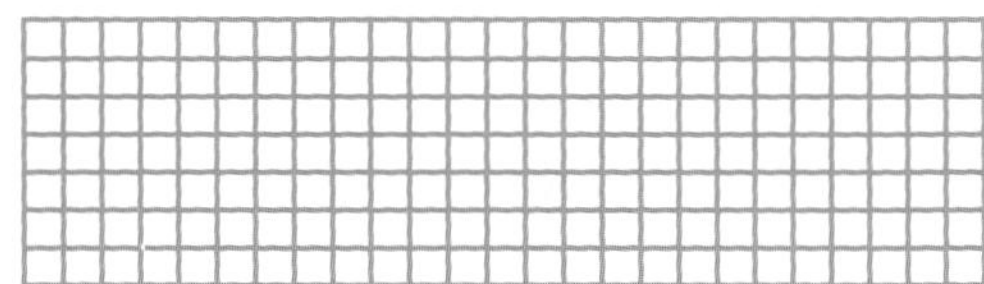

6 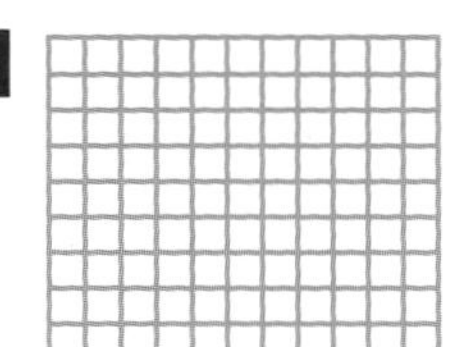

7 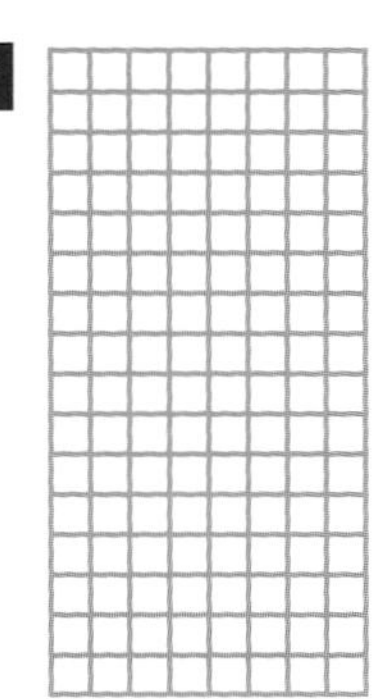

8 

Find the product using any method.

9 6×38 **10** 6×37 **11** 9×42 **12** 8×60

13 8×13 **14** 9×16 **15** 8×23 **16** 4×60

17 28×2 **18** 17×5 **19** 36×7 **20** 40×5 **21** 33×8

22 23×4 **23** 25×8 **24** 60×8 **25** 18×4 **26** 90×6

27 33×7 **28** 44×5 **29** 19×3 **30** 17×4 **31** 15×9

32 Tell which of ex. 9–31 you can solve mentally. Explain how you would solve three of them.

First, make one card for each of the numbers 1 through 9.

Next, on a sheet of paper, make a game sheet like the one at the right for each player.

You will need
- *index cards*
- *sheet of paper*

Play the Game

- ▶ Mix up the cards. Place them facedown. Pick two cards. Each player writes the two numbers in the boxes for Round 1.
- ▶ Find your product. Compare it with your partner's. The player with the greater product scores 1 point.
- ▶ If the products are the same, no one gets any points.
- ▶ Mix up the cards. Play Round 2 and Round 3. Continue playing, starting again at Round 1 until a player has 5 points.

What strategies did you use to place the numbers?

Game Sheet

Round 1: $1\square \times \square$ Round 2: $2\square \times \square$ Round 3: $3\square \times \square$

Score Chart

Round	1	2	3
Afi			
Tom			

Sample

Round 1: 9 3

Afi: $13 \times 9 = 117$

Tom: $19 \times 3 = 57$

Score: 1 point

mixed review • test preparation

1 7×9

2 6×3

3 $544 + 183$

4 $947 - 540$

5 $239 + 403$

6 $49 \div 7$ **7** $32 \div 4$ **8** $45 \div 5$ **9** $72 \div 9$

Extra Practice, page 501

Multiply 2-Digit Numbers

Cultural Note
Piñatas are used in birthday celebrations in Mexico.

A factory worker is packing piñatas to be shipped. There are 18 piñatas on each of 3 shelves. How many piñatas are there?

To estimate the number of piñatas, use mental math.

Think: $3 \times 20 = 60$

In the last lesson, you used this method to find the exact answer.

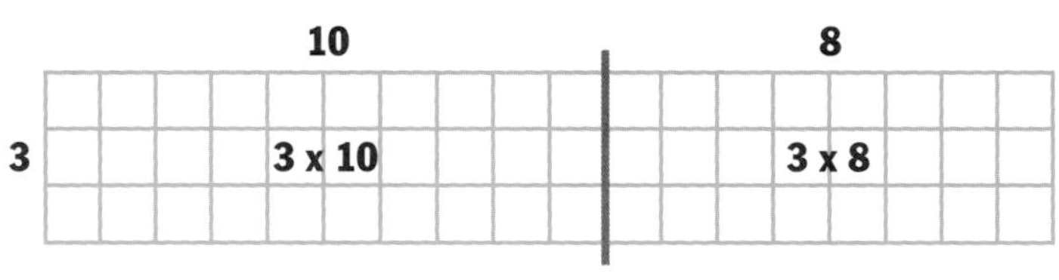

$$\begin{array}{r} 18 \\ \times\ 3 \\ \hline 24 \leftarrow 3 \times 8 \\ +\ 30 \leftarrow 3 \times 10 \\ \hline 54 \end{array}$$

Here is another method. Place-value models help you see the **regrouping.**

Step 1	Step 2
Multiply the ones. **Regroup if necessary.**	**Multiply the tens.** **Add all the tens.**
$\begin{array}{r} ^{2}\ \ \\ 18 \\ \times\ 3 \\ \hline 4 \end{array}$	$\begin{array}{r} ^{2}\ \ \\ 18 \\ \times\ 3 \\ \hline 54 \end{array}$
Think: 3×8 ones = 24 ones 24 ones = 2 tens 4 ones	**Think:** 3×1 ten = 3 tens 3 tens + 2 tens = 5 tens

There are 54 piñatas.

Check Out the Glossary
regroup
See page 544.

Talk It Over

- In what ways are these two methods alike? In what ways are they different?
- How did estimation help you know that 54 piñatas is a reasonable answer?

Piñatas are filled with small candies. **What if** there are 18 fruit candies in each piñata. How many fruit candies are there in 9 piñatas?

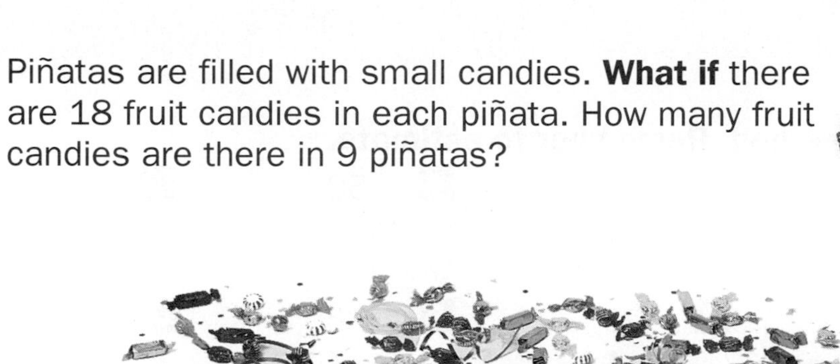

Multiply: 9×18

Step 1

Multiply the ones.
Regroup if necessary.

$$\begin{array}{r} {}^{7} \\ 18 \\ \times\ 9 \\ \hline 2 \end{array}$$

Think: 9×8 ones $= 72$ ones
72 ones $= 7$ tens 2 ones

Step 2

Multiply the tens.
Add all the tens
Regroup if necessary.

$$\begin{array}{r} {}^{7} \\ 18 \\ \times\ 9 \\ \hline 162 \end{array}$$

Think: 9×1 ten $= 9$ tens
9 tens $+ 7$ tens $= 16$ tens

There are 162 fruit candies.

More Examples

A $\begin{array}{r} 21 \\ \times\ 4 \\ \hline 84 \end{array}$ **B** $\begin{array}{r} {}^{3} \\ 16 \\ \times\ 5 \\ \hline 80 \end{array}$ **C** $\begin{array}{r} 63 \\ \times\ 3 \\ \hline 189 \end{array}$ **D** $\begin{array}{r} {}^{2} \\ 47 \\ \times\ 3 \\ \hline 141 \end{array}$

Check for Understanding

CHECK

Find the product using any method. Estimate to check the reasonableness of your answer.

1 $\begin{array}{r} 32 \\ \times\ 3 \\ \hline \end{array}$ **2** $\begin{array}{r} 20 \\ \times\ 6 \\ \hline \end{array}$ **3** $\begin{array}{r} 24 \\ \times\ 4 \\ \hline \end{array}$ **4** $\begin{array}{r} 43 \\ \times\ 3 \\ \hline \end{array}$ **5** $\begin{array}{r} 38 \\ \times\ 5 \\ \hline \end{array}$

Critical Thinking: Generalize **Explain your reasoning.**

6 When you multiply 2-digit numbers, how do you use facts? How do you use place value?

7 If you multiply a 2-digit number by a 1-digit number, what is the least product you can get? What is the greatest product you can get?

Turn the page for Practice.

PRACTICE

Practice

Find the product using any method. Remember to estimate.

1 13×3 **2** 37×3 **3** 15×5 **4** 50×8 **5** 61×5

6 65×6 **7** 14×4 **8** 90×4 **9** 86×7 **10** 48×5

11 3×17 **12** 55×8 **13** 52×7 **14** 5×20 **15** 70×3

16 8×23 **17** 23×4 **18** 6×40 **19** 45×6 **20** 8×77

21 Find the product of 32 and 7.

22 Find the product of 8 and 46.

23 Find the product of 81 and 4.

24 Find the product of 7 and 30.

Find only the products that are greater than 400. Use estimation to help you decide.

25 36×4 **26** 67×8 **27** 97×8 **28** 89×9 **29** 75×6

30 49×4 **31** 69×3 **32** 45×7 **33** 35×8 **34** 63×7

ALGEBRA Find the missing digit.

35 $15 \times ■ = 45$

36 $■3 \times 4 = 132$

37 $2■ \times 7 = 1■2$

38 $■8 \times ■ = 882$

ALGEBRA Write >, <, or =.

39 7 × 34 ● 897

40 210 ● 45 × 8

41 47 × 5 ● 78

42 7 × 45 ● 588

43 774 ● 86 × 9

44 42 × 9 ● 353

Make It Right

45 Here is how John found 6 × 54. Explain to a third grader what the mistake is. Write steps that show the correct answer.

$$\begin{array}{r} 54 \\ \times\ 6 \\ \hline 3{,}024 \end{array}$$

MIXED APPLICATIONS
Problem Solving

46 There are 8 dolls in each box shown at the right. How many dolls are there altogether in the boxes?

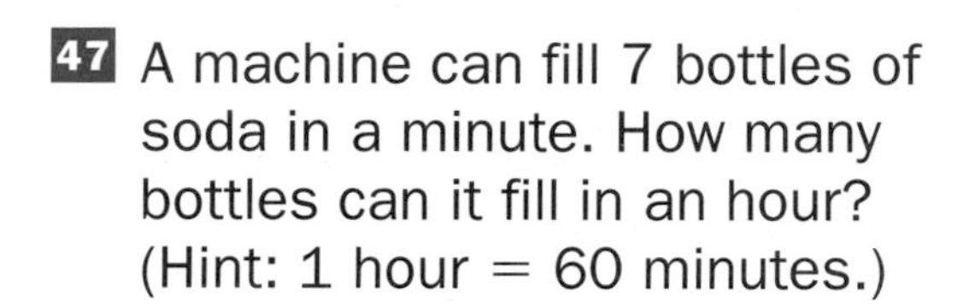

47 A machine can fill 7 bottles of soda in a minute. How many bottles can it fill in an hour? (Hint: 1 hour = 60 minutes.)

48 A factory packs 8 pencils in each package. How many packages will 24 pencils fill?

Use the information below to solve problems 49–51.

49 How much would it cost to buy 3 medium piñatas and 6 large piñatas?

50 How many different ways can you buy the piñatas to spend exactly $96? Show all your answers.

51 **Write a problem** about piñatas. Use the prices of the piñatas in the picture or make up prices of your own.

52 **Data Point** Survey your classmates about their favorite toy. Make a pictograph of the data. What is the most popular toy?

more to explore

Estimating Sums by Clustering

You can use multiplication to help you estimate certain sums.

Estimate: $62 + 57 + 52 + 66$ **Think:** Each number is about 60.

$4 \times 60 = 240$

Estimate the sum. Explain your thinking.

1 $43 + 39 + 41$ **2** $79 + 82 + 84 + 74$ **3** $311 + 295 + 308 + 302$

4 $58 + 62 + 63$ **5** $199 + 203 + 201$ **6** $88 + 91 + 86$

Extra Practice, page 501

Solve Multistep Problems

LEARN

Read **Action Sports makes skateboards with decks in 2 different patterns and 3 different colors. The company makes 8,000 of each type. How many skateboards will the company make?**

Plan You need to think about two questions to solve the problem.

Step 1 How many types of skateboards are possible?

Step 2 What is the total number of skateboards?

Solve **Step 1** Use a tree diagram.

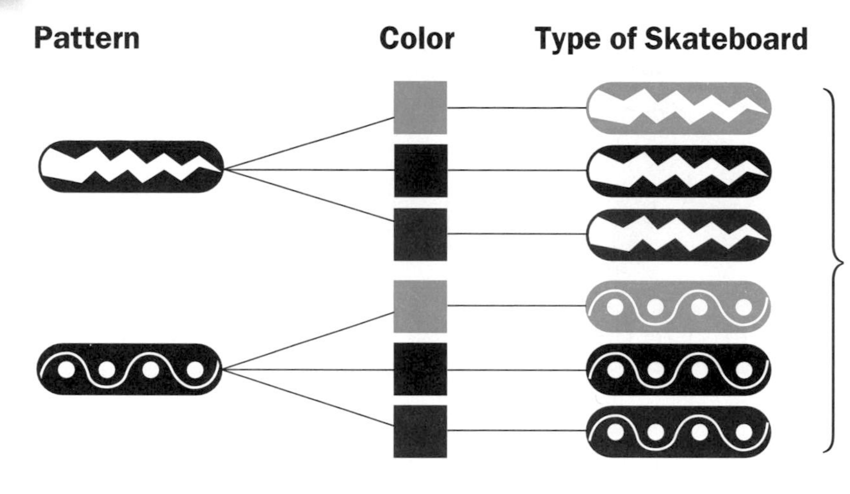

$3 \times 2 = 6$

There are 6 possible combinations.

Step 2 Use mental math. $6 \times 8{,}000 = 48{,}000$

The company will make 48,000 skateboards.

Look Back How could you solve this problem in a different way?

CHECK

Check for Understanding

Solve. Tell the steps you used.

1 **What if** Action Sports makes skateboard decks in 2 colors and 4 patterns. What is the cost to buy one of each type at $42 each? What two questions did you answer to solve the problem?

Critical Thinking: Analyze **Show two different ways to solve.**

2 Stickers for short decks sell for $1. Stickers for long decks sell for $2 more. How much does it cost to buy 2,000 stickers for short decks and 3,000 stickers for long decks?

MIXED APPLICATIONS

Problem Solving

PRACTICE

Yogurt Delight serves one fruit topping with either vanilla or chocolate yogurt on a sugar cone.

Yogurt Delight		
Flavors	**Prices**	
Vanilla Chocolate	2 scoops of one flavor and a fruit topping	$3.00
Fruit Toppings Strawberry Banana Mango Orange Pineapple	Additional scoops	$1.00 each
	Additional toppings	$1.00 each
	Sprinkles or chocolate chips	$1.00 extra
	Chocolate cones	$1.00 extra

1 Draw a diagram or make a table that shows all the possible flavor and fruit topping combinations at Yogurt Delight.

2 **Spatial reasoning** How many triangles can you find in Yogurt Delight's logo?

3 Yogurt Delight has 3 cases of 600 cones in the storeroom. The rest are behind the counter on 4 shelves. Each shelf holds 46 cones. How many cones are there in all?

4 **What if** your class saves $120 to buy yogurt for a class party. How many toppings, different-type cones, and scoops of each flavor can your class buy? How much money will be left over?

5 Caleb closes the Yogurt Delight store at 7 P.M. He works 8 hours a day. At what time does he start work?

6 **Write a problem** using the information from Yogurt Delight's price list. Solve it and have others solve it.

Use the graph for problems 7–10.

7 Which hobby is the most popular?

8 How many more students preferred baseball to reading?

9 How many students were surveyed?

10 **What if** listening to music was added to the survey. How do you think the graph would change?

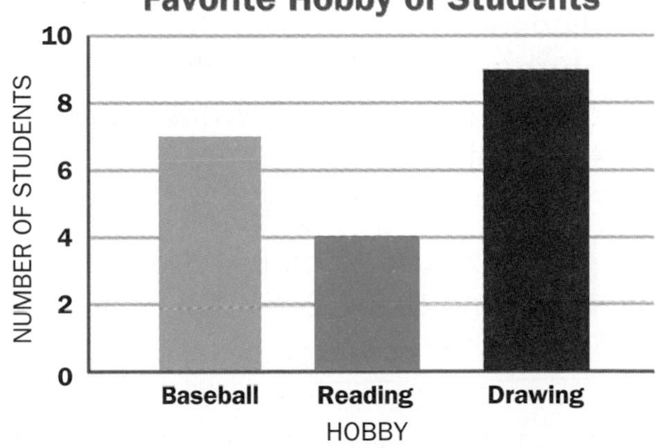

Extra Practice, page 502

midchapter review

ALGEBRA: PATTERNS Copy and complete.

1 $6 \times 9 = ■$
$6 \times 90 = ■$
$6 \times 900 = ■$
$6 \times 9{,}000 = ■$

2 $5 \times 60 = ■$
$5 \times 600 = ■$
$5 \times 6{,}000 = ■$

3 $8 \times 20 = ■$
$8 \times 200 = ■$
$8 \times 2{,}000 = ■$

Estimate how many.

4 4 cartons of tea bags

5 28 packs of soda

6 42 packs of hot dogs

7 6 cartons of eggs

Item	Number of Items in Each Pack/Carton
Soda	6
Eggs	12
Hot dogs	8
Tea bags	18

Find the product. Use graph paper if you need to.

8 22×4

9 34×5

10 37×6

11 $\$29 \times 7$

12 38×8

13 3×200

14 $5 \times \$8{,}000$

15 7×78

16 65×3

17 8×500

18 $3{,}000 \times 8$

19 4×83

20 52×7

Solve. Use mental math when you can.

21 A machine can make 3 doll arms each minute. How many can it make each hour? (Hint: 1 hour = 60 minutes.)

22 A machine makes about 26 earrings each hour. Estimate how many earrings 6 of the machines can make in an hour.

23 A toy company is selling stuffed dinosaurs for $7 each. A toy store orders 500. The shipping charge is $60 extra. What is the total bill?

24 A factory packs 9 model space shuttles in each box. How many models are in 54 boxes?

25 

Explain as many different ways as you can how to find 8×62.

developing number sense

MATH CONNECTION

Compare Exact Answers to Estimates

García Products donated 8 boxes of games to the hospital. Each box holds 32 games. Are there enough games to give as gifts to the 240 children at the hospital?

Estimate to solve this problem.

Exact Numbers	Estimate	Compare
8 x 32	$8 \times 30 = 240$	Since $32 > 30$, the exact answer is greater than 240.

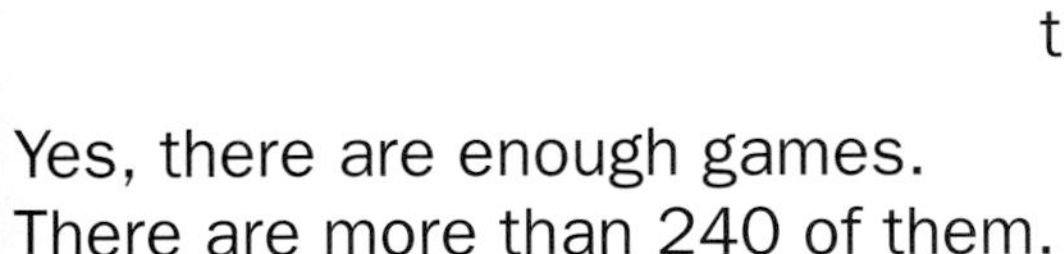

Yes, there are enough games.
There are more than 240 of them.

Estimate to solve the problem. Is the exact answer greater than or less than your estimate? Tell why.

1 The 4-H Club has $3,000. The members voted to buy 3 wheelchairs that cost $1,397 each for a hospital. Does the club have enough money?

2 The Key Club is saving money to donate a $320 television set to charity. The club can save $42 a month for 8 months. Will it have enough money?

3 Circus Products packs 48 toy clowns in a box. Are there enough clowns in 6 boxes for 300 children if each child gets 1 toy?

4 Ms. Kim works 34 hours a week in a hospital lab. She earns $9 an hour. Does she earn at least $270 a week?

5 The Computer Club washes cars to raise money for charity. They charge $7 for each car. They wash 47 cars. Have they raised $350?

6 The Grin and Bear It Company packages 24 toy bears in a box. Are there enough bears in 5 boxes for 100 children?

7 The staff at Children's Hospital is having a holiday party for 180 children. They have 9 packages of cups with 15 cups in a package. Are there enough cups for the party?

8 Kareem works 12 hours a week as a hospital volunteer. If he volunteers at least 40 hours a month, he gets free lunch in the cafeteria. Does he get his lunch free?

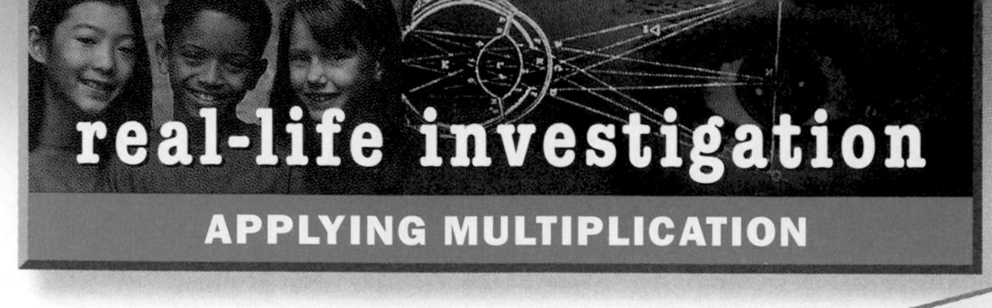

Paper Plane Factory

Assembly lines are used in many factories to increase the number of items made. In this activity, you will work in teams to make paper planes on an assembly line.

You will need

- *sheets of paper*

steps:

1 ***Fold in half and open up.***

2 ***Fold corners A and B to center line.***

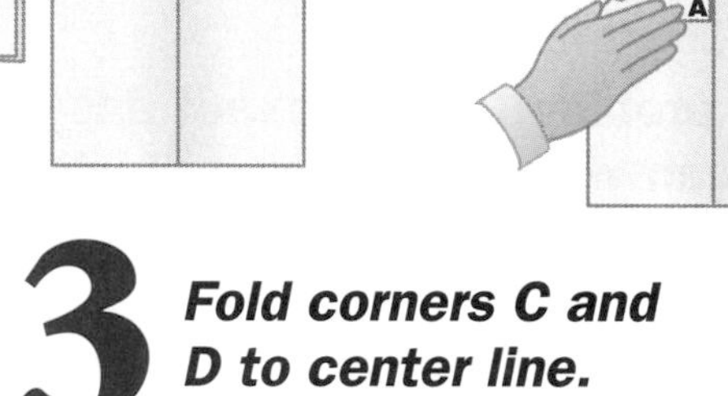

3 ***Fold corners C and D to center line.***

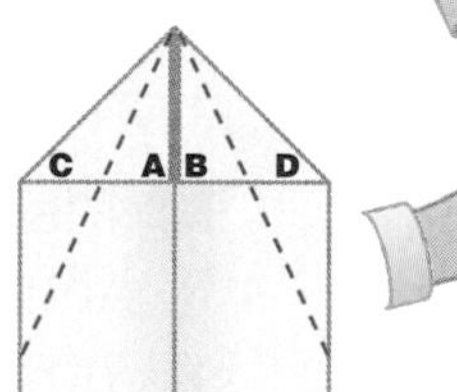

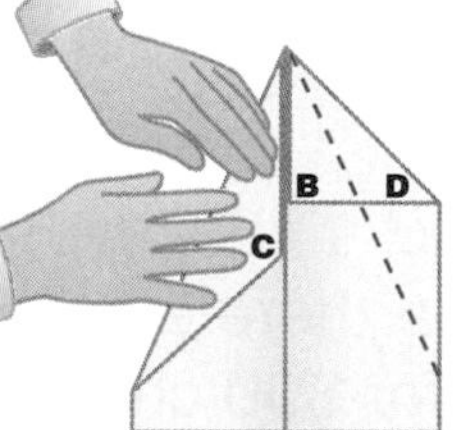

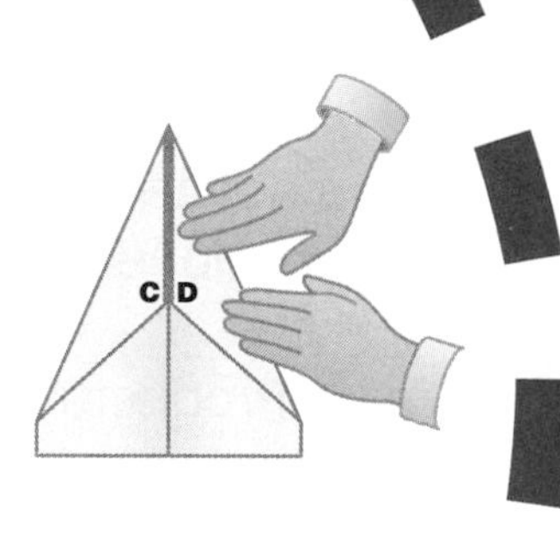

4 ***Fold in half.***

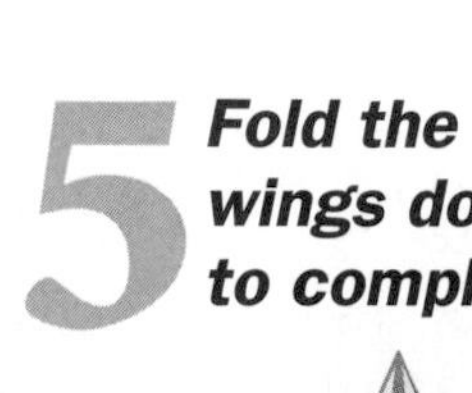

5 ***Fold the wings down to complete.***

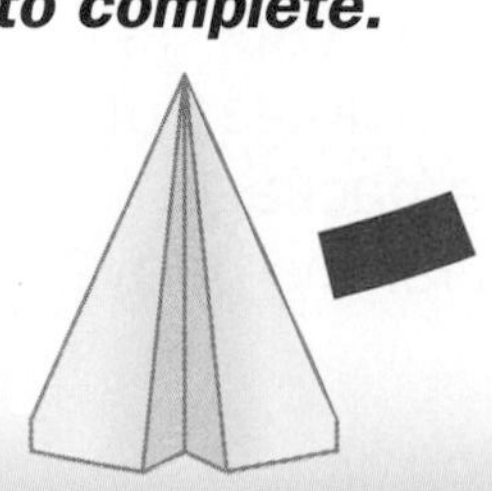

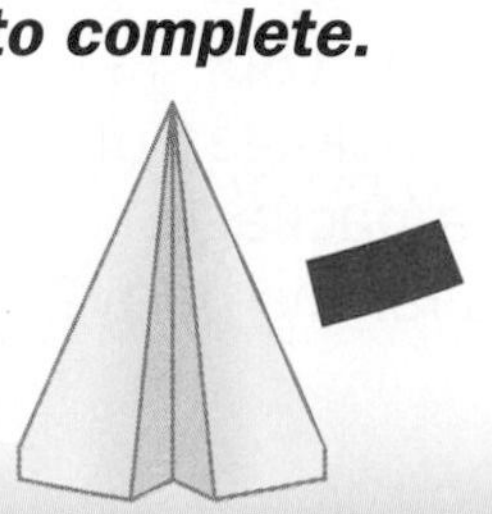

DECISION MAKING

Making Paper Airplanes

1 Form an assembly line to make paper planes. Use the design on the left. Let each person make one or more of the folds.

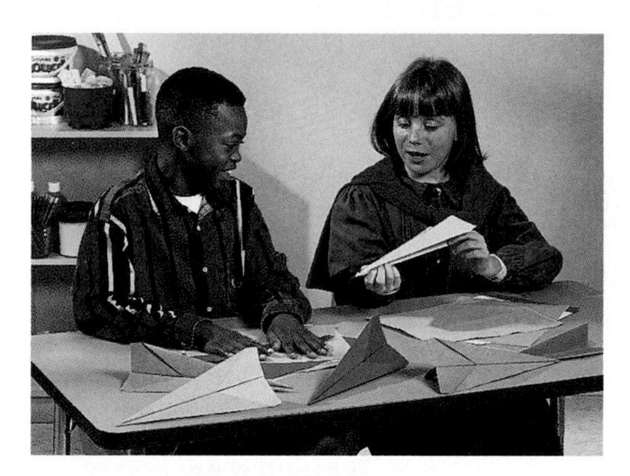

2 How many planes can your assembly line make in 5 minutes?

3 Decide how you can change your assembly line to increase the number of planes that your team can make. Make a new plan for your assembly line.

4 How many planes can your new assembly line make in 5 minutes?

Reporting Your Findings

5 Portfolio Prepare a report on what you learned. Include the following:

- ▶ Describe how your assembly line workers shared the work to make the greatest number of planes in 5 minutes. How many planes did they make?
- ▶ Predict the number of planes your assembly line could make in:
  a. one hour.
  b. a 7-hour day.
  c. a 5-day week.

Explain how you made your predictions.

6 Compare your report with the reports of other teams.

Revise your work.

- ▶ Are your calculations correct?
- ▶ Is your report clear and organized?
- ▶ Did you proofread your work?

MORE TO INVESTIGATE

PREDICT the number of paper planes your entire class could make in a month.

EXPLORE why paper planes fly.

FIND the world record for indoor flight-time of a paper plane.

Multiply Greater Numbers

LEARN

An old bottling machine puts 163 caps on bottles each hour. The factory bought a new robot that is 2 times as fast. How many bottles can the robot cap each hour?

To estimate the number of bottles the robot can cap each hour, use mental math.

Think: $2 \times 200 = 400$

To find the exact answer, use pencil and paper.

Step 1

Multiply the ones.
Regroup if necessary.

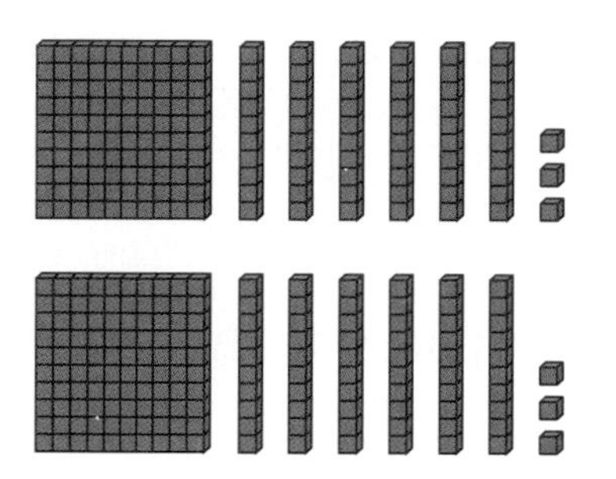

$$\begin{array}{r} 163 \\ \times \quad 2 \\ \hline 6 \end{array}$$

Think: 2×3 ones $= 6$ ones

Step 2

Multiply the tens.
Add all the tens.
Regroup if necessary.

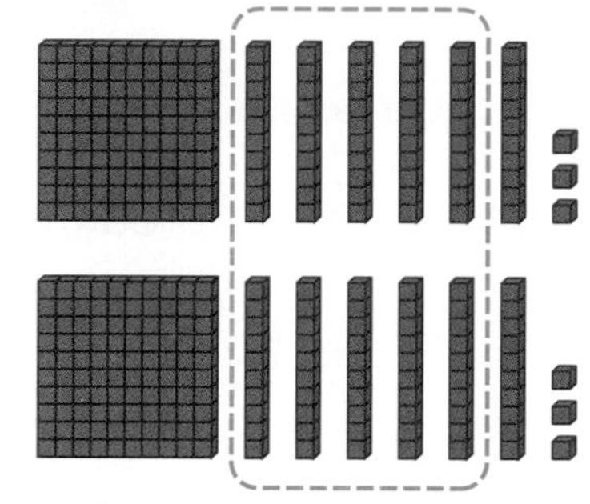

$$\begin{array}{r} {}^{1} \\ 163 \\ \times \quad 2 \\ \hline 26 \end{array}$$

Think:
2×6 tens $= 12$ tens
12 tens $= 1$ hundred 2 tens

Step 3

Multiply the hundreds.
Add all the hundreds.
Regroup if necessary.

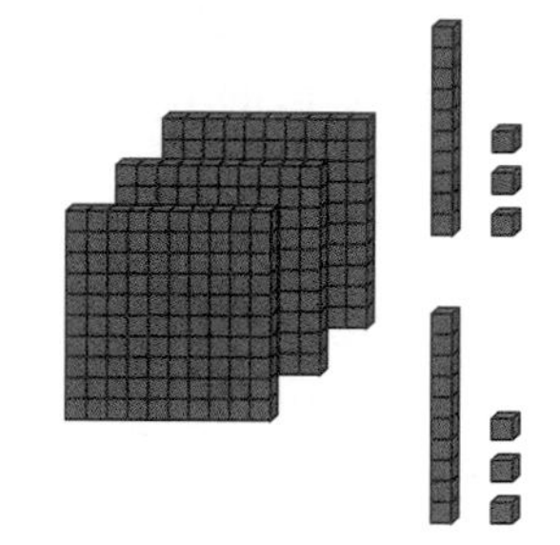

$$\begin{array}{r} {}^{1} \\ 163 \\ \times \quad 2 \\ \hline 326 \end{array}$$

Think:
2×1 hundred $= 2$ hundreds
2 hundreds $+ 1$ hundred $=$
3 hundreds

The robot can cap 326 bottles each hour.

Talk It Over

- How do you know the answer is reasonable?
- How is multiplying 3-digit numbers the same as multiplying 2-digit numbers? How is it different?

What if a machine labels 2,342 bottles each hour. How many bottles can the machine label in 7 hours?

Multiply: $7 \times 2,342$

You can use paper and pencil to find the product.

$$\begin{array}{r} {}^{2\ 21} \\ 2,342 \\ \times \quad 7 \\ \hline 16,394 \end{array}$$

You can also use a calculator.

$7 \times 2,342 =$ 16394.

The machine can label 16,394 bottles in 7 hours.

More Examples

A
$$\begin{array}{r} {}^{4} \\ 107 \\ \times \quad 6 \\ \hline 642 \end{array}$$

B
$$\begin{array}{r} {}^{34} \\ 356 \\ \times \quad 7 \\ \hline 2,492 \end{array}$$

C
$$\begin{array}{r} {}^{2} \\ 1,009 \\ \times \quad 3 \\ \hline 3,027 \end{array}$$

D
$$\begin{array}{r} {}^{2\ 21} \\ \$2,563 \\ \times \quad 4 \\ \hline \$10,252 \end{array}$$

Check for Understanding

Multiply. Estimate to check that your answer is reasonable.

1 401×3 **2** 997×2 **3** $1,750 \times 4$ **4** $8,002 \times 3$ **5** $\$8,765 \times 9$

6 6×115 **7** 9×843 **8** $2 \times 3,472$ **9** $8 \times 7,183$

Critical Thinking: Analyze **Explain your reasoning.**

10 You use a calculator to find $6 \times 3,336$. The display reads 200016. . Is the answer reasonable?

11 What error might you make when using a calculator to multiply?

12 Is it always easier to multiply greater numbers using a calculator? Give an example to support your reasons.

CHECK

Turn the page for Practice.

PRACTICE

Practice

Multiply. Remember to estimate.

1 123 × 4

2 181 × 5

3 206 × 7

4 437 × 2

5 $961 × 3

6 1,172 × 3

7 $4,903 × 8

8 5,092 × 4

9 9,238 × 9

10 $8,246 × 2

11 4 × 612 **12** 7 × 333 **13** 6 × 6,553 **14** 8 × $7,423

15 Find the product of 3 and 278.

16 What is 7 times 5,594?

Find only products between 1,200 and 60,000.
Use estimation to help you decide.

17 2 × 415 **18** 2 × 671 **19** 5 × 222 **20** 5 × 4,716

21 8 × 5,864 **22** 9 × 6,018 **23** 4 × 1,789 **24** 3 × 4,881

25 7 × 9,011 **26** 9 × 6,123 **27** 2 × 4,760 **28** 8 × 8,007

ALGEBRA Complete the table.

29

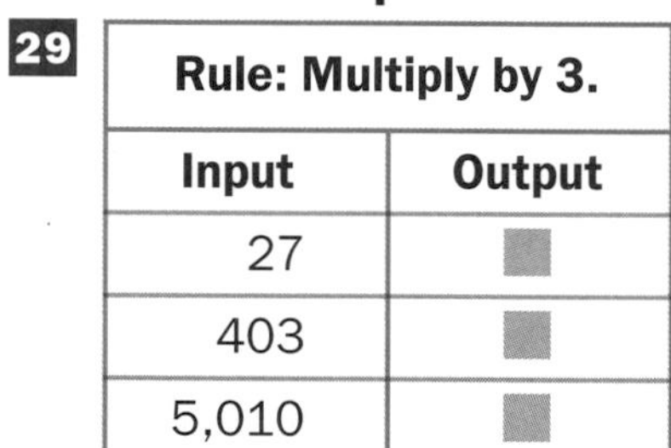

Rule: Multiply by 3.	
Input	**Output**
27	■
403	■
5,010	■

30

Rule: ■	
Input	**Output**
91	182
403	806
3,700	7,400

31

Rule: ■	
Input	**Output**
21	105
301	1,505
2,005	10,025

ALGEBRA Write the letter of the missing number.

32 ■ × 333 = 999 **a.** 5 **b.** 3 **c.** 2 **d.** 4

33 ■ × 876 = 3,504 **a.** 9 **b.** 2 **c.** 4 **d.** 5

34 5 × ■ = 11,110 **a.** 2,222 **b.** 2,111 **c.** 1,111 **d.** 5,555

Make It Right

35 Helen multiplied 2,135 by 4. Explain what the mistake is. Show the correct answer.

1 22
2,135
× 4
12,200

MIXED APPLICATIONS

Problem Solving

Use the graph and table for problems 36–39.

36 About how many umbrellas does the Adams machine make in 7 weeks?

37 How much would it cost to buy 3 Adams machines and 6 Croy machines?

38 READING ARITHMETIC WRITING **Draw Conclusions** The factory manager replaces a Becker machine with a Croy machine. What conclusions can you draw about her decision?

39 **Write a problem** that can be solved using the graph or the table. Solve it and have others solve it.

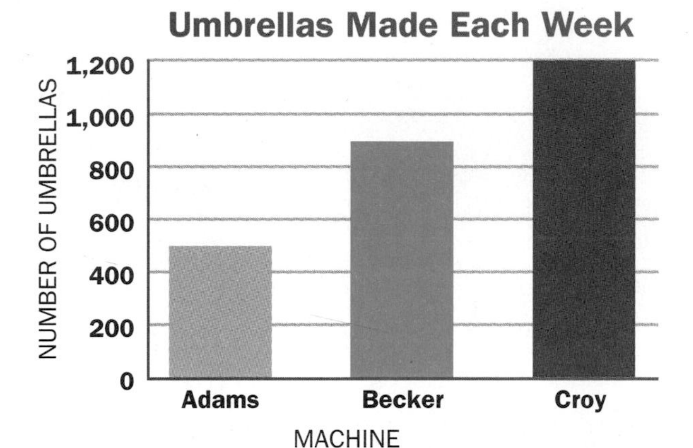

Machine	Cost of Each Machine
Adams	$798
Becker	$1,453
Croy	$5,674

Cultural Connection Egyptian Multiplication

In ancient Egypt, scribes kept records of payments to the pharaoh. As we do today, they used a number system based on ten. They used a line for 1 (|), a heel bone for 10 (∩), and a snare for 100 (𓍢). The number 142 would be 𓍢∩∩∩∩||.

What if 5 people paid 142 baskets of wheat each? Here is how the scribes would find 5 × 142.

Start with 142 and double the numbers.	Find those that add up to 5.	Add their products.
1 × 142 = 142	1 × 142 = 142	142
2 × 142 = 284	2 × 142 = 284	+ 568
4 × 142 = 568	4 × 142 = 568	5 × 142 = 710

Use this method to solve ex. 17, 19, 23, and 25 on page 190.

Multiply Money

Cultural Note

A chipao is a traditional Chinese dress.

Mrs. Hsu is buying 6 yards of silk to make *chipaos* (CHEE-powz). Each yard costs $52.99. What will be the total cost?

To estimate the cost, use mental math.

Think: $6 \times \$50 = \$300$

You can use pencil and paper to find the exact answer.

$$\begin{array}{r} {}^{1\,5\,5} \\ \$52.99 \\ \times \quad 6 \\ \hline \$317.94 \end{array}$$

Write a dollar sign and a decimal point in the product.

You can also use a calculator. $6 \times 52.99 =$ 317.94

The total cost will be $317.94.

More Examples

A

$$\begin{array}{r} {}^{5} \\ \$0.47 \\ \times \quad 8 \\ \hline \$3.76 \end{array}$$

$8 \times 0.47 =$ 3.76

B

$$\begin{array}{r} {}^{2} \\ \$6.03 \\ \times \quad 9 \\ \hline \$54.27 \end{array}$$

$9 \times 6.03 =$ 54.27

CHECK

Check for Understanding

Multiply. Estimate to check the reasonableness of your answer.

1 $\begin{array}{r} \$0.09 \\ \times \quad 8 \\ \hline \end{array}$ **2** $\begin{array}{r} \$0.64 \\ \times \quad 7 \\ \hline \end{array}$ **3** $\begin{array}{r} \$7.32 \\ \times \quad 5 \\ \hline \end{array}$ **4** $\begin{array}{r} \$36.12 \\ \times \quad 2 \\ \hline \end{array}$ **5** $\begin{array}{r} \$94.75 \\ \times \quad 4 \\ \hline \end{array}$

Critical Thinking: Analyze **Explain your reasoning.**

6 You use a calculator to find 6 x $3.45. The display reads 20.7 . What is the answer? Explain how you know that the answer is reasonable.

7 

How does finding the product of 4 and 5,482 help you to find the product of 4 and $54.82?

Practice

Multiply. Remember to estimate.

1. $\begin{array}{r} \$0.52 \\ \times \quad 6 \\ \hline \end{array}$
2. $\begin{array}{r} \$0.93 \\ \times \quad 7 \\ \hline \end{array}$
3. $\begin{array}{r} \$7.04 \\ \times \quad 8 \\ \hline \end{array}$
4. $\begin{array}{r} \$0.58 \\ \times \quad 4 \\ \hline \end{array}$
5. $\begin{array}{r} \$6.19 \\ \times \quad 5 \\ \hline \end{array}$
6. $\begin{array}{r} \$3.19 \\ \times \quad 2 \\ \hline \end{array}$
7. $\begin{array}{r} \$63.86 \\ \times \quad 5 \\ \hline \end{array}$
8. $\begin{array}{r} \$0.76 \\ \times \quad 3 \\ \hline \end{array}$
9. $\begin{array}{r} \$9.84 \\ \times \quad 3 \\ \hline \end{array}$
10. $\begin{array}{r} \$0.47 \\ \times \quad 9 \\ \hline \end{array}$
11. $\begin{array}{r} \$43.09 \\ \times \quad 6 \\ \hline \end{array}$
12. $\begin{array}{r} \$8.21 \\ \times \quad 9 \\ \hline \end{array}$
13. $\begin{array}{r} \$56.87 \\ \times \quad 2 \\ \hline \end{array}$
14. $\begin{array}{r} \$25.49 \\ \times \quad 8 \\ \hline \end{array}$
15. $\begin{array}{r} \$81.79 \\ \times \quad 6 \\ \hline \end{array}$

16. $3 \times \$0.84$
17. $8 \times \$24.95$
18. $2 \times \$4.75$
19. $\$70.39 \times 6$
20. $7 \times \$6.93$
21. $9 \times \$0.19$
22. $\$7.82 \times 5$
23. $4 \times \$85.85$

MIXED APPLICATIONS

Problem Solving

Use the table to solve problems 24–27.

COST OF HANDMADE CLOTHING			
Item	Fabric	Yards Needed	Cost for each Yard
Chipao	Silk	2	\$52.99
Vest	Satin	1	\$23.45
Jacket	Cotton	2	\$31.89

24. How many chipaos can Mrs. Hsu make with the 6 yards of silk fabric that she bought?

25. Mr. Li plans to make 5 vests and 4 jackets. How much fabric does he need?

26. **Make a decision** Mei can make a chipao in 6 hours. How much should she charge for selling it? Explain your reasoning.

27. **Write a problem** that uses information from the table. Solve it and ask others to solve it.

28. A weaver uses the silk from 10 cocoons. About how many yards of thread does the weaver use? **SEE INFOBIT.**

INFOBIT
Silk fabric is made from the cocoons of silkworms. When unwrapped, the strands from each cocoon yield up to 1,000 yards of silk thread.

mixed review • test preparation

1. $\begin{array}{r} 13{,}467 \\ -12{,}836 \\ \hline \end{array}$
2. $\begin{array}{r} 5{,}083 \\ +\quad 795 \\ \hline \end{array}$
3. $\begin{array}{r} 19{,}726 \\ +53{,}027 \\ \hline \end{array}$
4. $\begin{array}{r} 3{,}791 \\ -\quad 298 \\ \hline \end{array}$
5. $\begin{array}{r} \$4{,}002 \\ -\ 3{,}145 \\ \hline \end{array}$

Extra Practice, page 503

Problem Solvers at Work

Read
Plan
Solve
Look Back

LEARN

Part 1 Interpret Data

To keep the mill open for visitors, the mill needs to collect $500 worth of tickets each day.

The spreadsheet shows the number of people who bought tickets during a week in February.

	Adult	Children
Sunday	94	63
Monday	105	39
Tuesday	35	46
Wednesday	63	79
Thursday	34	65
Friday	57	86
Saturday	74	88

Mill Tour Admissions	
Adults	$6.00
Children (under 18)	$4.00

Work Together

Solve. Use the information in the spreadsheet.

1. On which days were more than $500 worth of tickets sold?

2. READING ARITHMETIC WRITING **Draw Conclusions** On what two days did most adults and children visit the mill? What conclusion can you draw about these days?

3. On which day did the mill collect the most money for tours?

4. What if one day the mill sells 150 tickets and collects between $700 and $750. How many children's tickets are sold? How many adult tickets are sold? Is there more than one possible answer?

5. **Make a decision** The mill manager can offer a discount either to schools or to senior citizens on Sundays, Mondays, or Tuesdays. Which discount would you choose? Which day? Why?

Part 2 Write and Share Problems

Josh used the data in the pictograph to write a problem.

Attendance for School Play, *WILLIE WONKA AND THE CHOCOLATE FACTORY*	
Wednesday	
Thursday	
Friday	
Saturday	

Key: Each stands for 50 people.

Tickets cost $3.

On which day did they make $450?

CHECK

Josh Watson
Hawthorne School
Indianapolis, IN

6 Solve Josh's problem.

7 Change Josh's problem so that it is either easier or harder to solve. Do not change any of the data in the pictograph.

8 Solve the new problem and explain why it is easier or harder to solve than Josh's.

9 Write a problem of your own that uses information from the graph.

10 Trade problems. Solve at least three problems written by your classmates.

11 What was the most interesting problem that you solved? Why?

Turn the page for Practice Strategies.

PRACTICE

Menu

Choose five problems and solve them. Explain your methods.

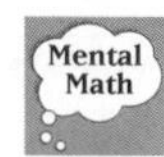

1 Barth, Inc., makes mouse pads. They pack 400 in each box. How many are in 6 boxes?

2 Betty Sue bought a CD that cost \$13.95. She paid \$14.79, with tax. How much was the tax?

3 Steve earns \$8.50 an hour for baby-sitting. How much does he earn by baby-sitting for 5 hours?

4 You are about 12 miles from your camp. You can walk about 2 miles every hour. How long will it take you to get back to your camp?

5 There are 24 dolls in each carton. There are 4 cartons in each box. How many dolls are in 5 boxes?

6 Joe has about 150 baseball cards. His sister has twice as many. About how many cards do they have altogether?

7 Each day an airline has 8 flights from Chicago to Dallas. Each flight can carry up to 245 people. What is the greatest number of people that can fly each day on those flights?

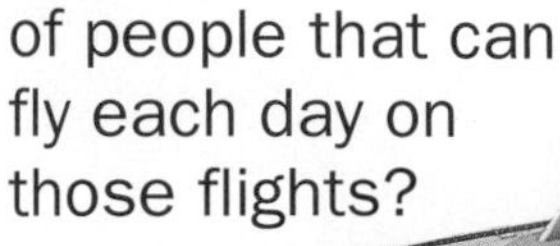

8 Ana wants to roast a 9-pound turkey. The directions say that the turkey must be roasted 20 minutes for each pound. If she wants the turkey to be done at 2:00 P.M., what time should she start roasting?

Choose two problems and solve them. Explain your methods.

9 Use the charts to plan two different breakfasts that have between 500 and 700 calories.

Food	Calories
Orange juice (8 ounces)	110
Grapefruit (half)	45
Orange	65
Egg	80
Whole milk (8 ounces)	160

Food	Calories
Toast (one slice)	
unbuttered	70
buttered	105
Muffin	120
Cornflakes (1 cup, with no milk)	100

10 **Spatial reasoning** How many squares are there in this figure? Count squares of any size. (Hint: There are more than 10.)

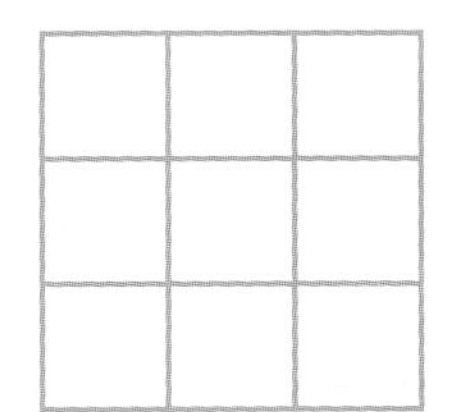

11 Write at least three possible newspaper headlines for the graph below.

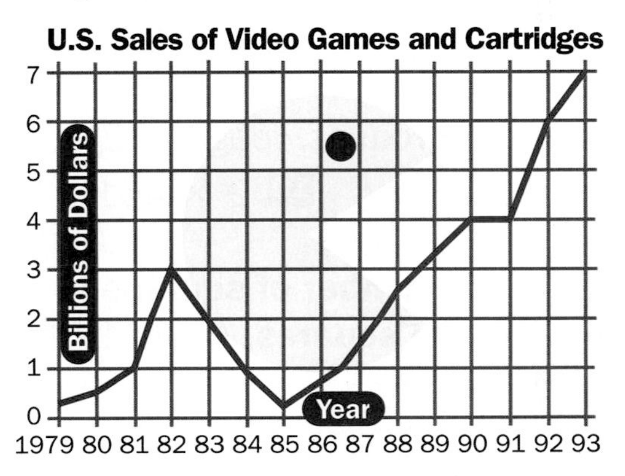

12 **At the Computer** You can make two different rectangles containing 4 squares, as shown on the right.

Use a drawing program to build as many different rectangles as possible containing 5 squares, 6 squares, and so on, up to 14 squares. Make a table like the one below. Then write about what you notice.

Number of Squares	1	2	3	4
Number of Different Rectangles	1	1	1	2

Extra Practice, page 503

chapter review

Language and Mathematics

Complete the sentence. Use a word in the chart. (pages 170–193)

1 The ■ of 8 and 20 is 160.

2 You ■ to change 36 tens to 3 hundreds 6 tens.

3 To estimate 7 × 382, you can ■ 382 and then multiply.

4 If you change the order of the ■, the product is the same.

Vocabulary
regroup
factors
product
sum
round
addends

Concepts and Skills

Find the product mentally. (page 170)

5 7 × 1,000 **6** 6 × 900 **7** 4 × $70 **8** 5 × 8,000

Estimate the product. (page 172)

9 5 × 63 **10** 9 × 451 **11** 8 × 1,070 **12** 4 × $6,553

Find the total number of squares in the rectangle without counting all the squares. (page 174)

13

14

15

Multiply. (pages 178, 188, 192)

16 97 × 3 **17** 406 × 5 **18** $6.34 × 2 **19** 8,199 × 4 **20** 58 × 8

21 18 × 5 **22** 6 × 22 **23** 3 × 514 **24** 7 × 399 **25** 9 × $857

26 9 × 82 **27** 4 × 79 **28** 2 × $56.05 **29** 8 × 634 **30** 6 × 2,007

Think critically. (page 178)

31 Analyze. Explain what mistake was made, then correct it.

$$\begin{array}{r} {}^{11} \\ 456 \\ \times \quad 3 \\ \hline 1{,}588 \end{array}$$

32 Generalize. Write *always, sometimes,* or *never.* Give examples to support your answer.

a. When you round one of two factors to find an estimate, the exact product is less than the estimate.

b. A 3-digit number multiplied by a 1-digit number has a 4-digit product.

c. The product of an odd number and an even number is an even number.

MIXED APPLICATIONS
Problem Solving

(pages 182, 194)

33 A machine prints 8,000 copies of a newsletter each hour. How many copies can the machine print in 7 hours?

34 Ruth's company needs to buy 3 printers. Each printer costs $1,876. Estimate about how much they will cost altogether.

35 What is the mystery number? It is an even number. The number is greater than the product of 2 and 23 but less than the sum of 17 and 33.

36 *The Sun Newspaper* has 758 more workers this year than last year. There are 1,742 workers this year. How many workers were there last year?

Use the information in the ad to solve problems 37–40.

37 What is the cost of 2 floor lamps and 3 sets of cookware?

38 What is the cost of 3 bunk beds?

39 Pedro has $2,200. Does he have enough to buy 3 bunk beds and 3 chairs?

40 You give the clerk $50 to pay for a chair and a radio alarm. How much change should you get?

Karen's Home Center

Radio Alarm	$18.59
Floor Lamp	$119
Cookware	$87
Bunk Bed	$669
Chair	$27

(Prices include tax)

Estimate the product.

1 8×72
2 $4 \times \$19$
3 9×37
4 6×832
5 3×154
6 7×108
7 $5 \times 2{,}089$
8 $3 \times \$5{,}592$

Multiply. Use mental math when you can.

9 38×4
10 80×9
11 57×6
12 $\$29 \times 4$
13 60×8
14 712×5
15 $\$300 \times 8$
16 406×2
17 600×7
18 $\$25.88 \times 3$

19 8×86
20 $5 \times \$60$
21 9×72
22 6×324
23 8×100
24 $7 \times \$6.39$
25 2×700
26 3×228
27 $9 \times 4{,}000$
28 $4 \times \$3{,}856$
29 $5 \times 5{,}983$

Solve. Use the table for problems 30–33.

Companies in Rogerville

Company	Number of Employees
Big Foot Shoe Co.	3,246
Rainbow Mfg.	876
Big Top Tents	2,141
Sean's Bicycles	8

30 Sean's employees each work 2,000 hours each year. How many hours each year do all Sean's employees work?

31 The Rainbow factory in Mayville has 3,600 employees. How many more employees are in the Mayville factory than in the Rogerville factory?

32 The total number of employees in all locations of these companies is 19,835. How many people outside Rogerville are employed at these companies?

33 Big Top Tents needs to buy 6 computers that cost $2,164 each and 4 printers that cost $420 each. How much will they spend in all?

What Did You Learn?

Lynne and Zoe calculated how many loaves of wheat bread to order for February. Lynne usually orders 8 loaves a day. There are 28 days in February.

Lynne		Zoe	
	28		28
	× 8		× 8
	64		160
	+160		+ 64
	224		224

- ▶ Explain how the multiplication methods are different and why the answers are the same. Use graph paper or pictures of place-value models if you wish.

You will need
- *graph paper*
- *place-value models*

- ▶ Explain how you could find the answer a different way.

A Good Answer
- clearly explains Lynne's and Zoe's methods and how they are alike and different
- tells another way to find the answer

You may want to place your work in your portfolio.

What Do You Think 

1. Can you recognize problems that can be solved by multiplying? Why or why not?
2. List all the ways you might use to solve a multiplication problem:
   - Use place-value models.
   - Use paper and pencil.
   - Use mental math.
   - Other. Explain.
3. Explain when you would use mental math to solve a multiplication problem.

math science technology
CONNECTION

FROM SEED TO CANDY BAR

The Cocoa Story

Cultural Note

Over 500 years ago, the Maya of Central America and the Aztec of Mexico cultivated cocoa beans. Now the Ivory Coast in Africa and Brazil in South America produce the most cocoa beans.

1 Today, cocoa beans come from South America, Central America, and Africa. They grow in pods on cacao trees.

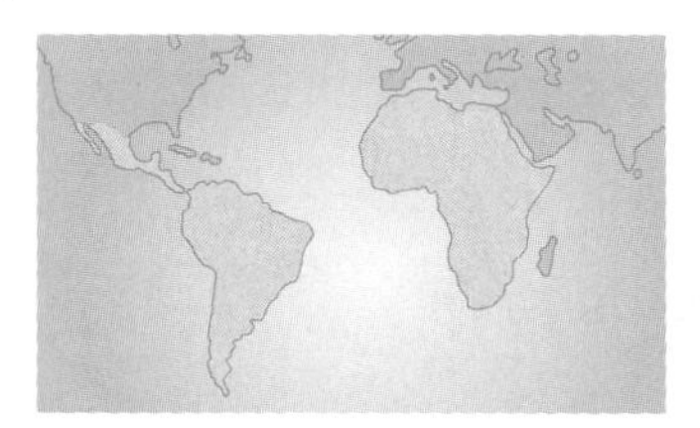

2 The cacao tree has 20 to 40 pods growing on it at a time. The trees are 25 to 40 feet tall.

3 Each pod has about 50 beans.

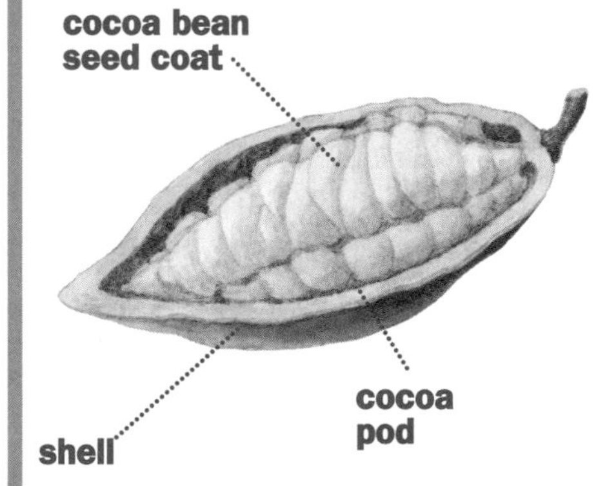

4 The beans are separated from the pod, kept dry, and shipped to the chocolate factory.

5 At the factory, the beans are roasted, ground, and made into a liquid.

6 About 9 beans are needed to make a 2-ounce milk chocolate bar.

- ► What other seeds do we use for food?
- ► Why are cocoa beans kept dry? What happens to seeds when they are kept moist?

How Much Chocolate Do You Eat?

Survey your class to find out how much chocolate is eaten in a month.

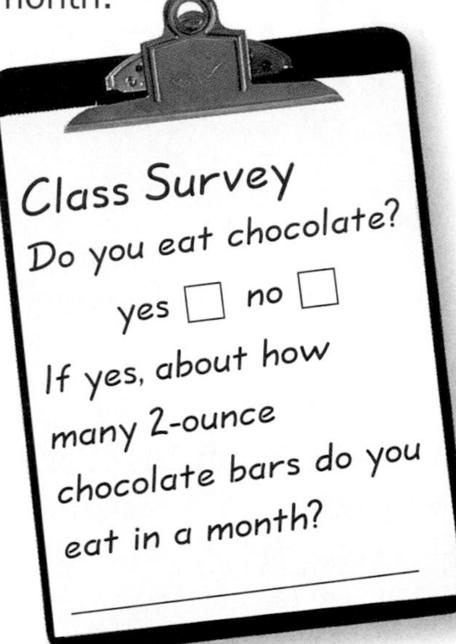

Use the data you collected to answer these questions.

1 About how many 2-ounce milk chocolate bars does your class eat in a month? in 6 months?

2 About how many cocoa beans would it take to supply milk chocolate bars to your class for 6 months?

3 **What if** you grow your own cocoa beans. About how many cacao trees will you need to supply chocolate bars to your class for 6 months?

At the Computer

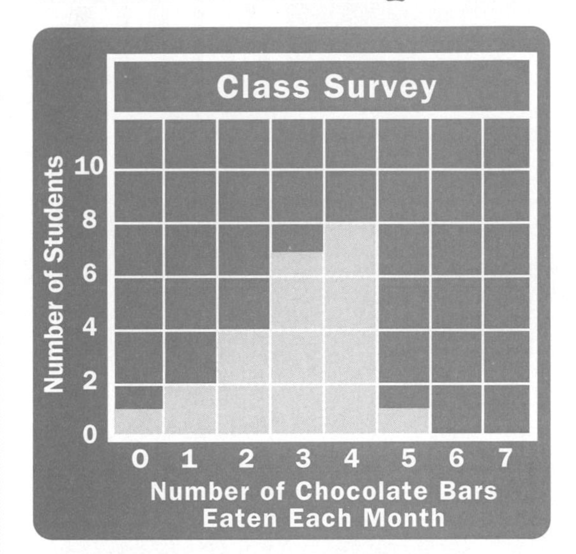

4 Use graphing software. Enter data from your survey to make a bar graph.

5 Write three statements about your class based on the graph.

CHAPTER 6

MULTIPLY BY 2-DIGIT NUMBERS

THEME

Earth Watch

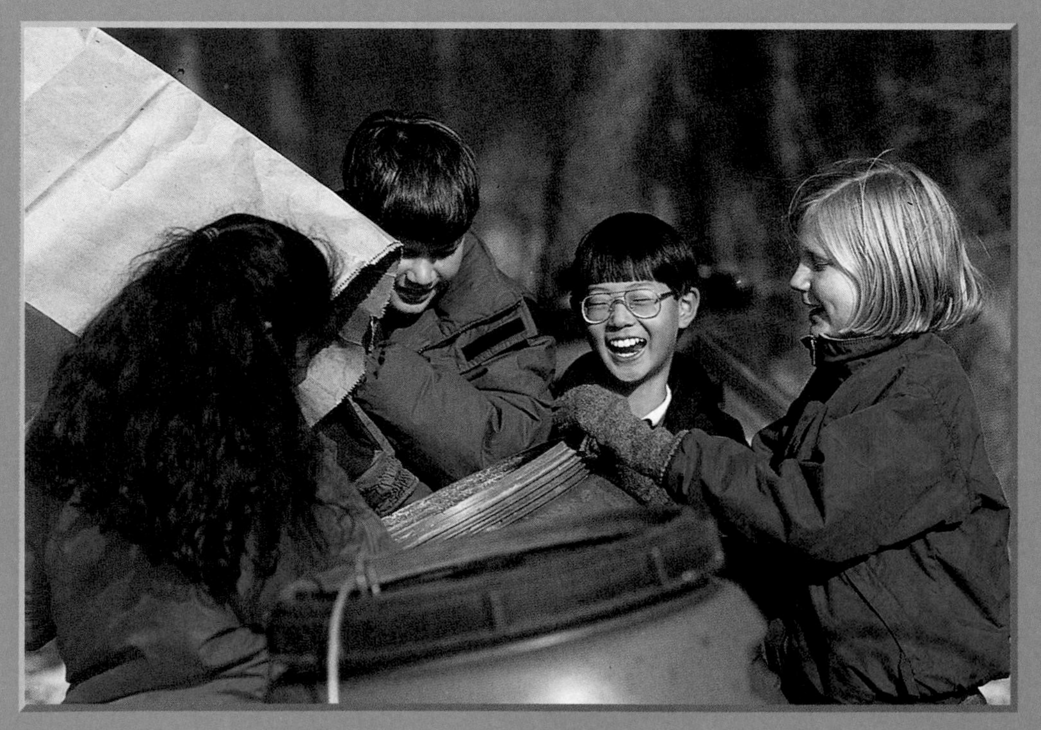

Have you ever thought of what you can do to help the environment? In this chapter you will see how multiplication is used in projects that help keep our earth a nice place to live.

What Do You Know ?

1 Maxie's class collected 5 pounds of aluminum. How many cans were collected? Explain how you found the number of cans.

2 **What if** the class collected 10 pounds of aluminum. Find the number of cans collected. Then tell how you found this number.

3 Portfolio Copy and complete the table. Explain how you completed it.

Weeks	1	2	3	4	5	6
Pounds	2	4	6			
Cans	46	92				

Steps in a Process

Maxie's class plans to recycle. They need boxes for cans and plastic bottles. They also need to take the cans to the recycling center once a month.

Paying attention to each step in a process helps you follow directions or understand information you read.

1 List the steps Maxie's class can follow to recycle their cans.

2 Why does it help to list the steps in order?

Vocabulary

product, p. 206
estimate, p. 208
round, p. 208
factor, p. 208
regroup, p. 214

Multiplication Patterns

LEARN

Have you ever seen signs along a highway or beach that say that the land is adopted? This means volunteers care for it by cleaning up the litter.

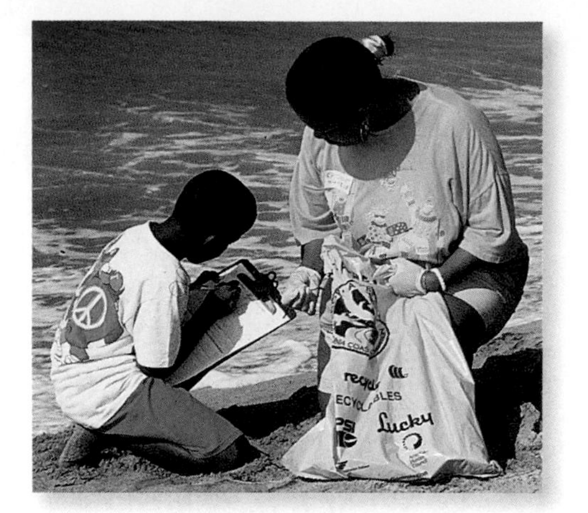

Suppose volunteers fill 30-gallon bags at a beach. If they fill 1,000 bags each day, how many gallons of trash do they pick up in 1 day? in 4 days?

ALGEBRA: PATTERNS Use patterns to find **products** mentally.

$3 \times 1 = 3$
$30 \times 1 = 30$
$30 \times 10 = 300$
$30 \times 100 = 3,000$
$30 \times 1,000 = 30,000$

They pick up 30,000 gallons of trash in 1 day.

$3 \times 4 = 12$
$30 \times 4 = 120$
$30 \times 40 = 1,200$
$30 \times 400 = 12,000$
$30 \times 4,000 = 120,000$

They pick up 120,000 gallons of trash in 4 days.

Check Out the Glossary
For vocabulary words
See page 544.

More Examples

A
$2 \times 3 = 6$
$20 \times 3 = 60$
$20 \times 30 = 600$
$20 \times 300 = 6,000$
$20 \times 3,000 = 60,000$

B
$4 \times 5 = 20$
$4 \times 50 = 200$
$40 \times 50 = 2,000$
$40 \times 500 = 20,000$
$40 \times 5,000 = 200,000$

C
$6 \times 8 = 48$
$60 \times 8 = 480$
$60 \times 80 = 4,800$
$600 \times 80 = 48,000$
$6,000 \times 80 = 480,000$

CHECK

Check for Understanding

Multiply mentally.

1 60×10 **2** $80 \times 1,000$ **3** 50×30 **4** 100×40

5 60×50 **6** 300×30 **7** 700×40 **8** $2,000 \times 50$

9 90×700 **10** $20 \times 8,000$ **11** $5,000 \times 80$ **12** $6,000 \times 90$

Critical Thinking: Summarize **Explain your reasoning.**

13 How does knowing basic facts help you find a product mentally?

Practice

Multiply mentally.

1. 20×100
2. 70×50
3. 60×300
4. $1,000 \times 90$
5. 80×60
6. 600×40
7. $50 \times 5,000$
8. $80 \times 4,000$
9. 900×30
10. $2,000 \times 40$
11. 70×600
12. $5,000 \times 40$
13. $8,000 \times 30$
14. $60 \times 1,000$
15. 800×50
16. 90×90
17. 70×70
18. $2,000 \times 90$
19. 70×800
20. $5,000 \times 60$

ALGEBRA Find the missing number.

21. $40 \times ■ = 1,200$
22. $60 \times ■ = 36,000$
23. $■ \times 1,000 = 80,000$
24. $80 \times ■ = 80,000$
25. $■ \times 200 = 8,000$
26. $■ \times 50 = 40,000$

MIXED APPLICATIONS
Problem Solving

27. Every month the Carter twins collect 30 newspapers from each of 70 homes. How many newspapers do they collect in a month?

28. On the average each American throws out over 1,000 pounds of trash each year. How much trash has just one American thrown out on the average in your lifetime?

29. **Logical reasoning** The product of two numbers is 2,400. One number is 20 more than the other. What are the two numbers?

30. A restaurant threw out 143 pounds of garbage in Week 1 and 251 pounds in Week 2. How many more pounds of garbage were thrown out in Week 2?

mixed review • test preparation

Complete. Write >, <, or =.

1. $685 + 34$ ● 719
2. $456 - 189$ ● 377
3. $549 ● $246 + $189
4. $3,004 - 95$ ● $2,119$
5. 2×8 ● 3×9
6. 6×4 ● 5×5
7. $24 \div 4$ ● $36 \div 6$
8. $40 \div 5$ ● $63 \div 9$

Estimate Products

LEARN

On the average, an American family throws out about 88 pounds of plastic each year. About how many pounds of plastic would a town of 765 families throw out?

You can **estimate** to solve this problem.

Round each **factor** so you can multiply mentally.

Estimate: 88×765 **Think:** 88×765

$\downarrow \quad \downarrow$

$90 \times 800 = 72{,}000$

A town of 765 families throws out about 72,000 pounds of plastic each year.

More Examples

A Estimate: 73×89 **Think:** $70 \times 90 = 6{,}300$

B Estimate: $24 \times \$3{,}258$ **Think:** $20 \times \$3{,}000 = \$60{,}000$

CHECK

Check for Understanding

Estimate the product. Tell how you rounded.

1 56×18 **2** 84×371 **3** $49 \times \$5{,}250$ **4** 91×645

5 65×14 **6** 372×48 **7** 754×34 **8** $\$1{,}925 \times 25$ **9** $5{,}289 \times 78$

Critical Thinking: Generalize **Explain your reasoning.**

10 How can multiplying tens, hundreds, and thousands mentally help you estimate?

11 How does an exact product compare to an estimate found by rounding *up* both factors? rounding *down* both factors?

Check Out the Glossary
For vocabulary words
See page 544.

Practice

Estimate the product.

1. 72 × $39
2. 55 × 63
3. 85 × $29
4. 12 × 18
5. 91 × 432
6. 83 × 654
7. 35 × 950
8. 49 × $519
9. 45 × $1,925
10. 28 × 5,250
11. 74 × 2,915
12. 17 × 6,643
13. 58 × 92
14. 15 × $712
15. 44 × 854
16. 75 × $3,925

Estimate to solve. Write > or <.

17. 47 × 35 ● 2,000
18. 57 × 32 ● 1,500
19. 756 × 89 ● 72,000

MIXED APPLICATIONS
Problem Solving

Use the pictograph for problems 20–22.

20. About how many pounds of food/yard, glass, metal, paper, and plastic garbage does one person produce in one year?

21. About how many times more paper is produced than plastic?

What is in your garbage?
Amount produced by one person in one year

Food Yard
Glass
Metal
Paper
Plastic

Key: Each stands for 30 pounds

22. **Write a problem** using information from the pictograph. Exchange problems with a classmate and solve each other's problems.

23. **Data Point** For one day, count the plastic items you throw away and recycle. Combine your list with others to make a class graph.

more to explore

Use Front Digits to Estimate

Here is another way to estimate a product.

Use the first digit of each factor. Replace the other digits with zeros. Then multiply.

Estimate: 66 × 4,710
↓ ↓
Think: 60 × 4,000 = 240,000

Estimate the product using front digits.

1. 24 × 356
2. 745 × 87
3. 55 × 860
4. 46 × 7,457

Extra Practice, page 504

Multiply 2-Digit Numbers

Grids can help you to multiply. You used grids to multiply by 1-digit numbers. You can also use grids to multiply by 2-digit numbers.

You will need
- *1–4 spinner*
- *centimeter graph paper*

Work Together

Work with a partner. Spin the spinner twice to get the tens and ones digits of a factor. Spin twice again to get the second factor.

Cut out or draw a rectangle on graph paper. Use the first factor as the length and the second factor as the width.

Find the product of the two factors without counting all the squares in the grid.

Record your results in a table.

Repeat with two new factors.

First Factor	Second Factor	Product
32	14	

Talk It Over

- Explain the method you used to find the totals.
- How can you tell if your products are reasonable?

Make Connections

Brian and Eileen broke the factors apart to find their products.

Brian

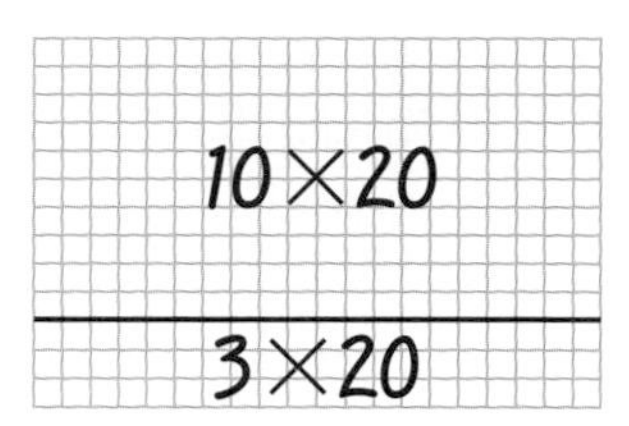

$$\begin{array}{rl} 20 & \\ \times\ 13 & \\ \hline 60 & \leftarrow 3 \times 20 \\ +\ 200 & \leftarrow 10 \times 20 \\ \hline 260 & \end{array}$$

Eileen

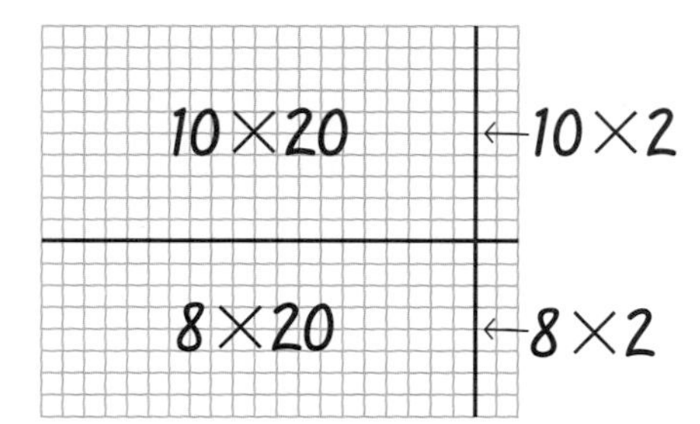

$$\begin{array}{rl} 22 & \\ \times\ 18 & \\ \hline 200 & \leftarrow 10 \times 20 \\ 20 & \leftarrow 10 \times 2 \\ 160 & \leftarrow 8 \times 20 \\ +\ 16 & \leftarrow 8 \times 2 \\ \hline 396 & \end{array}$$

- ▶ How did Brian and Eileen use place value when they broke apart the factors to find the products?
- ▶ Why does Brian's method have two addends while Eileen's method has four addends?

Check for Understanding

Multiply using any method.

1 25×30

2 36×23

3 17×20

4 12×38

5 21×35

CHECK

Critical Thinking: Analyze **Explain your reasoning.**

6 Another way to find 13×20 is by thinking: 13×2 tens $= 26$ tens $= 260$. How is this method like Brian's method? How is it different?

Turn the page for Practice.

PRACTICE

Practice

Find the total number of squares in the rectangle without counting.

1

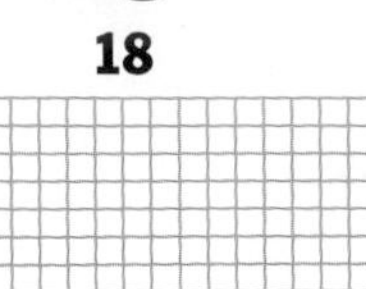

2

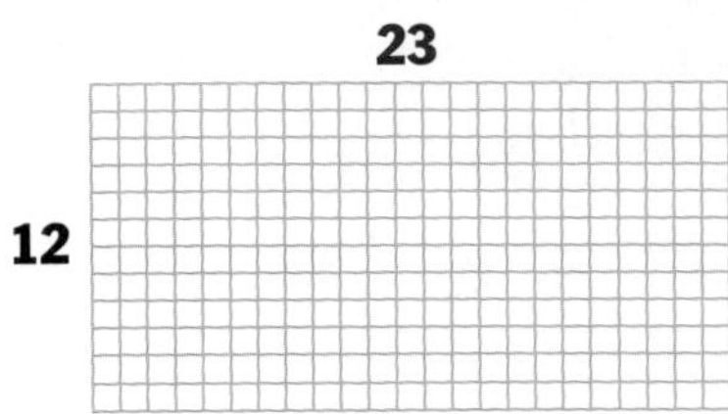

3

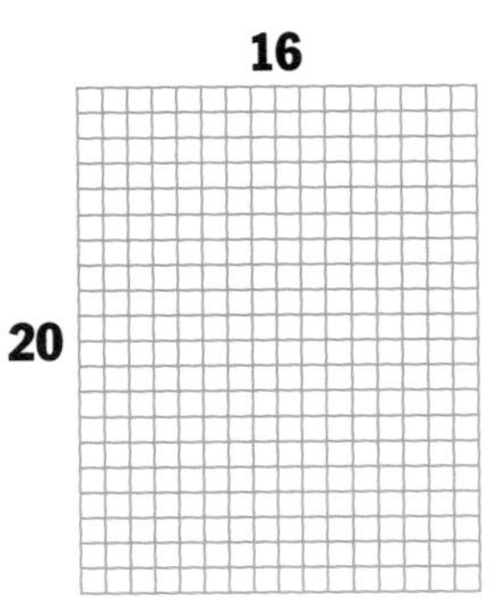

4

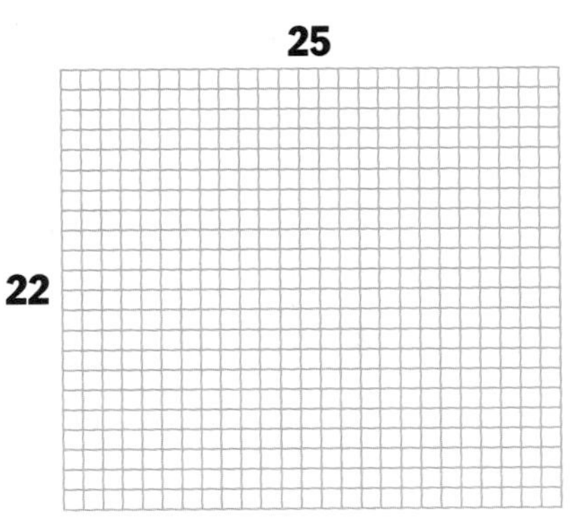

Multiply using any method.

5 42×20

6 23×30

7 19×33

8 44×29

9 20×18

10 34×26

11 28×41

12 33×30

13 35×40

14 13×38

15 24×42

16 41×25

17 39×39

18 28×19

19 35×21

Make It Right

20 Here is how Riley found 20×37. Explain what the mistake is, then correct it.

```
   37
 × 20
  140 ← 20 × 7
+  60 ← 20 × 3
  200
```

MIXED APPLICATIONS
Problem Solving

21 List all the pairs of numbers that have a product of 400.

20	10	25	15
18	24	100	20
5	16	40	80

22 How many tiles will you use if you cover a floor that is 18 feet by 11 feet using 1-foot by 1-foot tiles?

INFOBIT
In 1976, the people who live in Beijing, China, planted more than 300 million fast-growing trees along the path of the Great Wall of China.

23 **Make a decision** A carton of computer paper has 10 packages of paper in it and sells for $28. You can also buy the same paper for $3 per package. You have $35 to buy computer paper. What would you buy? Explain.

24 Were the trees along the path of the Great Wall of China planted more than or less than 25 years ago? **SEE INFOBIT.**

Cultural Connection
Hindu Lattice Multiplication

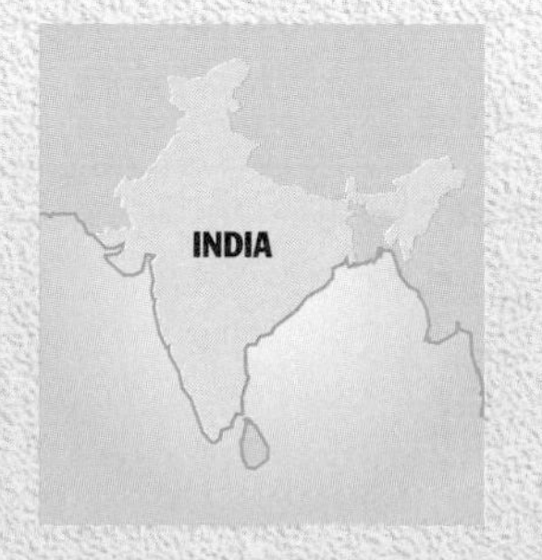

Here is a method Hindu mathematicians in India used to multiply two numbers.

To find 45×36:

Step 1 **Write the factors on the outside of the lattice as shown.**

Step 2 **Write the product of each pair of digits inside the lattice as shown.**

Step 3 **Starting at the top right, add the numbers in each diagonal row. Be careful to carry numbers to the next row if needed.**

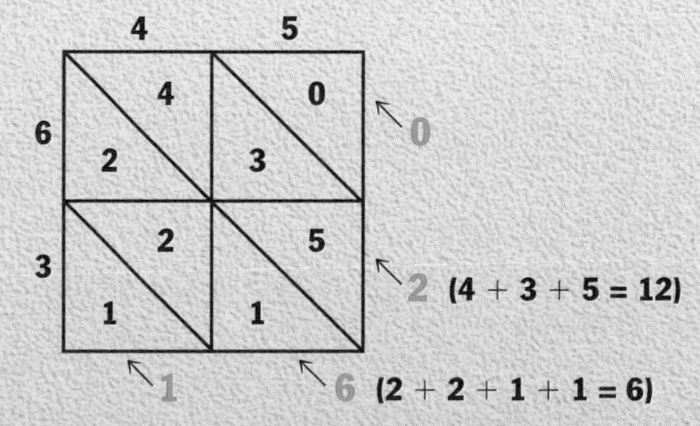

$45 \times 36 = 1,620$

Use lattice multiplication to find the product.

1 12×18 **2** 25×19 **3** 32×39 **4** 56×44

Extra Practice, page 505

Multiply 2-Digit Numbers

LEARN

To protect an acre of rain forest, you can send $35 to the Nature Conservancy's Adopt-an-Acre Program.

A fourth-grade class would like to protect 20 acres of rain forest. How much money does the class need to raise?

Estimate: $20 \times \$35$ **Think:** $20 \times \$40 = \$800$

In the last lesson you used this method to find the exact answer.

$$\begin{array}{rl} \$35 & \\ \times\ \ 20 & \\ \hline 100 & \leftarrow 20 \times 5 \\ +\ \ 600 & \leftarrow 20 \times 30 \\ \hline \$700 & \end{array}$$

Cultural Note

Tropical rain forests are located to the north and south of the equator in Central America, South America, Asia, and Africa. Over 750,000 different kinds of plants grow in Earth's rain forests.

Here is another method.

Step 1	Step 2
Multiply by the ones.	**Multiply by the tens. Regroup if necessary.**
$\begin{array}{r} \$35 \\ \times\ \ 20 \\ \hline 0 \end{array}$	$\begin{array}{r} {}^{1}\ \ \\ \$35 \\ \times\ \ 20 \\ \hline \$700 \end{array}$
Think: 0 ones $\times$ 35 = 0	**Think:** 2 tens $\times$ 35 = 70 tens

The class needs to raise $700.

Talk It Over

- How are the two methods similar? different?
- Describe how you could use a grid to find the product of 20 and 35.

What if the class plants 26 rows of 35 seedlings. How many seedlings does the class plant?

Estimate: 26×35 **Think:** $30 \times 40 = 1,200$

Multiply: 26×35

Step 1	Step 2	Step 3
Multiply by the ones.	**Multiply by the tens.**	**Add the products.**
$\begin{array}{r} {}^{3} \\ 35 \\ \times 26 \\ \hline 210 \end{array}$	$\begin{array}{r} {}^{1} \\ \not{3} \\ 35 \\ \times 26 \\ \hline 210 \\ 700 \end{array}$	$\begin{array}{r} {}^{1} \\ \not{3} \\ 35 \\ \times 26 \\ \hline 210 \\ +700 \\ \hline 910 \end{array}$
Think: 6 ones $\times$ 35 = 210 ones	**Think:** 2 tens $\times$ 35 = 70 tens	

The class plants 910 seedlings in all.

Check Out the Glossary
For vocabulary words
See page 544.

You can multiply money amounts the same way you multiply whole numbers.

Multiply: $26 \times \$0.35$

$$\begin{array}{r} \$0.35 \\ \times \quad 26 \\ \hline \$9.10 \end{array}$$

Insert the dollar sign and decimal point.

Check for Understanding

CHECK

Multiply using any method. Estimate to see if your answer is reasonable.

1. $\begin{array}{r} 13 \\ \times 12 \\ \hline \end{array}$
2. $\begin{array}{r} 25 \\ \times 11 \\ \hline \end{array}$
3. $\begin{array}{r} 34 \\ \times 22 \\ \hline \end{array}$
4. $\begin{array}{r} 29 \\ \times 30 \\ \hline \end{array}$
5. $\begin{array}{r} 48 \\ \times 12 \\ \hline \end{array}$

Multiply.

6. $16 \times \$53$
7. $25 \times \$0.67$
8. $34 \times \$0.59$
9. $41 \times \$0.82$

Critical Thinking: Compare

10. If $5 \times 45 = 225$, what is 50×45? Explain your reasoning.

11. Journal Why are there three steps to multiply 35 by 26 and only two steps to multiply 27 by 30?

Turn the page for Practice.

Practice

Multiply using any method. Remember to estimate.

1. $\begin{array}{r} 17 \\ \times 11 \\ \hline \end{array}$
2. $\begin{array}{r} 31 \\ \times 22 \\ \hline \end{array}$
3. $\begin{array}{r} 26 \\ \times 23 \\ \hline \end{array}$
4. $\begin{array}{r} \$25 \\ \times 40 \\ \hline \end{array}$
5. $\begin{array}{r} 32 \\ \times 19 \\ \hline \end{array}$
6. $\begin{array}{r} 24 \\ \times 35 \\ \hline \end{array}$
7. $\begin{array}{r} \$28 \\ \times 12 \\ \hline \end{array}$
8. $\begin{array}{r} 47 \\ \times 30 \\ \hline \end{array}$
9. $\begin{array}{r} 44 \\ \times 29 \\ \hline \end{array}$
10. $\begin{array}{r} \$60 \\ \times 18 \\ \hline \end{array}$
11. $\begin{array}{r} 24 \\ \times 18 \\ \hline \end{array}$
12. $\begin{array}{r} 35 \\ \times 23 \\ \hline \end{array}$
13. $\begin{array}{r} 29 \\ \times 42 \\ \hline \end{array}$
14. $\begin{array}{r} 35 \\ \times 58 \\ \hline \end{array}$
15. $\begin{array}{r} 64 \\ \times 19 \\ \hline \end{array}$
16. 25×50
17. 54×12
18. $36 \times \$27$
19. 73×11
20. 73×44
21. $63 \times \$92$
22. $88 \times \$0.36$
23. $52 \times \$0.84$

ALGEBRA Write the letter of the missing number.

24. $\blacksquare \times 50 = 1{,}000$ **a.** 6 **b.** 20 **c.** 15 **d.** 16
25. $40 \times \blacksquare = 2{,}400$ **a.** 60 **b.** 70 **c.** 6 **d.** 12
26. $\blacksquare \times 32 = 960$ **a.** 20 **b.** 15 **c.** 30 **d.** 25

Use the bar graph on the right for problems 27–28.

27. How many pounds of a person's trash are:
   a. recycled?
   b. placed in a landfill?

28. How many more pounds of trash are burned than are recycled? Explain how you know.

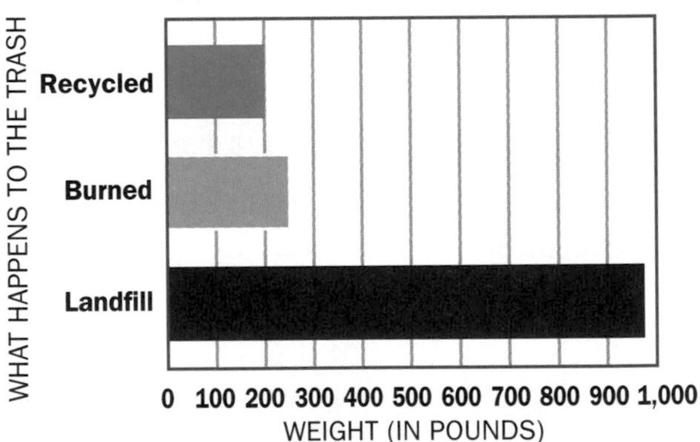

Make It Right

29. Here is how Luellen found 37×63. Tell what the mistake is, then correct it.

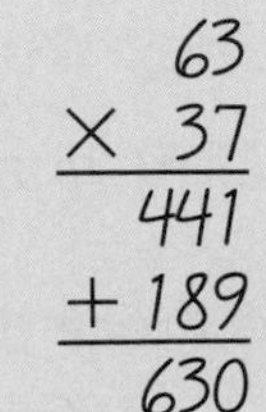

Hopscotch for Factors Game!

You will need
- *16 two-color counters*

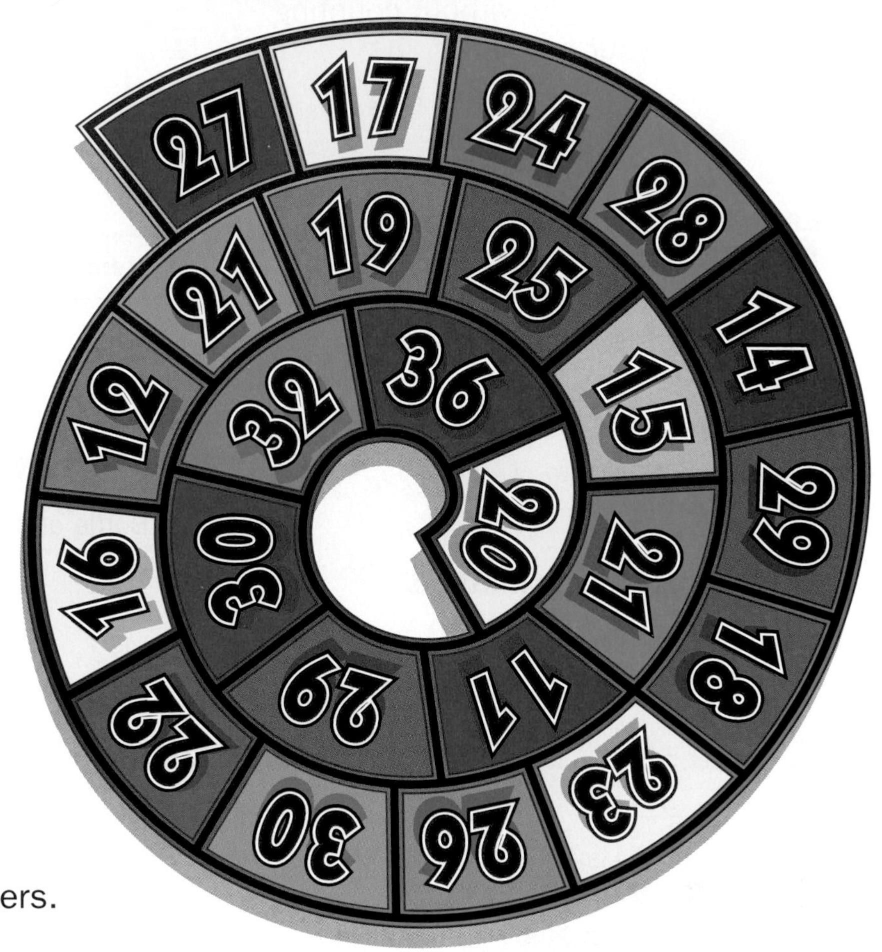

Play the Game

- ▶ Work with a partner to find pairs of numbers whose product is between 400 and 500.
- ▶ As you pick two numbers, place a counter on each one. Check the product. If it is between 400 and 500, keep both counters on the board. If not, remove the counters.
- ▶ The object of the game is to cover all of the numbers with counters.

How did you decide if a product is between 400 and 500?

Cultural Note

This game board is similar to the hopscotch pattern used in France. The pattern is called escargot (es-kahr-GOH), which is the French word for "snail."

mixed review • test preparation

Estimate.

1 $785 + 452$

2 $\$4,612 - 2,589$

3 $25,460 + 72,950$

4 $90,180 - 84,300$

5 $\$31,980 + 19,652$

6 4 × 12 **7** 8 × 68 **8** 9 × $25 **9** 8 × $72 **10** 9 × 45

Extra Practice, page 505

midchapter review

1 Find the total number of squares in the rectangle without counting. Explain your method.

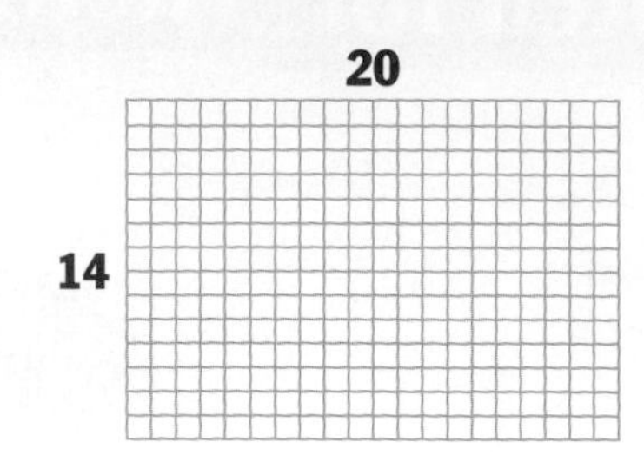

Estimate the total amount.

2 12 boxes of detergent

3 376 bottles of shampoo

4 23 bags of dog food

5 32 bottles of soda

Item	Amount
Bag of dog food	40 pounds
Bottle of soda	64 fluid ounces
Box of detergent	128 ounces
Bottle of shampoo	22 fluid ounces

Multiply using any method.

6 12×14

7 40×30

8 24×42

9 51×37

10 $2{,}000 \times 60$

11 $\$0.52 \times 38$

12 25×39

13 $\$0.46 \times 16$

14 $8{,}000 \times 70$

15 900×60

Complete. Write >, <, or =.

16 4,000 ● 75 × 67

17 8,000 × 90 ● 500,000

18 38 × 45 ● 63 × 24

19 99 × 39 ● 97 × 72

Solve. Use mental math when you can.

20 A class collects 43 bags of litter. Each bag weighs 18 pounds. How many pounds of litter does the class collect?

21 A club needs 17 gallons of paint. Each gallon costs $23. About how much will the club pay?

22 In a Green Day Parade, 30 groups marched through the town. Each group had about 40 people. About how many people marched in all?

23 On Earth Day, each group of volunteers collected 14 pounds of litter. If there were 23 groups, about how much litter was collected?

24 Banners for Earth Day cost $4.75 each. How much do 25 banners cost?

25 Journal Explain two different methods you can use to find 28 × 34.

Use Calculators and Spreadsheets

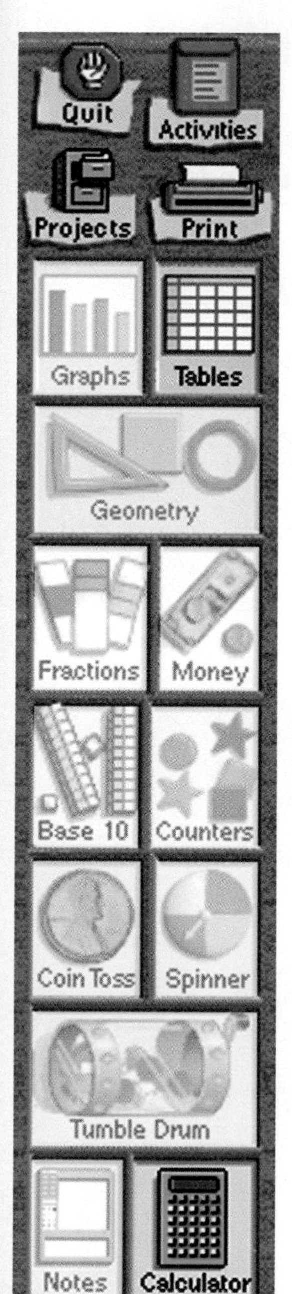

You can use a calculator to add, subtract, multiply, and divide. You can also use computer spreadsheets.

Suppose 19 students signed up for a recycling drive in the first week of January, 14 in the second week, 23 in the third week, and 16 in the fourth week. To find the total number of students, you could use a calculator to add the four numbers.

Suppose you wanted to do this calculation every month for a year. It would be better to use a spreadsheet.

Number of Students

Month	Week 1	Week 2	Week 3	Week 4	Total
Jan.	19	14	23	16	Col 1+Col 2+ Col 3+Col 4
Feb.	6	12	15	0	
Mar.	18	7	9	14	

▶ Col 2 Row 2 has 12 in it. Why can you add Col 1, 2, 3, and 4 in each row to get the total?

▶ Work with a partner and enter the data above in a computer table to see what your Total column looks like.

1 Create some other operations for your spreadsheet besides addition. Try the same operations on your calculator. Compare your work.

Critical Thinking: Generalize

2 When might it be better to use a spreadsheet than a calculator?

Be a Paper Recycler

Each year, paper is used to make millions of books, magazines, and newspapers and is turned into thousands of pounds of tissue paper in the United States.

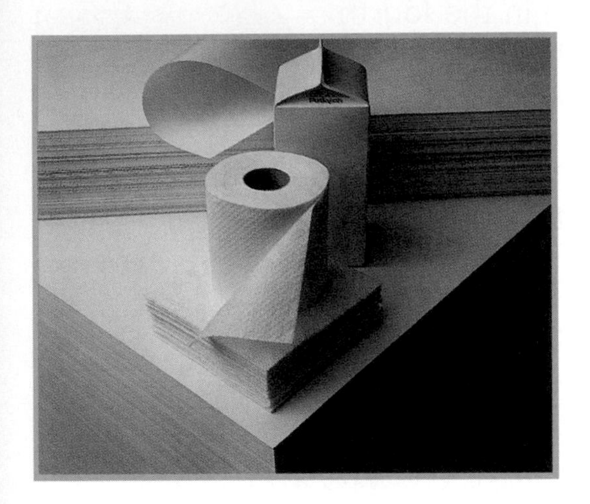

How do you use paper?

Paper and paperboard make up most of the material that is thrown out as waste.

Why is recycling paper so important?

In 1994, more paper was recycled than was sent to landfills. The recycled paper could fill a space about one yard deep and the area of Washington, D.C.

What can you do to help reach the recycling goal?

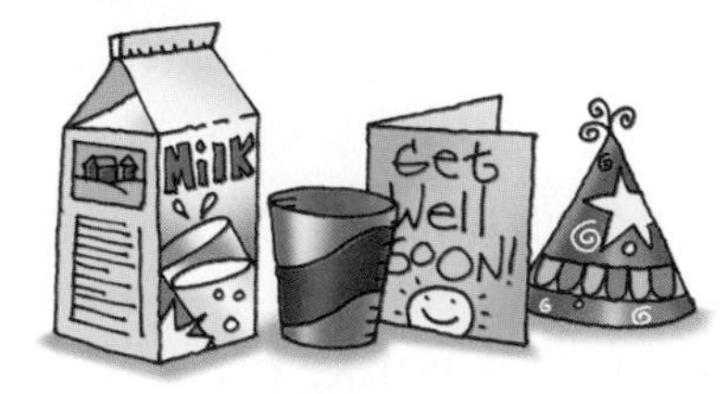

Goal for the Year 2000
The nation's paper industry has set a goal. By the year 2000, they will recycle half of all the paper Americans use each year.

DECISION MAKING

Reusing and Recycling

1 Work in a group. Choose where you will collect data on paper use—at home, in your classroom, in the school office or cafeteria.

2 Decide on the following things:
a. what kinds of paper to include
b. how long you will collect paper
c. how you will measure the paper
d. how you will record your findings

3 Use your data to estimate how much paper is thrown away in a month and in a year.

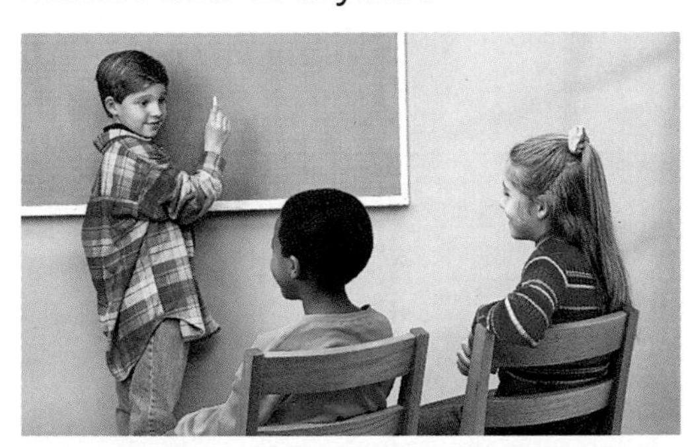

4 Estimate how much paper and paper products are recycled in a week and in a month.

Reporting Your Findings

5 Portfolio Prepare a report to persuade people to recycle paper. Include the following things:

- Explain why you think recycling is important.
- READING ARITHMETIC WRITING **Steps in a Process** List the steps your group followed to collect data.
- Write statements about the amount of paper your group members discovered was thrown away. Describe the math that you used.

6 Compare your report with the reports of other groups.

Revise your work.
- Have you checked your calculations?
- Is your report clear and organized?
- Did you proofread your work?

MORE TO INVESTIGATE

PREDICT how much paper and paper products your neighborhood or community throws away in a week.

EXPLORE what kinds of paper are recycled by a recycling center near you. Find out how the paper is then used.

FIND projects that you can take part in to help increase the recycling of paper.

Read | Plan | Solve | Look Back

Use Alternate Methods

LEARN

Read **Matt wants to solve this riddle.**

I am a number between 108 and 115. Each of my digits is less than 5. If you subtract 1 from me, you get a multiple of 5. What number am I?

Plan To find the answer, you can guess and test, or make a table.

Solve Guess and test.

Think: 110, 111, 112, 113, 114 have digits less than 5.

Try 110.
110 − 1 = 109 ← **not a multiple of 5**

Try 111.
111 − 1 = 110 ← **a multiple of 5**
So "I am 111."

Make a table.

Numbers from 109 to 114	109	110	111	112	113	114
Digits less than 5?	no	yes	yes	yes	yes	yes
Subtract 1. Multiple of 5?	no	no	yes	no	no	no

The only column in which each row has a *yes* is 111. So "I am 111."

Look Back How are the two methods alike? How are they different?

CHECK

Check for Understanding

1 **What if** the riddle did not say that all the digits are less than 5. How would the two methods change?

Critical Thinking: Analyze

2 Rewrite the riddle so that there is extra information but the answer is the same.

MIXED APPLICATIONS

Problem Solving

1 On one field trip, students rode their bikes 4 miles to the beach, biked 7 miles along the beach path, and then rode 5 miles back to their starting positions. How far did they travel? Explain your method. What other method could you have used?

2 A town recycles 30 tons of newspapers a week. About how many kilowatt-hours of electricity does the town save each week? **SEE INFOBIT.**

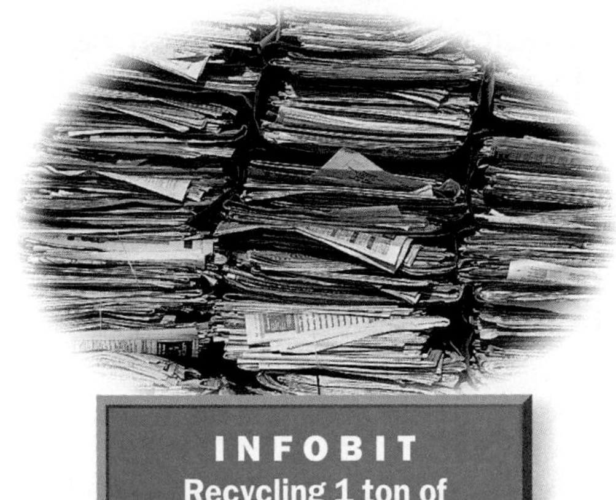

INFOBIT
Recycling 1 ton of newspapers saves about 4,000 kilowatt-hours of electricity.

3 A recycling center received 81 pounds of cans the first day. The next day it received 72 pounds. The cans are packed in 9-pound bags. How many bags did the center receive altogether?

4 **Logical reasoning** Helen rolls two number cubes to get a 2-digit factor. She rolls the cubes again to get another 2-digit factor. What is the greatest product she can get?

Use the price list for problems 5–9.

5 The Center offers 3 one-hour field trips every week. Twelve people go on each trip. How much money do the field trips earn each week?

6 A club has 28 members. Is it cheaper for them to go to the Ecology Center or watch the wildlife film?

7 Twenty-nine students plan to take a gardening class. What is the total cost for these students?

8 **Make a decision** Your class has $48 to spend at the Center. What should the class see and do? How much money will you spend?

9 **Write a problem** using at least two prices from the price list. Solve it using two methods. Have others solve it.

WETLANDS CENTER

Ecology Center	$3 for each person
Birdwatching Tour	**FREE**
Gardening Class (6 week session)	$90 for each group* $12 for each person
Wildlife film	$75 for each group*
Field trip	$9 for each person

* A group has ten or more people.

Multiply 3-Digit Numbers

LEARN

A leaky faucet dripping 1 drop each second wastes 879 gallons of water in a year. How many gallons of water can be saved by fixing the 23 faucets in an apartment building?

Estimate: 23 × 879

Think: 20 × 900 = 18,000

You can use pencil and paper to find the exact answer.

Multiply: 23 × 879

Step 1	Step 2	Step 3
Multiply by the ones.	**Multiply by the tens.**	**Add the products.**
$\begin{array}{r} {}^{2\,2} \\ 879 \\ \times\ 23 \\ \hline 2637 \end{array}$ ← 3 × 879	$\begin{array}{r} {}^{1\,1} \\ {}^{2\,2} \\ 879 \\ \times\ 23 \\ \hline 2637 \\ 17580 \end{array}$ ← 20 × 879	$\begin{array}{r} {}^{1\,1} \\ {}^{2\,2} \\ 879 \\ \times\ 23 \\ \hline 2637 \\ +17580 \\ \hline 20{,}217 \end{array}$

You can also use a calculator. 23 × 879 = 20217.

20,217 gallons of water can be saved.

CHECK

Check for Understanding

Multiply using any method. Estimate to see if your answer is reasonable.

1 $\begin{array}{r} 153 \\ \times\ 12 \\ \hline \end{array}$ **2** $\begin{array}{r} 281 \\ \times\ 23 \\ \hline \end{array}$ **3** $\begin{array}{r} \$3.54 \\ \times\ 17 \\ \hline \end{array}$ **4** $\begin{array}{r} 359 \\ \times\ 44 \\ \hline \end{array}$ **5** $\begin{array}{r} 825 \\ \times\ 51 \\ \hline \end{array}$

Critical Thinking: Generalize **Explain your reasoning.**

6 If 23 × 879 = 20,217, what is 23 × $8.79 equal to?

Practice

Multiply. Remember to estimate.

1. $\begin{array}{r} 158 \\ \times\ 85 \\ \hline \end{array}$
2. $\begin{array}{r} 430 \\ \times\ 42 \\ \hline \end{array}$
3. $\begin{array}{r} 608 \\ \times\ 28 \\ \hline \end{array}$
4. $\begin{array}{r} 706 \\ \times\ 28 \\ \hline \end{array}$
5. $\begin{array}{r} 414 \\ \times\ 92 \\ \hline \end{array}$
6. $\begin{array}{r} \$525 \\ \times\ \ \ 32 \\ \hline \end{array}$
7. $\begin{array}{r} \$0.87 \\ \times\ \ \ \ 35 \\ \hline \end{array}$
8. $\begin{array}{r} \$5.63 \\ \times\ \ \ \ 25 \\ \hline \end{array}$
9. $\begin{array}{r} \$8.47 \\ \times\ \ \ \ 39 \\ \hline \end{array}$
10. $\begin{array}{r} \$3.18 \\ \times\ \ \ \ 42 \\ \hline \end{array}$

11. 11 × 117
12. 45 × $0.88
13. 13 × 239
14. 22 × $3.01
15. 64 × $7.86
16. 39 × $0.07
17. 31 × 754
18. 37 × 306
19. 10 × 211
20. 19 × $1.99
21. 41 × 309
22. 45 × 807
23. 92 × 816
24. 53 × 635
25. 81 × $0.04
26. 74 × 560

MIXED APPLICATIONS

Problem Solving

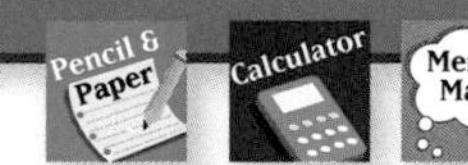

27. **What if** a toilet is flushed 25 times a day. An old toilet uses 5 gallons for each flush. A new toilet uses 2 gallons. How much water does the new toilet save?

28. **Data Point** Use the Databank on page 537. How long do you take to shower? About how much water do you use? Is this more than or less than the amount of water for a bath?

more to explore

Changing Units of Time

This chart can help you show times in different units.

1 minute (min) = 60 seconds (s)
1 hour (h) = 60 minutes
1 day (d) = 24 hours
1 week (wk) = 7 days
1 year (y) = 52 weeks
1 year = 12 months (mo)
1 year = 365 days

To show 3 hours in minutes, multiply.

Think: 1 hour = 60 minutes

3 hours = 3 × 60 minutes
= 180 minutes

Complete.

1. 4 minutes = ■ seconds
2. 5 days = ■ hours
3. 14 hours = ■ minutes
4. 7 hours = ■ seconds
5. 15 weeks = ■ days
6. 6 years = ■ days

Extra Practice, page 506

PRACTICE

Multiply Greater Numbers

LEARN

To help fight pollution, some students decided to recycle rings from a six-pack of soda. So far they have 4,278 bundles of rings with 25 rings in each bundle. How many rings do they have?

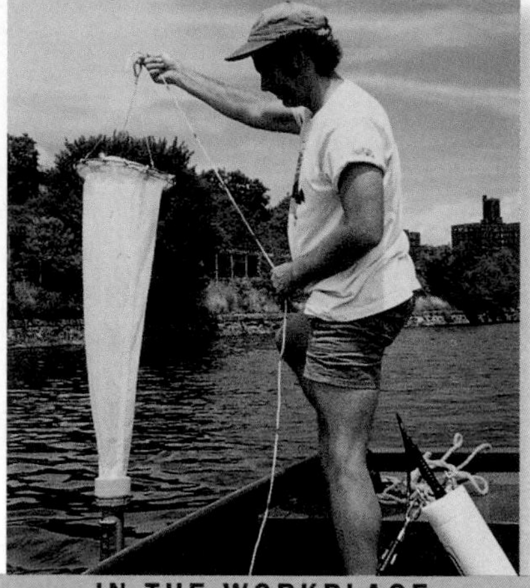

IN THE WORKPLACE
Field ecologist, Jeff Mantus, at work

Estimate: $25 \times 4{,}278$

$\downarrow \quad \downarrow$

Think: $30 \times 4{,}000 = 120{,}000$

You can use pencil and paper to find the exact answer.

Multiply: $25 \times 4{,}278$

Step 1	Step 2	Step 3
Multiply by the ones.	**Multiply by the tens.**	**Add the products.**
$\begin{array}{r} {}^{1\,3\,4} \\ 4{,}278 \\ \times \quad 25 \\ \hline 21390 \end{array} \leftarrow 5 \times 4{,}278$	$\begin{array}{r} {}^{1\,1} \\ \not{1}\,\not{3}\,\not{4} \\ 4{,}278 \\ \times \quad 25 \\ \hline 21390 \\ 85560 \end{array} \leftarrow 20 \times 4{,}278$	$\begin{array}{r} {}^{1\,1} \\ \not{1}\,\not{3}\,\not{4} \\ 4{,}278 \\ \times \quad 25 \\ \hline 21390 \\ +85560 \\ \hline 106{,}950 \end{array}$

You can also use a calculator. $25 \times 4{,}278 =$ 106950.

They have 106,950 rings.

CHECK

Check for Understanding

Multiply using any method. Estimate to see if your answer is reasonable.

1 $\begin{array}{r} 2{,}584 \\ \times \quad 37 \\ \hline \end{array}$ **2** $\begin{array}{r} 3{,}285 \\ \times \quad 21 \\ \hline \end{array}$ **3** $\begin{array}{r} \$42.25 \\ \times \quad 17 \\ \hline \end{array}$ **4** $\begin{array}{r} 12{,}045 \\ \times \quad 62 \\ \hline \end{array}$ **5** $\begin{array}{r} 51{,}009 \\ \times \quad 29 \\ \hline \end{array}$

Critical Thinking: Generalize

6  Explain how you would find the following products.

a. $40 \times 30{,}000$ **b.** $23 \times 3{,}216$ **c.** $72 \times 4{,}896$

Practice

Multiply mentally.

1 $5{,}000 \times 60$

2 $\$7{,}000 \times 30$

3 $8{,}000 \times 80$

4 $46{,}000 \times 20$

5 $25{,}000 \times 30$

Multiply using any method. Remember to estimate.

6 957×98

7 $\$6.78 \times 19$

8 $\$18.15 \times 75$

9 $6{,}029 \times 19$

10 $7{,}060 \times 87$

11 $\$25.72 \times 24$

12 $\$84.14 \times 9$

13 $3{,}225 \times 51$

14 $21{,}400 \times 33$

15 $43{,}200 \times 42$

MIXED APPLICATIONS
Problem Solving

16 **What if** there is an average of 4,074 students in each grade from kindergarten to grade 8 in your state. If each student recycles 5 bottles, how many bottles are recycled?

17 **Spatial reasoning** Which would be easiest to move if you were inside the shape? How would you move the shape?
a. a box **b.** a ball **c.** a cone

18 Tom and his brother spent $7.50 altogether at the Earth Day Festival. Tom spent $3.00 more than his brother. How much did each spend at the festival?

19 **Data Point** Survey your classmates to see how many times they open the refrigerator in one day. Make a table of your results. What does your survey show?

more to explore

Using Variables

ALGEBRA You can use a letter to show an unknown amount. Instead of writing $5 \times ■ = 45$, you can write $5 \times a = 45$.

A Find $9 + b$ when $b = 7$.
Since $b = 7$, $9 + 7 = 16$.

B Find the value of z in $4 \times z = 12$.
Since $4 \times 3 = 12$, $z = 3$.

Find the answer when $a = 6$.

1 $5 + a$ **2** $10 - a$ **3** $7 \times a$ **4** $a + 32$

Find the value of the letter.

5 $6 + z = 25$ **6** $8 \times b = 32$ **7** $49 - t = 35$ **8** $y - 12 = 25$

Extra Practice, page 507

Problem Solvers at Work

Read
Plan
Solve
Look Back

LEARN

Part 1 Use an Estimate or an Exact Answer

A recycling center trucks the materials it collects each week to a local processing plant. A truck can carry only 8,500 pounds of material on each trip.

Materials Collected at Recycling Center		
Materials	**Weight (in pounds)**	**Price (for each pound)**
Glass	2,700	3¢
Copper	600	1¢
Batteries	2,700	50¢
Aluminum	1,800	25¢
Steel	700	15¢
Cardboard/ newspaper	8,750	1¢

Work Together

Use the table to solve. Tell whether you can solve without finding the exact answer. Explain your reasoning.

1. Will a truck be able to carry all the glass, cardboard, and newspaper in one truckload?

2. If a truck takes all the glass, batteries, and aluminum in one truckload, how much cardboard can it take on the trip?

3. The price list shows what the plant pays the center for each pound of material. Will the center get more than $1,500 for the batteries?

4. **What if** the center can get $400 for this week's aluminum by sending it out of the state. Should the center agree to this offer? Why or why not?

5. **Make a decision** The truck can only make one trip this week. Which materials should the truck carry to earn the most money?

Part 2 Write and Share Problems

Megan used the information on page 228 to write a problem.

6. Solve Megan's problem. Can you use estimation to solve the problem? Why or why not?

7. **Steps in a Process** How can you solve the problem in another way? List the steps you would follow.

8. **Write a problem** about recycling. You should be able to solve the problem using estimation.

9. Write a similar problem that needs an exact answer to solve it.

10. Solve both problems. Explain why they can or cannot be solved using estimation.

11. Trade problems. Solve at least three problems written by your classmates.

12. What was the most interesting problem that you solved? Why?

CHECK

How much would the center get for all the copper?

Megan McAtee
Mandarin Oaks Elementary School
Jacksonville, FL

Cultural Note

Paul Revere used recycled copper metal to make parts of the dome of the Massachusetts statehouse.

Turn the page for Practice Strategies.

PRACTICE

Menu

Choose five problems and solve them. Explain your methods.

1 It takes 1,022 milk jugs to make a plastic park bench. Can you make 45 park benches from 42,000 milk jugs? How do you know?

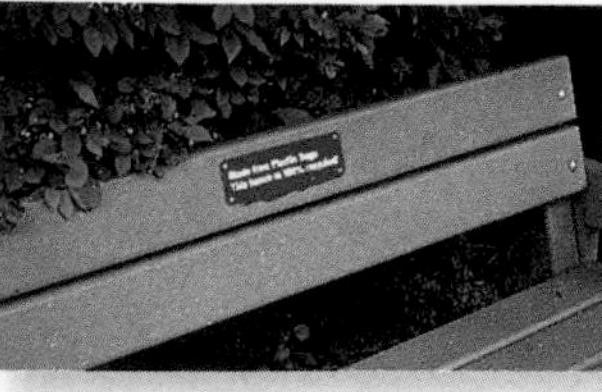

2 Use only the following keys on a calculator: 0 2 3 4 + − × ÷ . Give five different ways to find 120.

3 A school has 16 classes. Each class has 20 to 25 students. Each student writes a letter to a state representative about a recycling problem. What is the greatest number of letters that could be written?

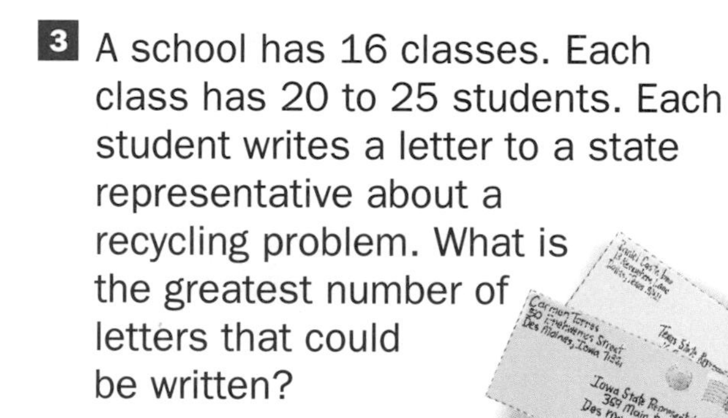

4 **Write a problem** about recycling that can be solved by finding 365 × 25. Did you use mental math, pencil and paper, or a calculator? Why?

5 What is your age in weeks? How did you find the answer?

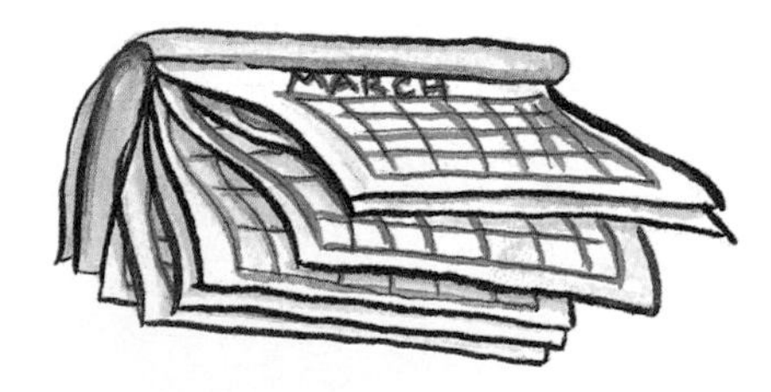

6 A town gives each of 6,750 families a recycling bin for aluminum cans. Each recycling bin costs 95¢. What is the total cost of the recycling bins?

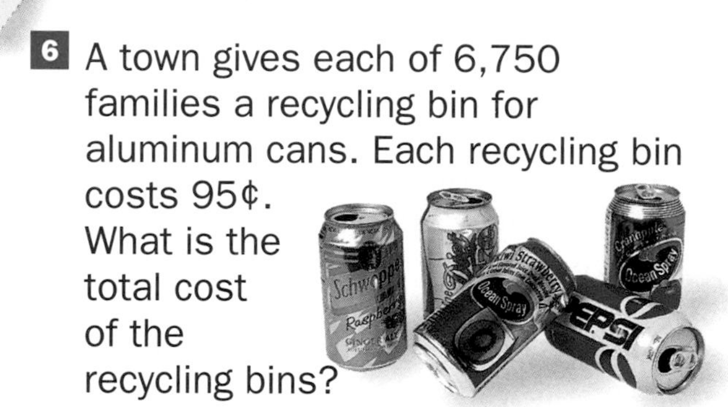

7 **Logical reasoning** Our class has two rules.
Rule 1: Use the computer if you finish your classwork early.
Rule 2: Use the computer for problems where you show graphs.
Today I used the computer. Did I finish my classwork early? Explain.

8 The Morales family recycled 750 pounds of garbage last year. This year they recycled 3 times as much garbage. How many pounds of garbage did they recycle for both years?

Choose two problems and solve them. Explain your methods.

9 **ALGEBRA: PATTERNS** Describe the pattern. Then find the next pair of factors.

16×20
18×18
20×16
22×14

10 Think of a number other than 0. Compare adding 10 repeatedly to that number to multiplying that number by 10 repeatedly. Which way makes the number grow faster? Why?

11 How many different rectangular grids can you make using 240 square units? What are they?

12 **At the Computer** About how much water do you use in a day? in a week? Make a spreadsheet like the one shown here to estimate the amount of water you use each day, week, month, and year.

- Estimate the amount of water you use in a day for each entry.
- Have the computer calculate an estimate of the amount you use each week, month, and year.
- Predict the total amount of water you use in one month before using the computer to find it. Write about what you notice.

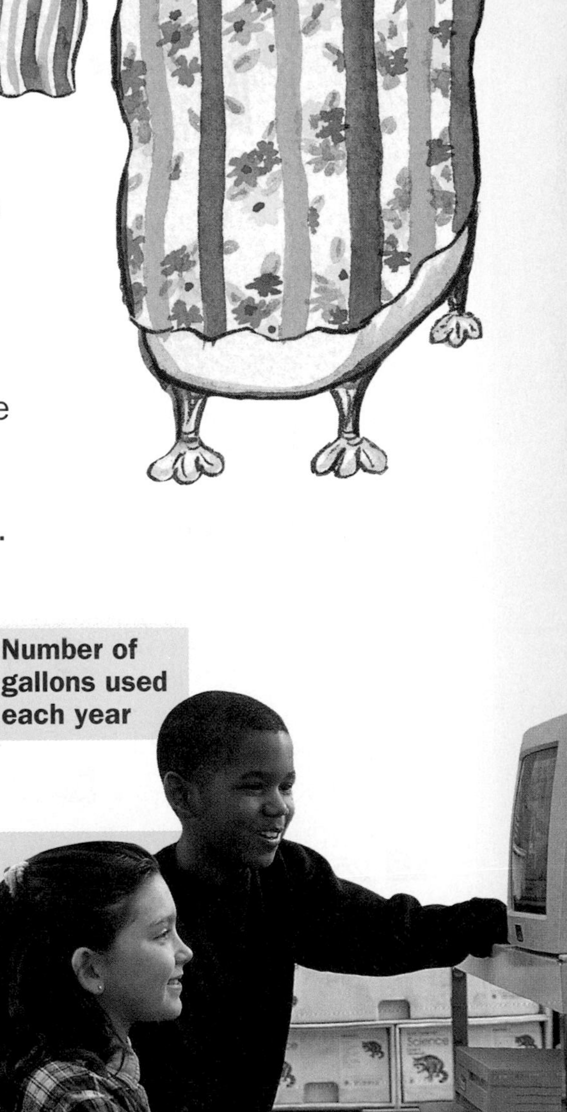

How water is used	Number of gallons used each day	Number of gallons used each week	Number of gallons used each month	Number of gallons used each year
Flushing toilet				
Shower				
Brushing teeth				
Washing dishes				
Total				

Extra Practice, page 507

Language and Mathematics

Complete the sentence. Use a word in the chart. (pages 206–227)

1 The numbers 12 and 40 are ■ of 480.

2 To estimate 45 × 2,689, you can round 45 and 2,689 and then ■.

3 If you change the order of the factors, the ■ is still the same.

4 If you round up both factors, the exact product is less than the ■.

Vocabulary
multiples
multiply
regroup
product
factors
estimate

Concepts and Skills

Without counting, find the total number of squares in the rectangle. (pages 210, 214)

5

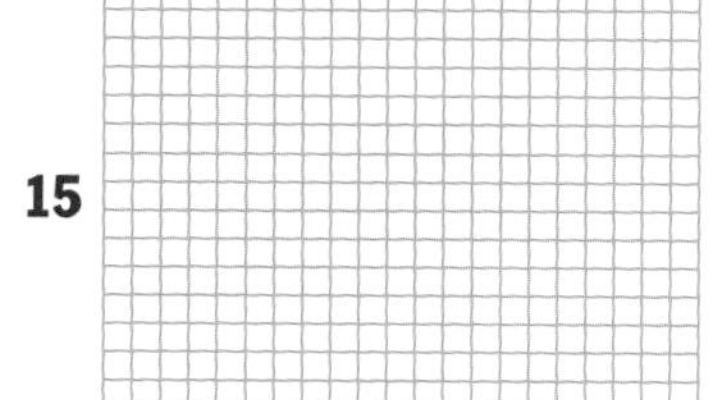

6

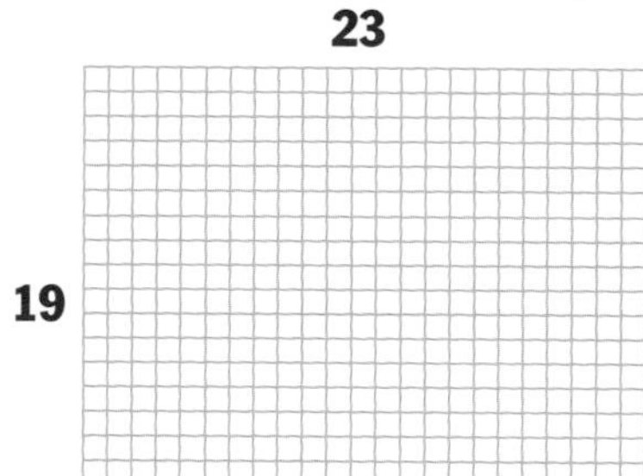

Estimate the product. (page 208)

7 78 × 17 **8** 51 × $98 **9** 52 × $0.44 **10** 47 × 113

11 52 × $621 **12** 11 × $7.81 **13** 81 × 2,079 **14** 19 × 27,100

Multiply. (pages 214, 224, 226)

15 20 × 30 **16** 32 × 11 **17** 50 × 30 **18** 35 × $0.27

19 25 × $0.13 **20** 42 × 98 **21** 56 × $0.90 **22** 77 × 58

23 30 × 90 **24** 48 × 42 **25** $0.45 × 62 **26** 570 × 37 **27** 400 × 50

28 6,098 × 81 **29** $42.50 × 25 **30** 9,604 × 54 **31** 13,800 × 70 **32** 52,000 × 18

Think critically. (page 224)

33 Analyze. Explain what mistake was made, then correct it.

$$\begin{array}{r} {}^{1\,2} \\ {}^{2\,3} \\ 247 \\ \times\ \ 35 \\ \hline 1235 \\ +74100 \\ \hline 75{,}335 \end{array}$$

34 Analyze. How can you find the product of 79 × 245 if you know that the product of 80 and 245 is 19,600?

MIXED APPLICATIONS
Problem Solving

Pencil & Paper | Calculator | Mental Math

(pages 222, 228)

35 The Interact Club has 8 members. Each member invited two friends to Cleanup Day. Each of these two friends invited one friend. How many people attended Cleanup Day? Explain your reasoning.

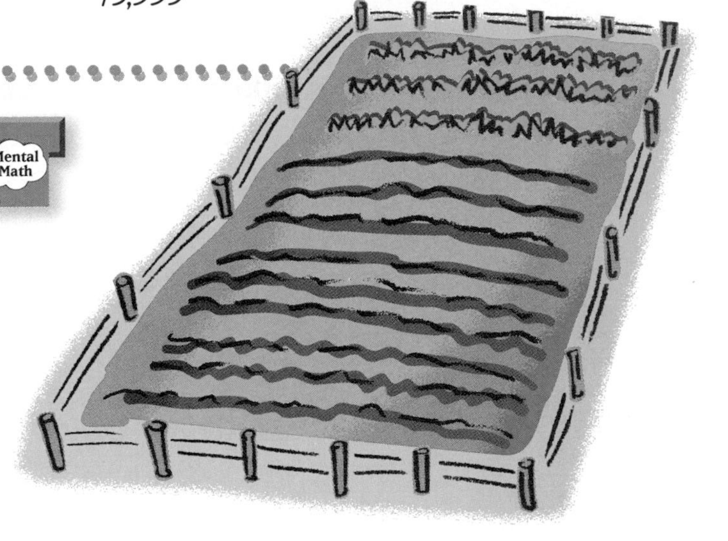

36 The product of two mystery numbers is 280. The sum of these numbers is 47. What are the mystery numbers?

37 **Spatial reasoning** Students planted 91 tree seedlings in the field as shown above. About how many seedlings will fill the field?

Use the sign for problems 38–40.

38 A large company recycles 19 tons of newspaper. How much money will the company collect? (Hint: 1 ton = 2,000 pounds)

39 A community collects 14 tons of aluminum cans and 24 tons of computer paper. How much money will the community collect?

40 How many tons of newspaper would a community have to collect to earn $100?

Sal's Recycling Center

WE BUY FROM YOU!

newspaper	**25¢ for 100 pounds**
computer paper	**7¢ for 100 pounds**
aluminum cans	**42¢ for 1 pound**
cast aluminum	**22¢ for 1 pound**
copper	**75¢ for 1 pound**
steel	**15¢ for 1 pound**

HOURS

Monday through Friday	7 A.M. to 5 P.M.
Saturday	8 A.M. to 3 P.M.

Estimate the product.

1 58×286 2 94×103 3 37×91 4 $48 \times \$0.88$

5 $18 \times \$7.66$ 6 68×335 7 $22 \times \$58.19$ 8 $28 \times 3{,}053$

9 81×457 10 77×915 11 $43 \times \$629$ 12 $15 \times 2{,}057$

Multiply. Use mental math when you can.

13 90×70 14 $\$5.73 \times 28$ 15 305×46 16 $\$72.60 \times 92$ 17 $8{,}005 \times 80$

18 68×41 19 50×20 20 $68 \times \$0.41$ 21 81×11

22 56×35 23 80×500 24 20×80 25 $30 \times \$100$

26 63×67 27 $48 \times 7{,}052$ 28 $87 \times \$39.04$ 29 $85 \times \$74{,}009$

Solve. Use the table for problems 30–31.

Average Content of 100 Pounds of Trash	
Paper products	39 pounds
Glass	6 pounds
Metals	8 pounds
Plastics	9 pounds
Wood	7 pounds
Food	7 pounds
Yard waste	15 pounds
Other	9 pounds

30 About how many times as many pounds of paper products are in the trash as metals?

31 If your family throws away 5,200 pounds of trash, do you expect to throw away more than 550 pounds of plastics? Show two ways to solve. Explain your reasoning.

32 Tara recycled 23 cans on Wednesday and 36 cans on Friday. She got 15¢ for each can. Does she have enough money to buy a book for $12.95? Explain.

33 Elle is 6 years older than Ming. The product of their ages is 216. Faye is 2 years younger than Elle. How old are Elle, Faye, and Ming?

What Did You Learn?

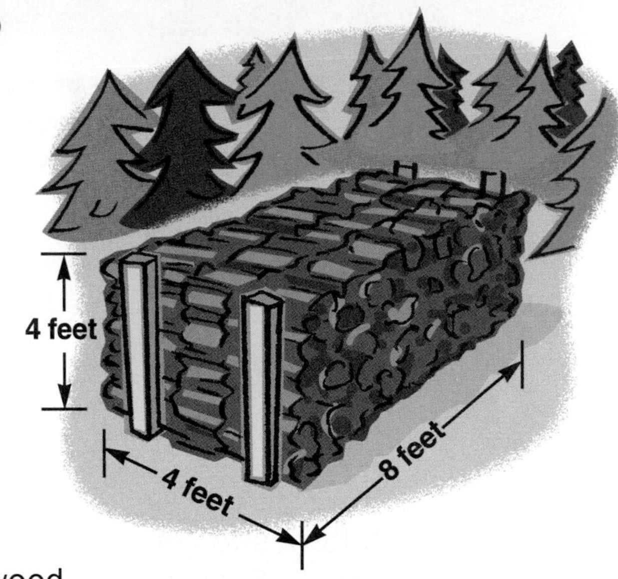

A Cord of Wood

A cord of wood can make about:

a. 1,500 pounds of paper.
b. 942 books, each weighing a pound.
c. 61,370 business envelopes.
d. 1,200 copies of *National Geographic.*
e. 2,700 copies of a daily newspaper.

Tell what you can make with the following. Give one example for each. Show the method you used to solve each problem and explain why you chose that method.

1 10 cords of wood

2 17 cords of wood

3 29 cords of wood

4 40 cords of wood

A Good Answer

- includes a response for each amount of wood
- shows accurate calculations and clearly explains why each method was chosen

You may want to place your work in your portfolio.

What Do You Think

1 How do you use what you know about multiplying by 1-digit numbers to help you multiply by 2-digit numbers?

2 If you want to find 13×356, which methods would you use?

- Use pencil and paper.
- Use a calculator.
- Use a diagram or model.
- Other. Explain.

3 Are you always able to break apart numbers to help you multiply? Why or why not?

math science technology

CONNECTION

BATTERY POWER

Cultural Note

An Italian scientist, Alessandro Volta, is believed to have invented the first working battery in 1800. The volt, a unit used to measure electrical energy, is named after him.

People use batteries to get energy for many things including radios, CD players, flashlights, cameras, and toys. The materials in batteries, however, can be harmful to the environment.

After they have been used up, regular and heavy-duty flashlight batteries are useless.

Alkaline batteries last longer than regular batteries but they also cost more.

Rechargeable batteries are expensive but they can be recharged as many as 1,000 times.

Batteries also come in different sizes.

Which battery is best? You can conduct your own experiment.

Place two of each type of size-D battery in a flashlight. Record the time it takes to use up each type of battery. Show your results in a table.

Type of Battery	Time	Cost of 2 Batteries
Regular	1 hour	$ 1.00
Heavy duty		
Alkaline		
Rechargeable		

You will need
- *4 flashlights*
- *2 of each kind of size-D battery: regular, heavy duty, alkaline, rechargeable*

▶ How else can you get energy besides using a battery?

Comparing Batteries

Use the results from your experiment.

1. How many regular batteries would it take to get the same amount of energy as two alkaline batteries?

2. Find and compare the prices of the different types of size-D batteries. Is it better to buy the heavy-duty batteries or the regular batteries? Explain your reasoning.

3. **What if** a charger for the rechargeable batteries costs $20.00. What is the cost of two rechargeable batteries and the charger? How many alkaline batteries could you buy for this amount? How long would this number of alkaline batteries last?

4. Which battery would be best to use in a flashlight? Explain your reasoning.

At the Computer

5. Write a lab report describing the experiment and the results. Tell which type of battery you think is best and explain why.

6. Make a spreadsheet that calculates the cost for each hour of use of each different type of battery.

Type of Size-D Battery	Cost of Batteries	Time Taken (hours)	Cost for each Hour of Use
Regular			
Alkaline			

cumulative review: chapters 1–6

TEST PREPARATION

Choose the letter of the best answer.

1 Choose the best estimate.
Seth can walk 4 miles in about ■ minutes.

A 8
B 80
C 800
D 8,000

2 All these numbers are:

24	32	16	8	48	3

F even.
G less than 50.
H odd.
J 2-digit numbers.

3 Find the missing number.

$$\begin{array}{r} 685 \\ +\ 14■ \\ \hline 831 \end{array}$$

A 0
B 4
C 6
D 7

4 Which names the same number as $(4 \times 5) \times 7$?

F $4 + (5 \times 7)$
G $4 \times (5 \times 7)$
H $4 \times (5 + 7)$
J $(4 \times 5) + 7$

5 If June 8th is on a Saturday, on which day is June 27th?

A Saturday
B Tuesday
C Sunday
D Thursday

6 Which statement is true?

F $34 \times 2{,}454 < 54 \times 1{,}414$
G $\$6{,}000 - \$189 > \$5{,}900 - \$99$
H $435 + 1{,}254 + 23 = 1{,}919$
J $9 \times 689 > 6{,}300$

7 If one number is removed from each box and placed in the other, the products would be equal. Which numbers should be moved?

10×60	5×30

A 60, 30
B 10, 30
C 60, 5
D 1, 3

8

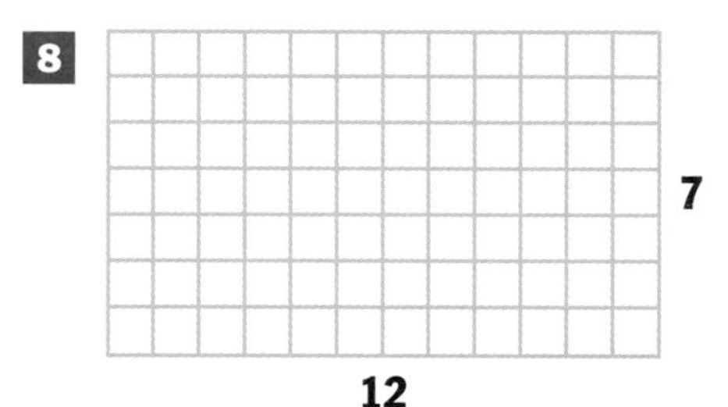

Find the total number of squares in the rectangle.

F 84
G 72
H 70
J 80

9 Find the difference between these products:
21 × 37 and **27 × 31**

A 0
B 40
C 60
D 80

10 You have saved $13 toward a bicycle. What other information is needed to determine how much more money you need to buy it?

- **F** The amount you save each day
- **G** The cost of the bicycle
- **H** The amount you started with
- **J** The cost of a helmet

11 What is shown on the number line?

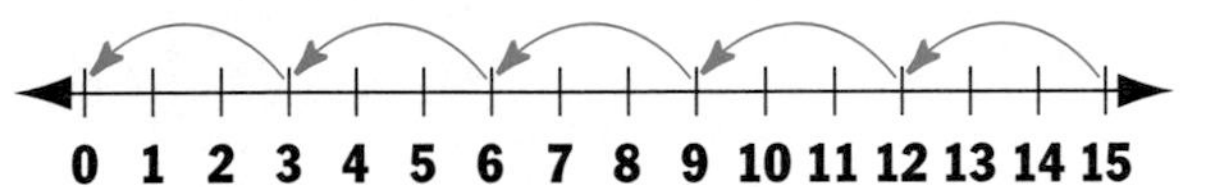

- **A** $15 - 5 = 3$
- **B** $5 + 3 = 15$
- **C** $15 \div 3 = 5$
- **D** $15 - 5 = 10$

12 Mike leaves home at 3:00 P.M. and shops for 1 hour and 45 minutes. If he needs 40 minutes to travel to the school play, which starts at 5:30 P.M., he will be ■.

- **F** 10 minutes early
- **G** on time
- **H** 5 minutes early
- **J** not given

13 Louise spent $4.60 on juice and fruit rolls. Juice sells for 35¢ a can, and fruit rolls are 50¢ each. How many of each did she buy?

- **A** 2 juice cans and 8 fruit rolls
- **B** 5 juice cans and 6 fruit rolls
- **C** 6 juice cans and 5 fruit rolls
- **D** 8 juice cans and 2 fruit rolls

14 Which number comes next?
8, 16, 24, 32, ■

- **F** 48
- **G** 8
- **H** 40
- **J** 34

15 Ty pays for a $0.99 book cover and a $1.75 pen with 3 one-dollar bills. What change might she get back?

- **A** a quarter and a dime
- **B** three dimes and a penny
- **C** five nickels and a penny
- **D** four nickels and four pennies

Use the graph for problem 16.

Paper Products at the Recycling Center	
Cardboard	■■■■■■ ■■■■■■■
Newspaper	■■■■■■
Office/ commercial paper	■■■
Books and magazines	■■■

Key: Each ■ stands for 2,000 pounds.

16 Which statement does *not* describe the data in the graph?

- **F** The recycling center has 50,000 pounds of paper products.
- **G** There is more cardboard than any other kind of paper material at the recycling center.
- **H** Recycling paper saves trees!
- **J** There is about twice as much cardboard as newspaper at the recycling center.

CHAPTER

7 MEASUREMENT

THEME

Under the Sea

Have you ever wanted to explore the oceans and seas? In this chapter, you will measure things from underwater as well as things around you. You will see why measurement is important.

What Do You Know ?

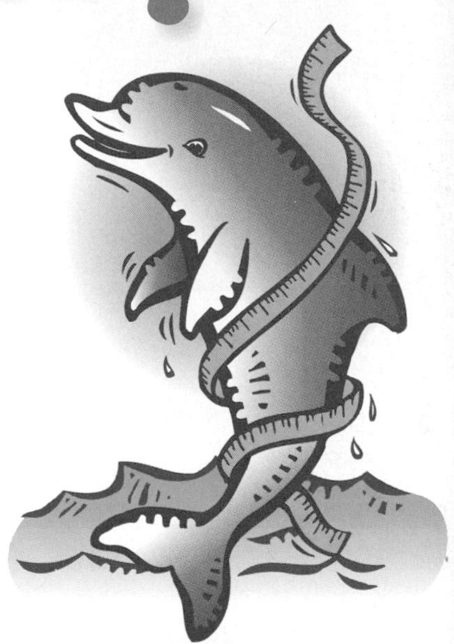

Scientists measure dolphins to learn more about how they grow.

1 What different types of measurements could scientists take on a dolphin?

2 What measurement tools could scientists use to collect dolphin measurements?

3 Portfolio A 15-foot-long bottle-nosed dolphin gave birth to a 5-foot-long calf. Name at least three things that are about the same length as each of the dolphins. Explain how you made your choices.

Use Diagrams A dolphin is 14 feet long and a fourth-grade student is 4 feet tall. Draw a diagram to compare the length of the dolphin to the height of the student.

A diagram is a drawing that shows what something looks like or how it works. A diagram often has labels to help you understand it better.

1 What does the diagram show you about the length of the dolphin and height of the student?

Vocabulary*

*partial list

length, p.242
customary, p.242
inch (in.), p.242
foot (ft), p.243
yard (yd), p.243
mile (mi), p.243
metric, p.246
meter (m), p.246
kilometer (km), p.246
perimeter, p.250
capacity, p.256
cup (c), p.256
pint (pt), p.256
quart (qt), p.256
gallon (gal), p.256
weight, p.257
ounce (oz), p.257
pound (lb), p.257
milliliter (mL), p.262
liter (L), p.262
mass, p.263
gram (g), p.263
kilogram (kg), p.263
degrees Fahrenheit (°F), p.266
degrees Celsius (°C), p.266

Length in Customary Units

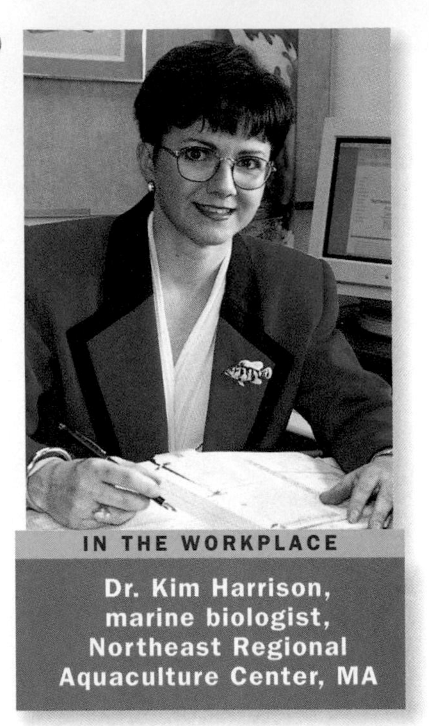

IN THE WORKPLACE

Dr. Kim Harrison, marine biologist, Northeast Regional Aquaculture Center, MA

Marine biologists estimate the lengths of sea animals to see if they are healthy.

An **inch (in.)** is a **customary** unit used to measure short lengths.

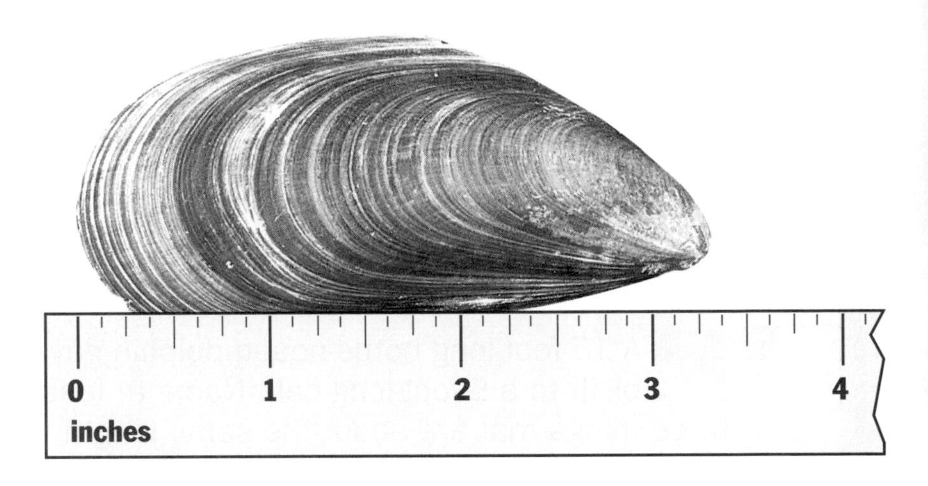

Check Out the Glossary
For vocabulary words
See page 544.

The length of the mussel is:

3 in. to the nearest inch.

$3\frac{1}{2}$ in. to the nearest half inch.

$3\frac{1}{4}$ in. to the nearest quarter inch.

Work Together

Practice estimating. Work with a partner to estimate and then measure some lengths in your classroom to the nearest inch.

You will need
- *inch ruler or measuring tape*

Record your work in a table.

Object	Estimate	Actual Measurement

▶ What methods did you use to estimate? Compare your methods with those of other pairs.

▶ How did your estimates compare to your actual measurements?

Make Connections

You can also use the units **foot (ft), yard (yd),** and **mile (mi)** to measure longer lengths.

▶ Name three things you would measure with each unit—inch, foot, yard, mile.

Item	Unit Used for Measurement
Carpet	foot
Football field	yard
Distance between cities	mile

CHECK

Check for Understanding

Estimate and then measure. Explain your methods.

1 your stride

2 your height

3 your arm span

Critical Thinking: Generalize **Explain your reasoning.**

4 When would you choose to measure to the nearest half or quarter inch?

PRACTICE

Practice

Estimate and then measure.

1 length of your foot

2 distance you can reach

3 How could you use this part of your thumb to estimate length?

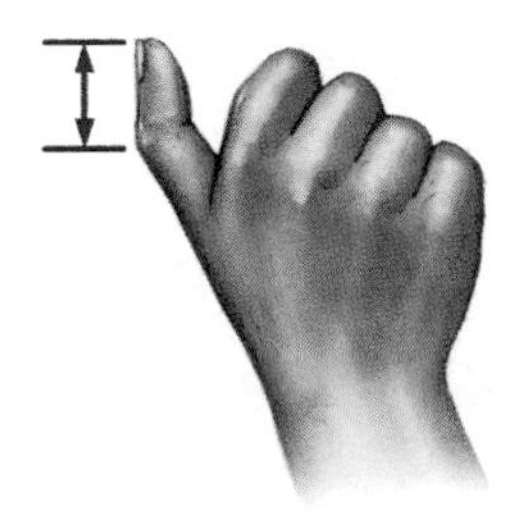

4 How could you use this part of your arm to estimate length?

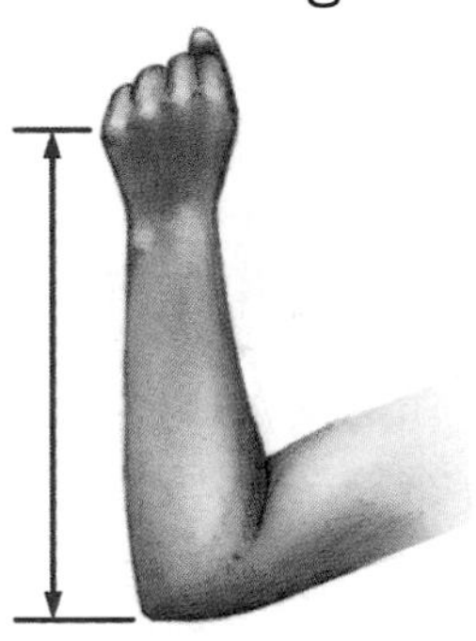

Write the letter of the best estimate.

5 height of a classmate
a. 50 ft **b.** 50 in. **c.** 50 yd

6 length of a school bus
a. 27 ft **b.** 27 in. **c.** 27 yd

7 distance around a running track
a. 440 mi **b.** 440 ft **c.** 440 yd

8 depth out in the middle of the ocean
a. 6 ft **b.** 6 mi **c.** 6 in.

9 Marta measured the length of her book to the nearest quarter inch as $13\frac{3}{4}$ in. What does it measure to the nearest inch?

10 Don can walk a mile in about 15 minutes. About how long would it take him to walk 5 miles?

Rename Customary Units of Length

LEARN

A carp is about 29 in. long. The giant American lobster is about 3 ft long. Which is longer?

You can rename units to make comparisons.

Cultural Note
In Japan, families fly carp-shaped wind socks as a wish for healthy children.

Compare 3 ft and 29 in.

Think: 1 ft = 12 in.

3 ft = 3 × 12 in. = 36 in.

36 in. > 29 in.

The lobster is longer.

12 **inches (in.)**	= 1 **foot (ft)**
3 **feet (ft)**	= 1 **yard (yd)**
1,760 **yards (yd)**	= 1 **mile (mi)**

More Examples

A Complete.
The marlin is about 8 ■ long.
a. inches
b. feet
c. yards

Think: The marlin is a little larger than the man.

The marlin is about 8 ft long.

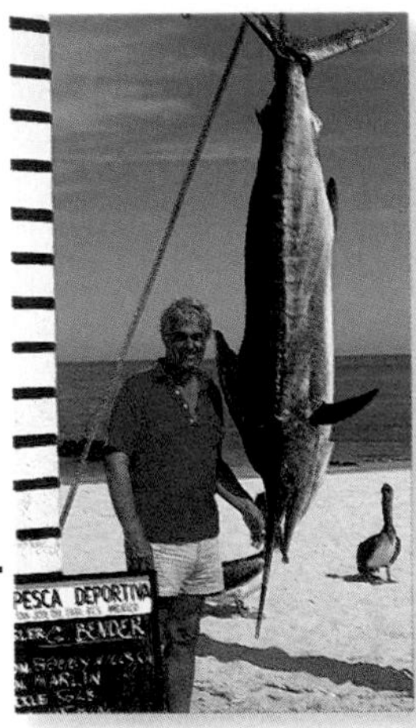

B Suppose you take 1-yd-long strides as you walk. How many strides will you take to walk 3 mi?

Complete.

3 mi = ■ yd **Think:** 1 mi = 1,760 yd

3 mi = 3 × 1,760 yd
= 5,280 yd

You will take 5,280 strides to walk 3 mi.

CHECK

Check for Understanding

Complete.

1 4 ft = ■ in. **2** 12 ft = ■ yd **3** 4 mi = ■ yd **4** 7 yd = ■ ft

Critical Thinking: Analyze

Explain your reasoning.

5 A Nassau grouper is 48 in. long. Tell how you would find this length in feet.

Practice

ALGEBRA: PATTERNS **Copy and complete the table.**

1

Feet	1	2	3	4	■
Inches	12	■	■	■	60

2

Yards	1	2	3	■	5
Feet	3	■	■	12	■

Complete.

3 12 yd = ■ ft **4** 5 mi = ■ yd **5** 9 ft = ■ in. **6** 96 in. = ■ ft

7 15 ft = ■ yd **8** 120 in. = ■ ft **9** 300 ft = ■ yd **10** 3,520 yd = ■ mi

Compare lengths. Write >, <, or =.

11 6 ft ● 60 in. **12** 4 yd ● 12 ft **13** 5,000 yd ● 5 mi **14** 20 ft ● 6 yd

15 Write the names in order from shortest to tallest height.

Jo Beth—70 in.	Kay—5 ft
Marcus—59 in.	Tom—6 ft

MIXED APPLICATIONS
Problem Solving

Use the pictures of the sea animals for problems 16–17.

Porkfish
13 in.

Atlantic Albacore Tuna
45 in.

Deep Sea Shrimp
6 in.

16 Which animals are longer than 1 foot?

17 Which animal is longer than 1 yard? How much longer?

18 Patti buys 2 *Great Underseas* videos. Each one costs $19.95 including tax. She gives the clerk 2 twenty-dollar bills. What is the cost? How much change will Patti get?

19 **Make a decision** You want to buy a rug. Your floor space is 108 in. wide and 144 in. long. What other units could you use? Which unit will you use when you go to buy the rug?

mixed review • test preparation

1 $925 + 637

2 $6.51 − 0.67

3 $60.71 × 9

4 $45 × 30

5 $9.99 × 12

Length in Metric Units

Centimeter (cm) and **millimeter (mm) are metric units of length.**

The length of this Siamese fighting fish is:
6 cm to the nearest centimeter.
64 mm to the nearest millimeter.

You can also use the units **decimeter (dm), meter (m),** and **kilometer (km)** to measure longer lengths.

10 **millimeters (mm)**	= 1 **centimeter (cm)**
10 **centimeters (cm)**	= 1 **decimeter (dm)**
10 **decimeters (dm)**	= 1 **meter (m)**
1,000 **meters (m)**	= 1 **kilometer (km)**

Work Together

Choose one object in your classroom that you would use each unit to measure—millimeter, centimeter, decimeter, and meter. Work with a partner to estimate and then measure those lengths and heights.

You will need
- *centimeter ruler*
- *meterstick or measuring tape*

Record your work in a table.

Object	Estimate	Actual Measurement

Talk It Over

- Explain your choices.
- How did your estimates compare to your actual measurements?

Check Out the Glossary
For vocabulary words
See page 544.

Make Connections

You can rename units to make comparisons.

Sam found that the width of the classroom doorway is 95 cm. Could a sea turtle that is 9 dm wide fit through the doorway?

Compare 9 dm and 95 cm.

Think: 1 dm = 10 cm

$9 \text{ dm} = 9 \times 10 \text{ cm} = 90 \text{ cm}$
$90 \text{ cm} < 95 \text{ cm}$

The sea turtle will fit through the doorway.

More Examples

A Complete.
The neon goby is about 5 ■ long.
a. mm **b.** cm **c.** dm

Think: The neon goby is about the length of your little finger.

A neon goby is about 5 cm long.

B How many kilometers do you walk if you walk 3,000 m?

Complete.
3,000 m = ■ km

Think: 1,000 m = 1 km
2,000 m = 2 km
3,000 m = 3 km

You walk 3 km.

Check for Understanding

CHECK

Estimate and then measure. Explain your methods.

1 your hand **2** your height **3** length of your stride

Complete.

4 7 m = ■ cm **5** 4 m = ■ dm **6** 60 mm = ■ cm **7** 8,000 m = ■ km

Critical Thinking: Analyze

Explain your reasoning.

8 Could a 4-dm-wide giant spider crab fit on a table or desk in your classroom? If so, where?

PRACTICE

Practice

Estimate and then measure.

1 your arm span

2 distance you can reach

Write the letter of the best estimate.

3 length of your stride — **a.** 7 dm **b.** 7 cm **c.** 7 m

4 distance a plane travels — **a.** 7 dm **b.** 70 cm **c.** 700 km

5 length of a baseball bat — **a.** 100 cm **b.** 80 km **c.** 3 m

6 distance around a running track — **a.** 4 km **b.** 400 km **c.** 400 m

ALGEBRA: PATTERNS Copy and complete the table.

7

Meters	1	2	■	4	5	■
Decimeters	10	■	■	40	■	60
Centimeters	100	■	300	■	■	■

Complete.

8 7 cm = ■ mm

9 8 m = ■ cm

10 4 km = ■ m

11 10 m = ■ dm

12 50 dm = ■ m

13 9,000 m = ■ km

14 100 cm = ■ dm

15 100 mm = ■ cm

16 12 dm = ■ cm

17 200 mm = ■ cm

18 15 m = ■ dm

19 1,000 cm = ■ m

Write >, <, or =.

20 60 cm ● 6 m

21 90 cm ● 9 dm

22 732 cm ● 7 m

23 15 cm ● 150 mm

24 32 cm ● 300 mm

25 4,000 m ● 40 km

26 50 dm ● 500 cm

27 250 mm ● 20 cm

28 40 cm ● 5 dm

Make It Right

29 Cory wrote these names in order from shortest to tallest. Tell what the error is and correct it.

Mark	2 m
Roy	14 dm
Sandra	165 cm

MIXED APPLICATIONS

Problem Solving

Use the bar graph for problems 30–35.

30 Which sharks are less than 500 cm long?

31 Which sharks are longer than 1,000 cm?

32 Which shark is between 4 m and 6 m in length?

33 Which shark is about as tall as your classroom?

34 Which shark is about as long as your classroom?

35 How many meters longer is a whale shark than a basking shark? How many centimeters is that?

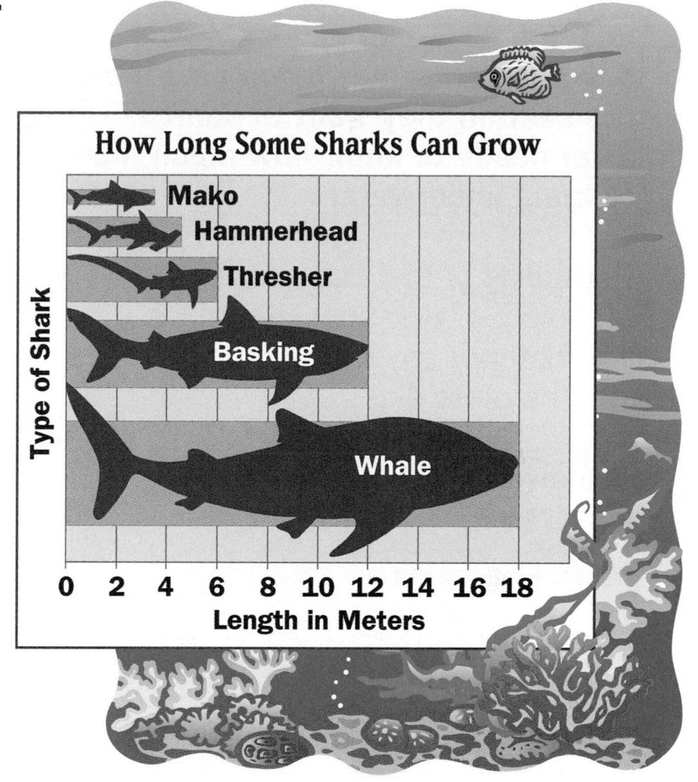

36 Alicia took a half hour to walk to the aquarium. She watched a dolphin show for 45 minutes. The show ended at 6:15 P.M. When did she leave to go to the aquarium?

37 A ticket to the Shark Show costs $1.50. Orlando buys 15 tickets. He pays with a twenty-dollar bill and a ten-dollar bill. How much change does he get?

38 Dom can walk 1 km in about 20 minutes. About how long would it take him to walk 5 km? Explain your reasoning.

39 **Write a problem** about buying a shark poster for your classroom. Pick a place for it. Ask a classmate to pick a size for the poster.

mixed review • test preparation

1 $\$14.25 + \$8.30$

2 $\$180 - \$120.95$

3 $\$17.45 - \$9.80$

4 $4 \times \$12.50$

5 $32 \times \$75$

6 $20 \times \$9.95$

7 $56 \div 7$

8 $36 \div 4$

9 $72 \div 8$

Extra Practice, page 509

Perimeter

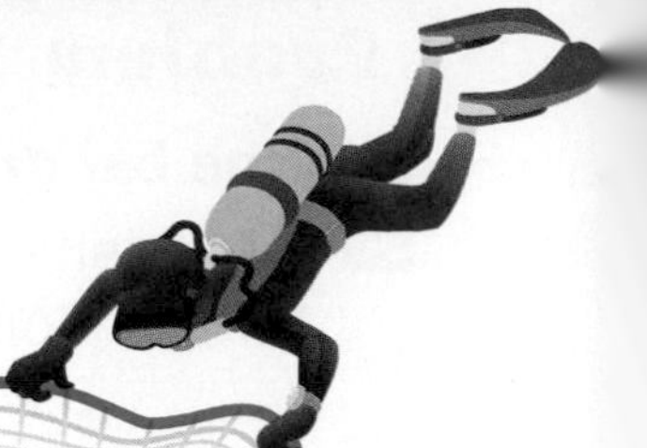

LEARN

Imagine diving for lost treasure! Salvage companies place rope around the section they want to search. The diver needs to know how much rope to bring underwater.

Perimeter is the distance around an object or a shape. To find the perimeter of any shape, add the lengths of its sides.

The diver needs 380 ft of rope.

Search Section	
Length	140 ft
Length	140 ft
Width	50 ft
Width	+ 50 ft
Perimeter	380 ft

More Examples

A

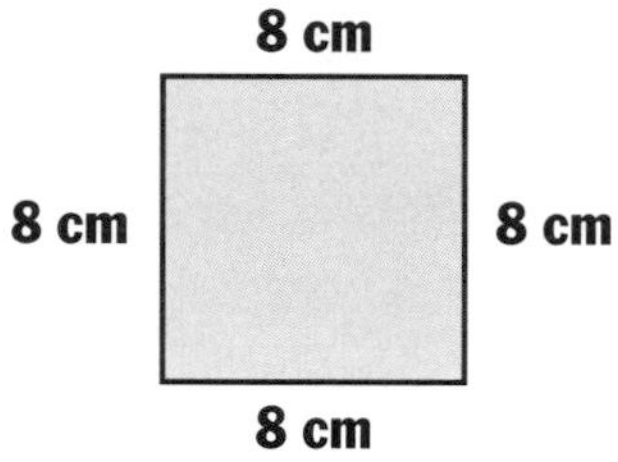

8 cm
8 cm
8 cm
+ 8 cm
32 cm

Perimeter = 32 cm

B

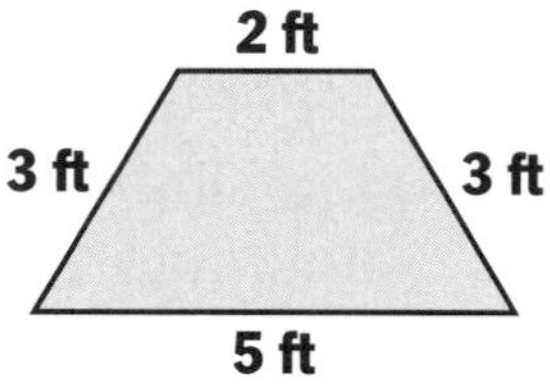

2 ft
3 ft
3 ft
+ 5 ft
13 ft

Perimeter = 13 ft

CHECK

Check for Understanding

Find the perimeter.

1

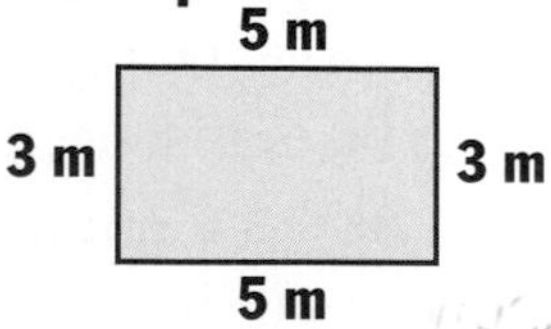

2

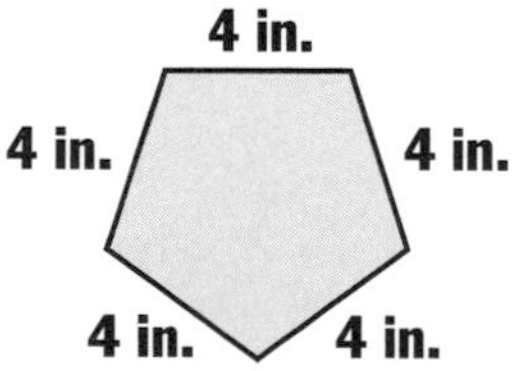

3

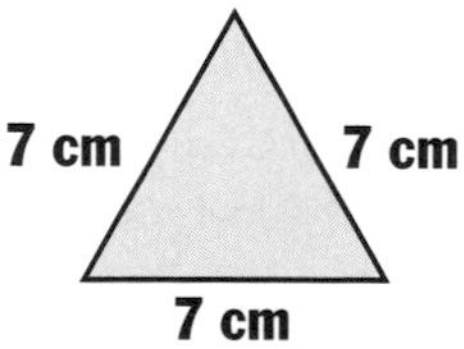

Critical Thinking: Analyze **Explain your reasoning.**

4 How many addends will you have if you find the perimeter of:
a. a figure with 6 sides?
b. a figure with 18 sides?

5 Journal Measure the perimeter of your classroom. Explain how you did it.

Check Out the Glossary
For vocabulary words
See page 544.

Practice

Find the perimeter.

1

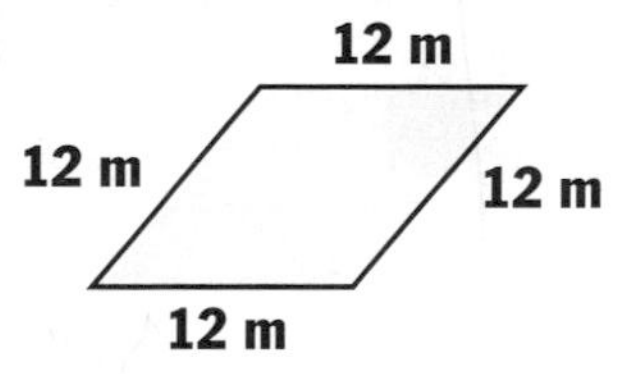

2

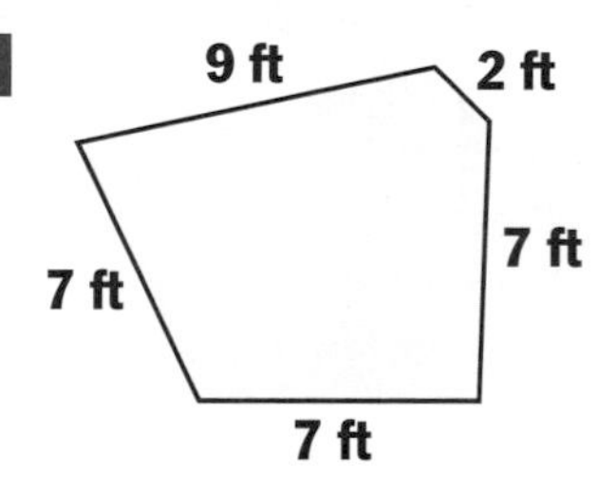

3

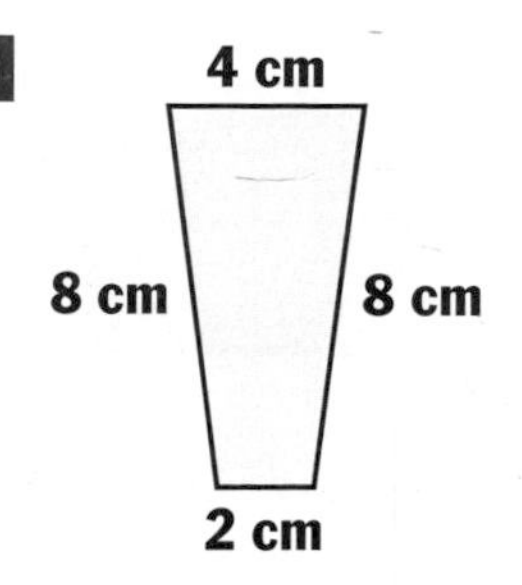

ALGEBRA **Find the length of the missing side.**

4

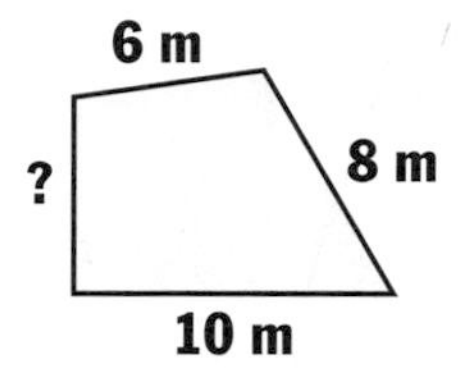

Perimeter = 30 m

5

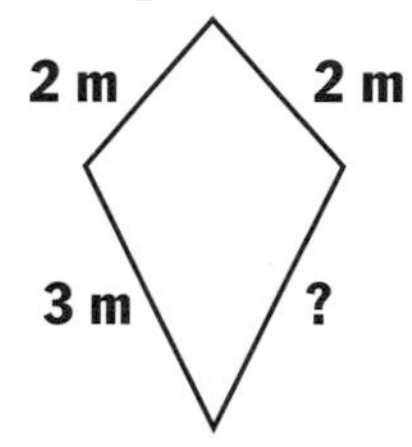

Perimeter = 10 m

6

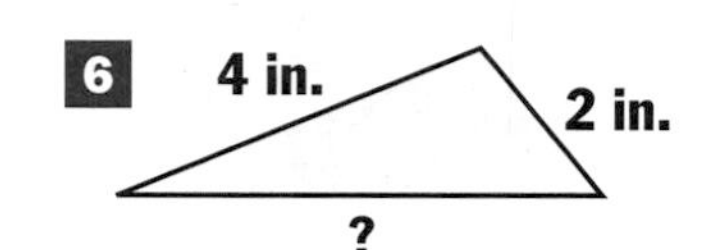

Perimeter = 11 in.

MIXED APPLICATIONS

Problem Solving

7  READING ARITHMETIC WRITING **Use Diagrams** Draw a diagram to show the distance a diver swims around a rectangular pool measuring 12 m by 8 m.

8 Divers use a waterproof notepad that costs $12.50. How much would it cost to supply a diving crew with 4 notepads?

more to explore

Squares and Rectangles

You can find the perimeter of a square if you know the length of one side.

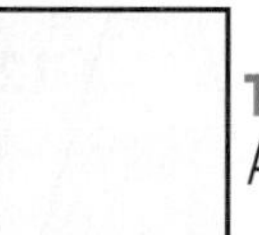

Think:
A square has four equal sides.

Perimeter = (3 + 3 + 3 + 3) yd
= 12 yd

You can also find the perimeter of a rectangle if you know the length of two sides that touch.

2 m
7 m

Think:
The opposite sides of a rectangle are equal.

Perimeter = (7 + 2 + 7 + 2) m
= 18 m

▶ How can you use multiplication to find the perimeters of the square and rectangle shown?

Extra Practice, page 509

midchapter review

Estimate and ther. measure the length to the nearest inch.

1 

2 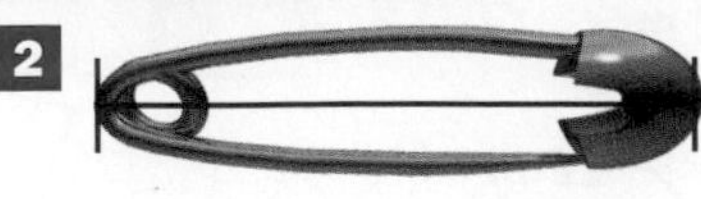

Estimate and then measure the length to the nearest centimeter.

3 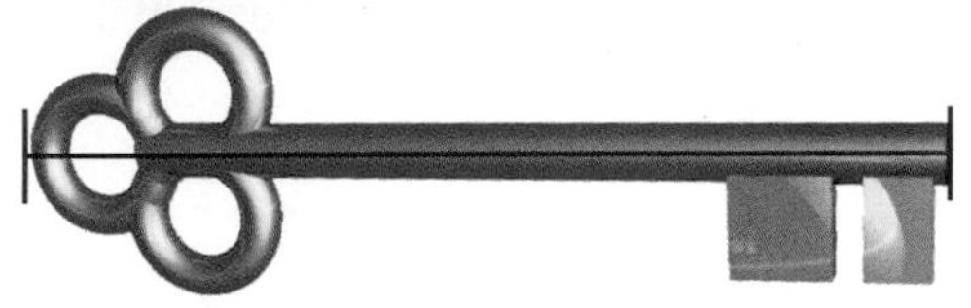

4 

Write the letter of the best estimate.

5	length of tennis racket	**a.** 3 ft	**b.** 3 in.	**c.** 3 mi
6	height of fire hydrant	**a.** 80 mm	**b.** 80 m	**c.** 80 cm
7	width of baseball card	**a.** 3 ft	**b.** 3 in.	**c.** 3 yd
8	distance between two cities	**a.** 250 cm	**b.** 250 dm	**c.** 250 km

Compare lengths. Write >, <, or =.

9 480 cm ● 4 m

10 4 ft ● 72 in.

11 18 ft ● 6 yd

Complete.

12 6 ft = ■ in.

13 18 ft = ■ yd

14 50 cm = ■ dm

Write the measurements in order from shortest to longest.

15 13 in., 5 ft, 1 yd, 2 ft, 72 in.

16 2 km, 19 cm, 25 mm, 3 dm, 1 m

Find the perimeter.

17 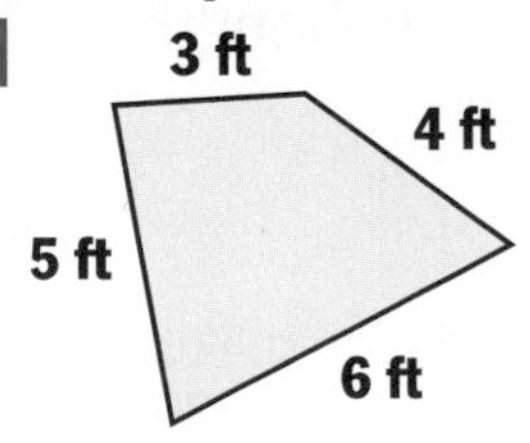

18 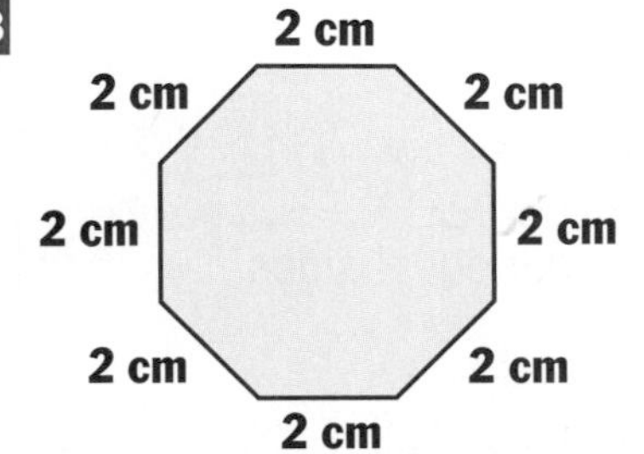

19 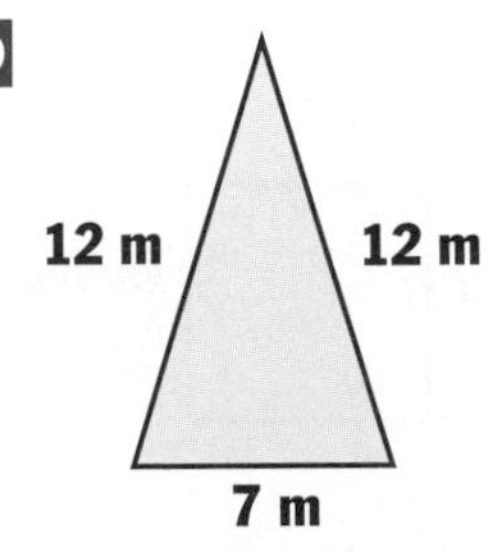

20 Journal Suppose you want to hang four posters evenly on your wall. What measuring tools would you use? What units of measurement would you use? Tell why.

Use Benchmarks

You can use common objects or parts of your body as *benchmarks* to help you estimate the lengths of other things.

Some Metric Benchmarks	
Length, Width, or Height	**About . . .**
Width of little finger	1 cm
Width of hand	1 dm
Width of cassette tape	1 dm
Length of stride	1 m
Height from toes to waist	1 m
Width of door	1 m
Height of door from floor to doorknob	1 m

1 Which of the body benchmarks in the table work for your body? Which do not?

2 Choose three objects whose lengths you would estimate by using the width of your little finger. Why did you choose them?

3 Choose three objects whose lengths you would estimate by using the width of your hand and three by using the length of your stride. Are these the same or different from the lengths you chose in problem 2? Why?

Tell how you would estimate the length, width, or height by using one or more of the benchmarks in the table.

4 height of a chair

5 length of a dining room table

6 length and width of a telephone book

7 width of a room

8 length of a hallway

9 height of a piano

TEMPERATURE AND SALINITY OF Ocean Water

Did you know that the temperatures of oceans in different parts of the world vary? How do you think the temperature affects how well salt dissolves in the water?

Make your own samples to test how well salt dissolves in water of different temperatures.

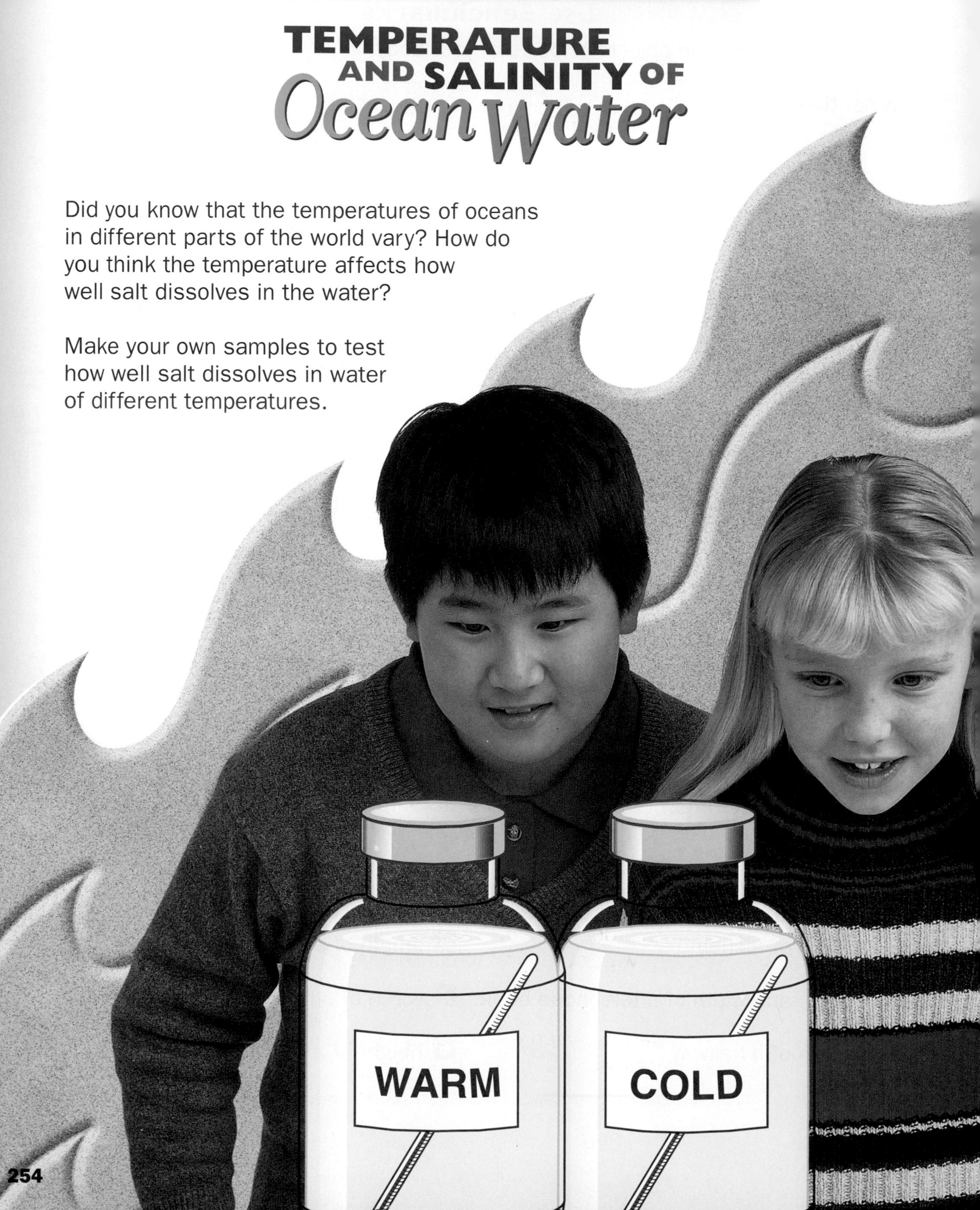

DECISION MAKING

Dissolving Salt

1. Work with a small group. Decide what each member of your group will do during the activity and how you will measure the temperature and amount of salt that is dissolved.

2. Fill two containers with 1 quart of water. Let one container sit in the classroom in sunlight. Place the other container in the refrigerator.

3. After 20 minutes, record the temperature of each container of water.

4. Add 1 level teaspoon of salt to each container at the same time. After each teaspoon is added, stir the containers until the salt is completely dissolved.

5. Continue to add 1 teaspoon of salt to each container until the salt will not dissolve completely. Record your findings in a table.

Reporting Your Findings

6. Portfolio Prepare a poster of the data and conclusions from your experiment. Include:

   - a description of the experiment and an explanation of how to tell when salt is completely dissolved.
   - a bar graph that shows the amount of salt that was dissolved in each container of water.
   - a sentence that describes how water temperature affects how well salt dissolves in water.

7. Compare your results with those of other groups. Did they draw the same conclusion?

Revise your work.

- Does your display include all three parts?
- Is your graph clearly labeled and easy to read?
- Did you proofread your work?

MORE TO INVESTIGATE

PREDICT whether more or less salt will dissolve in a container of boiling water than in the containers in your experiment.

EXPLORE how salt affects the freezing temperature of water.

FIND how salt in ocean water helps fish on sunny but very cold winter days.

Capacity and Weight in Customary Units

LEARN

If you fill a 5-gallon aquarium with only 1 gallon of water, it is still a 5-gallon aquarium. The size of the aquarium has not changed. It can still hold 5 gallons.

Capacity is the amount of liquid a container can hold.

1 fluid ounce (fl oz)

1 cup (c)

1 pint (pt)

1 quart (qt)

1 gallon (gal)

8 **fluid ounces (fl oz)**	= 1 **cup (c)**
2 **cups (c)**	= 1 **pint (pt)**
2 **pints (pt)**	= 1 **quart (qt)**
4 **quarts (qt)**	= 1 **gallon (gal)**

Check Out the Glossary
For vocabulary words
See page 544.

How many quart containers of water does it take to fill a 5-gal aquarium?

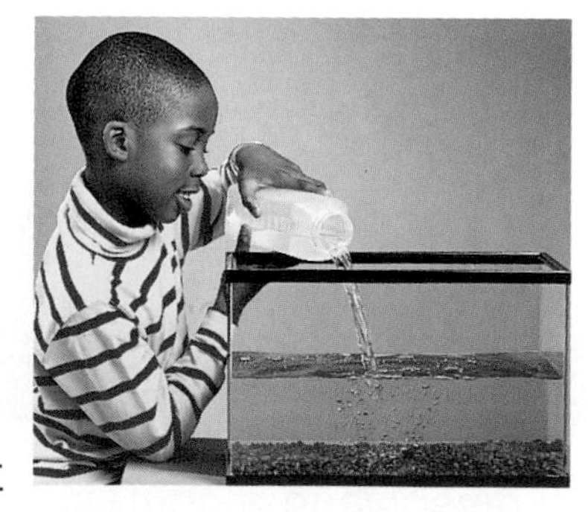

Think:
1 gal = 4 qt

$$5 \text{ gal} = 5 \times 4 \text{ qt}$$
$$= 20 \text{ qt}$$

It would take 20 quart containers.

A juice container holds 64 fl oz. How many cups is that?

Think: 1 c = 8 fl oz

$$64 \text{ fl oz} = 64 \text{ fl oz} \div 8$$
$$= 8 \text{ c}$$

That is 8 cups of juice.

▶ What do you think is the capacity of a drinking glass? a kitchen sink? Why? Measure and compare to your estimate.

A gallon of water weighs 8 pounds.

16 **ounces (oz)** = 1 **pound (lb)**

Weight is the amount of heaviness of an object.

1 ounce (oz)

1 ounce (oz)

1 pound (lb)

40 pounds (lb)

- Hold a dime in your hand. What items can you find in your classroom that weigh about the same as the dime? Use a scale to measure.

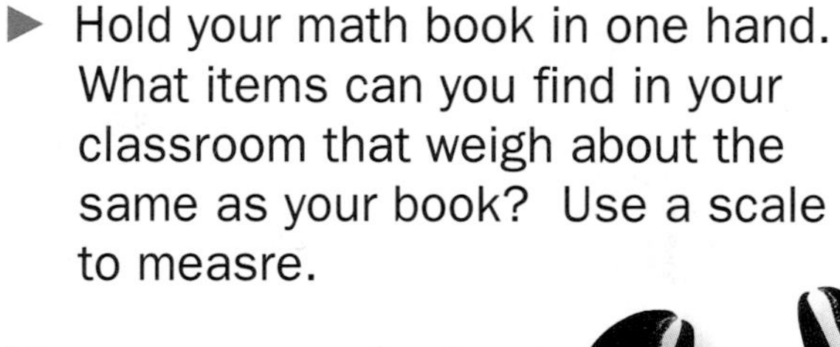

- Hold your math book in one hand. What items can you find in your classroom that weigh about the same as your book? Use a scale to measre.

How many ounces does a 5-lb bag weigh?

Think: 1 lb = 16 oz

5 lb = 5 × 16 oz
  = 80 oz

The bag weighs 80 oz.

How many pounds does a 48-oz lobster weigh?

Think: 1 lb = 16 oz
  2 lb = 32 oz
  3 lb = 48 oz

48 oz = 3 lb

The lobster weighs 3 lb.

Check for Understanding

CHECK

Write the letter of the best estimate.

1 amount of water in a bathtub
a. 50 fl oz **b.** 50 c **c.** 50 gal

2 weight of a school bus
a. 6,000 lb **b.** 6,000 oz **c.** 95 lb

Complete.

3 3 qt = ■ pt **4** 4 pt = ■ qt **5** 4 lb = ■ oz **6** 80 oz = ■ lb

Critical Thinking: Analyze **Explain your reasoning.**

7 Journal Marcy says that an aquarium filled with 10 gallons of water weighs 90 pounds. Is her statement reasonable? Why or why not?

Turn the page for Practice.

Practice

Choose the most reasonable estimate. Explain your reasoning.

1 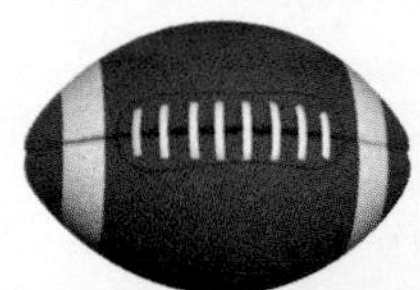

1 oz or 1 lb

2 

4 qt or 4 c or 4 fl oz

3 

30 fl oz or 30 qt or 30 gal

4 

2 gal or 2 pt or 2 c

5 

30 oz or 30 lb

6 

14 oz or 14 lb

Choose the best unit. Write *cup, pint, quart,* or *gallon.*

7 bucket
8 bud vase
9 milk pitcher
10 garbage can

Choose the better unit. Write *ounce* or *pound.*

11 eraser
12 hiking boots
13 computer
14 compact disc

ALGEBRA: PATTERNS **Copy and complete the table.**

15

Gallons	Quarts	Pints	Cups
1	4	■	16
2	■	16	32
3	12	24	■
■	16	■	64
5	■	40	■

16

Pounds	Ounces
1	■
■	32
■	48
4	■
5	■

Solve.

17 **What if** your full name is spelled out with wood blocks. Each block is 1 letter and weighs 1 lb. Name an object that weighs about the same as your name.

18 How would you balance the second scale? (Hint: You can mix containers.)

MIXED APPLICATIONS

Problem Solving

19 Brad decides to measure the school driveway with an inch ruler. Is this the best tool for him to use? Explain.

20 Manuel needs 3 gal of juice for a party. He has 7 quart bottles and 8 pint containers of juice. Does he have enough? Why or why not?

21 Jill's stride measures 26 in. She figures she can walk from her classroom door to the school library by taking no more than 15 steps. Her friend Aura measured the distance to be 32 ft. Is Jill's figuring correct? Explain your thinking.

22 **Logical reasoning** The largest whale ever recorded is a blue whale measuring just over 110 ft long. This whale was about as long as:
a. a cafeteria.
b. a classroom.
c. a school bus.

23 Yoko brings enough juice so that each person at a party gets 4 c. There are 24 people at the party. How much juice does Yoko bring?

24 Josie caught three fish that weighed 25 oz, 14 oz, and 34 oz. Which one weighed under 1 lb? over 2 lb?

25 Estimate and then measure the length of your hand span. What unit of measure did you use? Why? How could you use your hand span to estimate other lengths?

26 Dave starts a fishing trip at 6:30 A.M. If the trip lasts 4 hours 55 minutes, when does it end?

27 **Write a problem** about the perimeter of something in your classroom. Ask others to solve it.

mixed review • test preparation

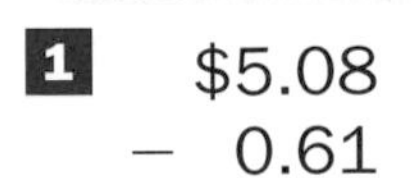

1 $\$5.08 - 0.61$	**2** $\$9.27 + 5.37$	**3** $\$34.92 + 72.64$	**4** $\$78.00 - 62.83$	**5** $\$35.16 - 26.99$
6 $\$2.83 \times 7$	**7** $\$17.52 \times 3$	**8** $\$9.41 \times 23$	**9** $\$4.99 \times 15$	**10** $\$37.65 \times 50$

Read
Plan
Solve
Look Back

Use Logical Reasoning

LEARN

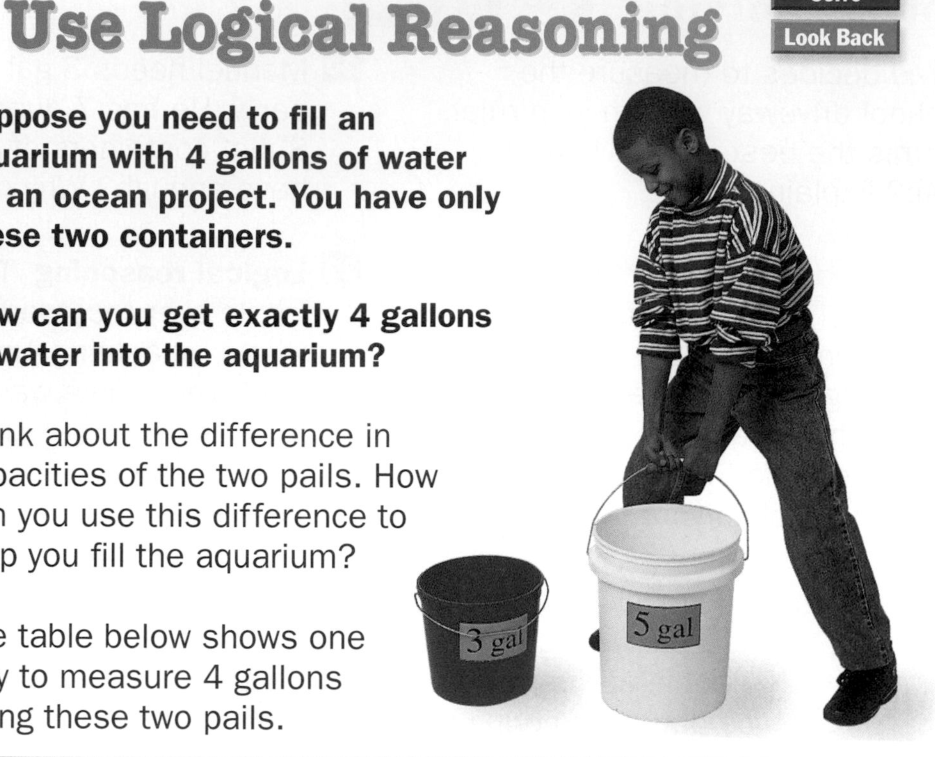

Read **Suppose you need to fill an aquarium with 4 gallons of water for an ocean project. You have only these two containers.**

How can you get exactly 4 gallons of water into the aquarium?

Plan Think about the difference in capacities of the two pails. How can you use this difference to help you fill the aquarium?

Solve The table below shows one way to measure 4 gallons using these two pails.

Steps	Water in 3-gal Pail	Water in 5-gal Pail	Water in Aquarium
1. Fill the 5-gallon pail.	0	5	0
2. Pour water from the 5-gallon pail into the 3-gallon pail to fill it.	3	2	0
3. Pour what remains in the 5-gallon pail into the aquarium.	3	0	2

You now have 2 gallons in the aquarium. Empty the 3-gallon pail and repeat steps 1–3 to get 4 gallons in the aquarium.

Look Back Why is making a table helpful for solving this problem?

CHECK

Check for Understanding

1 How would you use the 3-gallon and the 5-gallon pails to fill the aquarium with each amount of water?

a. 7 gal **b.** 8 gal **c.** 9 gal

Critical Thinking: Analyze **Explain your reasoning.**

2 Which amount (7, 8, or 9 gallons) requires the least number of steps? Which requires the greatest number of steps?

MIXED APPLICATIONS

Problem Solving

1 Suppose you have 3 glasses—2 fl oz, 5 fl oz, and 8 fl oz. What is the easiest way to fill a pitcher with 6 fl oz? 1 fl oz?

2 Dante collects seashells. At last count, he had 156 shells. His goal is to collect 500 shells. If he collects 8 shells each week, how long will it be before he reaches his goal?

3 The tanks at the Monterey Bay Aquarium in California hold 750,000 gal of water. Is this about 2 times, 4 times, or 6 times the capacity of the ocean tank at the New England Aquarium? SEE INFOBIT.

INFOBIT
The giant ocean tank at the New England Aquarium is 23 feet deep and holds almost 190,000 gal of water.

4 Calvin has 3 times as many fish in his aquarium as Sheila. Sheila has 2 more than Cameron, who has twice as many as Carlos. If Carlos has 4 fish in his aquarium, how many do the others have in theirs?

5 Amber presented her project 2 days after Brian. Marlene presented hers 4 days before Amber. The presentations were given on Monday, Wednesday, and Friday. When did each person present?

Use the table for problems 6–10.

Lengths of Coastal Waters Fish	
Atlantic manta ray	22 ft
Bluefish	2 ft
Blue marlin	10 ft
Ocean sunfish	11 ft
Smalltooth sawfish	16 ft
Swordfish	7 ft

6 About how many swordfish would you need to equal the length of 3 smalltooth sawfish?

7 How long are the three shortest fish? the three longest? What is the difference between the shortest and the longest?

8 Which fish measures 132 in.?

9 What object is about as long as an Atlantic manta ray? a bluefish? a swordfish? Explain your answers.

10 **Write a problem** using any of the information from the table. Solve it. Then have a friend solve it.

LEARN

Capacity and Mass in Metric Units

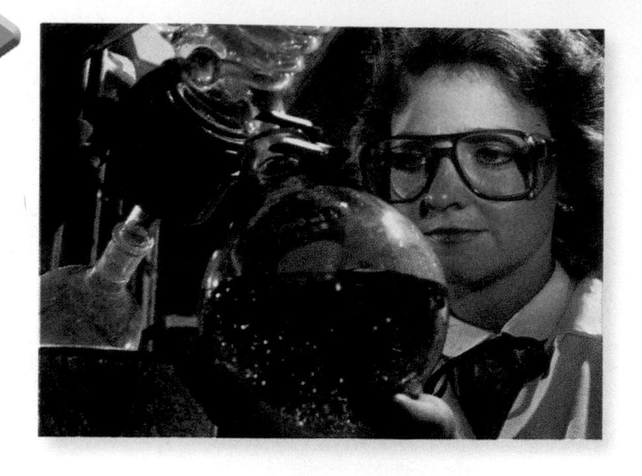

Most scientists measure capacity with metric units.

Capacity is the amount a container can hold.

1 **milliliter (mL)**

1 **liter (L)**

5 **liters (L)**

1,000 **milliliters (mL)** = 1 **liter (L)**

How many 1-mL eyedroppers would it take to fill a 2-L bottle with water?

Think: 1 L = 1,000 mL

2 L = 2 × 1,000 mL
= 2,000 mL

It would take 2,000 eyedroppers.

There are 50,000 mL of water in an aquarium. How many liters is that?

Think: 1,000 mL = 1 L
50,000 mL = 50 L

The aquarium has 50 L of water.

► What do you think is the capacity of a drinking glass? a sink? Why? Measure and compare to your estimate.

Check Out the Glossary
For vocabulary words
See page 544.

Biologists measured the largest leatherback turtle at 500 kilograms!

1,000 **grams (g)** = 1 **kilogram (kg)**

Mass is the amount of matter that makes up an object.

1 gram (g) each

1 kilogram (kg)

Box Turtle
500 grams (g)

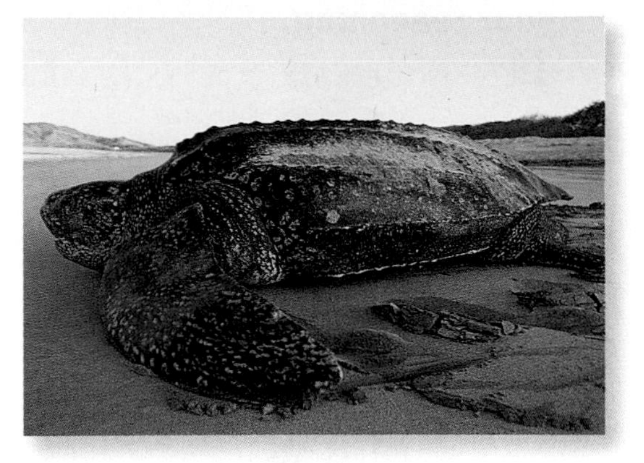

Leatherback Turtle
500 kilograms (kg)

The mass of a bicycle is 15 kg. How many grams is that?

Think: 1 kg = 1,000 g

$15 \text{ kg} = 15 \times 1{,}000 \text{ g}$
$= 15{,}000 \text{ g}$

The bicycle's mass is 15,000 g.

A bluefin tuna's mass is 454,000 g. Is it lighter or heavier than the leatherback turtle?

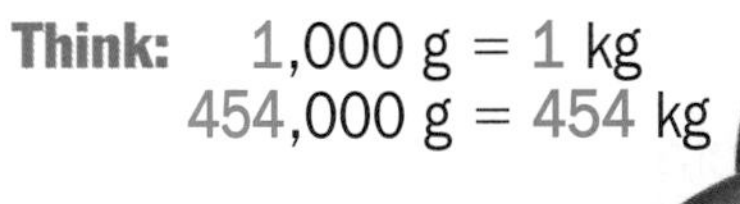

Think: 1,000 g = 1 kg
454,000 g = 454 kg

$454 \text{ kg} < 500 \text{ kg}$

The bluefin tuna is lighter than the leatherback turtle.

Check for Understanding

Write the letter of the best estimate.

1 capacity of a soup can
a. 350 mL **b.** 35 L **c.** 35 mL

2 mass of a television set
a. 25 g **b.** 250 kg **c.** 25 kg

Complete.

3 3 L = ■ mL **4** 7 kg = ■ g **5** 2,000 g = ■ kg **6** 10,000 mL = ■ L

Critical Thinking: Generalize **Explain your reasoning.**

7 Look at some 1-L containers. Are they all the same shape?

8 What would you measure using grams? using kilograms?

CHECK

Turn the page for Practice.

Practice

Choose the more reasonable estimate. Explain your reasoning.

1

300 kg or 300 g

2

6 mL or 6 L

3

2,000 kg or 20 kg

4

50 L or 500 mL

5

130 g or 13 g

6

1 L or 10 mL

Choose the best unit. Write *milliliter, liter, gram,* or *kilogram.*

7 capacity of a milk jug

8 mass of a rowboat

9 mass of a clamshell

10 capacity of a straw

ALGEBRA: PATTERNS **Copy and complete the table.**

11

Liter	1	2	■	4	■
Milliliter	1,000	■	3,000	■	■

12

Kilogram	■	2	3	4	5
Gram	1,000	■	■	■	5,000

Complete.

13 10 L = ■ mL **14** 9 kg = ■ g **15** 8,000 mL = ■ L **16** 30,000 mL = ■ L

17 40 L = ■ mL **18** 20 kg = ■ g **19** 10,000 g = ■ kg **20** 292,000 g = ■ kg

21 15 kg = ■ g **22** 23 L = ■ mL **23** 11,000 g = ■ kg **24** 133,000 mL = ■ L

Make It Right

25 Manny says that he drinks a 250-L glass of orange juice every morning. Write to him and explain why this is impossible.

MIXED APPLICATIONS

Problem Solving

26 Rico knows that there are 4 qt in a gallon. He also knows that a liter is a little bit more than a quart. He says there are a little less than 4 L in a gallon. Explain why you agree or disagree with Rico.

27 How many centimeters long can kelp grow? **SEE INFOBIT.**

28 Could kelp grow to full length in 50 days? in 100 days? Show your work. **SEE INFOBIT.**

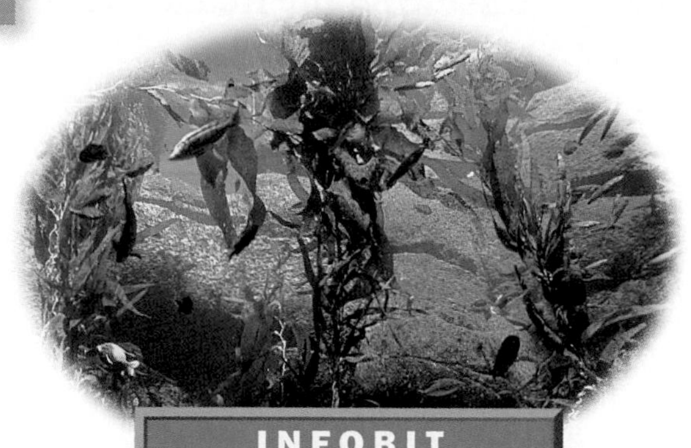

INFOBIT
Kelp is a large type of seaweed. It can grow up to 60 cm a day. It can reach 60 m in length.

29 **Data Point** Choose ten objects in your classroom. Survey your classmates to see how they estimate the mass of each object. Measure and compare. Show your results in a graph.

30 Margaret has 800 cm of tape. She gives Alex 4 m for his project, she gives Carey 2 m for her project, and she keeps the rest for her project. How much does she keep?

Cultural Connection

Measurement from the Middle East

In the Middle East, the first units of measure came from the plant world. The Arabic word *qirat* (KEE-raht) means the seed of the coral tree. Qirats were used for thousands of years to measure the weight of precious gems. Today, we say *carat.*

A carat is still the unit of measure by weight for jewels. In 1913, jewelers around the world agreed that a carat would be equivalent to 200 milligrams.

Find the weight in milligrams.

1 1 carat **2** 3 carats **3** 5 carats **4** 10 carats **5** 23 carats

Extra Practice, page 511

Problem-Solvers at Work

Read
Plan
Solve
Look Back

LEARN

Part 1 Checking for Reasonableness

Did you ever check a thermometer before deciding what to wear? Suppose you are going to the ocean for the day. The air temperature is 70°F. The water temperature is 55°F.

Temperature can be measured in **degrees Fahrenheit (°F)** or in **degrees Celsius (°C).**

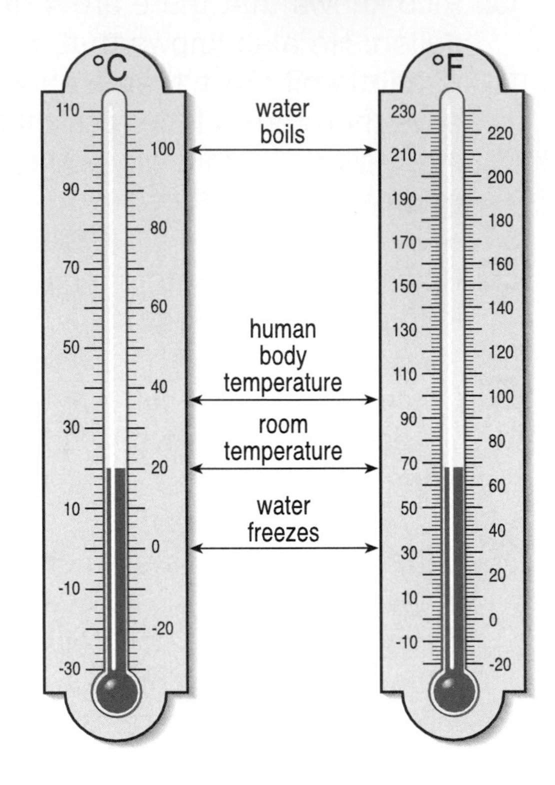

Work Together

For problems 1–3, decide if the statement is reasonable or unreasonable. Explain your answer.

1 You wear shorts and a sweatshirt while collecting seashells.

2 You go into the ocean for a swim.

3 You say that the water temperature is 13°C. If it rises about 8 degrees, you will go swimming.

4 **What if** the air temperature is 18°C. What should you wear?

5 **Make a decision** A thermometer reads ⁻6°C. What outdoor activities would it be reasonable for you to do?

Part 2 Write and Share Problems

Lisa wrote a problem about a book.

A book is 8 in. wide and 12 in. long. It weighs 48 lb. Which of these measurements is not reasonable? What should it be?

Lisa Lay
E. N. Rogers School
Lowell, MA

CHECK

6 Solve Lisa's problem.

7 **Write a problem** of your own about one of the items on the table. Or choose any other item you wish. Include as many measurements as you can. Make one of the measurements unreasonable.

8 Trade problems. Solve at least three problems written by your classmates.

9 READING ARITHMETIC WRITING **Use Diagrams** Describe one of the items on the table. Challenge another student to draw a diagram from your description.

10 What was the most interesting problem you solved? Why?

Check Out the Glossary
For vocabulary words
See page 544.

Turn the page for Practice Strategies.

PRACTICE

Menu

Choose five problems and solve them. Explain your methods.

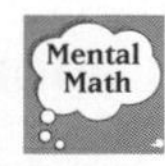

1 Robert and his family visited New York. The high temperature for their afternoon tour was 14°C. At night, the temperature dropped 20 degrees. How do you think they dressed in the afternoon? at night?

2 Karla has scored the most goals on her team. She has scored 4 times as many as Barb, who has scored 2 more than Tanya. Tanya has scored 2 goals for her team. How many goals has Karla scored?

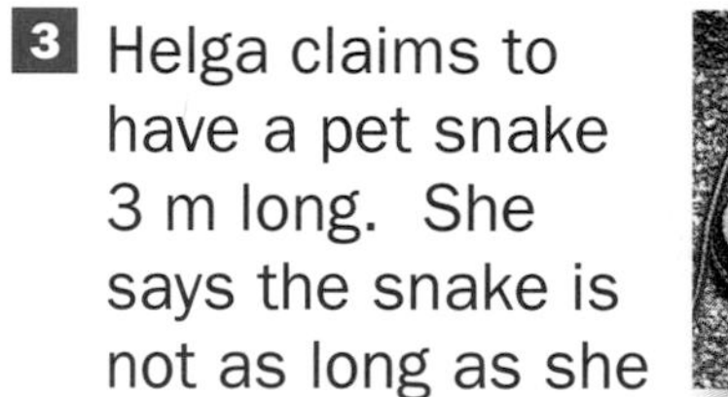

3 Helga claims to have a pet snake 3 m long. She says the snake is not as long as she is tall. Is this reasonable? Why or why not? What length might her snake be?

4 Glenn needs ribbon to wrap around the book he bought for a birthday present. The book measures 12 in. by 10 in. and is 1 in. thick. How much ribbon will he need?

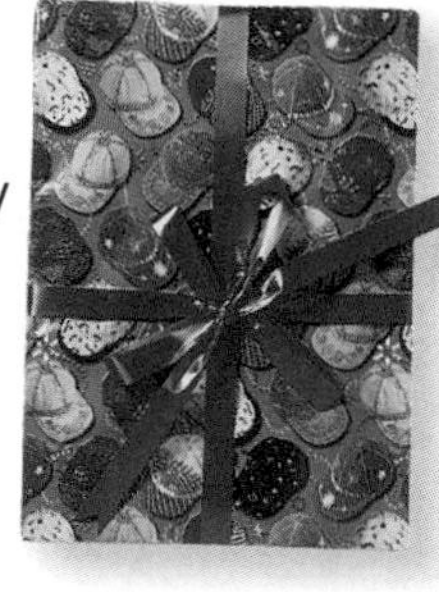

5 May recorded Tuesday's high and low temperatures as 100°F and 67°F. Estimate what these temperatures are in degrees Celsius.

6 A bullfrog is 20 cm long. It makes a leap that is 15 times its body length. Is the leap greater than or less than 2 m?

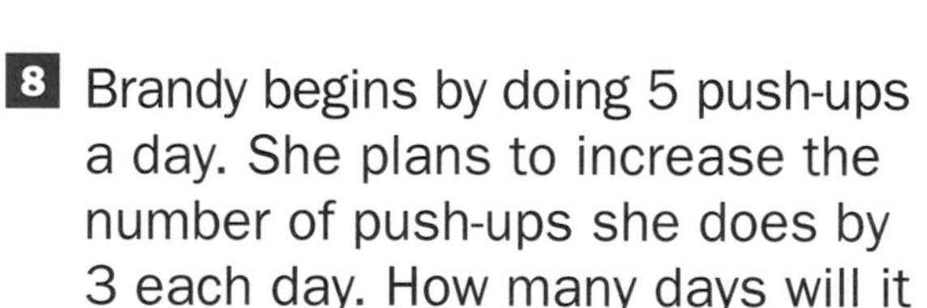

7 Brenda ate breakfast and got dressed in 30 minutes. She th watched television for 1 hour 15 minutes, read a book for 35 minutes, and worked on the computer for 40 minutes. She then played outside with her friends. If she started at 7:30 A.M., what time did she go outside?

8 Brandy begins by doing 5 push-ups a day. She plans to increase the number of push-ups she does by 3 each day. How many days will it take her to reach her goal—50 push-ups a day? **What if** she starts out doing only 3 the first day. How many more days would it take?

Choose two problems and solve them. Explain your methods.

9 Use the table. Identify and find at least three differences between depths.

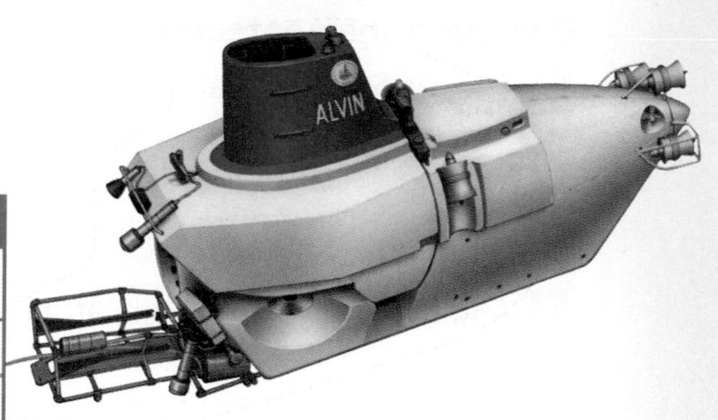

World's Three Largest Oceans			
Ocean	**Rank**	**Average Depth**	**Deepest Point**
Pacific	1	4,028 m	11,500 m
Atlantic	2	3,688 m	9,200 m
Indian	3	4,284 m	7,450 m

10 **Spatial reasoning** Jack built a large cube out of centimeter cubes. Each side is 4 cubes long. How many cubes did Jack use in all? What is the perimeter of each side of the large cube?

11 **Write a problem** that requires changing units of measure to find the answer. Solve your problem. Then give it to a classmate to solve. Compare your answers and the methods you used. How are they similar? How are they different?

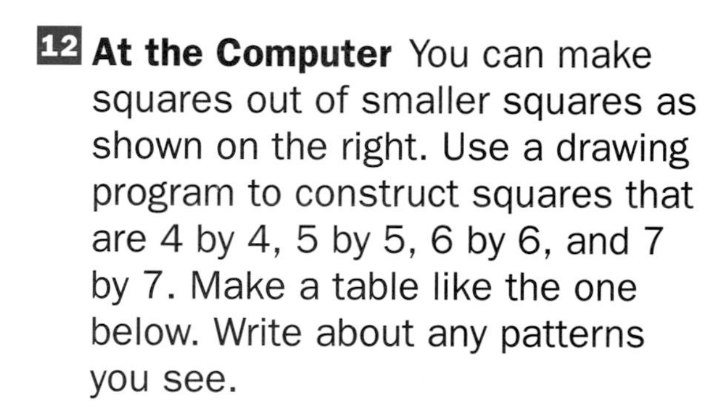

12 **At the Computer** You can make squares out of smaller squares as shown on the right. Use a drawing program to construct squares that are 4 by 4, 5 by 5, 6 by 6, and 7 by 7. Make a table like the one below. Write about any patterns you see.

Number of Squares	1	4	9
Dimensions	1 by 1	2 by 2	3 by 3
Perimeter	4	8	12

Extra Practice, page 511

Language and Mathematics

Complete the sentence. Use a word in the chart. (pages 242–265)

1 The ■ of an aquarium is usually found in gallons.

2 By knowing the length and width of a rectangle, you can find its ■.

3 A kilometer is a ■ unit of measure.

4 The ■ of a car would be measured in kilograms.

Vocabulary
metric
customary
perimeter
length
capacity
weight
mass

Concepts and Skills

Write the letter of the best estimate. (pages 242, 244, 246, 256, 262)

5 width of a street	**a.** 18 in.	**b.** 18 ft	**c.** 18 mi
6 length of a mouse	**a.** 4 in.	**b.** 4 ft	**c.** 4 yd
7 height of a flagpole	**a.** 8 dm	**b.** 8 m	**c.** 8 cm
8 length of an envelope	**a.** 25 cm	**b.** 25 km	**c.** 25 m
9 capacity of a teacup	**a.** 10 pt	**b.** 10 qt	**c.** 1 c
10 capacity of a glue bottle	**a.** 120 L	**b.** 12 L	**c.** 120 mL
11 weight of a kitten	**a.** 30 lb	**b.** 3 oz	**c.** 3 lb
12 mass of a city telephone book	**a.** 10 g	**b.** 100 g	**c.** 1 kg

Complete. (pages 242, 244, 246, 256, 262)

13 4 qt = ■ c	**14** 3 gal = ■ pt	**15** 4 lb = ■ oz	**16** 6 ft = ■ in.
17 2 yd = ■ in.	**18** 108 in. = ■ yd	**19** 5 L = ■ mL	**20** 4 cm = ■ mm
21 7 km = ■ m	**22** 25 dm = ■ cm	**23** 6,000 g = ■ kg	**24** 10 kg = ■ g

Write *reasonable* or *unreasonable* for the statement. If unreasonable, write the unit of measure that would make it reasonable. (page 266)

25 Eli wore his bathing suit when it was 32°F.

26 Lynda ran 2 mi after school.

27 Danica's little brother weighs 25 oz.

Think critically. (page 250)

28 Generalize. Is this statement *true* or *false*? Explain why.
The perimeter of a rectangle can always be found if you know the length and the width.

MIXED APPLICATIONS
Problem Solving

Pencil & Paper | Calculator | Mental Math

(pages 260, 266)

Use the table for problems 29–31.

29 Which sea monster is about half as long as a man is tall?

30 How many sea monsters (and which ones) could you line up to equal the length of your classroom?

31 What in your school would compare to the length of the loggerhead sponge? the giant earthworm? the giant squid?

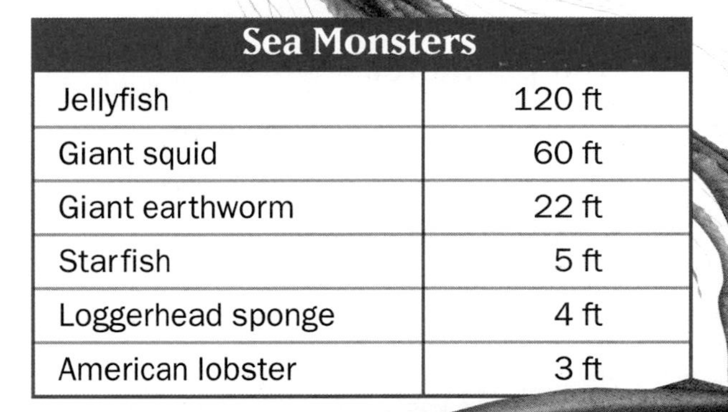

Sea Monsters	
Jellyfish	120 ft
Giant squid	60 ft
Giant earthworm	22 ft
Starfish	5 ft
Loggerhead sponge	4 ft
American lobster	3 ft

32 What is the easiest way to fill a 1-gal pitcher using only a 3-gal container and a 4-gal container?

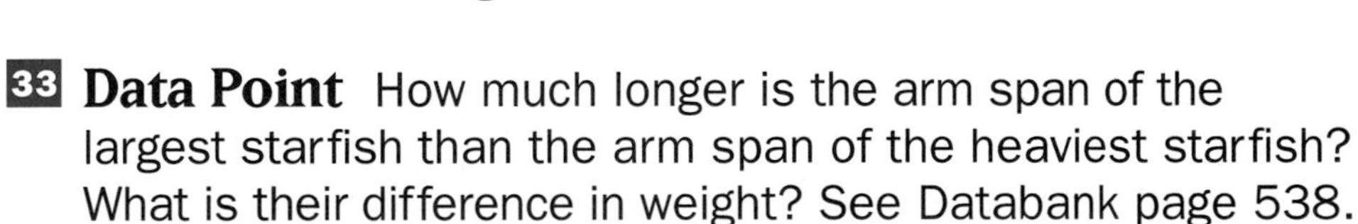

33 **Data Point** How much longer is the arm span of the largest starfish than the arm span of the heaviest starfish? What is their difference in weight? See Databank page 538.

Measure the line to the nearest inch.

1 ________

2 ____________________

3 ________________________________

4 ____________

Write the letter of the best estimate.

5 width of a window
a. 24 ft **b.** 24 in. **c.** 24 yd

6 length of a pen
a. 12 m **b.** 12 km **c.** 12 cm

7 mass of a paper clip
a. 1 g **b.** 10 g **c.** 1 kg

8 capacity of a soup bowl
a. 1 pt **b.** 1 gal **c.** 1 fl oz

9 capacity of a juice bottle
a. 2 mL **b.** 2 L **c.** 20 mL

10 weight of a textbook
a. 2 oz **b.** 20 lb **c.** 2 lb

Complete.

11 6 qt = ■ c

12 15 dm = ■ cm

13 4 kg = ■ g

14 6 yd = ■ ft

15 5,000 m = ■ km

16 4 yd = ■ in.

17 44 qt = ■ gal

18 2 L = ■ mL

Find the perimeter.

19

3 in.
3 in. 3 in.
3 in.

20

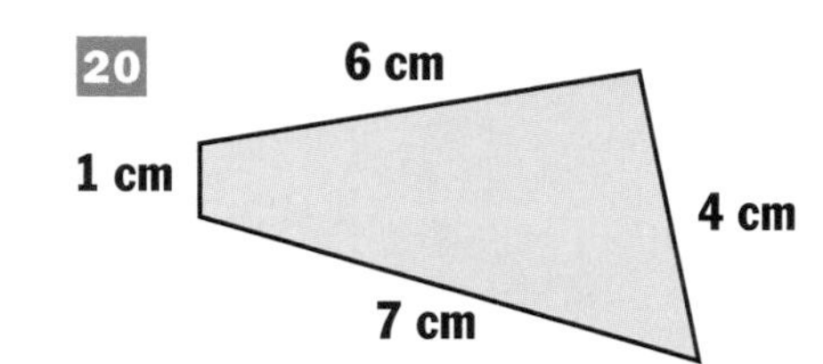

21

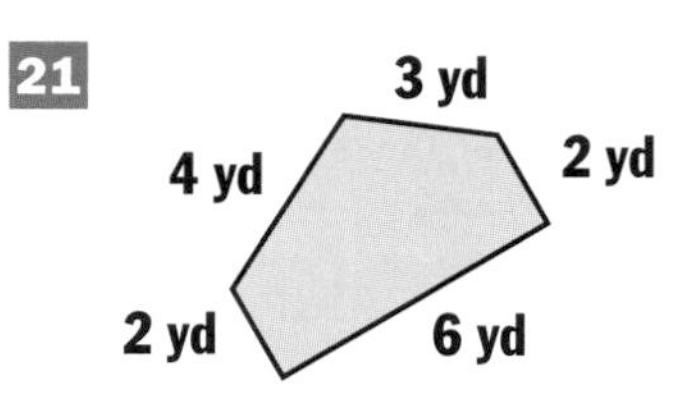

Solve.

22 A bicycle is 5 ft 4 in. long and 3 ft 9 in. high. Will it fit in a shipping box that is 66 in. long and $44\frac{1}{2}$ in. high? Why or why not?

23 Mel walks 75 ft in 1 min. His brother walks 21 yd in 2 min. If they start together, how far from Mel will his brother be in 6 min?

24 The art teacher wants to put a border around a rectangular bulletin board that is 7 ft long and 5 ft wide. How many yards of border will he need?

25 José is twice as old as Alberto. In 2 years, Alberto will be half Corey's age at that time. Corey is now 14. How old are José and Alberto?

What Did You Learn?

List three objects that can be found in your school. Choose objects whose length, mass, weight, or capacity you can measure.

- ▶ Describe the different ways that each object can be measured.
- ▶ For each object:
  - **a.** describe the most important measurement and explain why you chose it.
  - **b.** estimate the measurement.
  - **c.** explain how you arrived at your estimate and decided which unit to use.
  - **d.** find the actual measurement.

A Good Answer

- explains the reasoning involved in choosing how an object should be measured
- gives reasonable estimates and accurate measurements

You may want to place your work in your portfolio.

What Do You Think

1 Do you understand which tools and units to use when you measure? If not, what do you find most difficult?

2 What do you think is the most important step when measuring an object? Why?
- Finding an estimate first.
- Deciding what measurement to take.
- Choosing an appropriate unit and tool.
- Other. Explain.

math science technology CONNECTION

SONAR

Cultural Note

In 1793, the Italian biologist Lazzaro Spallanzani found that bats make very high squeaks that cannot be heard by people. These sounds bounce off objects and back to the bat like an echo.

By listening to echoes, a bat can tell the difference between an insect and the twig it is sitting on! People also are able to use sound to measure ocean depth.

In about 1920, scientists discovered how to bounce sounds off the ocean floor and record the echo. This is called *sonar*, which stands for **so**und **na**vigation and **r**anging.

Scientists measure the time it takes for the sound to travel from the ship to the floor of the ocean and back to the ship. Since we know how fast sound can travel in water, we can calculate the depth of the ocean by timing how long it takes for the echo to return to the ship.

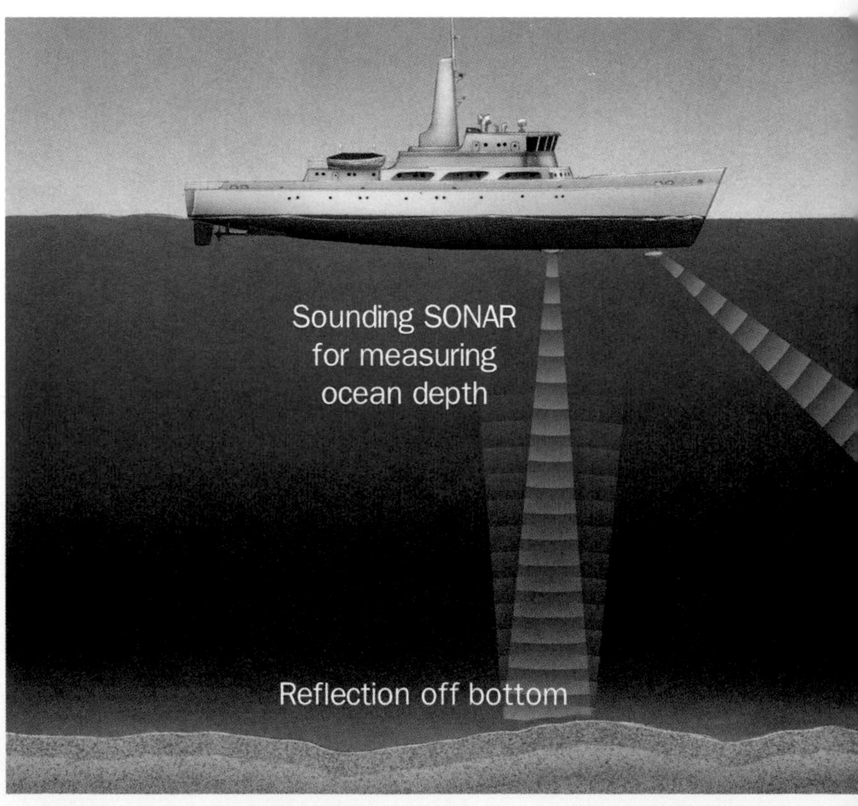

- ▶ What other animals do you know of that use sonar to tell distances?
- ▶ Harder surfaces reflect sound better. On the ocean floor, which would reflect sound better, rocks or mud?

Measuring Ocean Depth

Sound travels about 5,000 ft every second in water. Suppose it takes 2 seconds for an echo to travel from a ship to the ocean floor and back. That means that the sound travels for 1 second before hitting the ocean floor and another second to return to the boat. That means the ocean is 5,000 ft deep.

1 How deep is the ocean if it takes 4 seconds for the sound to return to the ship?

2 How long will it take the sound to return to a ship if the ocean is 20,000 ft deep?

3 Make a table of times for an echo to return to a ship, and then calculate the ocean depth in feet.

At the Computer

4 Use a graphing program to make a line graph for the distance traveled by sound and the time taken to return to the ship. What do you notice?

5 Use the line graph to estimate how many seconds it will take for sound to return to a ship if the ocean is 7,500 ft deep.

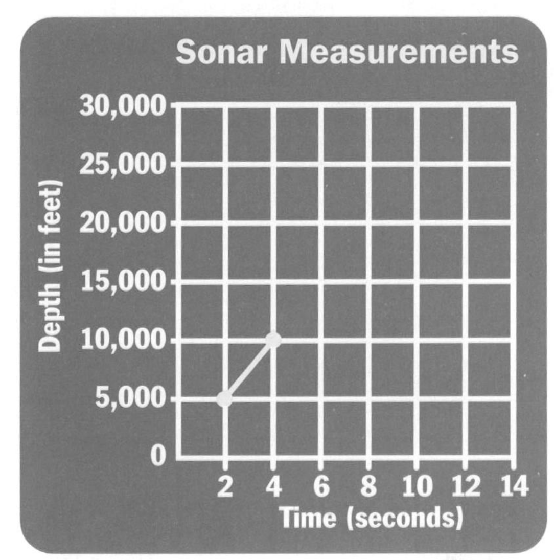

CHAPTER 8

DIVIDE BY 1- AND 2-DIGIT NUMBERS

THEME

Our States

How is your state interesting? In this chapter, you will learn many interesting facts about our states. Mathematics is used to describe the people and things that make each state special.

What Do You Know

The Travel Industry Association of America helps people arrange tours within the United States and abroad.

1 You want to visit all 50 states. If you visit 8 states a year, how many years will it take? Explain.

2 Portfolio Choose the states you want to visit and how many you would like to visit each year. Find how many years it would take to visit these states. Explain your reasoning.

Problem/Solution **A class wants to plan a trip to a new park in their city. The only map of their city they have is dated 1980.**

In stories, if you identify a problem, it helps you understand the situation and the solution.

1 What problem might the students have?

2 How could students solve the problem?

Vocabulary

estimate, p.280
compatible number, p.280
regroup, p.282
dividend, p.283
divisor, p.283
quotient, p.283
remainder, p.283
average, p.300

Division Patterns

LEARN

Maria loves her dogs—all 120 of them! She takes people on dogsledding tours. She also races. A 3-day race can cover 180 mi. If the same distance is covered each day, how many miles are traveled a day?

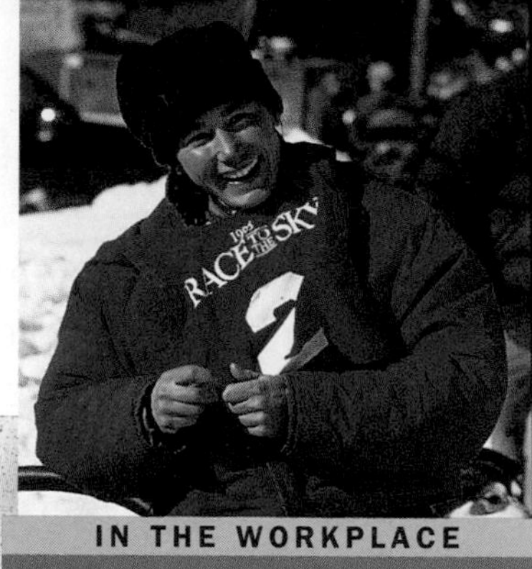

IN THE WORKPLACE

Maria Hayashida, musher from Jackson Hole, WY, competed in the 1996 Alaskan Iditarod Race

 **ALGEBRA: PATTERNS** You can use patterns to help you divide mentally. What patterns do you see below?

18 ÷ 3 = 6

180 ÷ 3 = 60

1,800 ÷ 3 = 600

60 miles are traveled each day.

Jackson Hole

WYOMING

More Examples

A 7 ÷ 1 = 7
70 ÷ 1 = 70
700 ÷ 1 = 700
7,000 ÷ 1 = 7,000

B 12 ÷ 3 = 4
120 ÷ 3 = 40
1,200 ÷ 3 = 400

C 20 ÷ 5 = 4
200 ÷ 5 = 40
2,000 ÷ 5 = 400

CHECK

Check for Understanding

ALGEBRA: PATTERNS Complete. Describe the pattern.

1 9 ÷ 3 = 3
90 ÷ 3 = ■
900 ÷ 3 = ■
9,000 ÷ 3 = ■

2 30 ÷ 6 = ■
300 ÷ 6 = ■
3,000 ÷ 6 = 500

3 72 ÷ 9 = ■
720 ÷ 9 = 80
7,200 ÷ 9 = ■

Divide mentally.

4 140 ÷ 2 **5** 240 ÷ 6 **6** 450 ÷ 9 **7** 480 ÷ 8

8 1,500 ÷ 3 **9** 1,800 ÷ 6 **10** 6,400 ÷ 8 **11** 8,100 ÷ 9

Critical Thinking: Analyze **Explain your reasoning.**

12 If you think of 180 as 18 tens, how can you find 180 ÷ 6 mentally? How can you find 1,800 ÷ 6?

Practice

Divide mentally.

1 $80 \div 2$ **2** $60 \div 3$ **3** $50 \div 5$ **4** $60 \div 6$

5 $320 \div 4$ **6** $160 \div 8$ **7** $400 \div 5$ **8** $490 \div 7$

9 $540 \div 9$ **10** $630 \div 9$ **11** $2{,}100 \div 3$ **12** $3{,}000 \div 6$

13 $2{,}700 \div 3$ **14** $3{,}600 \div 9$ **15** $4{,}800 \div 6$ **16** $7{,}200 \div 8$

ALGEBRA Find the missing number.

17 ■ $\div 2 = 300$ **18** $540 \div$ ■ $= 60$ **19** $720 \div$ ■ $= 90$

20 $3{,}500 \div$ ■ $= 700$ **21** ■ $\div 4 = 300$ **22** ■ $\div 7 = 700$

MIXED APPLICATIONS
Problem Solving

Price for each Pound of Fruit

Fruit	Price
Oranges	$1.29
Pineapples	$1.69
Grapes	$1.79
Apples	$0.99
Pears	$0.69

Use the sign for problems 23–25.

23 Which costs more, 4 pounds of pears or 2 pounds of grapes? How much more? Explain.

24 Gina buys a pound of apples and a pound of pineapples. She pays with a $5 bill. How much change will she get?

25 **Make a decision** You want to spend no more than $3 on fruit and get at least two different kinds of fruit. What would you buy? Why?

26 A musher switches her 120 dogs around for her trips. She has 4 dogs on each sled. How many sleds are pulled before the same dogs are used again?

mixed review • test preparation

1 $861 + 246$ **2** $947 - 540$ **3** 40×51 **4** $9{,}000 \times 40$

Find the perimeter.

5

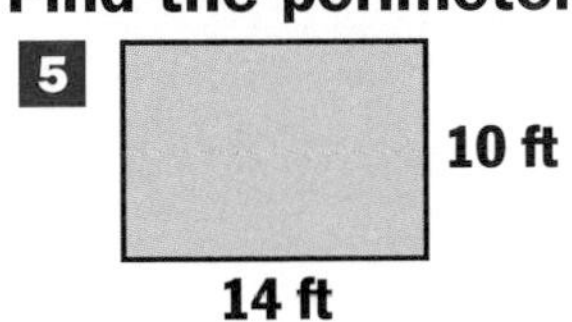

6

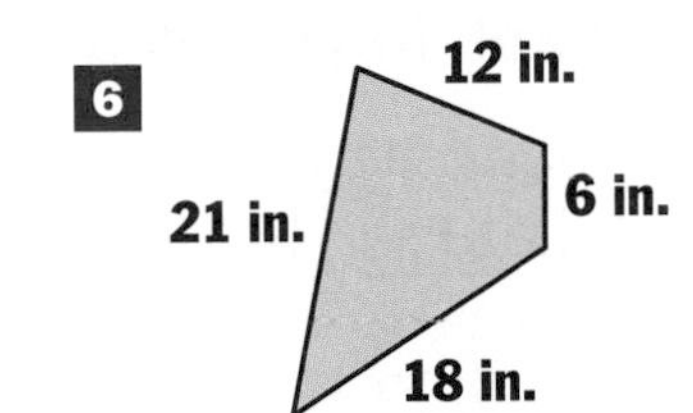

7

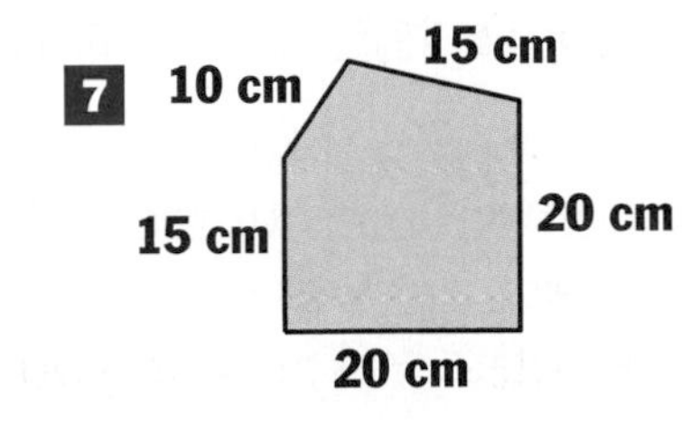

Extra Practice, page 512

LEARN

Estimate Quotients

A Little League team needs $752 for buses and tickets. The team decides to earn money by running a snack stand for 8 weeks. About how much do they need to earn each week?

You can **estimate** to solve this problem.

To estimate a quotient, you can use **compatible numbers** that help you divide mentally.

Estimate: 752 ÷ 8

Find a number close to 752 that you can divide by 8 mentally. You can use the fact 72 ÷ 8 = 9 to help you choose a compatible number.

Think: 720 ÷ 8 = 90

The team needs to earn about $90 each week.

Each year, the Little League World Series is held in South Williamsport, PA.

More Examples

A Estimate: 409 ÷ 6
Think: 42 ÷ 6 = 7
420 ÷ 6 = 70

B Estimate: 192 ÷ 5
Think: 20 ÷ 5 = 4
200 ÷ 5 = 40

Check Out the Glossary
estimate
compatible numbers
See page 544.

CHECK

Check for Understanding

Estimate. Show the compatible numbers you used.

1 57 ÷ 3
2 98 ÷ 5
3 312 ÷ 6
4 652 ÷ 8
5 787 ÷ 9
6 2,489 ÷ 4
7 4,809 ÷ 7
8 6,225 ÷ 9

Critical Thinking: Analyze **Explain your reasoning.**

9 Show two different ways to estimate the quotient using compatible numbers. Which estimate will be closer to the exact answer?
a. 292 ÷ 4 **b.** 2,234 ÷ 6

Practice

Estimate. Show the compatible numbers you used.

1 $79 \div 4$	**2** $80 \div 3$	**3** $37 \div 2$	**4** $96 \div 5$
5 $167 \div 2$	**6** $152 \div 8$	**7** $379 \div 4$	**8** $532 \div 6$
9 $1{,}578 \div 5$	**10** $2{,}654 \div 3$	**11** $4{,}512 \div 5$	**12** $7{,}465 \div 8$
13 $198 \div 4$	**14** $4{,}732 \div 6$	**15** $518 \div 7$	**16** $8{,}976 \div 9$

MIXED APPLICATIONS

Problem Solving

17 A town has $324 for supplies for its 4 Little League teams. About how much can each team spend for supplies? Explain your thinking.

18 What is the distance between the bases on a major league field? **SEE INFOBIT.**

19 Major league bats cannot be longer than 42 inches. How much longer is this than the longest Little League bat? **SEE INFOBIT.**

20 **Data Point** Survey your classmates to find their favorite baseball team. Show your results on a graph.

INFOBIT
Little League bats cannot be longer than 33 inches. The distance between the bases is 60 feet. This is 30 feet shorter than on a major league field.

21 Little League coaches try out 488 children during 8 sessions. About how many children try out at each session?

22 **Write a problem** that you can solve by estimating a quotient. Have others solve it.

mixed review • test preparation

1 $123.89 − $95.99	**2** $912.43 + $0.99 + $1.79	**3** 242×6
4 $3 \times 5{,}004$	**5** $904.00 + $19.36	**6** $307 − $249

Extra Practice, page 512

Divide by 1-Digit Numbers

LEARN

You can use cubes to help you divide by 1-digit numbers.

Work Together

Work in a group to share cubes.

Make a pile of 9 trains that each have 10 cubes. Make another pile with 9 single cubes.

Take a handful of trains and single cubes. Record the total in a table like the one below.

Spin the spinner to see how many people will share the cubes.

Divide the cubes so that each person gets an equal number. Record your work.

Repeat the activity five times.

You will need
- *1–9 spinner*
- *connecting cubes*

KEEP IN MIND
▶ Be prepared to explain your answers and methods.

Total	Number of People	Number of Cubes for Each Person	Number Left Over
59	4		

Talk It Over

▶ How did you decide when to **regroup?**

▶ How can you use estimation to check your answers?

Make Connections

Lena's group shared 59 cubes among 4 people. They recorded it like this.

Total	Number of People	Number of Cubes for Each Person	Number Left Over
59	4	14	3

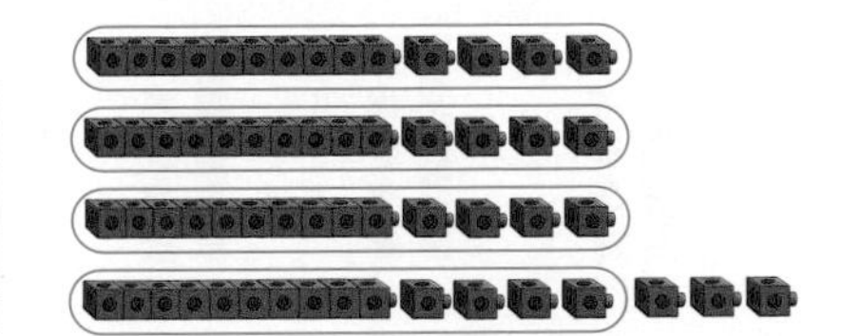

You can also use a division sentence to record what was done.

$$59 \div 4 = 14 \text{ R3}$$

59 ↑ **dividend** ÷ 4 ↑ **divisor** = 14 ↑ **quotient** R3 ← **remainder**

Check Out the Glossary
For vocabulary words
See page 544.

- ▶ Explain what each part of the division sentence means.
- ▶ Write division sentences for each row in your table.

Check for Understanding

CHECK

Complete the division sentence for the picture.

1 $36 \div 4 = \blacksquare$

2 $43 \div 6 = \blacksquare$

Divide. You may use place-value models.

3 $51 \div 3$ **4** $75 \div 5$ **5** $98 \div 7$ **6** $76 \div 4$

7 $39 \div 6$ **8** $88 \div 9$ **9** $87 \div 2$ **10** $99 \div 8$

Critical Thinking: Generalize **Explain your reasoning.**

11 How are remainders related to divisors in the division sentences below? What is the greatest possible remainder for a given divisor? the least possible remainder?

$26 \div 2 = 13$
$77 \div 2 = 38 \text{ R1}$

$57 \div 3 = 19$
$73 \div 3 = 24 \text{ R1}$
$29 \div 3 = 9 \text{ R2}$

$88 \div 4 = 22$
$61 \div 4 = 15 \text{ R1}$
$34 \div 4 = 8 \text{ R2}$
$55 \div 4 = 13 \text{ R3}$

Turn the page for Practice.

Practice

Write a division sentence to represent the models.

1

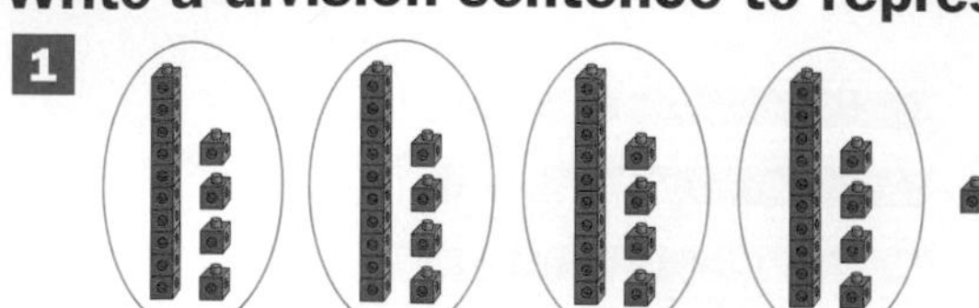

2

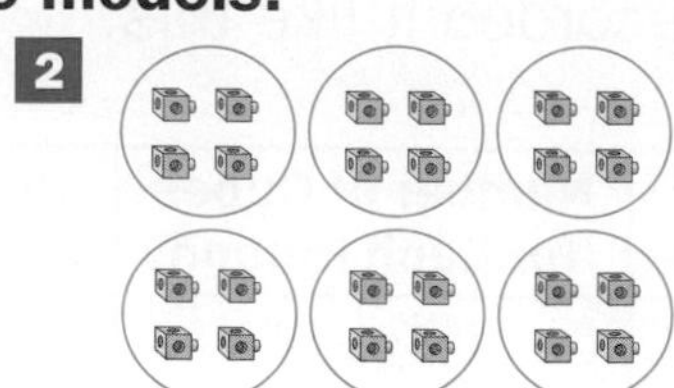

3

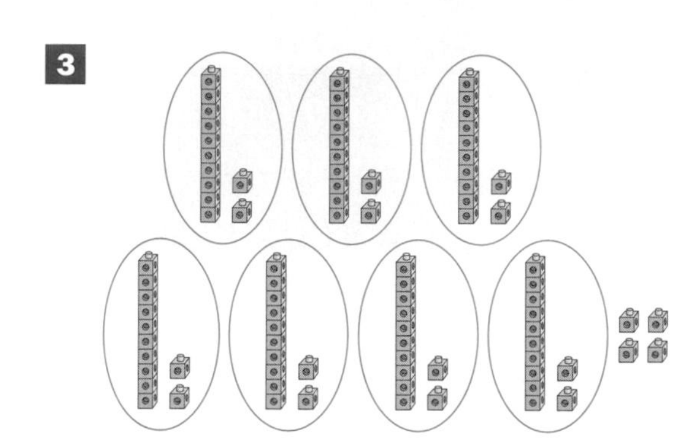

4

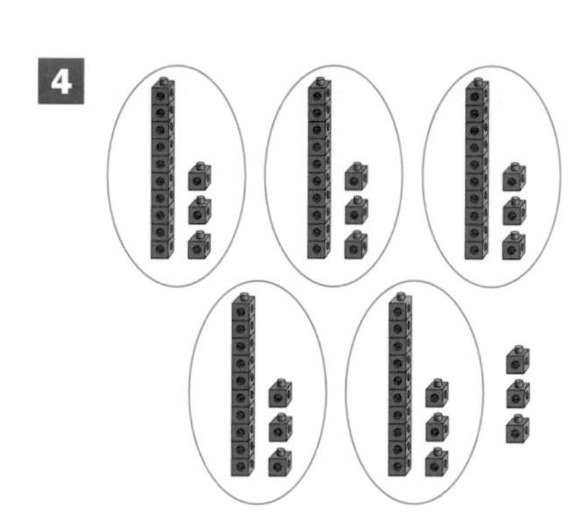

Divide. You may use place-value models.

5 $49 \div 8$ **6** $37 \div 4$ **7** $31 \div 7$ **8** $26 \div 3$

9 $91 \div 6$ **10** $70 \div 5$ **11** $49 \div 2$ **12** $77 \div 5$

13 $97 \div 7$ **14** $99 \div 2$ **15** $87 \div 8$ **16** $47 \div 3$

17 $89 \div 9$ **18** $84 \div 4$ **19** $69 \div 7$ **20** $46 \div 6$

21 $47 \div 5$ **22** $46 \div 2$ **23** $74 \div 3$ **24** $44 \div 8$

25 $64 \div 3$ **26** $19 \div 3$ **27** $91 \div 8$ **28** $67 \div 5$

Make It Right

29 Explain what mistake was made. Show the correct answer.

$58 \div 4 = 12\ R10$

MIXED APPLICATIONS

Problem Solving

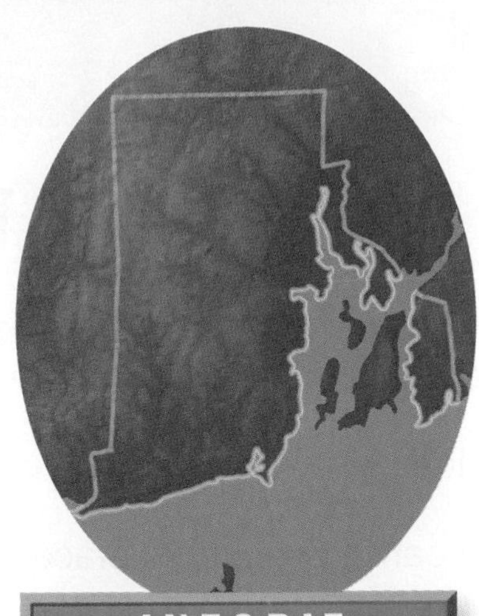

30 Each column in front of the Supreme Court in Washington, D.C., is 3 stories high. If you created a single column from all the columns, it would be 48 stories high! How many columns are there?

31 You can travel from Providence down Narragansett Bay and then to Block Island. A boat sails this distance in 2 hours. It covers a mile in about 3 minutes. How far is Block Island from Narragansett Bay? **SEE INFOBIT.**

INFOBIT
Rhode Island is the smallest state. Narragansett Bay reaches 30 miles inland to Providence.

Use the bar graph for problems 32–34.

32 In 1994, about how many more people lived in Washington, D.C., than Wyoming?

33 One of these states was the least populated state in the country in 1994. Which state was it?

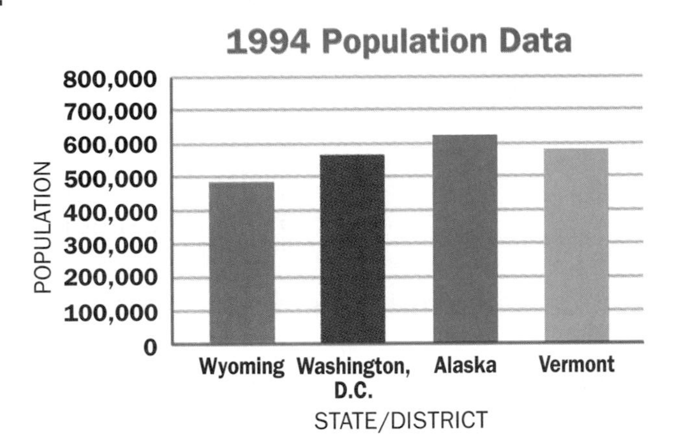

34 

Problem/Solution Write another problem based on the graph. Tell how to solve the problem.

more to explore

Prime and Composite Numbers

A *prime number* is divisible by only two numbers, itself and 1.

$17 \div 1 = 17$ $17 \div 17 = 1$

17 is divisible by 1 and 17 only.

A *composite number* is divisible by more than two numbers.

$15 \div 1 = 15$ $15 \div 5 = 3$
$15 \div 15 = 1$ $15 \div 3 = 5$

15 is divisible by 1, 3, 5, and 15.

Tell if the number is prime or composite.

1 13   **2** 21   **3** 19   **4** 7   **5** 14

Divide by 1-Digit Numbers

LEARN

In the last lesson, you snapped apart cubes to find the answer. Here are some other ways to divide.

Divide: $87 \div 7$

You can divide using place-value models.

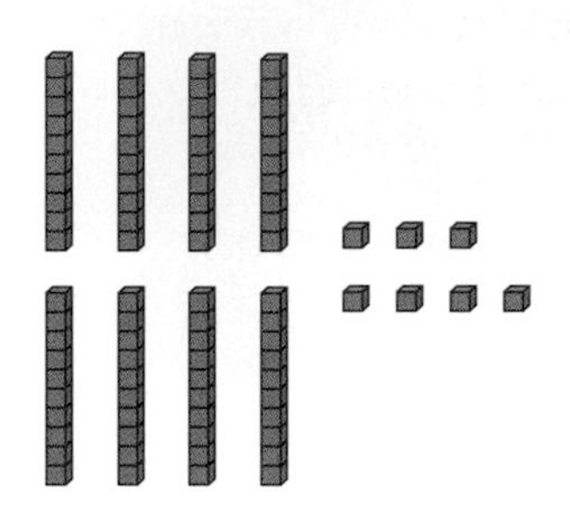

8 tens 7 ones

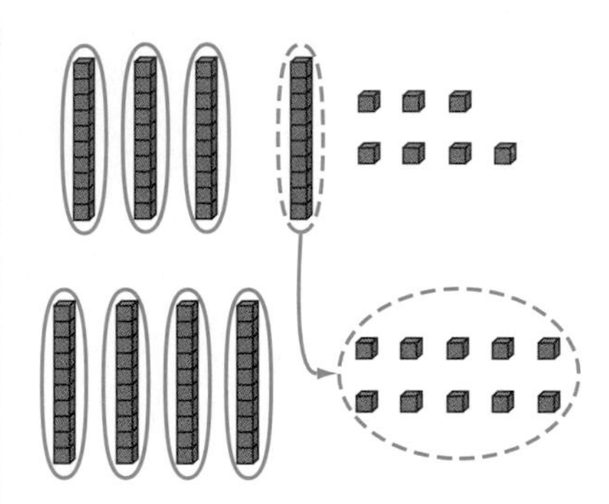

Put 1 ten in each of 7 groups. Regroup 1 ten as 10 ones. Divide the ones equally among the 7 groups.

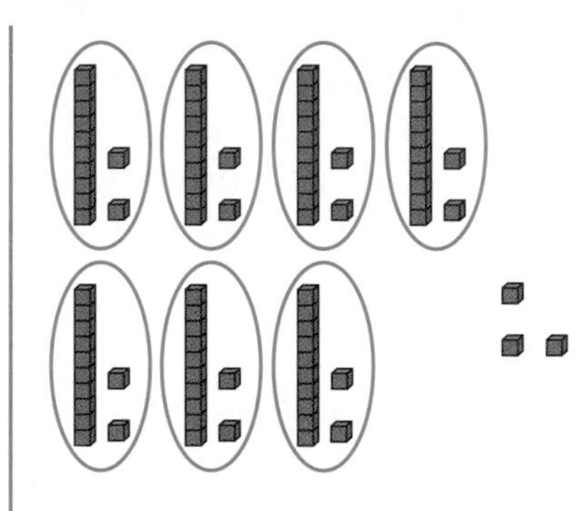

Each of the 7 groups has 1 ten 2 ones. There are 3 ones left over.

$87 \div 7 = 12$ R3

You can also divide using pencil and paper.

Estimate to place the first digit of the quotient.

Estimate: $87 \div 7$ **Think:** $70 \div 7 = 10$ The first digit is in the tens place.

Step 1

Divide the tens.

$$\begin{array}{r} 1 \\ 7\overline{)87} \\ -7 \\ \hline 1 \end{array}$$

Think: $7\overline{)8}$ with quotient 1

Multiply: $1 \times 7 = 7$

Subtract: $8 - 7 = 1$

Compare: $1 < 7$

So $87 \div 7 = 12$ R3.

Step 2

Bring down the ones. Divide the ones. Write the remainder.

$$\begin{array}{r} 12 \text{ R3} \\ 7\overline{)87}\phantom{\text{ R3}} \\ -7\downarrow\phantom{\text{ R3}} \\ \hline 17\phantom{\text{ R3}} \\ -14\phantom{\text{ R3}} \\ \hline 3\phantom{\text{ R3}} \end{array}$$

Think: $7\overline{)17}$ with quotient 2

Multiply: $2 \times 7 = 14$

Subtract: $17 - 14 = 3$

Compare: $3 < 7$

Check: $7 \times 12 = 84$

$84 + 3 = 87$

Talk It Over

▶ Why is it a good idea to compare the difference with the divisor at each step?

Divide: $800 \div 7$

Estimate to place the first digit of the quotient.

Estimate: $800 \div 7$ **Think:** $700 \div 7 = 100$
The first digit is in the hundreds place.

Step 1	Step 2	Step 3
Divide the hundreds.	**Bring down the tens.** **Divide the tens.**	**Bring down the ones.** **Divide the ones.** **Write the remainder.**
$7\overline{)800}$ quotient 1; -7; 1. **Think:** $7\overline{)8}$ = 1. Multiply: $1 \times 7 = 7$. Subtract: $8 - 7 = 1$. Compare: $1 < 7$	$7\overline{)800}$ quotient 11; $-7\downarrow$; 10; -7; 3. **Think:** $7\overline{)10}$ = 1. Multiply: $1 \times 7 = 7$. Subtract: $10 - 7 = 3$. Compare: $3 < 7$	$7\overline{)800}$ quotient 114 R2; $-7\downarrow$; 10; $-7\downarrow$; 30; -28; 2. **Think:** $7\overline{)30}$ = 4. Multiply: $4 \times 7 = 28$. Subtract: $30 - 28 = 2$. Compare: $2 < 7$. **Check:** $7 \times 114 = 798$; $798 + 2 = 800$

So, $800 \div 7 = 114$ R2.

More Examples

A $4\overline{)\$52}$ = \$13
$-4\downarrow$
12
-12
0

B $5\overline{)327}$ = 65 R2
$-30\downarrow$
27
-25
2

C $4\overline{)340}$ = 85
$-32\downarrow$
20
-20
0

Check for Understanding

Divide.

1 $6\overline{)82}$ **2** $4\overline{)97}$ **3** $5\overline{)429}$ **4** $4\overline{)\$692}$ **5** $8\overline{)968}$

6 $57 \div 5$ **7** $77 \div 3$ **8** $\$384 \div 4$ **9** $743 \div 6$ **10** $987 \div 7$

Critical Thinking: Summarize

11 

Write the steps you would use to find $794 \div 3$. Explain how to check the answer.

CHECK

Turn the page for Practice.

PRACTICE

Practice

Divide mentally.

1. $60 \div 6$
2. $80 \div 4$
3. $81 \div 9$
4. $90 \div 3$
5. $280 \div 7$
6. $720 \div 9$
7. $630 \div 7$
8. $350 \div 5$

Divide. Remember to estimate.

9. $5\overline{)85}$
10. $2\overline{)79}$
11. $8\overline{)89}$
12. $7\overline{)99}$
13. $6\overline{)74}$
14. $9\overline{)\$855}$
15. $2\overline{)38}$
16. $9\overline{)876}$
17. $6\overline{)68}$
18. $7\overline{)380}$
19. $4\overline{)948}$
20. $6\overline{)919}$
21. $8\overline{)914}$
22. $5\overline{)659}$
23. $4\overline{)944}$
24. $3\overline{)176}$
25. $5\overline{)96}$
26. $2\overline{)\$496}$
27. $3\overline{)95}$
28. $4\overline{)776}$

ALGEBRA Write the letter of the missing number.

29. $347 \div ■ = 69$ R2 — **a.** 4 **b.** 5 **c.** 6 **d.** 7
30. $158 \div ■ = 52$ R2 — **a.** 5 **b.** 6 **c.** 3 **d.** 8
31. $720 \div 8 = ■$ — **a.** 9 **b.** 80 **c.** 90 **d.** 8

Write the letter of the correct answer.

32. If the divisor is 7, which cannot be a remainder? — **a.** 0 **b.** 6 **c.** 9 **d.** 4
33. If the remainder is 6, which cannot be a divisor? — **a.** 4 **b.** 9 **c.** 7 **d.** 8
34. If the dividend is 42, the quotient will have 2 digits if the divisor is ■. — **a.** 6 **b.** 5 **c.** 7 **d.** 3
35. If the dividend is 42, the quotient will have 1 digit if the divisor is ■. — **a.** 6 **b.** 4 **c.** 3 **d.** 2

Make It Right

36. Here is how Karl divided 334 by 3. Explain what mistake was made. Show how to correct it.

```
   11 R1
3)334
 -3
  04
 - 3
   1
```

Remainders Game!

First, make three sets of index cards labeled 0 through 9. Create a scorecard like the one shown.

You will need
- *0–9 spinner*
- *index cards*

Next, mix up the cards and have each player choose three cards.

Play the Game

- ▶ Spin the spinner once to get a divisor.

  Note: Do not use zero as a divisor.

- ▶ Each player arranges the three cards as a 3-digit dividend and finds the quotient and remainder. Record the remainder as the player's score.

- ▶ Replace the cards and mix them up. Choose new cards and then spin the spinner for a new divisor.

- ▶ Continue playing until a player has 20 points.

What strategy could you use to help you get the most points?

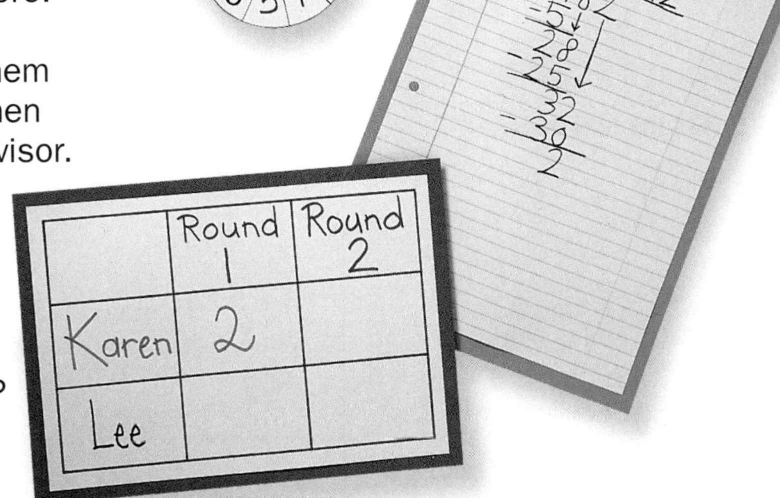

mixed review • test preparation

1 $752 + 0 = ■$

2 $(2 \times 9) \times 5 = (2 \times ■) \times 9$

3 $24 \times ■ = 24$

4 $68 + 57 = 57 + ■$

5 $■ \times 95 = 0$

6 $33 \times 56 \times ■ = 0$

7 $6 \times 26 = (3 \times 26) + (■ \times 26)$

8 $■ \times 42 = (4 \times 42) + (4 \times 42)$

Zeros in the Quotient

LEARN

The Statue of Liberty stands 305 feet high above New York Bay on Liberty Island. A class is making a 3-foot model of the statue. How many times larger is the actual statue?

Cultural Note

The words at the bottom of the Statue of Liberty were written by the American poet Emma Lazarus. They include, "Give me your tired, your poor, your huddled masses yearning to breathe free."

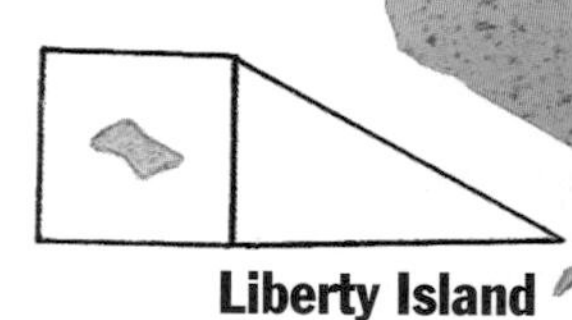

Estimate to place the first digit of the quotient.

Estimate: $305 \div 3$ **Think:** $300 \div 3 = 100$
The first digit is in the hundreds place.

Step 1

Divide the hundreds.

$$\begin{array}{r} 1 \\ 3\overline{)305} \\ -3 \\ \hline 0 \end{array}$$

Think: $3\overline{)3}$ (quotient 1)
Multiply: $1 \times 3 = 3$
Subtract: $3 - 3 = 0$
Compare: $0 < 3$

Step 2

Bring down the tens.
Divide the tens.

$$\begin{array}{r} 10 \\ 3\overline{)305} \\ -3\downarrow \\ \hline 00 \end{array}$$

Think: $0 < 3$
Not enough tens.
Write 0 in the quotient.

Step 3

Bring down the ones.
Divide the ones.
Write the remainder.

$$\begin{array}{r} 101 \text{ R2} \\ 3\overline{)305}\phantom{\text{ R2}} \\ -3\downarrow\downarrow\phantom{\text{ R2}} \\ \hline 005\phantom{\text{ R2}} \\ -3\phantom{\text{ R2}} \\ \hline 2\phantom{\text{ R2}} \end{array}$$

Think: $3\overline{)5}$ (quotient 1)
Multiply: $1 \times 3 = 3$
Subtract: $5 - 3 = 2$
Compare: $2 < 3$

The actual statue is over 101 times larger than the model.

CHECK

Check for Understanding

Divide.

1 $6\overline{)425}$ **2** $2\overline{)211}$ **3** $8\overline{)833}$ **4** $4\overline{)842}$ **5** $3\overline{)902}$

Critical Thinking: Analyze **Explain your reasoning.**

6 When do you get a zero in the tens place of a quotient?

Practice

Divide.

1 $4\overline{)429}$ **2** $3\overline{)614}$ **3** $8\overline{)322}$ **4** $5\overline{)518}$ **5** $7\overline{)146}$

6 $3\overline{)\$912}$ **7** $7\overline{)846}$ **8** $2\overline{)814}$ **9** $6\overline{)\$660}$ **10** $4\overline{)837}$

11 $62 \div 3$ **12** $82 \div 8$ **13** $153 \div 5$ **14** $732 \div 7$

15 $\$412 \div 4$ **16** $\$850 \div 5$ **17** $\$749 \div 7$ **18** $\$418 \div 2$

How many:

19 $5 are in $100? **20** $9 are in $954? **21** $1 are in $740?

MIXED APPLICATIONS
Problem Solving

Use this information for problems 22–23:

> 387 students and teachers want to visit the Statue of Liberty. The school hires buses that hold 44 people each.

22 Will 7 buses be enough to hold all of the students and teachers? Why or why not?

23 **What if** there are 9 buses. Will they be able to go to the Statue of Liberty? Why or why not?

24 In 1993, New York City was listed as the place where 128,434 immigrants intended to live. Los Angeles was listed by 106,703 immigrants. About how many more immigrants listed New York than Los Angeles? Explain your thinking.

25 The ferry to the Statue of Liberty has benches that each hold up to 8 people. If 240 people sit at the same time, what is the fewest number of benches needed to hold all the people? Explain your thinking.

mixed review • test preparation

1 $768 + 625$ **2** $6{,}520 - 129$ **3** $3{,}109 \times 5$ **4** $5{,}675 \times 27$

Choose the letter of the best unit to measure.

5 mass of a child
a. kg **b.** mg **c.** g

6 capacity of a bottle
a. mL **b.** g **c.** L

7 height of a door
a. cm **b.** m **c.** g

Extra Practice, page 513

Divide Greater Numbers

LEARN

In Florida, the Olympic torch was carried 1,148 miles in 9 days. If the same distance was covered each day, how many miles was the torch carried each day?

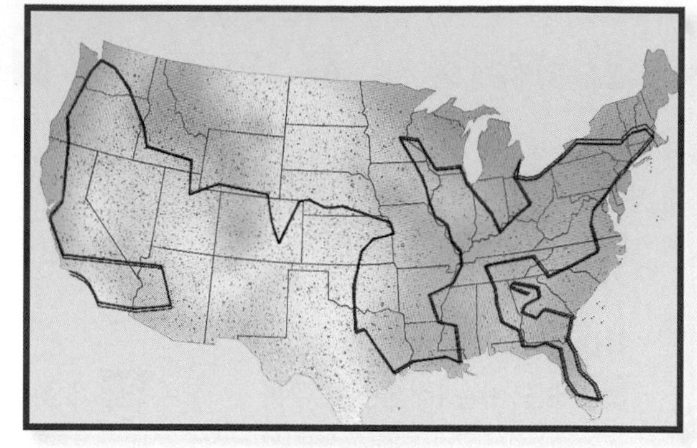

1996 Olympic Torch Route

Estimate to place the first digit of the quotient.

Estimate: $1{,}148 \div 9$

Think: $900 \div 9 = 100$
The first digit is in the hundreds place.

Step 1

Divide the hundreds.

$$\begin{array}{r} 1 \\ 9\overline{)1{,}148} \\ -\ 9 \\ \hline 2 \end{array}$$

Think: $9\overline{)11}$ with quotient 1
Multiply:
$1 \times 9 = 9$
Subtract:
$11 - 9 = 2$
Compare: $2 < 9$

Step 2

Bring down the tens.
Divide the tens.

$$\begin{array}{r} 12 \\ 9\overline{)1{,}148} \\ -\ 9\downarrow \\ \hline 24 \\ -\ 18 \\ \hline 6 \end{array}$$

Think: $9\overline{)24}$ with quotient 2
Multiply:
$2 \times 9 = 18$
Subtract:
$24 - 18 = 6$
Compare: $6 < 9$

Step 3

Bring down the ones.
Divide the ones.
Write the remainder.

$$\begin{array}{r} 127 \text{ R5} \\ 9\overline{)1{,}148} \\ -\ 9\downarrow \\ \hline 24 \\ -\ 18\downarrow \\ \hline 68 \\ -\ 63 \\ \hline 5 \end{array}$$

Think: $9\overline{)68}$ with quotient 7
Multiply:
$7 \times 9 = 63$
Subtract:
$68 - 63 = 5$
Compare: $5 < 9$

You can also use a calculator.
Press: 1,148 ÷R 9 = *127 R5*

The torch was carried about 128 miles each day.

CHECK

Check for Understanding

Divide.

1 $4{,}257 \div 3$ **2** $\$7{,}368 \div 8$ **3** $3{,}278 \div 3$ **4** $61{,}721 \div 2$

Critical Thinking: Analyze

5 Kim used a calculator to find $1{,}245 \div 7$. Her answer was 1,779 R2. Is this answer reasonable? How can you check? What might Kim have done?

PRACTICE

Practice

Divide.

1 $6\overline{)\$738}$ 2 $7\overline{)712}$ 3 $4\overline{)4,275}$ 4 $5\overline{)3,154}$

5 $7\overline{)75,214}$ 6 $6\overline{)14,567}$ 7 $8\overline{)\$320}$ 8 $9\overline{)\$9,144}$

9 912 ÷ 9   10 \$2,040 ÷ 4   11 1,054 ÷ 3   12 38,606 ÷ 6

Find only those quotients that are greater than 800.

13 \$3,975 ÷ 5   14 12,562 ÷ 7   15 4,175 ÷ 3   16 \$4,368 ÷ 6

17 7,345 ÷ 4   18 1,121 ÷ 3   19 1,605 ÷ 2   20 \$80,325 ÷ 9

MIXED APPLICATIONS
Problem Solving

21 In 1996, the Olympic torch was carried across California for 6 days and covered 1,208 miles. Did the torch travel more miles each day in Florida (see page 292) or California? Explain.

22 During the Summer Olympic Games, some athletes try to eat about 8,000 calories each day. If an athlete eats 6 meals each day, about how many calories should she eat at each meal?

Cultural Connection Babylonian Division

Ancient Babylonians divided by 5 using a method similar to the one shown below.

Divide: 340 ÷ 5

Step 1
Multiply the dividend by 2.
2 × 340 = 680

Step 2
Divide the product by 10.
680 ÷ 10 = 68

So 340 ÷ 5 = 68.

Use Babylonian division to divide. Then use a calculator to check your answer.

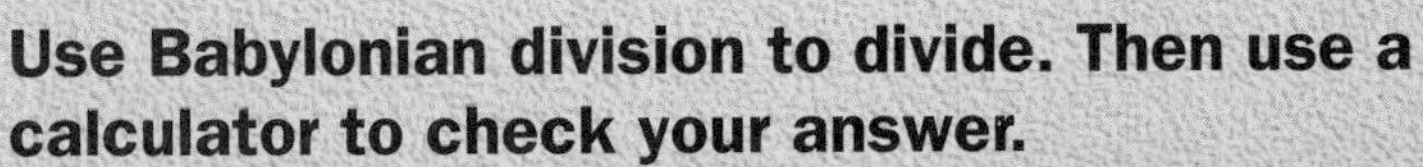

1 260 ÷ 5   2 720 ÷ 5   3 180 ÷ 5   4 370 ÷ 5   5 480 ÷ 5

Problem-Solving Strategy

Read
Plan
Solve
Look Back

Guess, Test, and Revise

LEARN

Read **Eighteen students in a science class sign up to collect water samples from Lake Michigan on one day during the week. A group of either 3 or 4 students will go each day. On how many days will a group of 3 collect samples? a group of 4?**

Plan List the information you know.

a. The sum of the days with 3 students and the days with 4 students is 5 days.

b. 18 students will collect samples.

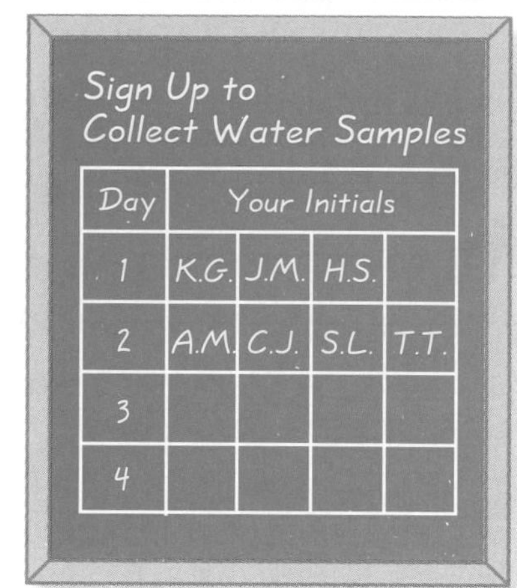

Solve

Guess Start with two numbers whose sum is 5. Try 1 and 4.

Then find the number of students.

Think: 1 day—3 students $1 \times 3 = 3$
4 days—4 students $4 \times 4 = 16$

Test $3 + 16 = 19$ students Too high

Revise Choose two other numbers whose sum is 5.

Guess Try 2 and 3. **Think:** $2 \times 3 = 6$
$3 \times 4 = 12$

Test $6 + 12 = 18$ students

There will be 3 students on 2 days and 4 students on 3 days.

Look Back How can you solve this problem a different way?

CHECK

Check for Understanding

1 **What if** there are 17 students instead of 18. On how many days will there be 3 students? 4 students?

Critical Thinking: Summarize

2 Explain how the test answer helps you to make a new guess.

PRACTICE

MIXED APPLICATIONS

Problem Solving

1 Jenny is making sand art. A bottle holds 8 inches of sand. Jenny wants to have 2 inches more of red sand than blue sand. How many inches of each color will she pour?

2 Nick got home from school at 3:00 P.M. He was in school for 6 hours and it took him a half hour to get to school and another half hour to walk home. At what time did Nick leave for school?

3 You have $80. You already have groceries that cost $73.22. Can you buy a box of cereal for $3.95? Explain.

4 **Spatial reasoning** The sheet of paper was folded in half and then holes were punched through it. Show what it will look like unfolded.

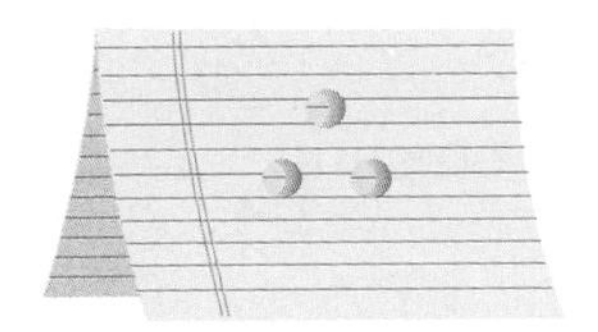

5 **Logical reasoning** If you multiply a number by 3, you get an even number. Can the number be odd? Explain.

6 Mrs. Kelly stores 120 books in a small bookcase and a large bookcase. Each large shelf holds 20 books. Each small shelf holds 15 books. How many shelves are in the large bookcase? The small bookcase?

7 Detroit, Michigan, is famous for making cars. One factory has 320 cars to deliver on trailers. Each trailer holds 8 cars. What is the least number of trailers needed to deliver the cars? Explain your answer.

8 In 1851, there were only 3 schools in Minnesota to teach 250 students. If each school had about the same number of students, were there more than 80 students in each school? Explain your answer.

Use the table for problems 9–11.

9 Which city has about five times as many people as Grand Rapids?

10 Which two cities are closest in population?

11 About how many more people does the most populated city have than Lansing?

Population of Largest Cities in Michigan	
Detroit	1,027,974
Grand Rapids	189,126
Warren	144,864
Flint	140,761
Lansing	127,321

Extra Practice, page 514

midchapter review

ALGEBRA: PATTERNS Complete.

1 $12 \div 4 = ■$
$120 \div 4 = ■$
$1{,}200 \div 4 = ■$

2 $■ \div 6 = 4$
$■ \div 6 = 40$
$■ \div 6 = 400$

3 $42 \div 7 = ■$
$420 \div 7 = ■$
$4{,}200 \div 7 = ■$

Estimate how many are in each box if the boxes have the same number in them.

4 7 boxes of pens

5 4 boxes of binders

6 6 boxes of scissors

7 8 boxes of erasers

Item	Total
Pens	287
Binders	121
Erasers	693
Scissors	321

Divide.

8 $4\overline{)532}$ **9** $7\overline{)95}$ **10** $3\overline{)314}$ **11** $6\overline{)61}$ **12** $8\overline{)29}$

13 $6\overline{)713}$ **14** $7\overline{)372}$ **15** $4\overline{)427}$ **16** $4\overline{)1{,}961}$ **17** $2\overline{)\$88{,}122}$

Find the missing number.

18 $3{,}200 \div 8 = ■$ **19** $900 \div ■ = 300$ **20** $2{,}000 \div ■ = 500$

Solve. Use mental math when you can.

21 The Rainbow Bridge in Utah is 200 ft high. It is the largest natural arch in the world. Gateway Arch in Missouri is the tallest monument in the United States. It is 430 ft taller than the Rainbow Bridge. How tall is it?

22 The Hershey Plant in Pennsylvania is the world's largest chocolate factory. Some candy bars are divided into 8 squares each. If you had 872 squares, how many chocolate bars would you have?

23 One of the largest stock exchanges in the United States is the Philadelphia Stock Exchange. Mr. Morris bought $552 worth of stocks. He paid $8 for each share. How many shares did he buy?

24 **What if** Mr. Morris paid $9 for each share and bought $864 worth of stocks. About how many shares would he own?

25 

Describe two different methods you could use to find $482 \div 4$.

Explore Divisibility Rules

Imagine hiking along a trail that starts in Maine and goes all the way to Georgia. You can do it on the Appalachian Trail. What if a hiker starts in New York and goes 135 miles in 9 days. Can she hike the same distance each day?

You can use a divisibility rule to find out if there will be a remainder when you divide.

Note: If there is no remainder when you divide, then the dividend is divisible by the divisor.

A number is divisible by 9 if the sum of its digits is divisible by 9.

$1 + 3 + 5 = 9$
9 is divisible by 9.

135 is divisible by 9.

Check your answer: $135 \div 9 = 15$.

She can travel 15 miles each day.

Here are more divisibility rules.
A number is divisible by:

2 if the ones digit is 0, 2, 4, 6, or 8.
3 if the sum of its digits is divisible by 3.
5 if the ones digit is 0 or 5.
6 if it is divisible by both 2 and 3.
10 if the ones digit is 0.

Use divisibility rules to answer mentally. Explain your thinking.

1 A tour company donates 1,456 tickets for a tour to Niagara Falls to 3 schools. Can you divide the tickets equally among the schools?

2 A group of 459 students are marching in a parade in New York City. Can they march in equal rows of 9 students across?

3 Can 657 jelly beans be divided equally among 2 classes?

4 Can you divide 354 counters into groups of 2? 3? 5? 6? 9? 10?

5 **Make a decision** You want to share cookies evenly among 6 of your friends. Would you buy a box that had 47, 98, or 102 cookies?

6 **Write a problem** that can be solved by using a divisibility rule. Solve it. Trade it with a classmate. Solve each other's problem.

real-life investigation

APPLYING DIVISION

A STATE TOUR

Where would you want to go if you and your friends could take a tour of places in your state? In this activity, you will work with a group to take a tour of special places you choose and then split the cost.

DECISION MAKING

Planning the Tour

1 Work with a group. Collect travel information about interesting places in your state.

2 Decide what places you and your group will tour and how long you will spend at each place.

3 Estimate the total amount you will spend for each of the following:
a. transportation from your home to the places on the tour. You need to decide if you will use an airplane, a train, or a bus.
b. motels or hotels for overnight stays.
c. meals.
d. tickets for the places on the tour.

4 Decide how you will split the cost of the tour among your group. For each person, find:
a. the cost each day.
b. the total amount to pay.

Reporting Your Findings

5 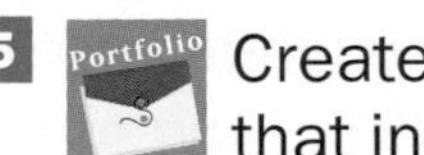

Create an advertisement that includes the following:

- ▶ a description of the tour and all that is included.
- ▶ a list that shows the cost each day for hotels, meals, entertainment, and transportation for one person.
- ▶ an explanation of how the costs were divided and why.

6 Compare your tour with those of other groups.

Revise your work.
- ▶ Did you provide all the information that was asked for?
- ▶ Is the advertisement organized and easy to follow?

MORE TO INVESTIGATE

PREDICT which locations in your state would be the most popular. Tell why.

EXPLORE other places of interest in your state and the distance you would travel each day to get there from your home.

FIND out how travel agents plan trips. Where do they get their information?

LEARN

Average

For three years in a row, Emmitt Smith of the Dallas Cowboys won the NFL rushing title. In addition, he scored 12 touchdowns in 1991, 18 in 1992, and 9 in 1993. How many touchdowns did he average each year?

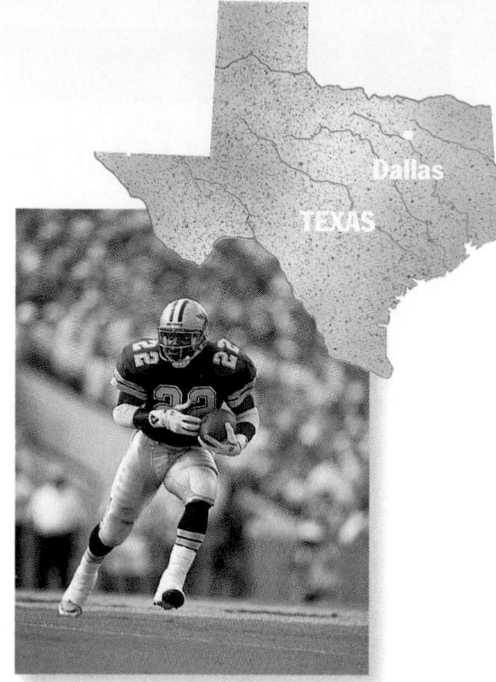

The Dallas Cowboys have won the Super Bowl five times.

Work Together

Work with a partner.

You will need
- *connecting cubes*

Let each cube stand for a touchdown. Use stacks of cubes to show the number of touchdowns for each year.

Rearrange the cubes so that the stacks are all the same height. The number of cubes in each stack is the **average** number of touchdowns.

Record your work in a table like the one shown.

Number of Stacks	Total Number of cubes	Average
3	12+18+9=39	

Use cubes to help you find the average of these numbers. Record your work.

a. 15, 8, 17, 9, 6
b. 17, 7, 12, 3, 2, 13
c. 25, 19, 16, 24

Make Connections

Look at your table. First, you found the total number of cubes. Then, you separated the total into equal groups.

Check Out the Glossary
average
See page 544.

You can use addition and division to find the average of any group of numbers, such as Emmitt Smith's touchdowns.

Step 1	Step 2
Add all the numbers.	**Divide the sum by the number of addends.**
12 + 18 + 9 = 39	39 ÷ 3 = 13

Emmitt scored an average of 13 touchdowns.

- **What if** Emmitt had scored only 3 touchdowns in 1993. How would this change the average? Explain.
- **What if** Emmitt had scored 13 touchdowns in 1994. How would this change the average? Explain.

CHECK

Check for Understanding

Find the average.

1 Number of minutes spent with each customer: 21, 42, 15, 35, 9, 22

2 Number of kilometers driven each day: 101, 92, 119, 85, 123

Critical Thinking: Analyze

Explain your reasoning.

3 Find the average math score for the first 4 weeks and then for the 5 weeks. How does one zero affect the grade? Explain.

Week	Math Quiz Scores
1	98
2	94
3	61
4	87
5	0

PRACTICE

Practice

Find the average.

1 Number of cousins each student has: 2, 6, 10

2 Number of tickets sold for each play: 11, 33, 55, 202, 101, 6

3 Number of umbrellas sold each day: 35, 46, 17, 119, 75, 14

4 Number of meals served at each school: 752, 864, 328

5 Find the average number of votes for each state.

6 Find the average number of tickets sold for the six months.

Where Would You Like to Live?

Class	Florida	California
4A	𝍸 III	𝍸 I
4B	𝍸 𝍸	𝍸 III
4C	IIII	𝍸 II
4D	𝍸 I	𝍸 𝍸 I

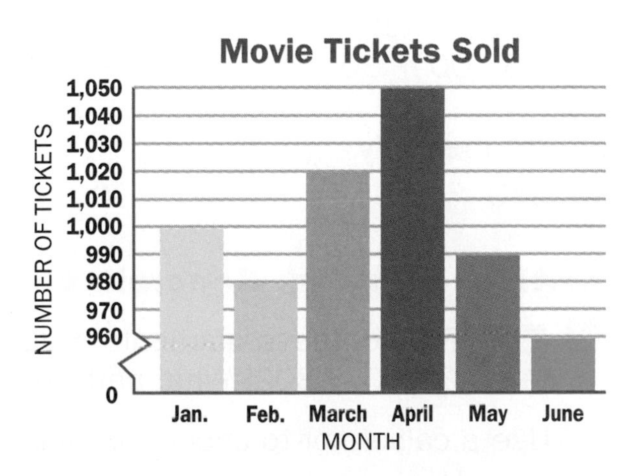

Extra Practice, page 514

Divide by Multiples of Ten

LEARN

Have you ever said, "Are we there yet?" when traveling with your family?

The distance from Toledo, Ohio, to Washington, D.C., is 450 miles. If you drive 50 miles each hour, how long will the trip take?

ALGEBRA: PATTERNS You can use patterns to help you divide mentally. What patterns do you see below?

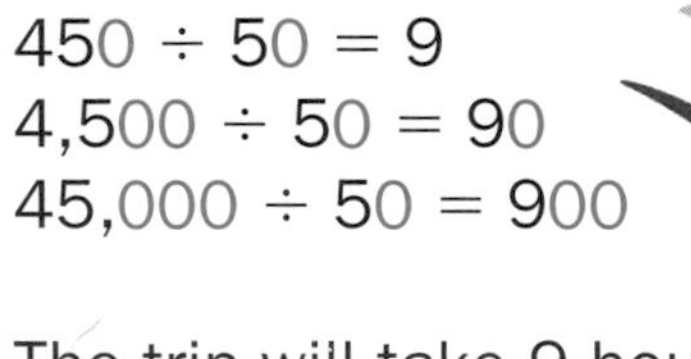

45 ÷ 5 = 9
450 ÷ 50 = 9
4,500 ÷ 50 = 90
45,000 ÷ 50 = 900

The trip will take 9 hours.

More Examples

A 6 ÷ 2 = 3
60 ÷ 20 = 3
600 ÷ 20 = 30
6,000 ÷ 20 = 300

B \$28 ÷ 7 = \$4
\$280 ÷ 70 = \$4
\$2,800 ÷ 70 = \$40
\$28,000 ÷ 70 = \$400

C 20 ÷ 5 = 4
200 ÷ 50 = 4
2,000 ÷ 50 = 40
20,000 ÷ 50 = 400

CHECK

Check for Understanding

ALGEBRA: PATTERNS Describe and complete the pattern.

1 15 ÷ 3 = 5
150 ÷ 30 = ■
1,500 ÷ 30 = ■
15,000 ÷ 30 = ■

2 40 ÷ 5 = 8
400 ÷ 50 = ■
4,000 ÷ 50 = ■
40,000 ÷ 50 = ■

3 8 ÷ 4 = 2
80 ÷ 40 = ■
800 ÷ 40 = ■
8,000 ÷ 40 = ■

Divide mentally.

4 540 ÷ 90 **5** 90 ÷ 30 **6** \$3,000 ÷ 60 **7** 64,000 ÷ 80

Critical Thinking: Generalize

8 

Predict these quotients: 600 ÷ 200 and 3,200 ÷ 800. What patterns with zeros did you use?

Use a calculator to check your answer.

Practice

PRACTICE

Divide mentally.

1. 630 ÷ 90
2. 160 ÷ 20
3. 600 ÷ 30
4. 300 ÷ 60
5. 1,000 ÷ 50
6. 2,100 ÷ 30
7. $8,000 ÷ 20
8. 4,000 ÷ 80
9. 5,600 ÷ 70
10. $7,200 ÷ 90
11. 4,000 ÷ 50
12. $3,600 ÷ 40
13. 16,000 ÷ 40
14. 35,000 ÷ 50
15. 32,000 ÷ 80
16. 21,000 ÷ 30

ALGEBRA **Find the missing number.**

17. 60 ÷ 30 = ■
18. 160 ÷ 20 = ■
19. 300 ÷ ■ = 60
20. 420 ÷ ■ = 6
21. 800 ÷ ■ = 40
22. ■ ÷ 70 = 70

MIXED APPLICATIONS

Problem Solving

23. **What if** your family wants to stay in a motel in Washington, D.C. The motel costs $720. The price is $90 each night. How many nights do you stay?
24. **What if** your family finds a motel in another part of town that charges $80 each night. How many nights can you and your family stay and still pay $720?
25. In three days, Mr. Garcia drove 305 miles, 450 miles, and 280 miles. What is the average number of miles that he drove each day? Explain your method.
26. Michael practiced for a marathon by running 37 miles each week for 7 weeks. He then ran 49 miles each week for 20 weeks. How many miles did he run altogether?
27. **Write a problem** that you could solve mentally using division. Solve it. Share your problem and solution with the class.
28. **Data Point** Use the Databank on page 539. Make a bar graph to show the populations. Add the population of your city.

mixed review • test preparation

1. 65,842 + 16,625
2. 138,615 − 56,124
3. 5,600 × 40
4. 15 × 249

Order the measurements from least to greatest.

5. 300 cm, 2 m, 100 cm, 250 mm
6. 50 in., 2 yd, 5 ft, 1 yd

Divide by Tens

LEARN

You used models to divide by 1-digit numbers. You can also use them to divide by tens.

Work Together

Work in a group.

Place 9 tens models in a pile. Place 9 ones models in another pile. Take a handful of models from each pile to get the total.

Spin the spinner to get the number in each group. Separate the total to find how many groups.

Record your results in a table. Write a division sentence or example.

Total	Number in Each Group	Number of Equal Groups	Number Left Over	Division Sentence or Example
97	40	2	17	$97 \div 40 = 2$ R17

You will need
- *spinner*
- *place-value models*

Take turns spinning, dividing, and recording.

Talk It Over

▶ Compare the remainder to the divisor in each line of your table. What do you notice?

Make Connections

Molly's group divided by tens this way and labeled their results.

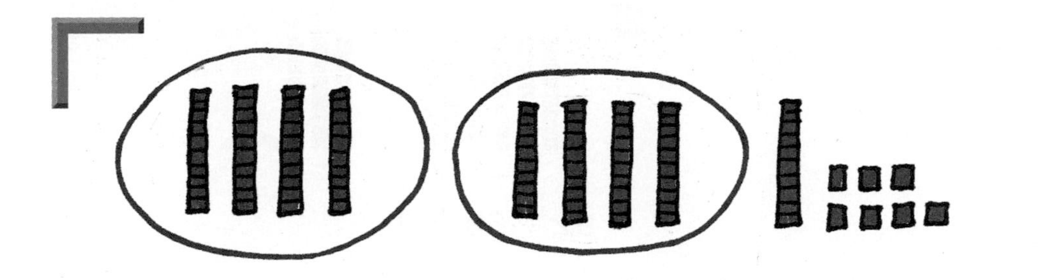

2 ← number of groups
number in each group → 40)97 ← total
−80
17 ← amount left over

You can also use this method with greater numbers.

Divide: $128 \div 40$

Estimate to place the first digit of the quotient.

Estimate: $128 \div 40$ **Think:** $120 \div 40 = 3$
The first digit is in the ones place.

```
    3 R8
40)128
 − 120
     8
```

Think: 40)128 (quotient 3)
Multiply: $3 \times 40 = 120$
Subtract: $128 - 120 = 8$
Compare: $8 < 40$

Check:
$3 \times 40 = 120$
$120 + 8 = 128$

Check for Understanding

Divide using any method.

1 10)79 **2** 20)60 **3** 30)167 **4** 20)109 **5** 80)488

CHECK

Critical Thinking: Generalize

6 Suppose you divide 158 by 40. What compatible numbers would you use to estimate? Would your estimate be the actual quotient? Why or why not?

Turn the page for Practice.

PRACTICE

Practice

Complete the division sentence.

1

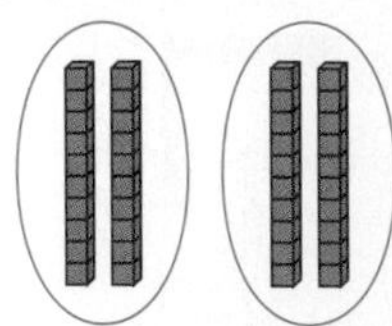

$55 \div 20 = ■$

2

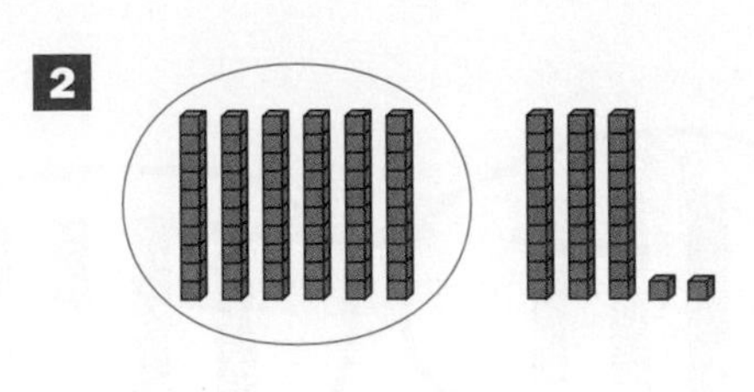

$92 \div 60 = ■$

Divide using any method.

3 $20\overline{)93}$	**4** $10\overline{)69}$	**5** $20\overline{)31}$	**6** $30\overline{)68}$	**7** $40\overline{)94}$
8 $20\overline{)183}$	**9** $70\overline{)222}$	**10** $50\overline{)265}$	**11** $60\overline{)\$240}$	**12** $80\overline{)588}$
13 $128 \div 30$	**14** $\$540 \div 60$	**15** $374 \div 50$	**16** $495 \div 80$	**17** $300 \div 70$
18 $468 \div 90$	**19** $354 \div 40$	**20** $\$630 \div 70$	**21** $356 \div 60$	**22** $859 \div 90$

ALGEBRA Use mental math to choose the letter of the correct answer. Explain your thinking.

23 $30\overline{)99}$ = ■   **a.** 3   **b.** 3 R9   **c.** 9 R9   **d.** 33

24 $■\overline{)584}$ = 7 R24   **a.** 12   **b.** 90   **c.** 80   **d.** 70

25 $20\overline{)■}$ = 6 R8   **a.** 12   **b.** 112   **c.** 120   **d.** 128

26 $387 \div 40 = ■$ **a.** 9   **b.** 9 R27   **c.** 27   **d.** 36

27 $247 \div ■ = 3$ R7   **a.** 60   **b.** 70   **c.** 80   **d.** 90

Make It Right

28 Here is how Kerry found $48 \div 30$. Tell what mistake was made. Explain how to correct it.

$48 \div 30 = 18$

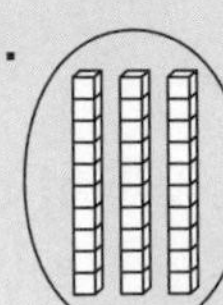

MIXED APPLICATIONS

Problem Solving

Use the table for problems 29–31.

29 There were 280 calories in the bread Leslie ate for breakfast. How many slices did she have?

30 Russell had a cup of milk, one egg, and some bread for breakfast. If he had a total of 515 calories for breakfast, how many slices of bread did he have? Explain your method.

Food	Quantity	Calories
Eggs	1	85
Milk	1 cup	150
Bread	1 slice	70

31 Kelly ate 600 calories at breakfast. What combination of eggs, milk, and bread could she have eaten?

32 A farmer buys chicken feed for $74.70. She gives the clerk four $20 bills and three quarters. What is her change?

33 Ralph's Breakfast Nook seats 16 people. Some booths seat 2 and others seat 3. How many of each type of booth does Ralph have in his restaurant?

34 **Write a problem** that can be solved using multiplication or division. Solve it both ways. Share your problem and solutions with the class.

more to explore

Changing Measurements

You can divide to change smaller units of measurement into larger units of measurement.

Change 108 inches into feet.

Think: 12 in. = 1 ft
108 in. ÷ 12 = ■ ft

108 ÷ 12 = 9

108 in. = 9 ft

Length	Weight	Capacity
12 in. = 1 ft 3 ft = 1 yd	16 oz = 1 lb	2 c = 1 pt 2 pt = 1 qt 4 qt = 1 gal

Convert the measurement to the unit shown. You may use a calculator.

1 1,728 ft = ■ yd **2** 72 in. = ■ ft **3** 128 oz = ■ lb **4** 1,236 qt = ■ gal

Extra Practice, page 515

Problem Solvers at Work

Read
Plan
Solve
Look Back

LEARN

Part 1 Interpret the Quotient and Remainder

Did you know that almost all the peanut butter that is made is eaten in the United States? You can make about 20 sandwiches from an 8-oz jar of peanut butter.

Work Together

Solve. Explain your thinking.

1. A student makes a total of 179 sandwiches for a picnic. Write the division sentence you would use to find the number of 8-oz jars of peanut butter he uses.

2. What information does the quotient give you? the remainder?

3. How many jars of peanut butter does he need to make 179 sandwiches?

4. **What if** the student packs the same number of sandwiches in each of 8 boxes. He then places the leftover sandwiches one at a time into some of the boxes. How many sandwiches will be in each box?

5. **Make predictions** Predict whether there will be more or fewer sandwiches in a box if the sandwiches are separated equally into 9 boxes with leftovers added one at a time to different boxes. Explain.

Cultural Note

George Washington Carver, born in Diamond, Missouri, was a chemist and inventor who developed hundreds of products from peanuts and other plants.

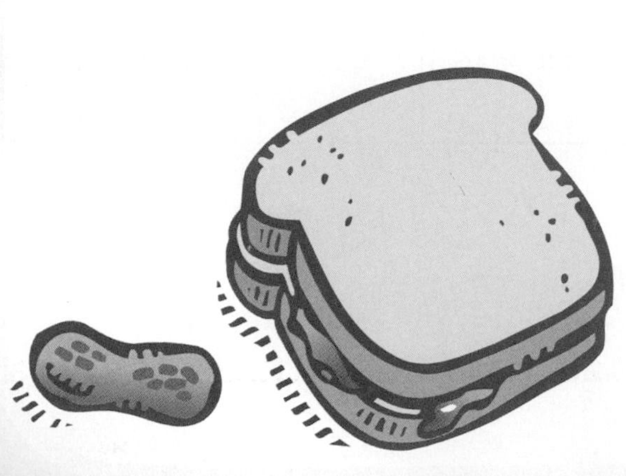

Part 2 Write and Share Problems

Scott used the information about the peanuts to write his own problem.

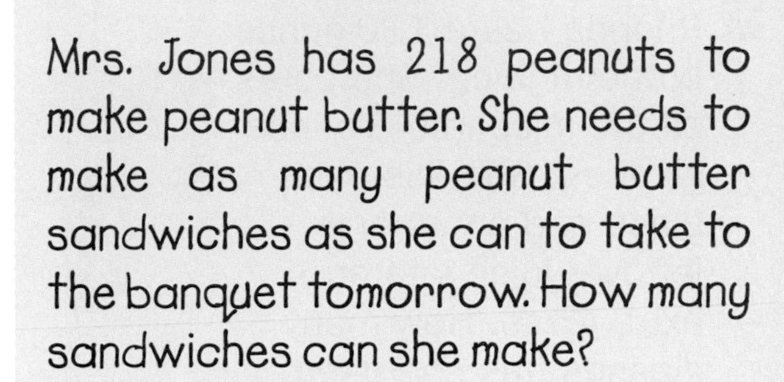

CHECK

6  Solve Scott's problem using division.

7 Change the information about peanuts that was used. Explain how the change affects the answer to Scott's problem.

8 **Write a problem** that uses division. The answer should include a remainder.

9 READING ARITHMETIC WRITING **Problem/Solution** Explain why the problem can be solved with division. Explain what the remainder means.

10 Trade problems. Solve at least three problems written by your classmates.

11 What was the most interesting problem that you solved?

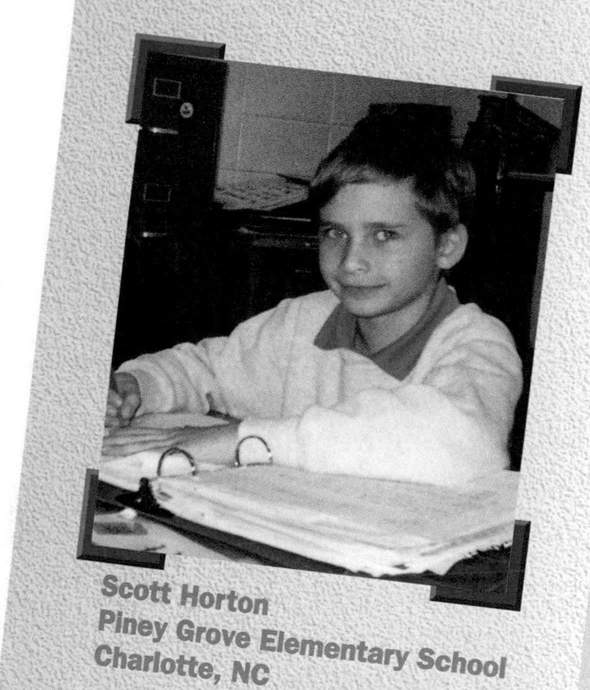
Scott Horton
Piney Grove Elementary School
Charlotte, NC

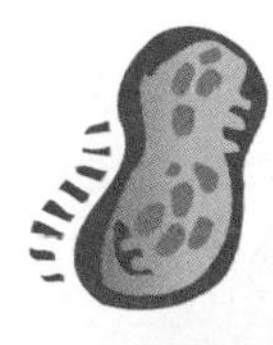

Turn the page for Practice Strategies.

Menu

Choose five problems and solve them. Explain your methods.

1 Rhonda has a 130-ounce bottle of juice. She pours all the juice into 8-ounce glasses. She fills each glass and pours any leftover juice into an extra glass. How many glasses has she used? Explain your answer.

2 There were 178 people at a restaurant. There were 16 more men than women. How many men were there? How many women were there?

3 **Logical reasoning** Eight fourth graders have been on the roller coaster and the parachute drop. Six have only been on the roller coaster. Five have only been on the parachute drop. How many have been on the parachute drop?

4 Mr. Ruiz must be at work by 9:00 A.M. Trains leave his town at 8:00, 8:15, 8:30, and 8:45 A.M. It takes him 5 minutes to walk from his house to the train. He rides the train a half hour and then walks another 5 minutes to his office. What is the latest time he can leave his house and get to work on time? Why?

5 Mrs. Teng wants to share 164 crayons equally among her 9 grandchildren. What is the greatest number of crayons she can give each grandchild? Explain how you got your answer.

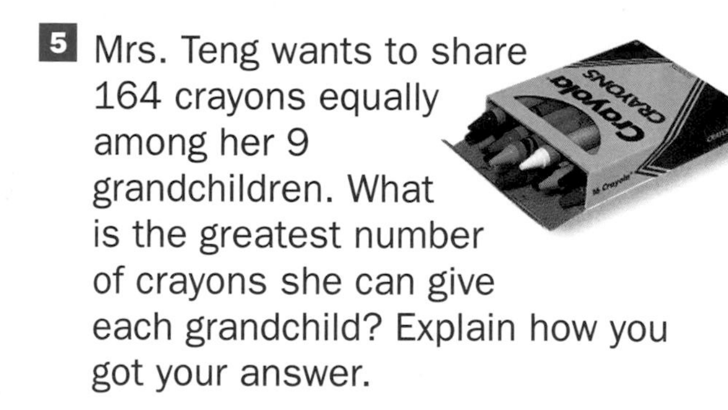

6 Sweaters sell for $25 each at Willmett's Department Store. At Buy Low's the same sweaters are 2 for $48. How much more do you have to spend at Willmett's for the 2 sweaters?

7 Cashews cost $4 a pound. Pistachios cost $3 a pound. Suppose you want to buy an equal number of pounds of each kind of nut. What is the greatest number of pounds of each you can buy for $30?

8 Alberto carries a 70-ounce container of water on a hike. If he drinks about 6 ounces every hour, will the water last for 14 hours? Explain.

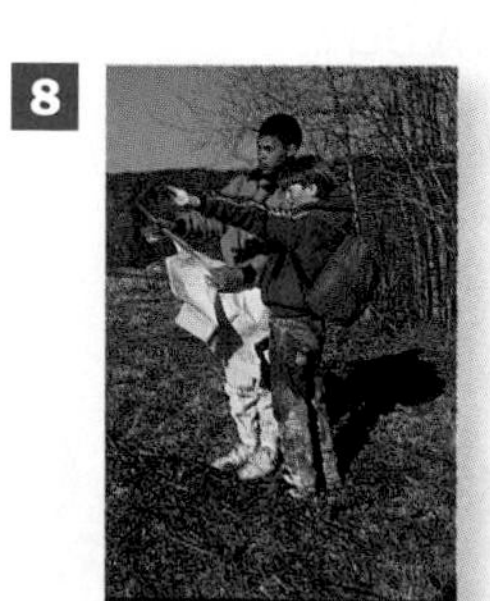

Choose two problems and solve them. Explain your methods.

9 **Data Point** Choose five classmates or family members. Find out how many hours each week they each watch television, do homework, talk on the phone, and sleep. Show the data in a table or graph. Use the data to find the average number of hours for each activity.

10 Tom will plant several rows of corn, beans, squash, and cabbage seeds. He plans to have corn in the 2nd row, 6th row, and 10th row. If he continues this pattern, how many rows of corn will he plant in the first 20 rows? Design a pattern to plant 30 rows while keeping the same pattern for corn.

11 Write a schedule for your next school day. Include information from the chart on the left.

Activity	Time Taken
Meals	45 minutes
Travel time	30 minutes
School	6 hours 15 minutes
Sleep	7 hours 45 minutes

School-Day Schedule

Activity	Time Start	Time End
Breakfast	6:30 A.M.	7:15 A.M.
Get Dressed	7:15 A.M.	7:30 A.M.
Travel to School		

12 **At the Computer** You can use a spreadsheet program to keep track of your grade averages.

a. Choose three subject areas. Keep track of the grades for your next four tests. Find the average for each subject.

b. Add a column for Test 5 and change the rule for the averages to include the new tests. Experiment to see what grade you need on Test 5 to raise the average for each subject by 1 point.

c. Predict how your next test will actually affect your average. Then check your prediction by putting your actual grade in the spreadsheet under Test 5.

Subject	Test 1	Test 2	Test 3	Test 4	Average
Math					
Science					
Music					

Extra Practice, page 515

Language and Mathematics

Complete the sentence.
Use a word in the chart. (pages 278–307)

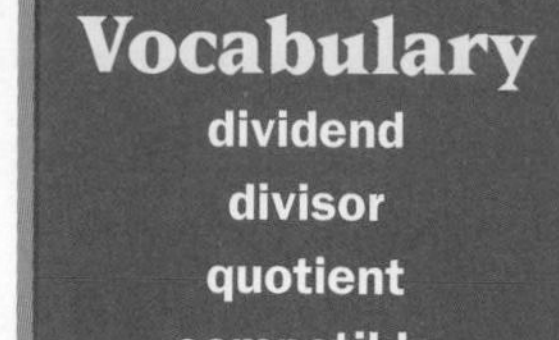

1 The ■ of 720 ÷ 8 is 90.

2 To estimate 169 ÷ 4, use ■ numbers.

3 The ■ of 35, 45, and 52 is 44.

4 If you divide 754 by 8, the ■ must be less than 8.

Concepts and Skills

Divide mentally. (page 278)

5 80 ÷ 2 **6** 420 ÷ 7 **7** 200 ÷ 4 **8** 3,000 ÷ 5

9 80 ÷ 20 **10** 350 ÷ 70 **11** 7,200 ÷ 90 **12** 4,000 ÷ 50

Estimate. Show the compatible numbers you used. (page 280)

13 827 ÷ 9 **14** 143 ÷ 5 **15** 1,427 ÷ 4 **16** 5,376 ÷ 8

17 209 ÷ 40 **18** 365 ÷ 60 **19** 484 ÷ 70 **20** 848 ÷ 90

Find the average. (page 300)

21 20, 32, 18, 26 **22** 231, 124, 512 **23** 154, 23, 129, 400, 9

Divide. (page 282)

24 9)37 **25** 4)85 **26** 2)75 **27** 5)94 **28** 8)93

29 9)$657 **30** 6)421 **31** 3)157 **32** 7)$742 **33** 3)765

34 2,437 ÷ 6 **35** 1,534 ÷ 8 **36** 4,578 ÷ 3 **37** 4,431 ÷ 4

38 360 ÷ 10 **39** 184 ÷ 20 **40** 227 ÷ 30 **41** 182 ÷ 40

Think critically. (page 282)

42 Analyze. Explain what the mistake is, then correct it.

```
   11 R3
 7)85
 − 75
   10
 −  7
    3
```

43 Generalize. Write *true* or *false*. Explain why.

a. To find the average of a set of data, add the data and divide the sum by 5.

b. When you divide a 4-digit number by a 1-digit divisor, the quotient must always have at least 4 digits.

c. When you use compatible numbers to estimate the quotient, the estimate is sometimes greater than the exact answer.

MIXED APPLICATIONS
Problem Solving

(pages 294, 308)

44 The longest river in the United States is the Mississippi River. It is 2,340 miles long. A family travels about 200 miles along the river each day. Can they travel along the entire length in 9 days? Explain.

45 A local charity group has collected 23 boxes of food to give to 5 shelters. Some of the shelters will get 4 boxes and others will get 5. How many shelters will get 4 boxes?

46 Use the diagram. How many feet of fencing are needed to go around the pool?

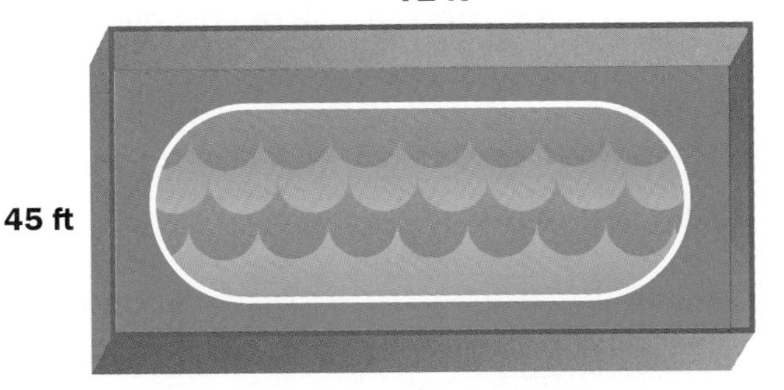

47 It takes Juan a half hour to clean each of the 8 kennels in the dog shelter. At what time must he begin cleaning the kennels in order to be done by 4:00 P.M.?

Use the table for problems 48–50.

48 How much larger is the largest state than the fifth-largest state?

49 About how many times larger than California is Alaska?

50 Which two states are about half as large as Texas?

Area of the Five Largest States

State	Area (in square miles)
California	163,707
Alaska	656,424
Texas	268,601
New Mexico	121,598
Montana	147,046

Estimate. Show the compatible numbers you used.

1 731 ÷ 90

2 308 ÷ 50

3 $417 ÷ 7

4 1,397 ÷ 20

5 $5,567 ÷ 9

6 6,298 ÷ 70

Find the average.

7 75, 54, 31, 12

8 312, 154, 89

9 19, 549, 32, 100

Divide. Use mental math when you can.

10 $312 ÷ 6

11 560 ÷ 7

12 3,457 ÷ 8

13 4,324 ÷ 3

14 638 ÷ 70

15 360 ÷ 6

16 291 ÷ 40

17 2,000 ÷ 50

18 $612 ÷ 6

19 $9\overline{)25}$

20 $7\overline{)82}$

21 $4\overline{)3{,}413}$

Solve.

22 Alley is standing in a line. She is 12th from the left and 17th from the right. How many people are standing in line?

23 Twice as many adults as children visited the town hall in one day. If 456 people visited the town hall, how many were children?

24 The road distance between Los Angeles and Atlanta is 2,182 miles. Suppose you average 50 miles an hour and you drive for 8 hours each day. Could you drive from Los Angeles to Atlanta in 5 days? Why or why not?

25 Los Angeles is 2,451 air miles from New York. If an airplane takes 5 hours to make this trip, about how many miles each hour is the plane flying?

performance assessment

What Did You Learn?

A town wants to hire buses to take 389 students to visit the dinosaur exhibits at the University of Nebraska State Museum. Each bus holds 45 students. Will 9 buses hold all the students?

▶ Solve this problem in two different ways. How could you use division? multiplication?

▶ Which way for solving this problem do you prefer? Explain why.

A Good Answer

- sets up the correct division and multiplication problems
- clearly shows the steps for solving the problem using each operation

You may want to place your work in your portfolio.

What Do You Think 

1 How do you know how to interpret the remainder when you solve a problem?

2 How have you used averages in the past? Was finding the average helpful? Explain.

3 What method do you use to find the first digit in the quotient?
- Estimate using compatible numbers.
- Choose a quotient by thinking of a fact and then use multiplication and addition to check your guess.
- Other. Explain.

math science technology

CONNECTION

Map It!

Cultural Note

The oldest known map of the world is from about 600 B.C. It was found in Mesopotamia, an area that held some of the earliest civilizations. It is carved on a clay tablet about the size of your hand. At the center of the map is the city of Babylon. The makers of most early maps put their own cities at the center.

Modern maps are very accurate because they are made from satellite photographs of Earth.

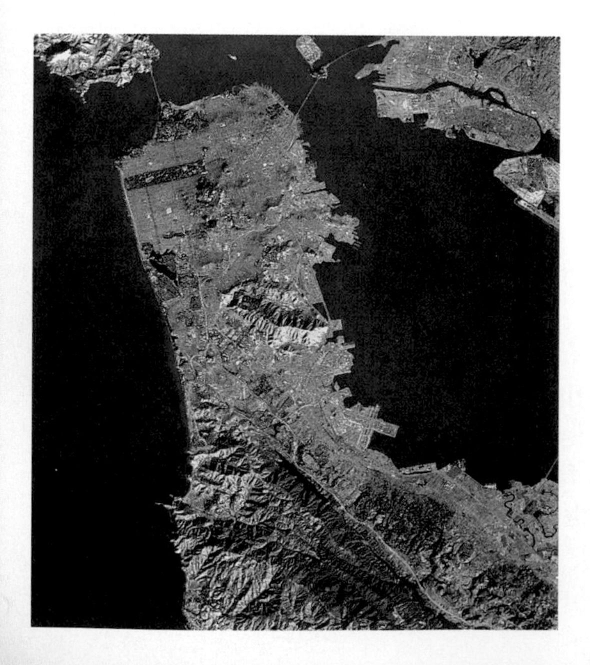

- What types of maps have you seen or made? What was shown on the maps? Why were they useful?
- Why do you think it might be helpful to place a landmark familiar to you as the center of your map? This is similar to what was done on the Babylonian map.
- How are the maps you have seen similar to each other? How are they different?

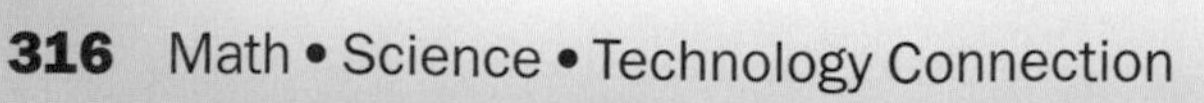

Classroom Maps

Maps show real objects or distances at a smaller size. They use scales to tell how much larger the real objects are.

The scale 1 cm = 1 m tells you that 1 centimeter stands for 1 meter. Anything shown on the map that is 1 cm long is actually 1 m long.

1 A map shows a chalkboard that is 2 cm long. The chalkboard is actually 4 m long. What scale is on the map?

2 **What if** you drew a similar map but used a different scale, 1 cm = 200 m. Would the chalkboard be smaller or larger on the map? Explain.

3 **What if** you wanted to draw a map of your state. Which scale would you use? Explain your reasoning.

a. 1 cm = 1 m **b.** 1 cm = 10 m
c. 1 cm = 1 km **d.** 1 cm = 10 km

At the Computer

4 Use a drawing or graphics program to make a map of your school, neighborhood, or town.

5 Change the scale of the map to see how this affects how clear and useful the map is. Tell what you notice.

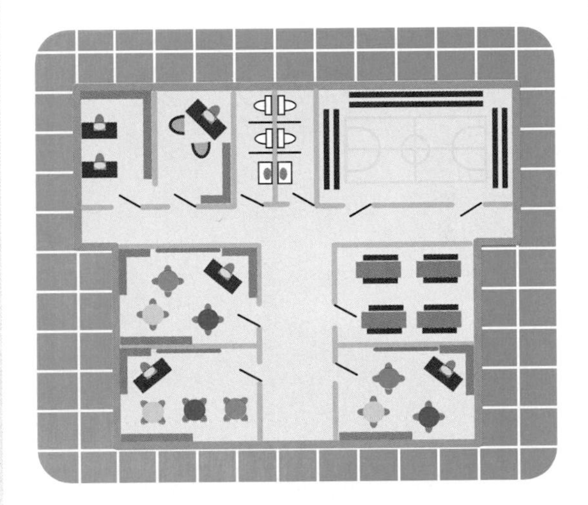

CHAPTER

9 GEOMETRY

THEME

Art and Nature

Do you know that geometry is almost everywhere you look? In this chapter you will look at geometric shapes in the art of various cultures and discover interesting shapes in nature.

What Do You Know ?

Look at the objects around you in the classroom. Create a table that lists the objects and their shapes.

1. What shapes did you list? Explain how you recognize each shape.

2. Describe at least three shapes that you can see in nature or in art.

3. Portfolio Choose three shapes from the list below. Sketch each shape. Then describe the shape in as many ways as you can.

triangle	**rectangle**	**circle**
kite	**cylinder**	**cube**

Categories

Look at the table you made of objects and their shapes. Put the shapes you found in your classroom into three different groups or categories.

Sometimes it is helpful to group things that are alike. When you put things into categories, you can compare them more easily.

1. What groups of shapes did you make? Describe each group.

2. How are your categories different?

Vocabulary*

*partial list

3-dimensional, p. 320
edge, p. 321
face, p. 321
vertex, p. 321
2-dimensional, p. 324
open, p. 324
closed, p. 324
polygon, p. 324
line, p. 328
ray, p. 328
intersecting, p. 328
parallel, p. 328
perpendicular, p. 328
diagonal, p. 330
angle, p. 332
congruent, p. 340
similar, p. 340
line of symmetry, p. 344
flip, p. 346
slide, p. 346
turn, p. 346
area, p. 348
square unit, p. 348
volume, p. 352
cubic unit, p. 352

3-Dimensional Figures

LEARN

You can fold paper to make 3-dimensional figures.

You can also fold patterns to make 3-dimensional figures. This pattern can be folded to make a box.

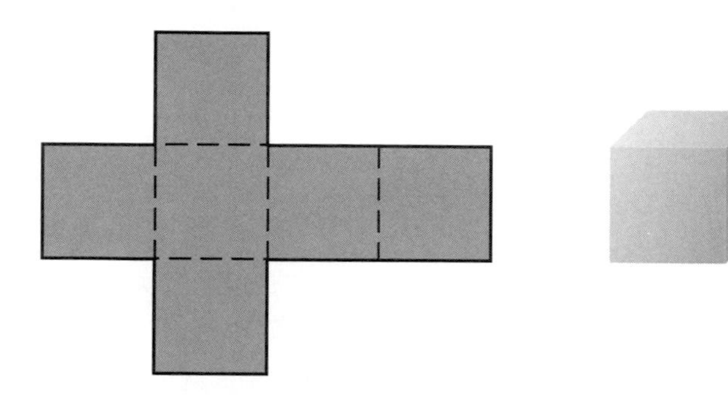

Work Together

Work with a partner. Copy and cut out patterns along the solid lines. Fold each pattern at the dotted lines. Tape the sides together to make the 3-dimensional figure.

You will need
- *patterns for figures*
- *scissors*
- *tape*
- *newspapers, magazines*

Note: A 3-dimensional figure has length, width, and height.

Look at newspapers and magazines. Find pictures of objects that have the same shape as your 3-dimensional figures.

Talk It Over

- How would you describe each figure? How are the figures alike? How are they different?
- What figure did you find the most pictures of? Why do you think this is so?

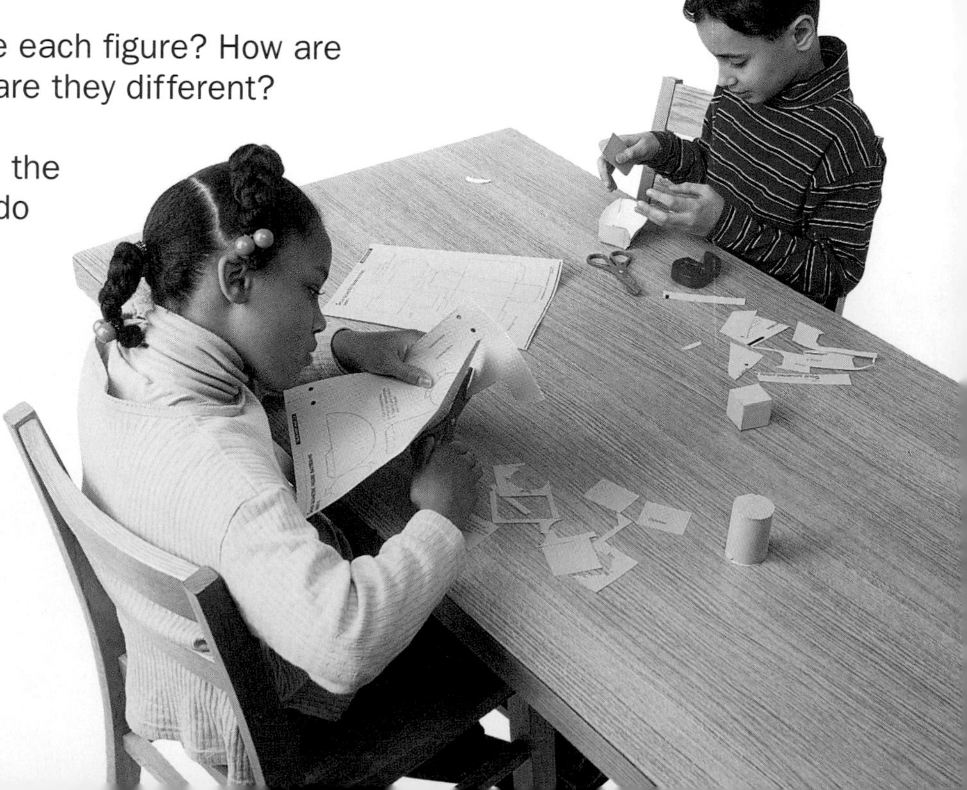

Check Out the Glossary
For vocabulary words
See page 544.

Make Connections

Look at these figures. Think about ways to describe them by describing their parts.

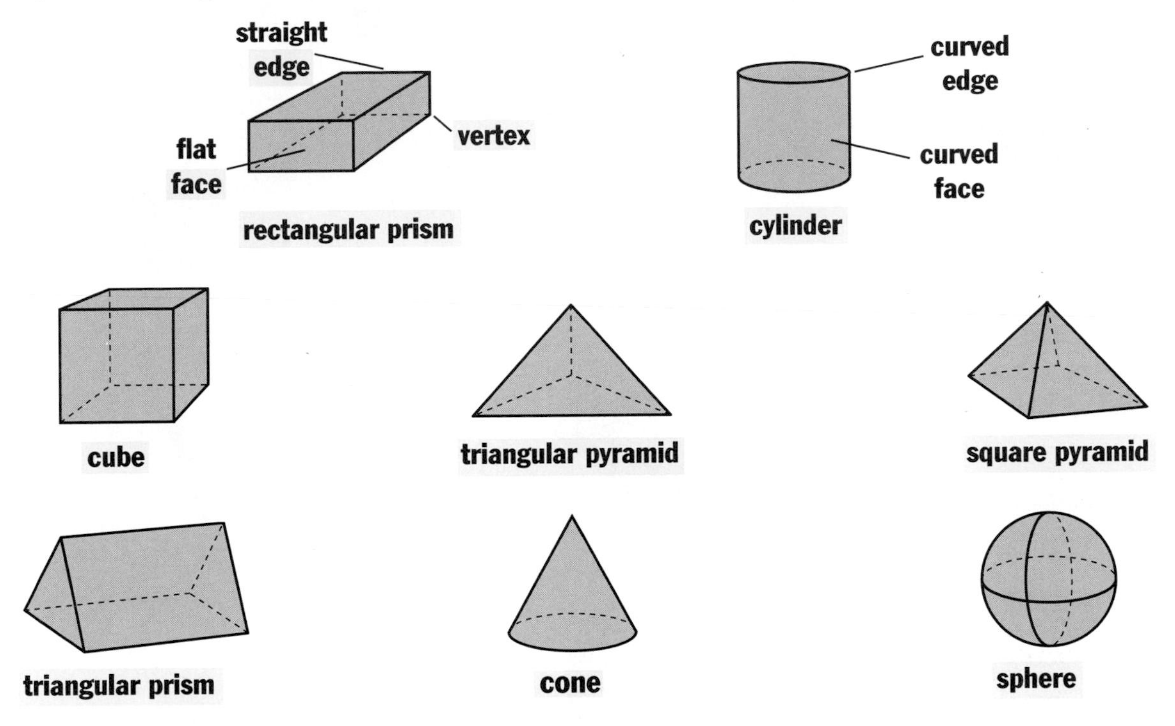

- ▶ Which figures have flat faces and straight edges? How many vertices, faces, and edges does each have?
- ▶ Which figures have curved edges or curved faces?

Check for Understanding

CHECK

Describe the figure. Tell how many vertices, edges, and faces it has.

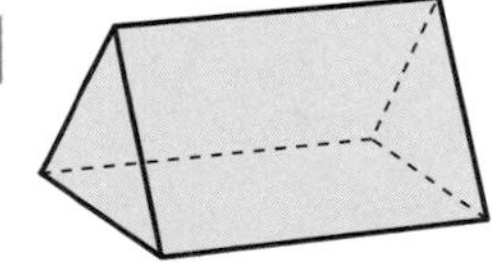

2

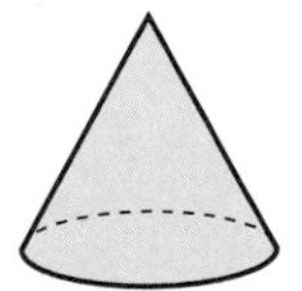

3

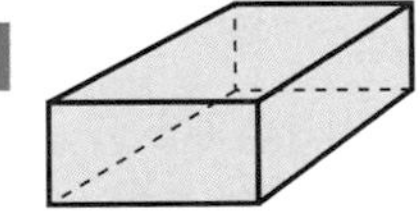

Critical Thinking: Analyze **Explain your reasoning.**

4 Look at the three patterns below. Which of these patterns can be folded to form a cube?

a.

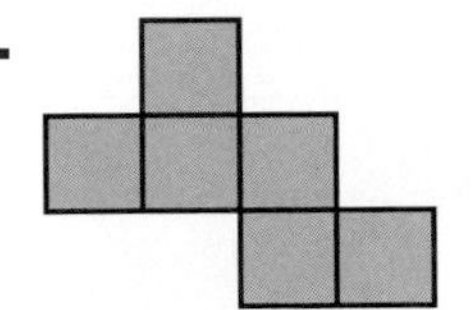

b.

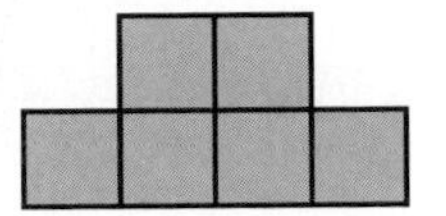

c.

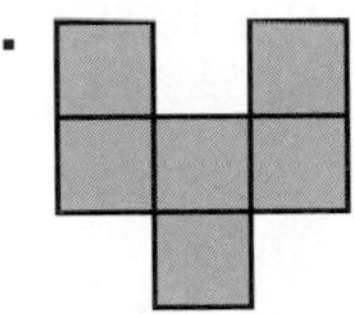

Turn the page for Practice.

PRACTICE

Practice

Name the 3-dimensional figure the object looks like.

1 

2 

3 

4 

5 

6 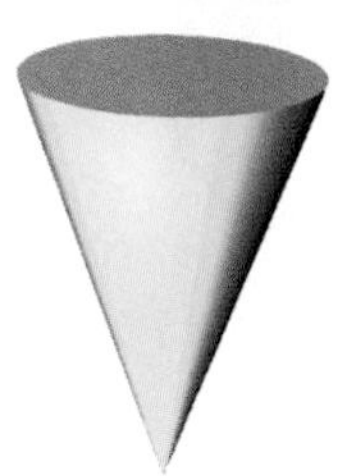

7 

8 

9 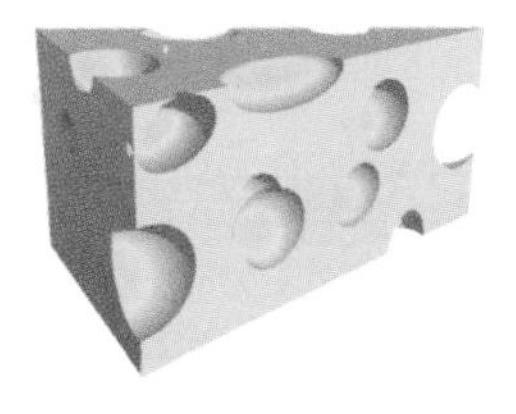

Describe the figure. Tell how many edges, vertices, and faces it has.

10 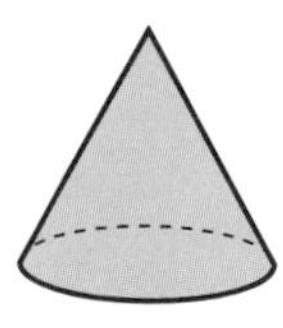

11 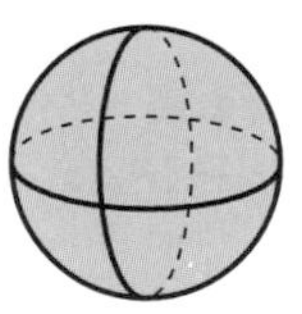

12 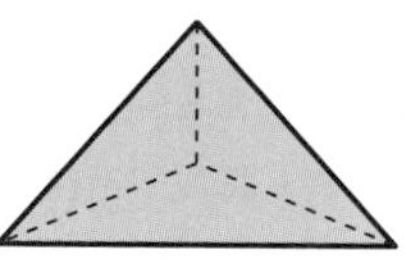

13 What figure has:

a. 1 curved face and no edges?

b. only 4 flat faces with all faces the same shape?

Make It Right

14 Marcos described this figure. Tell what the error is and correct it.

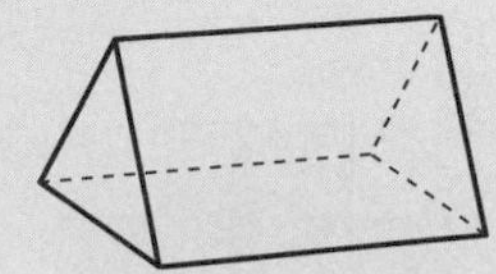

4 flat faces
6 vertices
9 edges

MIXED APPLICATIONS
Problem Solving

15 There are 35 marbles in a jar. If 8 children each get the same number of marbles, how many marbles will be left over?

16 **Spatial reasoning** Sheila built a 3-dimensional figure out of straws. Her figure has 8 vertices, 12 edges, and 6 faces. What could her figure be?

17 The Khufu pyramid is a square pyramid. About how long is the perimeter of its base? SEE INFOBIT.

INFOBIT
The pyramid of Khufu at Giza in Egypt is the world's largest pyramid. Each side of its base measures 230 m.

18 Marlena saved \$33.61 to buy her mother and father each a present. She bought her father a sweater for \$16.55 and her mother a purse for \$13.87. How much money does Marlena have left?

19 Rory bought a bicycle for \$160. He paid for it with 1 hundred-dollar bill and 6 ten-dollar bills. What other combinations of one-, ten-, and hundred-dollar bills could he have used? Explain.

20 Flyers from two different supermarkets were sent to you. Shopfirst has 9 oranges on sale for 81¢. ValueFood has 3 oranges on sale for 24¢. Where would you buy oranges? Explain why.

21 **Write a problem** about a 3-dimensional figure. Ask for a description of the figure, including the number of vertices, faces, and edges. Ask someone to solve the problem.

22 **Data Point** Copy the table on the right. Use the categories shown to sort the objects in the Databank on page 540. Complete the table by listing the object beside each category it fits.

Sorting Objects	
a. even number of faces	
b. odd number of faces	
c. straight edges only	
d. curved edges only	

mixed review • test preparation

1 2,623 + 845 **2** 983 − 217 **3** 286 × 419 **4** 17,600 × 43

5 84 ÷ 6 **6** 184 ÷ 30 **7** 288 ÷ 9 **8** 482 ÷ 70

2-Dimensional Figures and Polygons

LEARN

What 2-dimensional figures do you see when you look at these objects?

The sides of many 3-dimensional objects show 2-dimensional figures.

Note: A 2-dimensional figure has length and width only.

You can describe a 2-dimensional figure as **open** or **closed.**

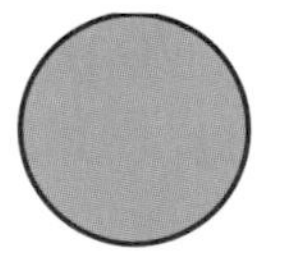 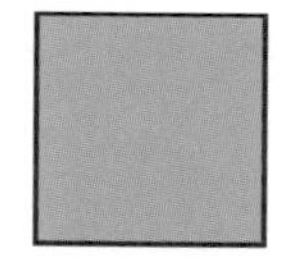 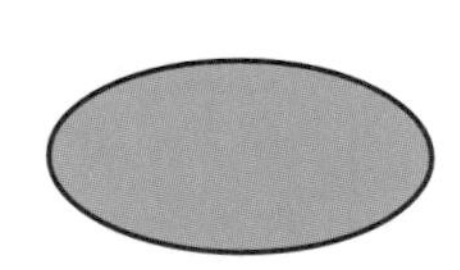

Closed figures

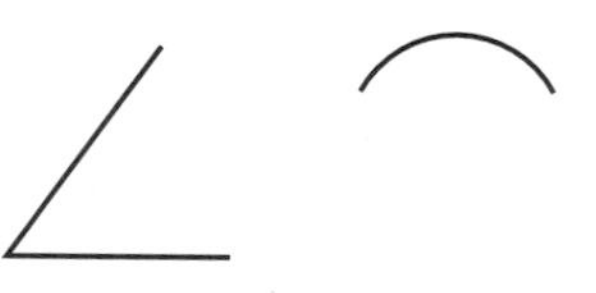

Open figures

Polygons are closed 2-dimensional figures made up of straight sides. A square is a polygon. A circle is not.

You can group polygons by the number of sides that they have.

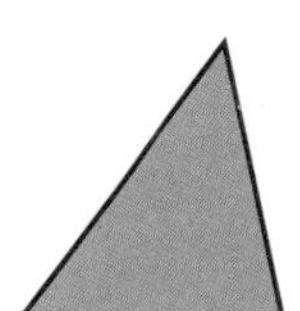

triangle
3 sides

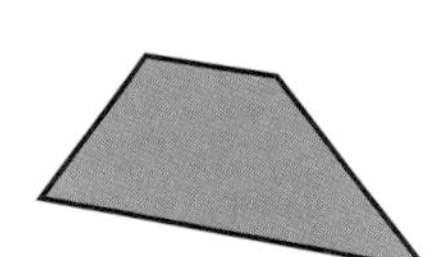

quadrilateral
4 sides

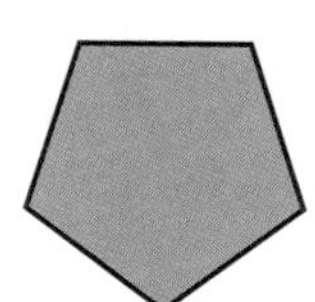

pentagon
5 sides

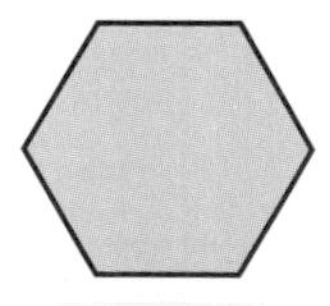

hexagon
6 sides

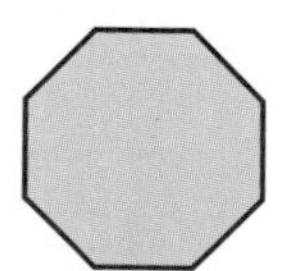

octagon
8 sides

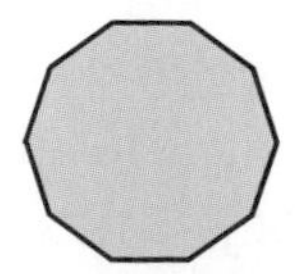

decagon
10 sides

Talk It Over

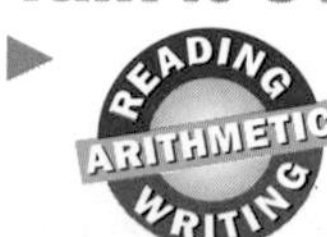

▶ **Categories** Do you think that you can group polygons by the number of corners that they have? Why or why not?

Check Out the Glossary
For vocabulary words
See page 544.

You can group quadrilaterals by the length of their sides and the size of their corners.

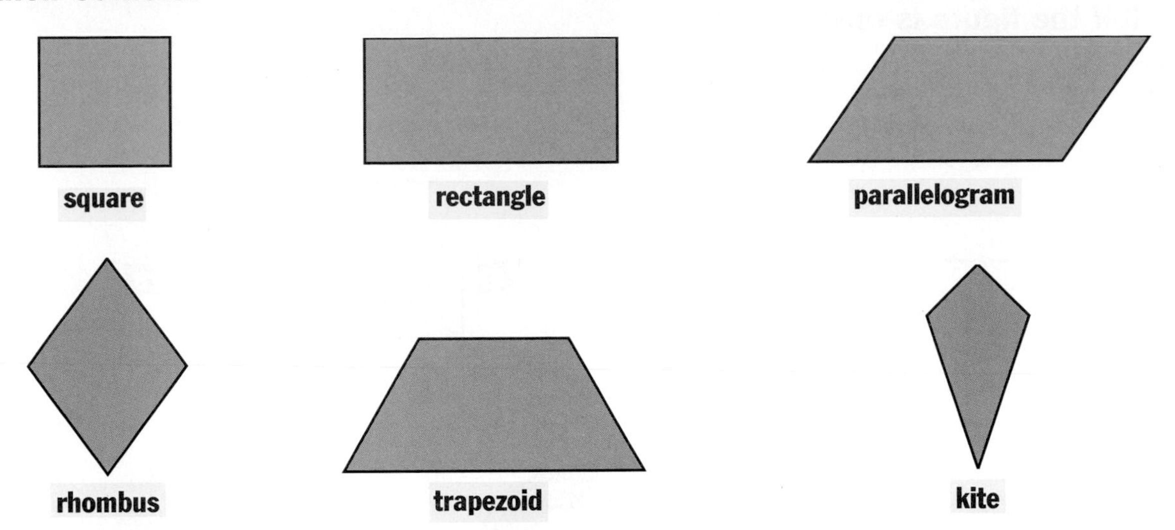

▶ How are all quadrilaterals alike? How is a square different from a trapezoid?

Check for Understanding

Is the figure open or closed? If closed, is it a polygon?

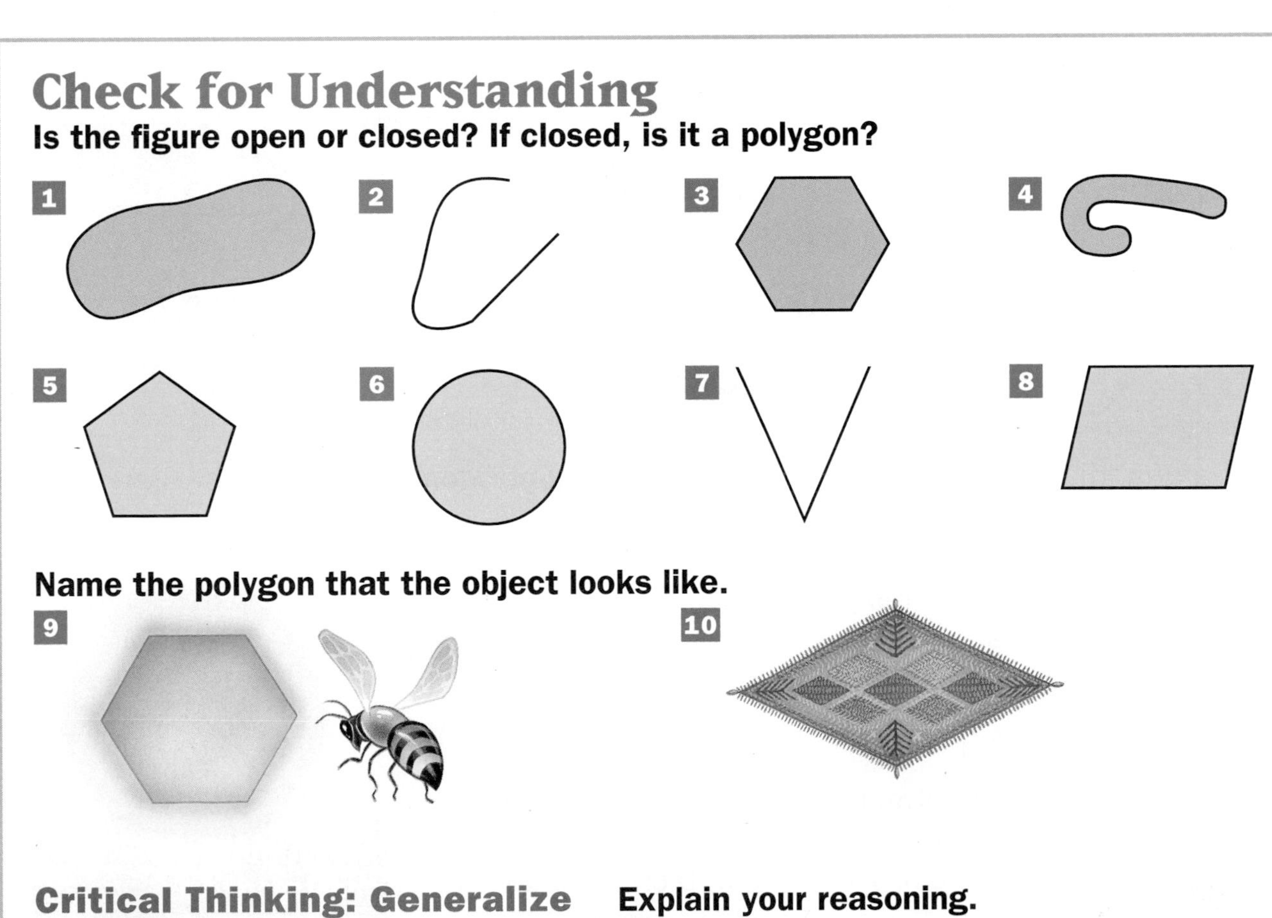

Name the polygon that the object looks like.

Critical Thinking: Generalize **Explain your reasoning.**

11 

Compare the square and the rhombus. Tell how they are alike and how they are different.

CHECK

Turn the page for Practice.

PRACTICE

Practice

Tell if the figure is open or closed.

1 

2 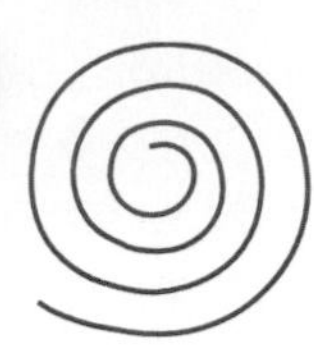

3 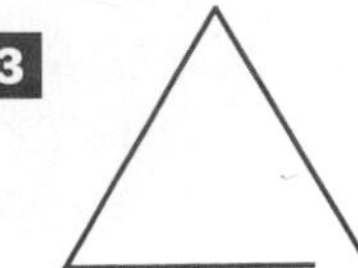

4 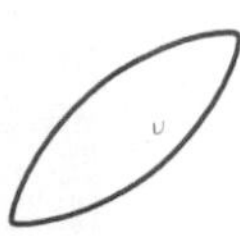

5 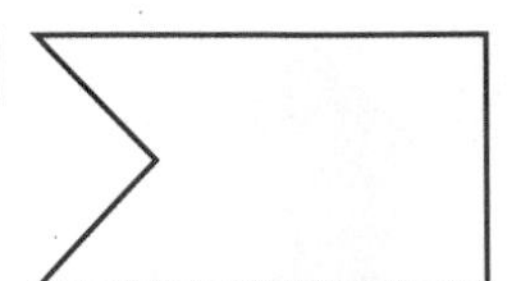

6 

7 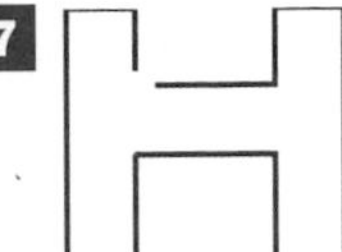

8 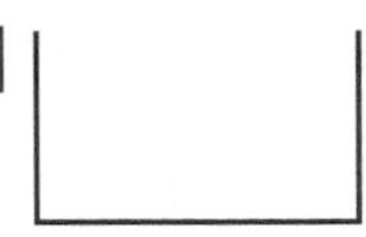

Tell if the figure is a polygon.

9 

10 

11 

Name the polygon.

12 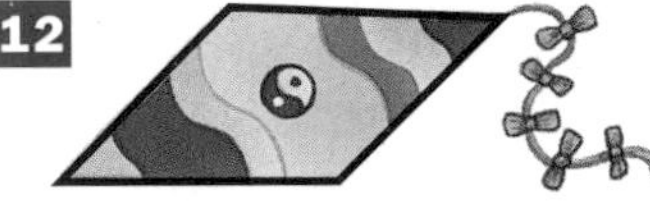

13 

14 

15 

16 

17 

18 a quadrilateral with all sides equal in length

19 a polygon with 10 sides

Draw the figure and name it.

20 a 4-sided polygon

21 a 3-sided polygon

22 a 6-sided polygon

23 a 5-sided polygon

24 A **heptagon** is a polygon with 7 sides. Draw one that has the following measurements for 6 of its sides: 3 cm, 5 cm, 7 cm, 7 cm, 9 cm, and 11 cm.

25 Draw a pentagon and label the sides to show a perimeter of 25 cm.

Check Out the Glossary
heptagon
See page 544.

MIXED APPLICATIONS
Problem Solving

26 Marla used toothpicks to make the following: 3 triangles, 4 pentagons, 5 quadrilaterals, and 6 octagons. How many toothpicks did she need?

27 **Spatial reasoning** When you look at the bottom of a 3-dimensional figure, you see a square. What figure could you be looking at?

28 **Logical reasoning** Use a table like the one at the right to help you find the shape.
Clue 1: I have four sides.
Clue 2: My sides are not equal lengths.
Clue 3: My opposite sides are equal.

	Clue 1	Clue 2	Clue 3
Square	Y		
Circle	N		
Parallelogram	Y		
Rhombus	Y		
Trapezoid	Y		

29 **Write a problem** similar to problem 27 for a shape other than a square. Solve it. Then have others solve it.

30 Each side of a hexagon measures 6 cm. What is the perimeter of the hexagon?

Cultural Connection
Kites from Different Cultures

The first kites were built over 3,000 years ago in China. These kites had bamboo pipes attached to them. The pipes made loud whistling sounds when the kites were flown. The Chinese used these kites in battle to frighten their enemies.

In Japan, there are kites that are larger than humans. They are flown in a festival that has been celebrated for over 400 years.

Modern kites are flown for fun by adults and children around the world.

▶ Name the shapes you see in the kite shown.

Line Segments, Lines, and Rays

Artists use geometric ideas in their work.

A **line** is a straight figure that goes on forever in both directions.

A **ray** is another straight figure. It has one **endpoint** and goes on forever in just one direction.

"METROPOLITAN PORT," 1935–1939, JOSEPH STELLA, NATIONAL MUSEUM OF AMERICAN ART

A **line segment** is a straight figure with two endpoints.

Line segments that stay the same distance apart from each other are **parallel.**

Line segments that meet or cross each other are called **intersecting** line segments.

Intersecting line segments that form square corners are **perpendicular** to each other.

Work Together

Work with a partner to sketch your own designs.

Without showing your partner, sketch a design using the geometric figures above. Then exchange sketches.

Describe your partner's sketch using the words *line*, *ray*, *line segment*, *parallel*, *intersecting*, and *perpendicular*.

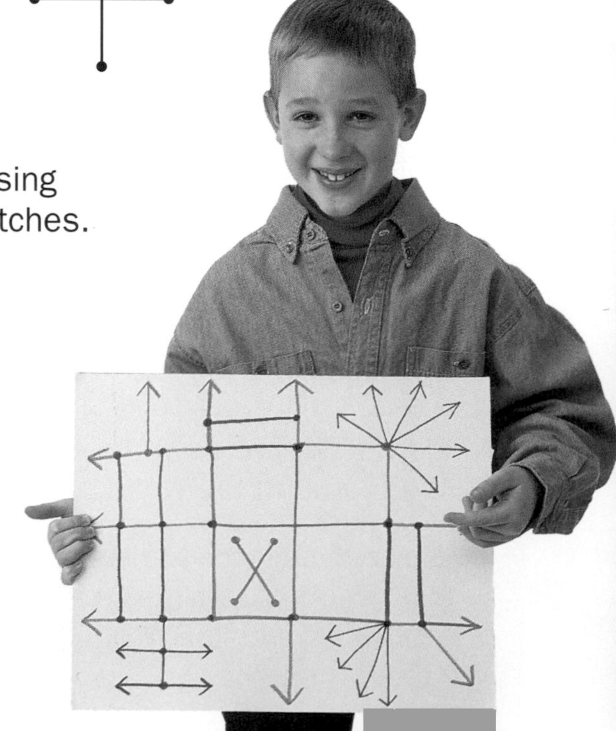

Talk It Over

- What could your partner have added to the sketch to make it easier to describe?

Check Out the Glossary
For vocabulary words
See page 544.

Make Connections

Sometimes it is easier to describe geometric figures if they are labeled.

Line segment *AB* is parallel to line segment *CD*.

Line *HJ* is perpendicular to line segment *AB*.

Ray *EG* intersects line *LH*.

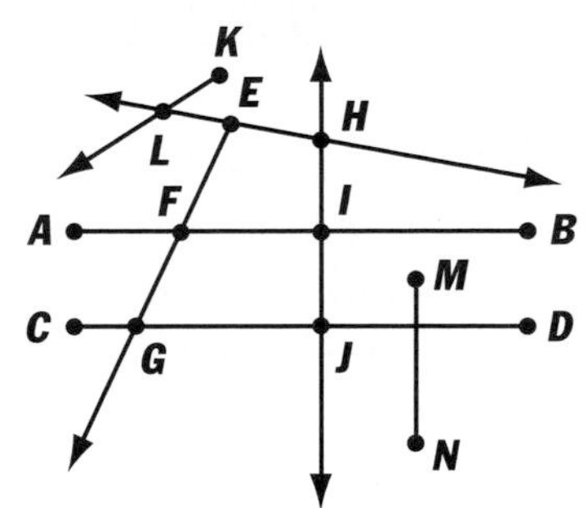

Check for Understanding

Use the figures above for ex. 1–4.

1 List three line segments.

2 List two rays.

3 List two intersecting line segments. Are they perpendicular? Explain.

4 Are there any parallel lines?

Critical Thinking: Generalize **Explain your reasoning.**

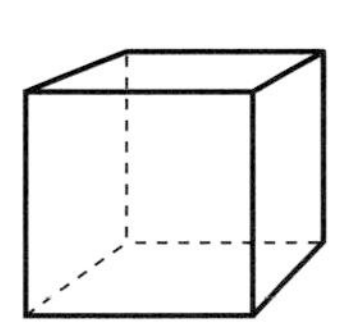

5 Explain why you agree or disagree with each statement.
a. The opposite faces of a cube are perpendicular.
b. The corner of a cube is formed by three perpendicular faces.

Practice

Describe the figure.

1

2

3

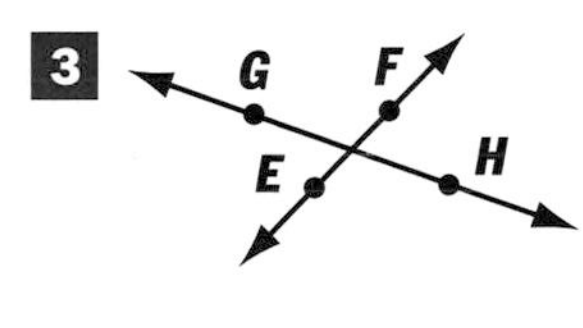

4

5 Sketch some polygons. Describe them using the terms *line segment*, *parallel lines*, and *intersecting lines*.

6 The horizon is a real-world example of a line. List a real-world example of a *line segment*, *parallel lines*, and *intersecting lines*.

Extra Practice, page 516

Read
Plan
Solve
Look Back

Make an Organized List

PRACTICE

Compare the number of sides and the number of diagonals of different polygons. How can you use this relationship to predict the number of diagonals in an octagon?

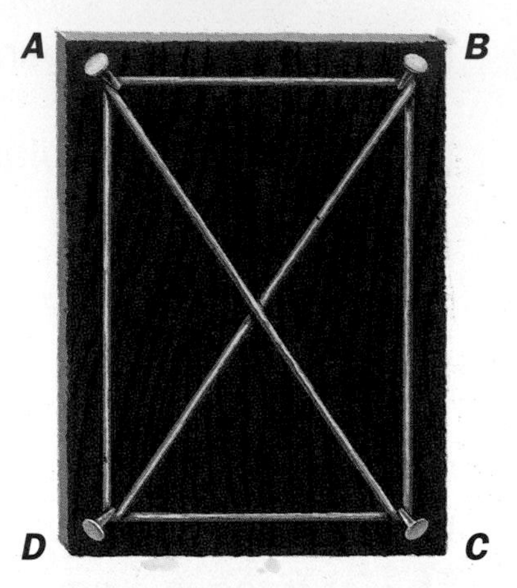

A line segment that connects two vertices, but is not a side, is called a **diagonal.** In the string figure on the right, a diagonal connects vertex *A* to vertex *C* and another diagonal connects vertex *B* to vertex *D*.

To solve the problem, you can make a table to help you compare the numbers and to find a pattern.

Check Out the Glossary
diagonal
See page 544.

Solve

Order the shapes in the table according to the number of sides. Look for a pattern.

Polygon	Number of Sides	Number of Diagonals
Quadrilateral	4	2
Pentagon	5	5
Hexagon	6	9
Heptagon	7	14

+ 3
+ 4
+ 5

An octagon has one more side than a heptagon, so the octagon has $14 + 6 = 20$ diagonals.

Look Back

Can the pattern be used to predict the number of diagonals in other polygons? Show an example.

CHECK

Check for Understanding

1 Alberto claims that a 12-sided polygon, or dodecagon, has 54 diagonals. Is he correct? How did he get this number without drawing the shape?

Critical Thinking: Summarize

2 Why is it a good idea to organize the shapes in the table according to the number of sides they have?

MIXED APPLICATIONS

Problem Solving

1 The first day of camp is July 8th, which falls on a Monday. The last day of camp is July 31. What day is that?

2 Melinda drew a polygon that had 27 diagonals. How many sides are in her polygon?

3 **Make a decision** You are visiting an amusement park. Use the list at the right. How would you spend 50 tickets at the park?

RIDES.................................4 TICKETS
TILT A WHIRL — FERRIS WHEEL

GAMES...............................5 TICKETS
BASKETBALL SHOOT — PONY RIDE
COIN TOSS — WATER RACE

FOOD/DRINK.................3 TICKETS
COTTON CANDY — HOT DOGS
SODA

4 A student saved the same amount of money each week for 7 weeks. He saved a total of $84.00. How much money did he save each week?

5 **ALGEBRA: PATTERNS** Terry's goal is to write to at least 50 sports stars to ask for autographs. She wrote 1 letter the first week, 3 letters the second, 6 letters the third, and 10 letters the fourth week. What pattern is Terry following?

6 Students are playing telephone. The first player gives a message to 2 other players. Each of these players gives the message to 2 other players. If each of these players then tells 3 more players, how many players will have given or heard the message?

7 An African giant frog measures 31 in. with its legs outstretched. When it is sitting normally, it measures 14 in. How much more does it measure when its legs are outstretched?

8 **Write a problem** that describes a polygon and asks for its name. Ask others to solve your problem. If they cannot, ask them to suggest ways you can describe the polygon more clearly.

Use the bar graph for problems 9–11.

9 How many students were surveyed?

10 Which flavor was the favorite of 4 students?

11 How many more students prefer vanilla to chocolate ice cream?

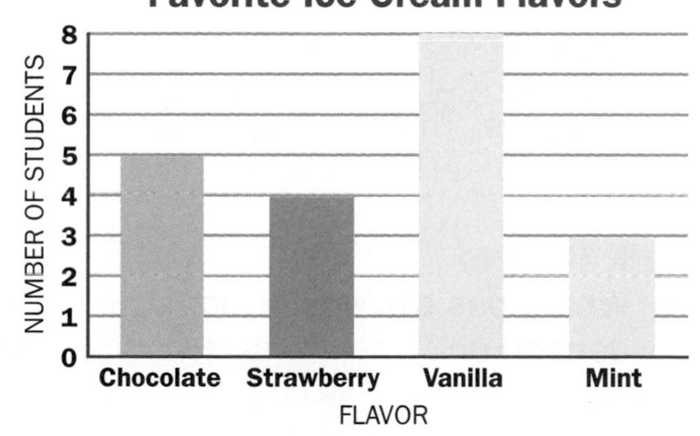

Extra Practice, page 517

Angles

LEARN

Maybe you noticed that if you draw intersecting line segments or rays, you create angles.

Check Out the Glossary
For vocabulary words
See page 544.

Angles are formed when two line segments or rays meet or cross.

You can find angles everywhere in nature.

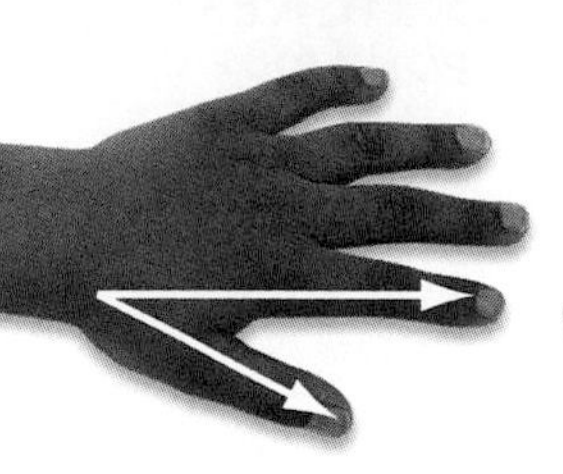

A **right angle** is formed by perpendicular lines. A right angle forms a square.

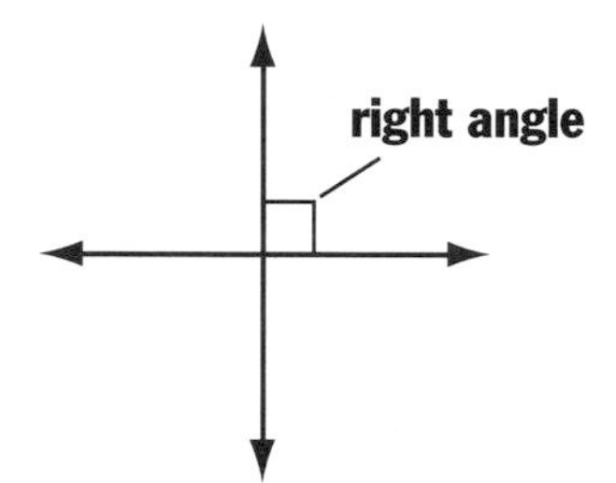

Artists, architects, and builders use a right angle as a measuring tool.

Work Together

Work with a partner to make a list of angles that you see in your classroom.

Use a corner of a sheet of paper, which is a right angle, as a paper tool.

Find angles in your classroom. Use your paper tool to compare them to right angles. Then choose categories to help you sort the angles.

Talk It Over

- ► What does the size of an angle depend on?

Make Connections

Angles that are not right angles also have special names.

An angle that is smaller than a right angle is called an **acute angle.**

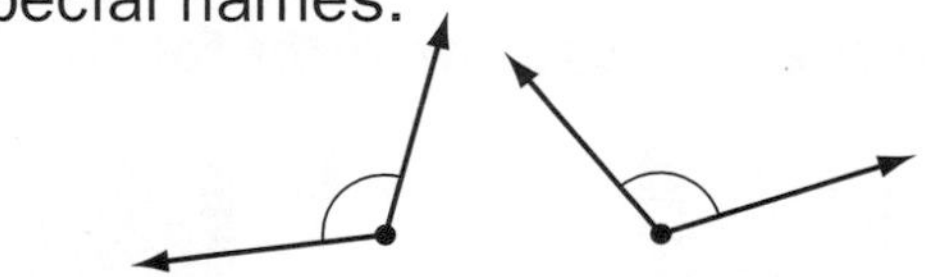

An angle that is larger than a right angle is called an **obtuse angle.**

► Use *acute*, *obtuse*, and *right angle* to describe the angles on your list.

Check for Understanding

CHECK

Write *acute, obtuse,* or *right* for the angle.

1 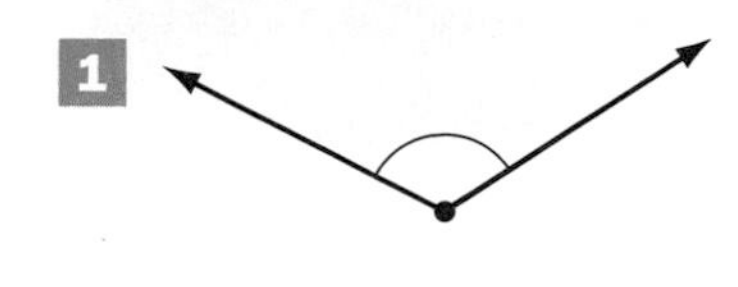

2 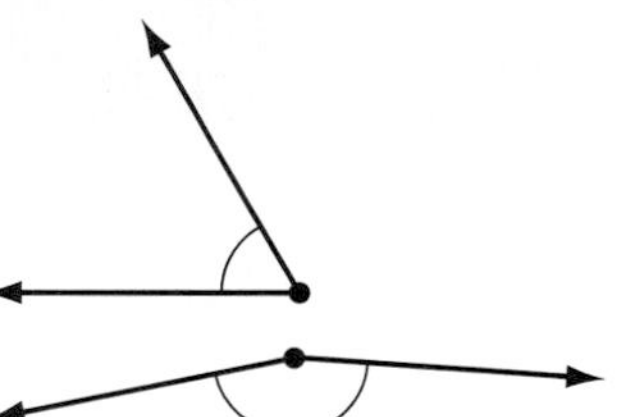

3 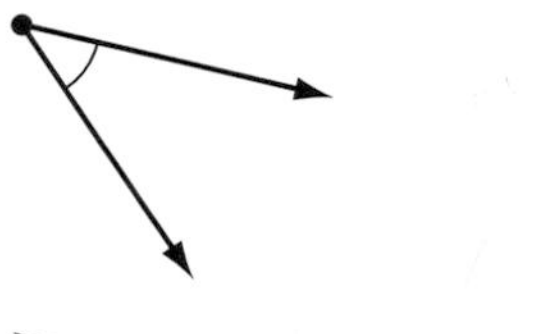

4 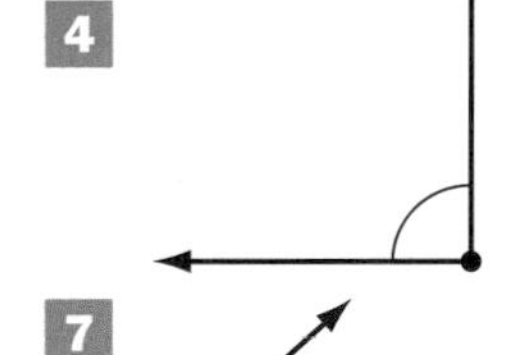

5

6

7

8 

9 

Write *acute, obtuse,* or *right* for each angle in the figure.

10 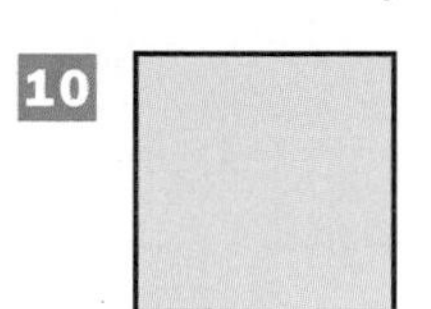

11 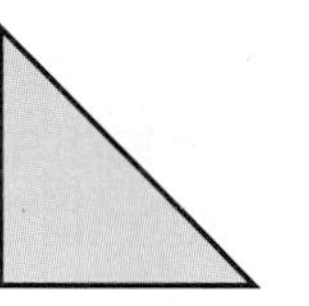

12 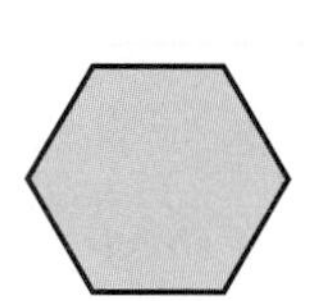

13 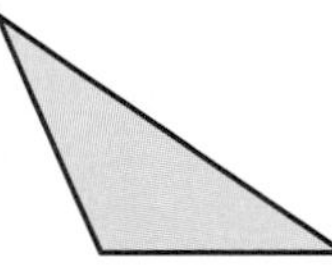

14 

15 

16 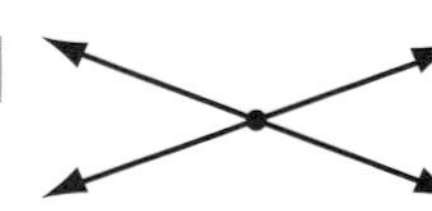

17 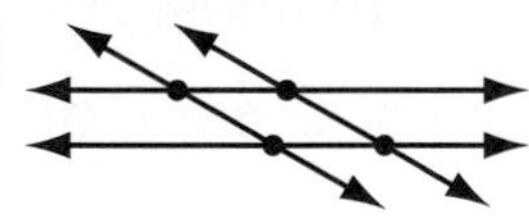

18 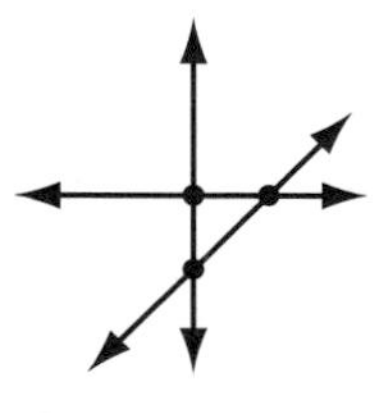

Critical Thinking: Summarize **Explain your reasoning.**

19 Explain the difference between a right angle, an acute angle, and an obtuse angle. Give an example of each.

Turn the page for Practice.

Practice

Write *acute*, *obtuse*, or *right* for the angle.

 1

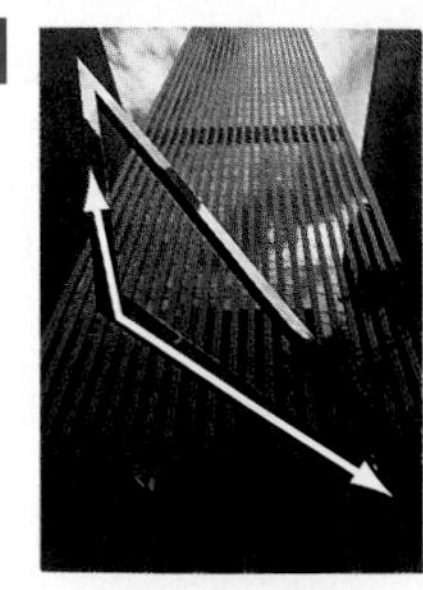

 2

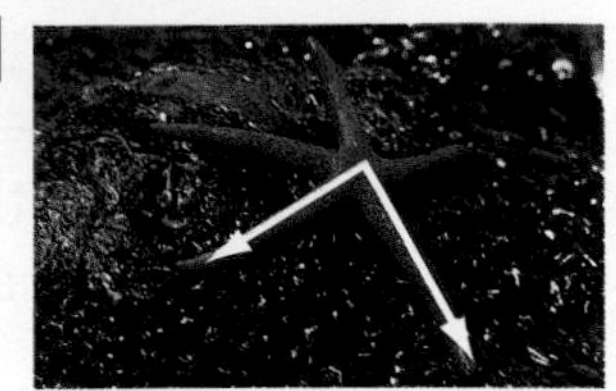

3

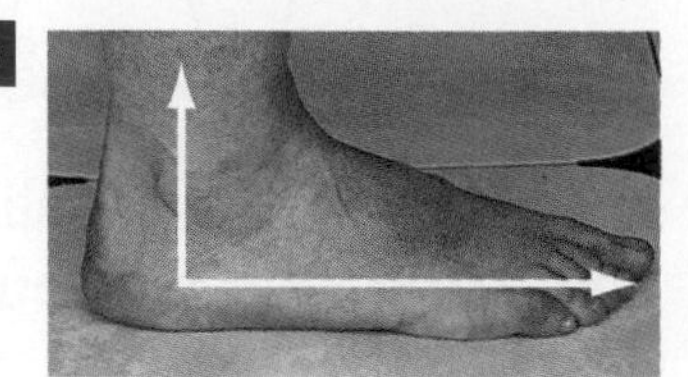

 4

5

6

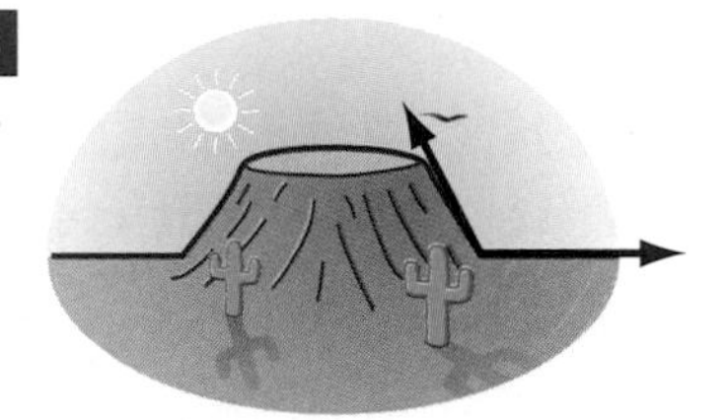

7 Which of these polygons have right angles?

a.

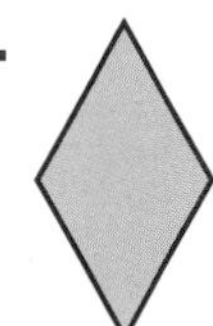

b.

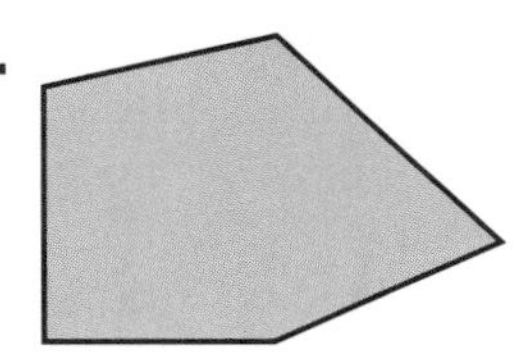

c.

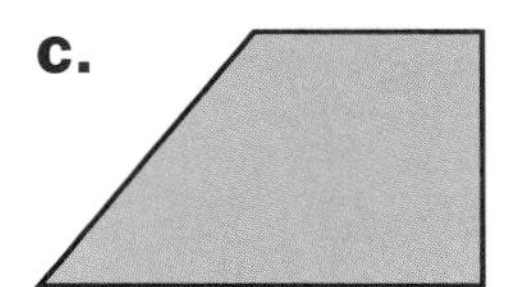

d.

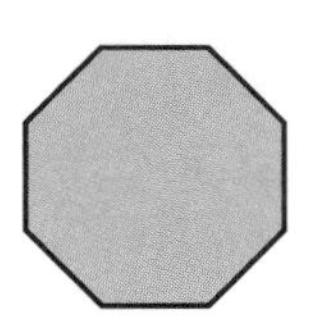

Write the letter that best describes the polygon.

 8

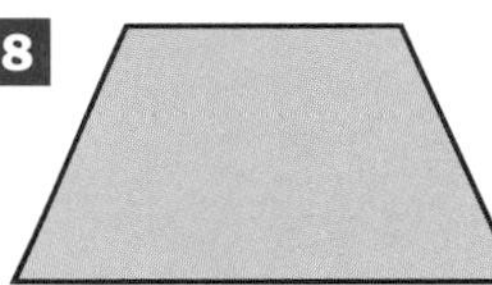

a. 2 acute angles, 2 right angles, 1 pair of parallel angles

b. 2 acute angles, 2 obtuse angles, 1 pair of parallel lines

9

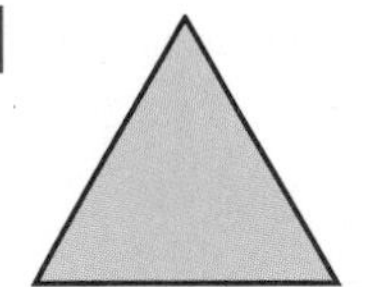

a. 2 acute angles, 1 obtuse angle, 2 pairs of perpendicular lines

b. 3 acute angles, 3 intersecting lines

10

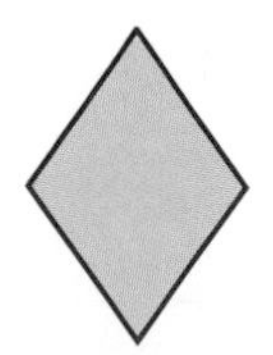

a. 2 acute angles, 2 obtuse angles, 2 pairs of parallel lines

b. 4 obtuse angles, 2 pairs of intersecting lines

Make It Right

11 Todd says this triangle has 3 right angles. Write him a note explaining what error was made.

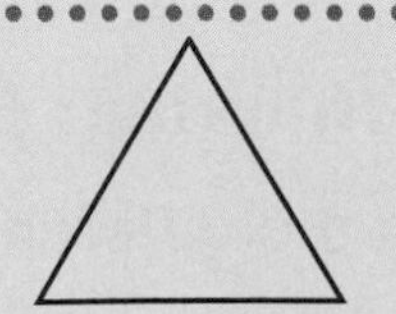

MIXED APPLICATIONS
Problem Solving

12 **Logical reasoning** What polygon am I?
I have 1 pair of parallel lines.
I have more than 1 right angle.
I have 2 obtuse angles and
1 acute angle.

13 **Spatial reasoning** Veronica bought a collector's doll. The box it came in was in the shape of a triangular prism. What are the shapes of the faces of the box?

Use the line plot for problems 14–15.

14 How many students spend 3 or more hours on the Internet each month?

15 Do more students spend 1 hour or 4 hours on the Internet each month? How many more students?

Hours Spent on the Internet Each Month by Students

0	1	2	3	4	5
x x	x x x x x	x	x x x	x x x	x

HOURS

more to explore

Triangles

You can describe triangles using their angle sizes.

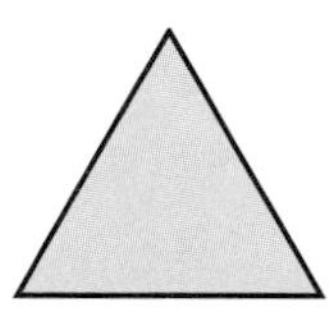

An acute triangle has all acute angles.

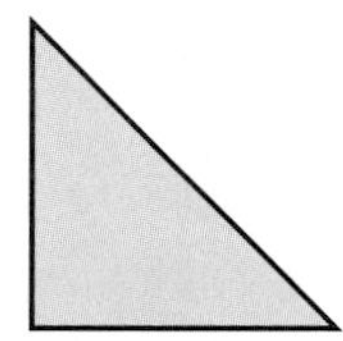

A right triangle has 1 right angle and 2 acute angles.

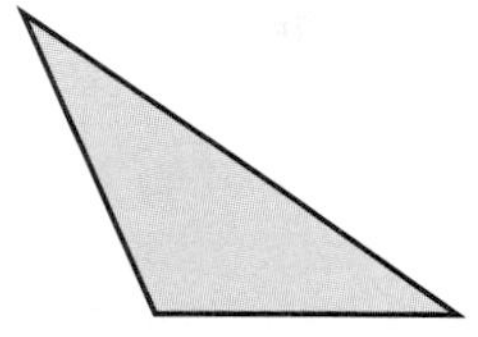

An obtuse triangle has 1 obtuse angle and 2 acute angles.

Tell if the triangle is an acute, obtuse, or right triangle.

1 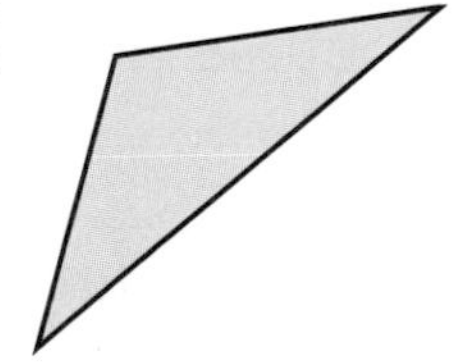

2 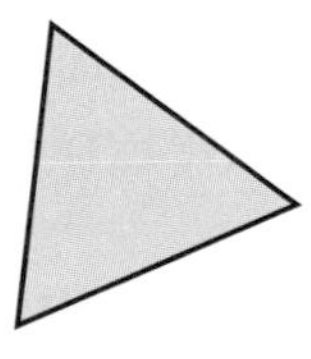

3 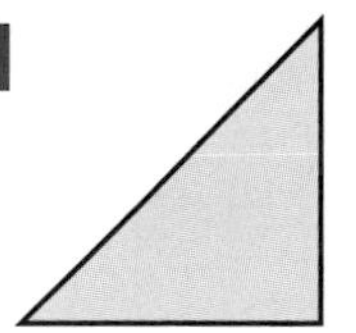

4 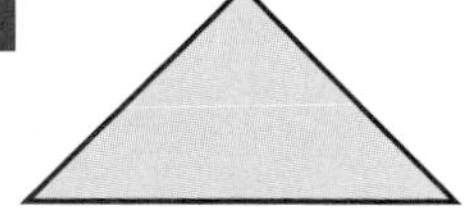

5 Can you have a triangle with all right angles? with all obtuse angles? with 2 obtuse angles? Why or why not?

Name the figure the object looks like.

1 

2 

3 

4 

Is the figure open or closed? If closed, is it a polygon?

5 

6 

7 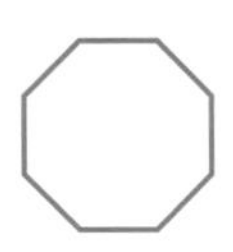

8 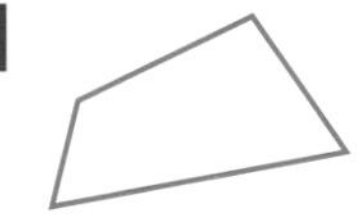

Name the figure.

9 

10 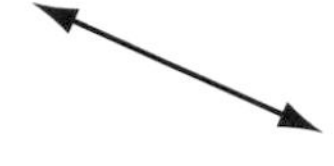

11 

12 

Write *acute, obtuse,* or *right* for the angle.

13 

14 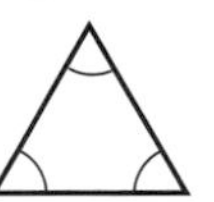

15 

16 

Solve.

17 Find the number of faces, edges, and vertices in a rectangular prism. Make an organized list to help find the answer.

18 I am a 4-sided polygon. All my sides are equal in length. I have 2 acute angles and 2 obtuse angles. What am I?

19 Compare the following:
a. a line and a line segment
b. a line and a ray
c. an acute and an obtuse angle

20 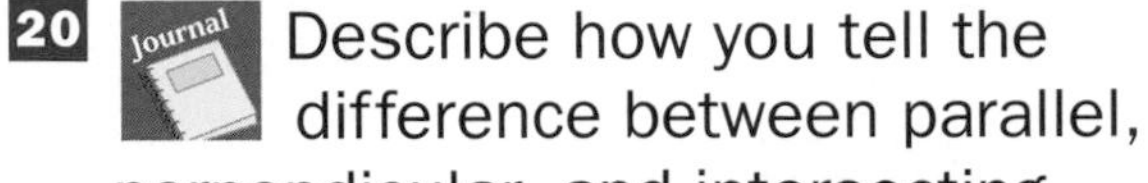
Describe how you tell the difference between parallel, perpendicular, and intersecting lines.

developing spatial sense

MATH CONNECTION

Use Perspective

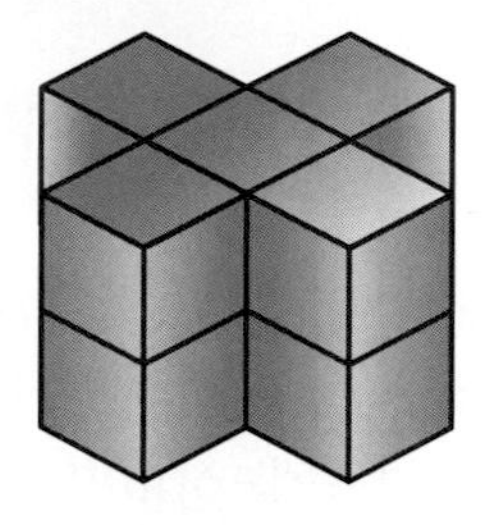

You can describe a stack of cubes using the top, front, and side views.

A stack of cubes is shown on the right. The top view, front view, and side view of the stack are shown below.

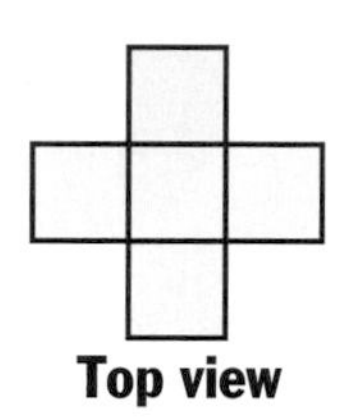

Top view

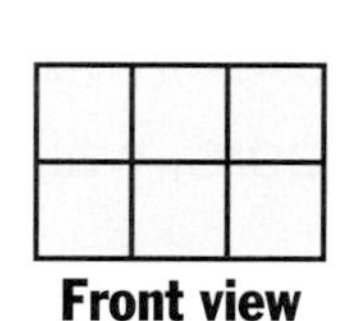

Front view

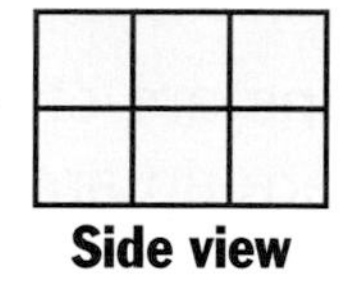

Side view

Use centimeter cubes to build each stack of cubes. Then draw the three views for each.

1 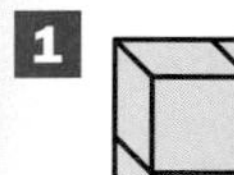 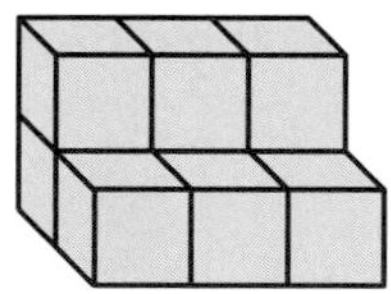

2 

3 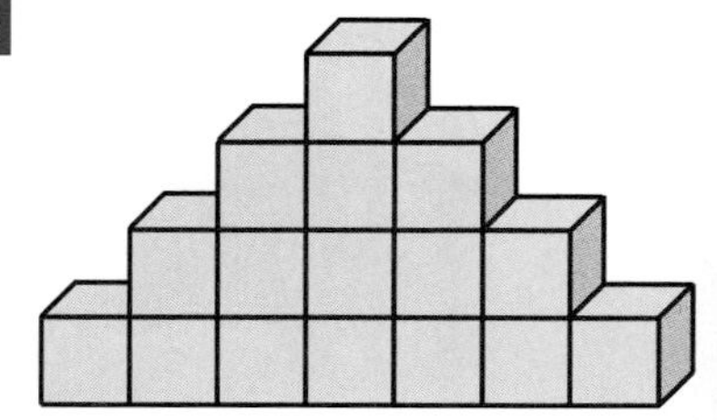

Use the three views to build the stack. Tell how many cubes you used.

4 

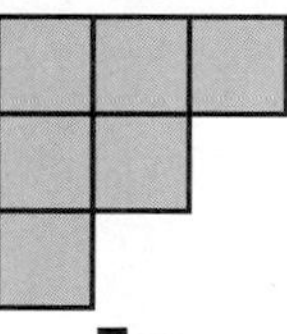

Top

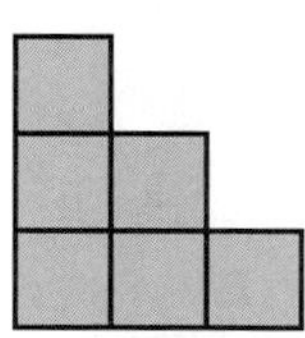

Front

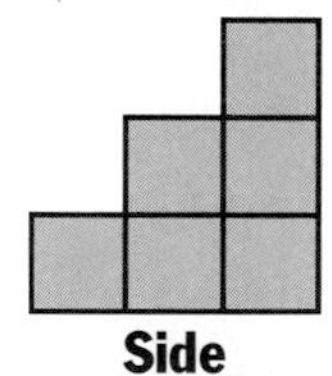

Side

5

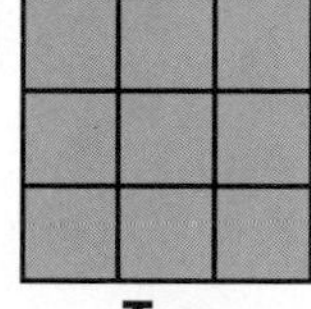

Top

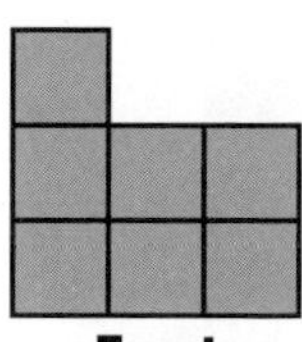

Front

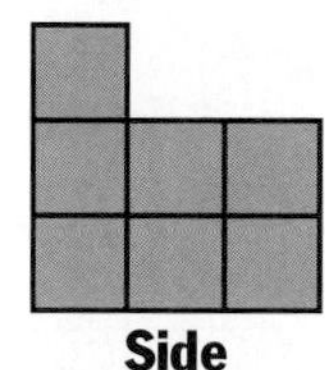

Side

real-life investigation

APPLYING GEOMETRY

TESSELLATIONS

In many cultures, people create designs by repeating one or more shapes many times. The shapes are fitted together without overlapping and with no spaces between them. The design is called a **tessellation.**

Tessellations are often used to cover floors, walls, or other places where an area has to be filled.

Find out what shapes you can tessellate to create your own designs!

Cultural Note

One of the most famous uses of tessellations in modern times was by M. C. Escher, a Dutch graphic artist. He tessellated animals in several works to create very interesting images.

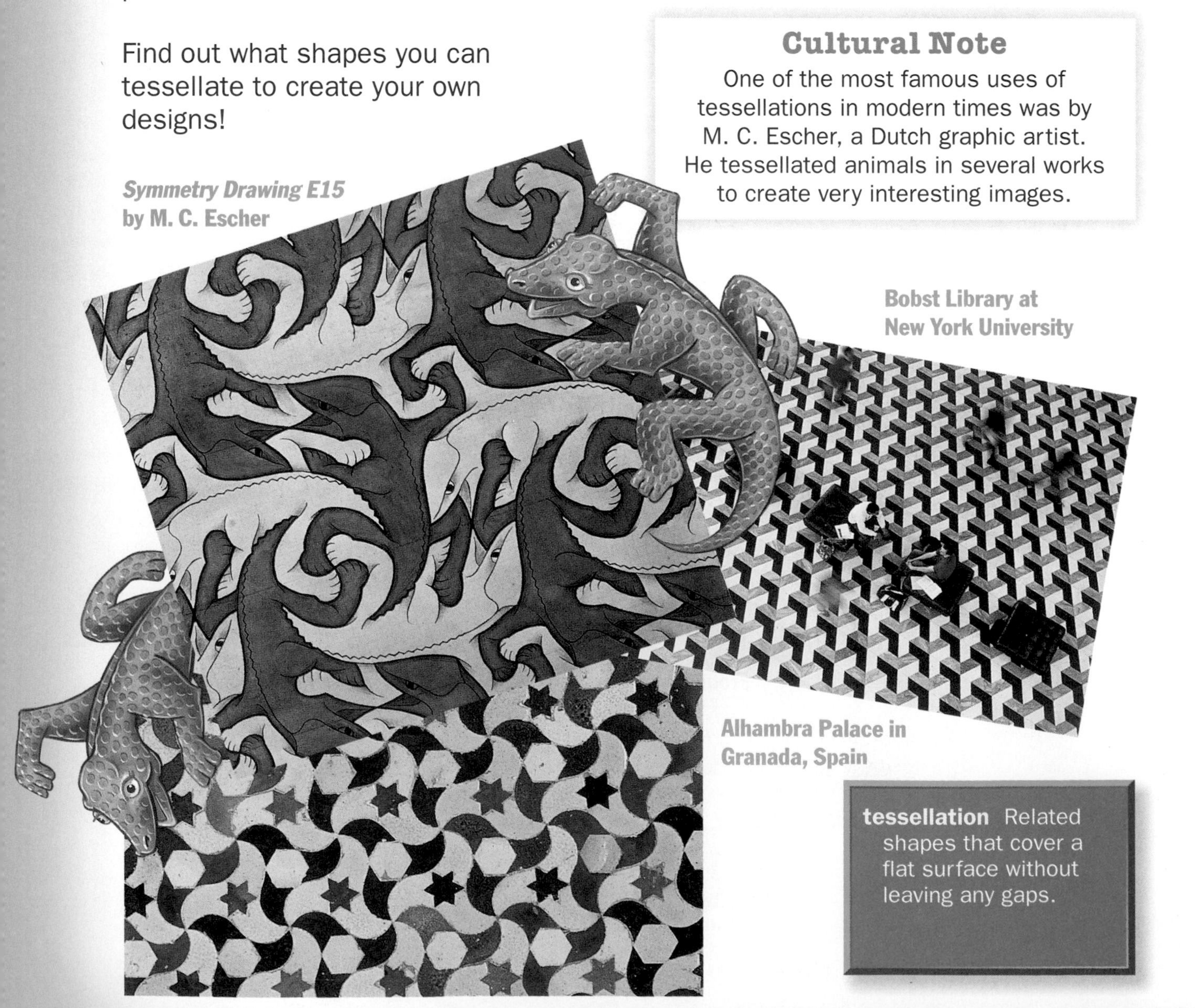

Symmetry Drawing E15 by M. C. Escher

Bobst Library at New York University

Alhambra Palace in Granada, Spain

tessellation Related shapes that cover a flat surface without leaving any gaps.

DECISION MAKING

Creating Tessellations

1 Work with a partner. You may use pattern blocks, objects that you trace, or cut out shapes.

2 Decide on the following things:
- what shape or shapes to use
- what kind of paper you will use
- where on the paper you will start the design
- what colors you will use

3 Create a tessellation. Lay out and trace the shapes until your entire paper is filled. Use at least three colors to complete your design.

4 Now try another one.

Reporting Your Findings

5 Portfolio Prepare to present your tessellations. Include the following:
- Finished tessellations.
- A description of the shapes you used.
- Any discoveries you made about the types of shapes that tessellate.

6 Compare your tessellations with those of your classmates. How are they alike? How are they different?

Revise your work.
- Do your shapes fit together without overlapping and without space between them?
- Did you include a written description of the polygons and colors you used?

MORE TO INVESTIGATE

PREDICT which polygons, other than the ones you used, will tessellate. Try them out.

EXPLORE ways to create a curved shape that will tessellate.

FIND more examples of tessellations in art and nature. What is tessellated on a snake?

Congruency and Similarity

LEARN

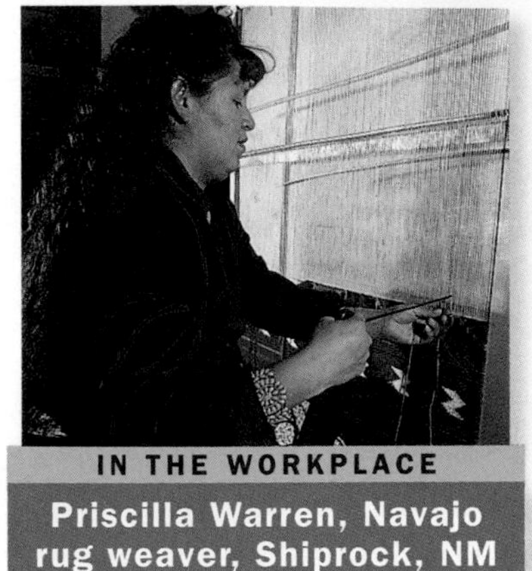

IN THE WORKPLACE
Priscilla Warren, Navajo rug weaver, Shiprock, NM

Many Navajo blankets have patterns that are made up of congruent and similar figures.

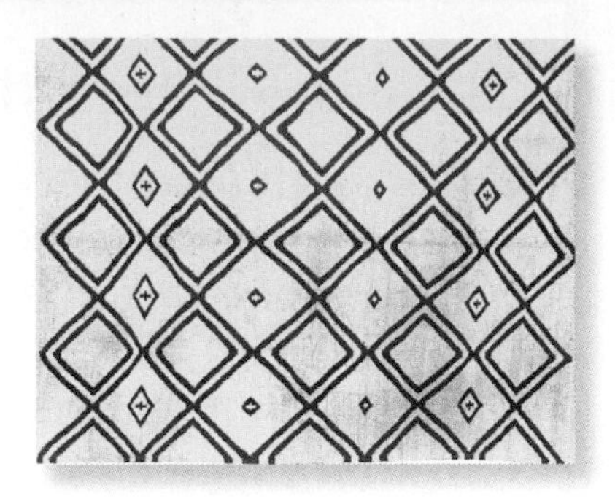

Figures that are the same shape and size are **congruent.**

Figures that are the exact same shape but are different sizes are **similar.**

Work Together

Work with a partner. One partner creates a figure on dot paper.

You will need
- *dot paper*

The other partner uses dot paper to create two figures, one that is congruent to the original figure and one that is similar.

Change roles and continue making new figures as well as figures that are similar and congruent.

Check Out the Glossary
congruent
similar
See page 544.

Talk It Over

► What method did you use to create a figure congruent to your partner's original figure? one that is similar?

► How could you prove the figures are congruent?

Make Connections

Jackie and Don used dot paper to show how they created a figure and then created congruent and similar figures.

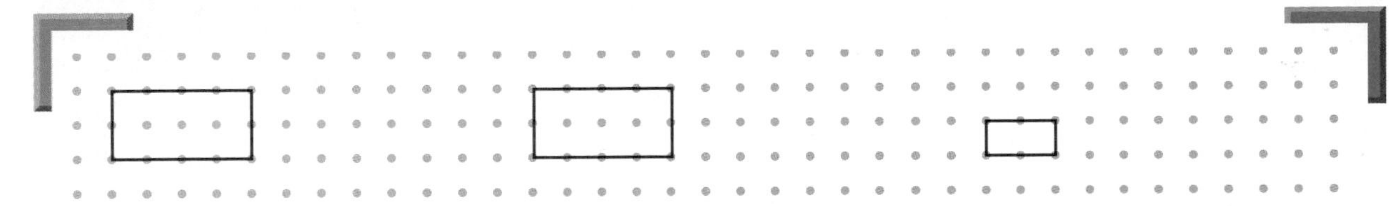

First, I make a drawing of a rectangle on my dot paper.

To make a congruent rectangle, I draw another rectangle with the same width and length.

To make a similar rectangle, I draw a rectangle that is half as wide and half as long.

- ▶ Would ink on a stamp create a congruent or similar shape to the figure on the stamp?

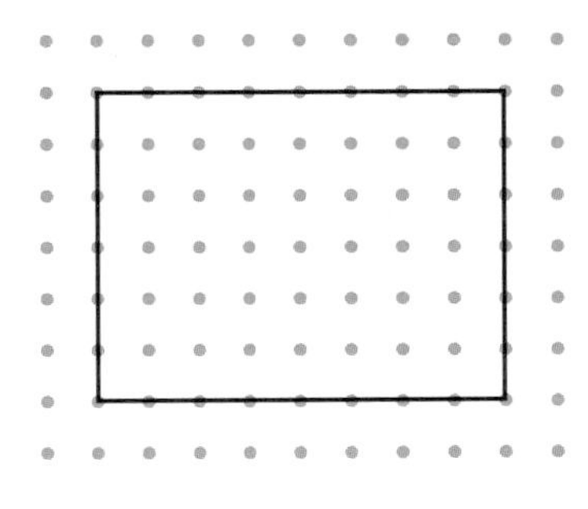

- ▶ The rectangle on the right is not similar to Jackie and Don's rectangle because the length is twice as long but the width is three times as wide. Draw another rectangle that is not similar to their rectangle.

Check for Understanding

Are the figures congruent? If not, are they similar?

1  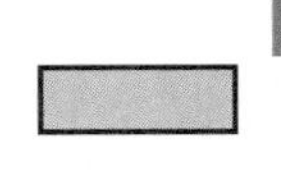

2 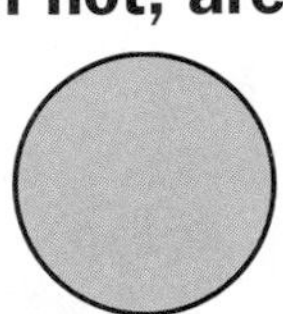 

3 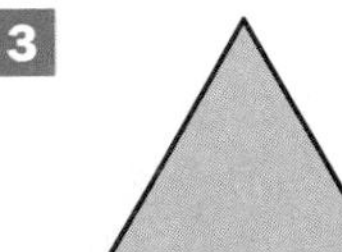 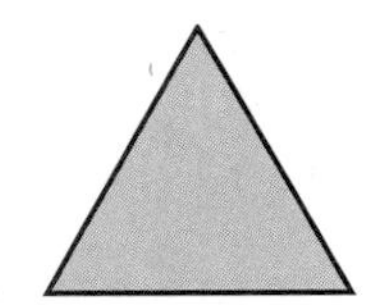

Draw the figure on dot paper. Then draw one figure that is congruent and one that is similar to the original.

4 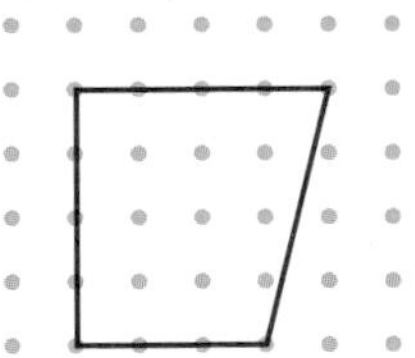

5 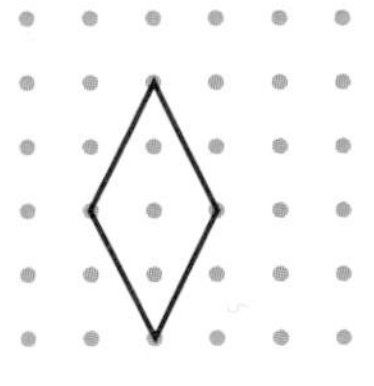

6 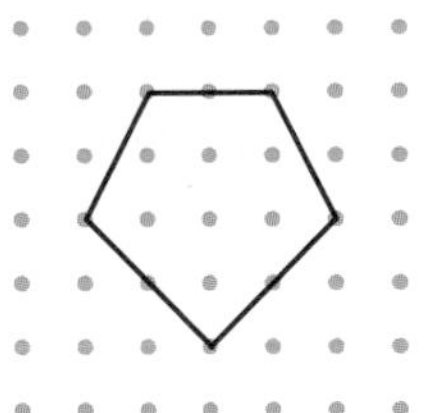

Critical Thinking: Analyze **Explain your reasoning.**

7  Hannah believes that if you can fit one shape exactly over another, then the shapes are congruent. Explain why you agree or disagree with her.

Turn the page for Practice.

Practice

Is each figure *congruent* or *similar* to the original figure?

1  **a.** 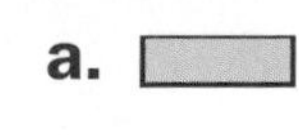 **b.** 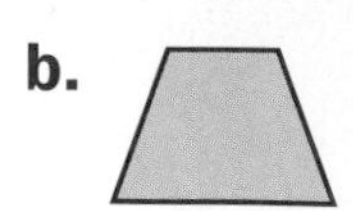 **c.** 

2  **a.** 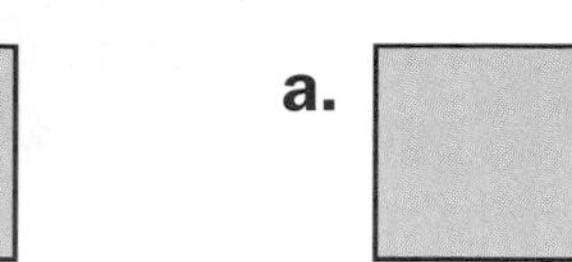 **b.** 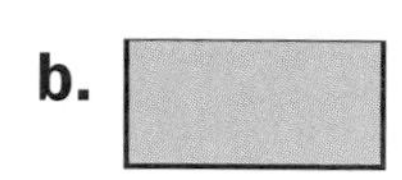 **c.** 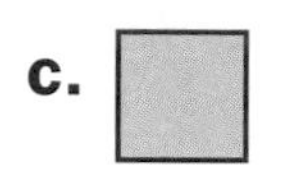

3 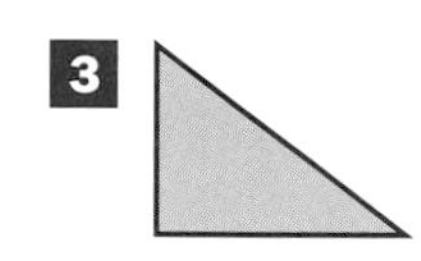 **a.** 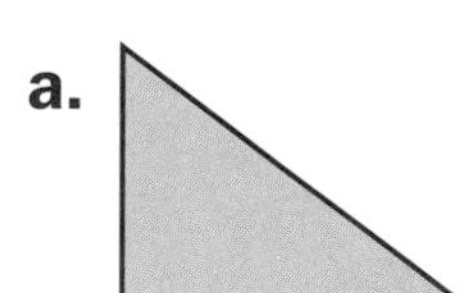 **b.** 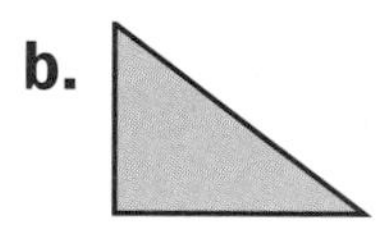 **c.** 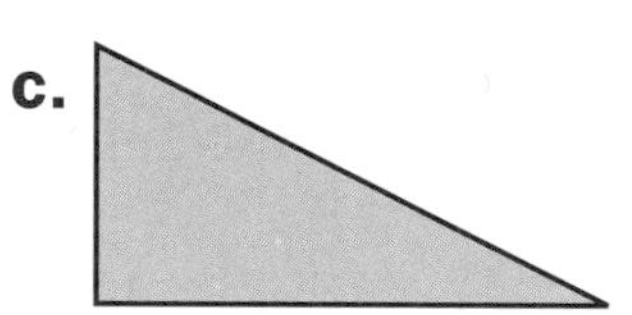

Are the figures congruent? If not, are they similar?

4  

5 

6  

7  

Draw the figure on dot paper. Then draw a congruent figure and similar figure.

8 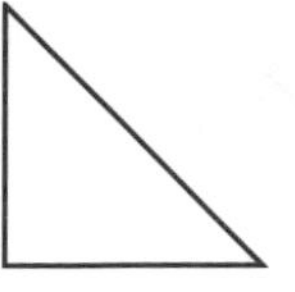

9 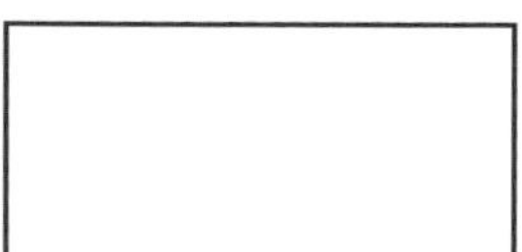

10 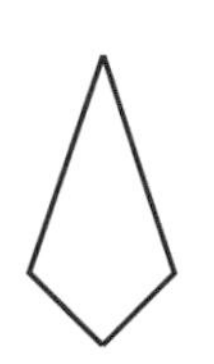

11 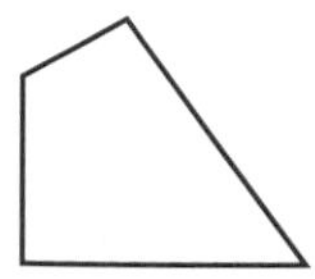

12 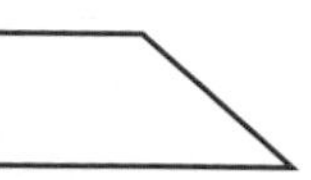

13 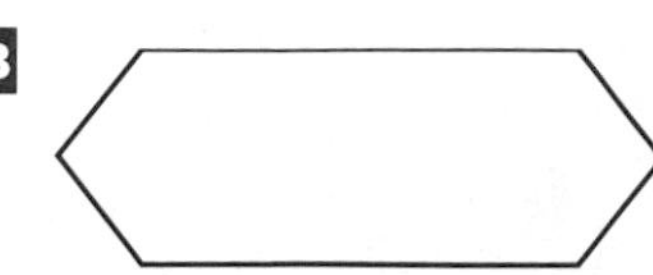

Play with a partner.
Place a barrier on a desk between the players. The barrier should be tall enough so that it prevents one player from seeing the other player's work.

You will need
- *dot paper*
- *rulers*
- *cardboard barriers*

Play the Game

- ▶ The first player draws a figure on dot paper and describes the figure.
- ▶ The second player follows the directions and draws a congruent figure. Then both players compare the figures to see if they are congruent.
- ▶ If the figures are not congruent, then the players discuss better ways to describe the figure.
- ▶ If the figures are congruent, the players trade roles and continue to play.

What strategies did you use to make sure your directions were complete and easy to follow?

more to explore

Circle Measurements

How does the distance around a circle compare to its diameter?

You will need
- *string*
- *scissors*

Measure the outside of the circle on the right with a piece of string. Then measure the diameter with a piece of string.

Compare the lengths. What do you notice?

Use circular containers or objects such as cans to draw circles of different sizes. Compare the distance around each circle to its diameter. What pattern do you see?

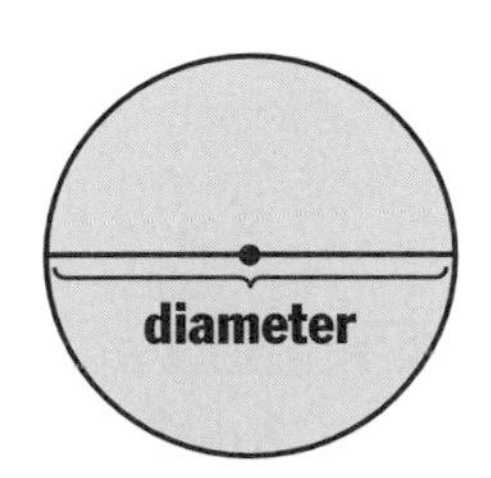

Extra Practice, page 517

Symmetry

LEARN

You probably see examples of symmetry every day. A figure that can be folded to make equal parts is a symmetrical figure. The fold line is a **line of symmetry.**

Check Out the Glossary
line of symmetry
See page 544.

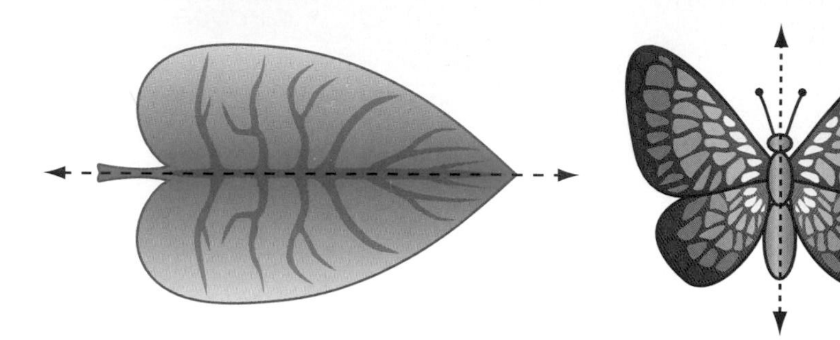

If you have ever seen a cell in a bee's honeycomb, you may have noticed that its shape is similar to a hexagon. A hexagon has 6 lines of symmetry.

Here are other pattern block shapes and their lines of symmetry.

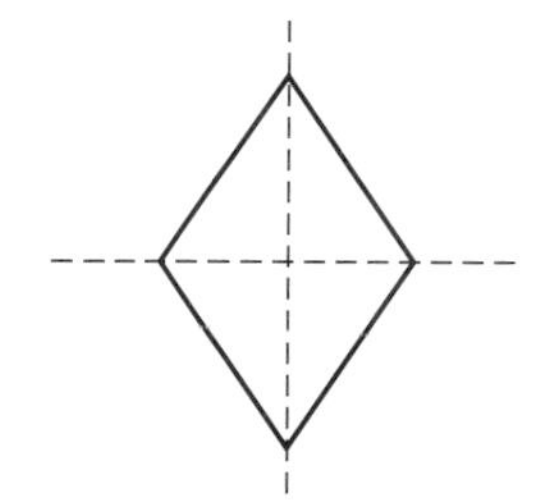

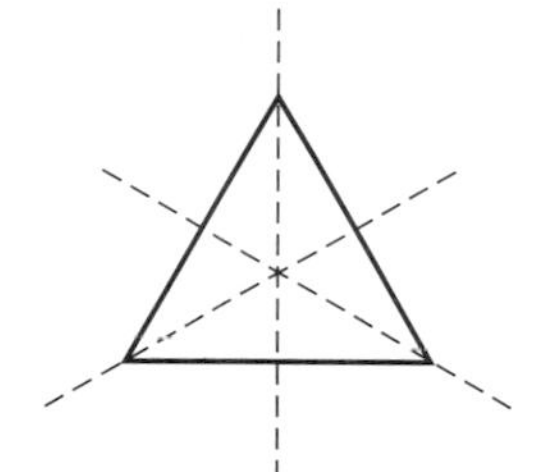

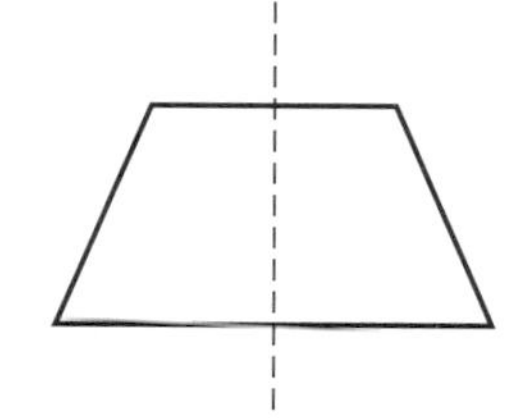

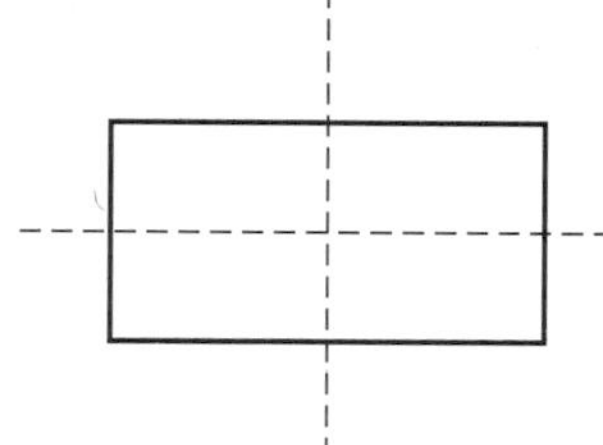

CHECK

Check for Understanding

1 Show eight capital letters that are symmetrical. How many lines of symmetry does each of them have?

2 Which numbers from 1 to 10 are symmetrical? How many lines of symmetry does each of these numbers have?

Critical Thinking: Analyze **Explain your reasoning.**

3 Can you find lines of symmetry for every shape? If not, show examples of figures that do not have lines of symmetry.

4 If a figure is flipped across a line of symmetry to get another figure, will the two figures be congruent?

Practice

Is the dashed line a line of symmetry?

1

2

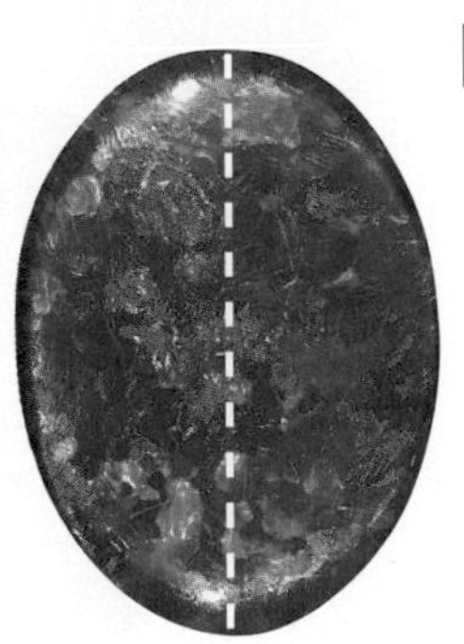

3

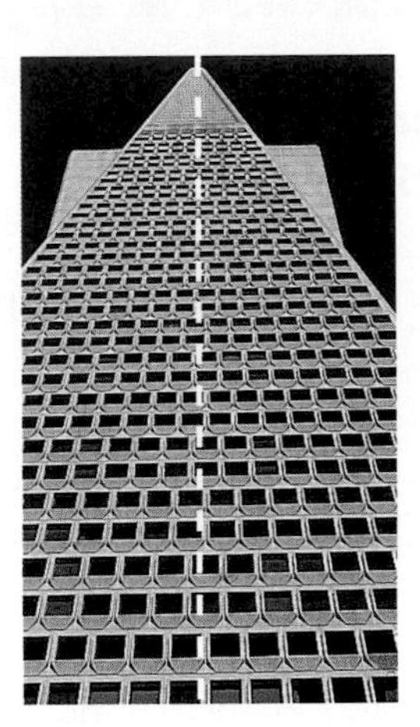

4

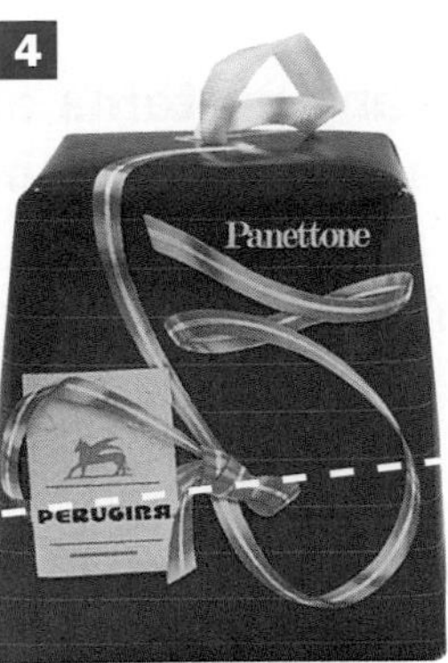

Is the figure symmetrical? If yes, draw its lines of symmetry.

5

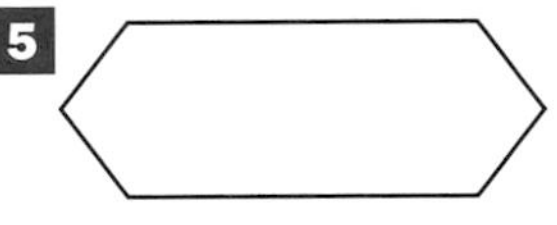

6

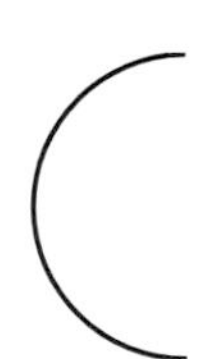

7

8

MIXED APPLICATIONS
Problem Solving

9 What are the missing digits?

$$\begin{array}{r} 304 \\ \times\ 1\blacksquare \\ \hline 21\blacksquare 8 \\ +\ 3\blacksquare 40 \\ \hline 5{,}168 \end{array}$$

10 **Logical reasoning** A rectangle drawn on graph paper has 63 small squares inside. The length of the rectangle is 2 squares longer than the width. Draw the rectangle on graph paper and label its sides.

11 Describe a seesaw using as many geometric and measurement words as you can. Include drawings with your descriptions.

12 **Write a problem** that asks for the lines of symmetry of a figure. Solve it and have others solve it.

mixed review • test preparation

Estimate.

1 $346 + 128$ **2** $163 \div 4$ **3** $781 - 302$ **4** 38×41

5 $536 - 292$ **6** 87×21 **7** $479 + 826$ **8** $3{,}402 \div 50$

Extra Practice, page 518

Slides, Flips, and Turns

LEARN

Many patterns are created by moving figures in different ways.

You can **flip (reflect)** a figure over a line.

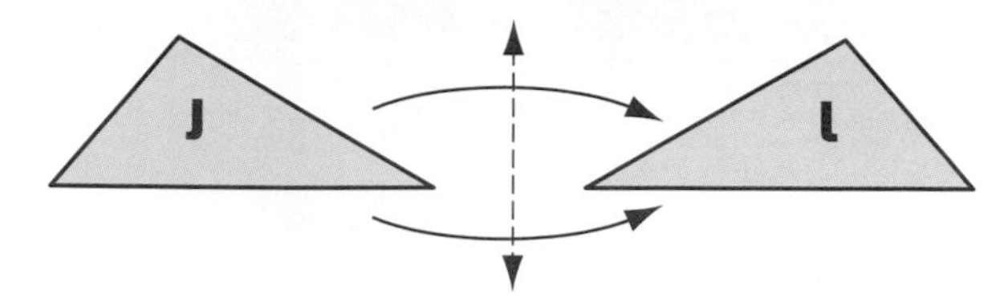

You can **slide (translate)** a figure across a line.

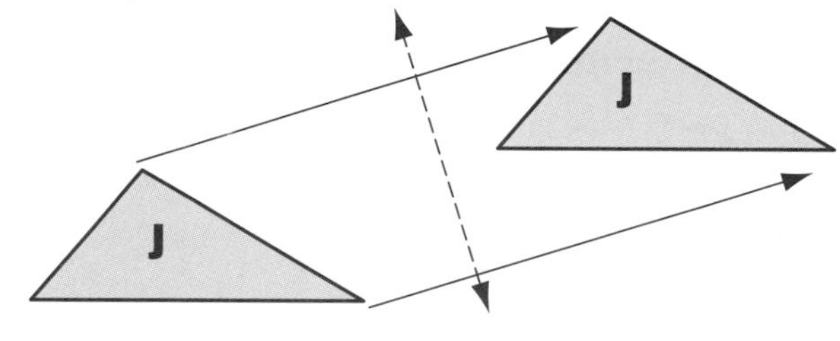

You can **turn (rotate)** a figure around a point.

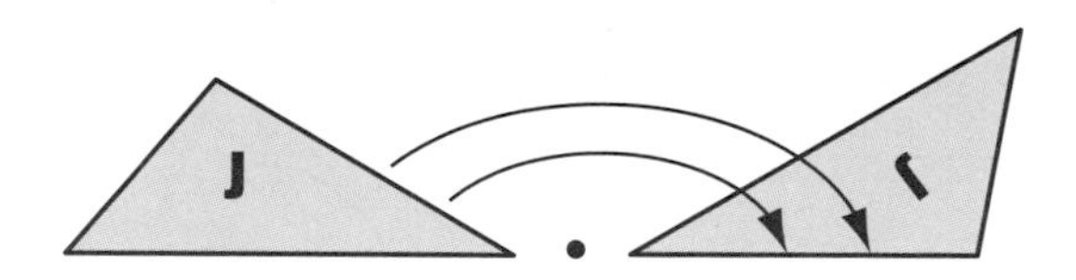

You will need
- *graph paper*
- *rulers*

Work Together

Work with a partner. Draw a geometric figure on graph paper. Now draw another figure by either flipping, sliding, or turning the original figure.

Exchange drawings with your partner. Have your partner check that the two figures are congruent and tell whether you used a flip, a slide, or a turn.

Change places and follow the same steps.

Check Out the Glossary
For vocabulary words
See page 544.

Make Connections

You can check that two figures are congruent by tracing one of the figures. If you can slide, flip, or turn the traced figure so that it fits exactly over the second figure, the two figures are congruent.

Check that the figures below are congruent using tracing paper. Tell whether you used a slide, flip, or turn.

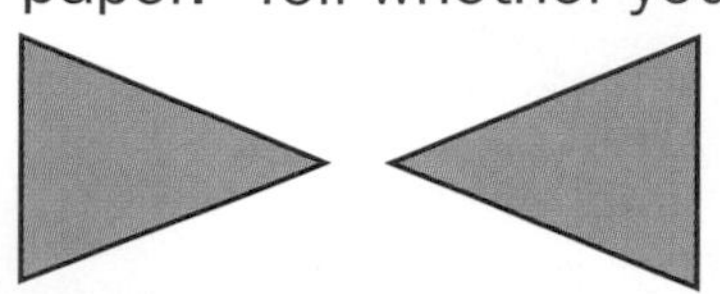

Check for Understanding

Tell how the figure on the left was moved to get the figure on the right. Explain how you can check that the two figures are congruent.

1

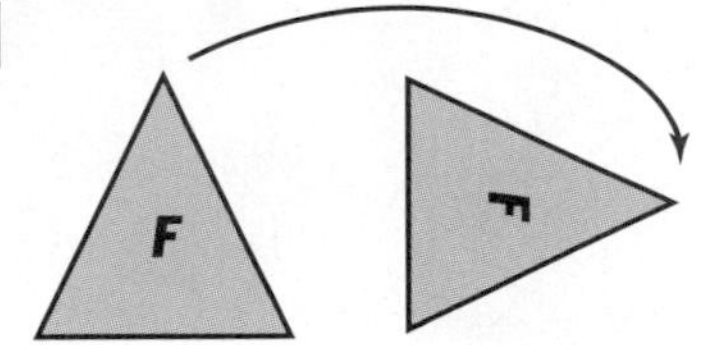

2

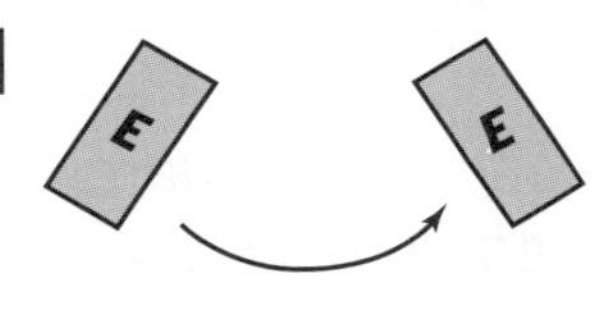

3

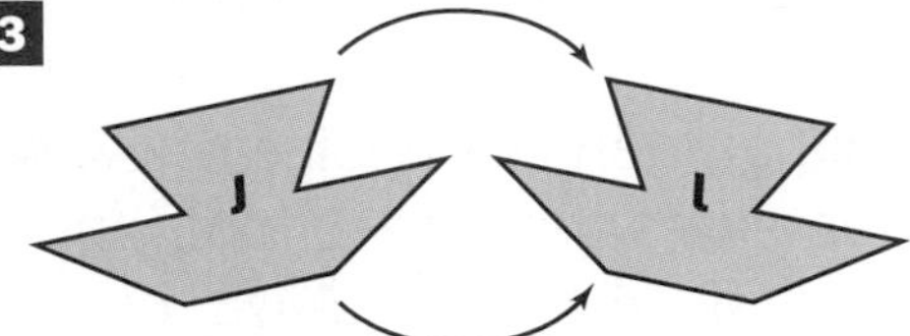

4

5

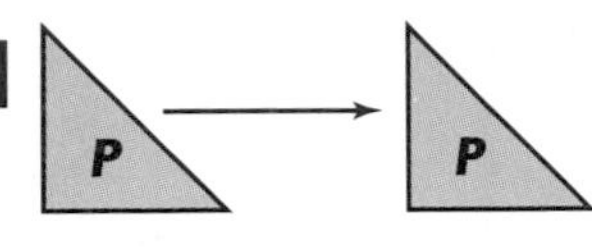

6

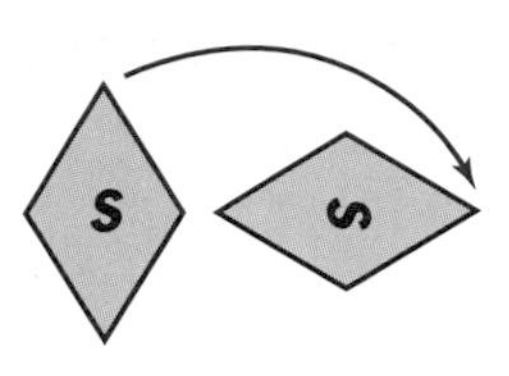

Critical Thinking: Analyze **Explain your reasoning.**

7 Can you slide, flip, or turn a figure to create a new figure that is not congruent to the original figure?

Practice

Tell how the figure on the left was moved to get the figure on the right. Explain how you can check that the two figures are congruent.

1

2

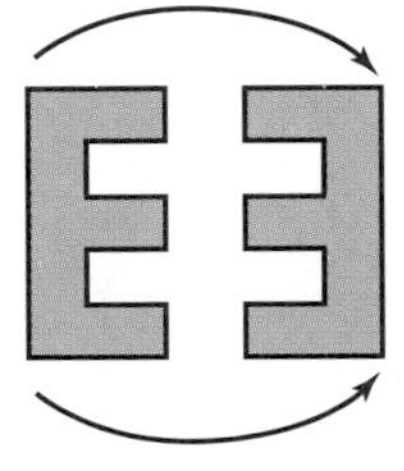

3

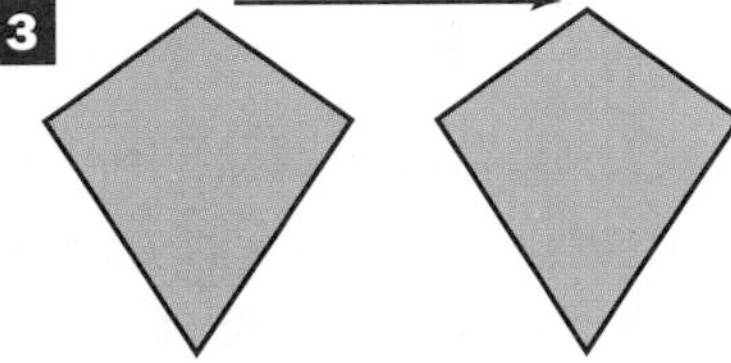

Copy the figure onto graph paper. Move the figure in the way indicated.

4

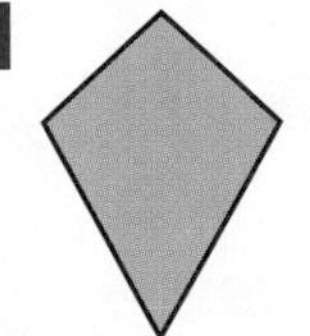

slide (translate)

5

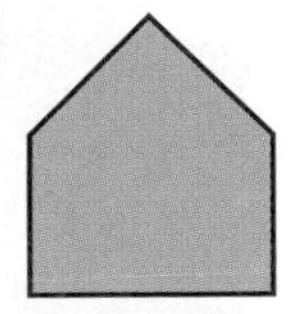

turn (rotate)

6

slide (translate)

7

flip (reflect)

8

slide (translate)

9

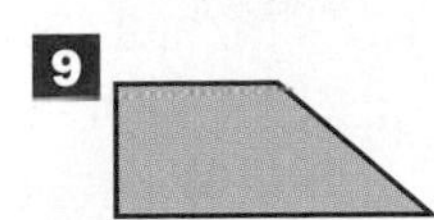

flip then slide
(reflect then translate)

Extra Practice, page 518

Area

The **area** of a 2-dimensional figure is the number of square units needed to cover the figure.

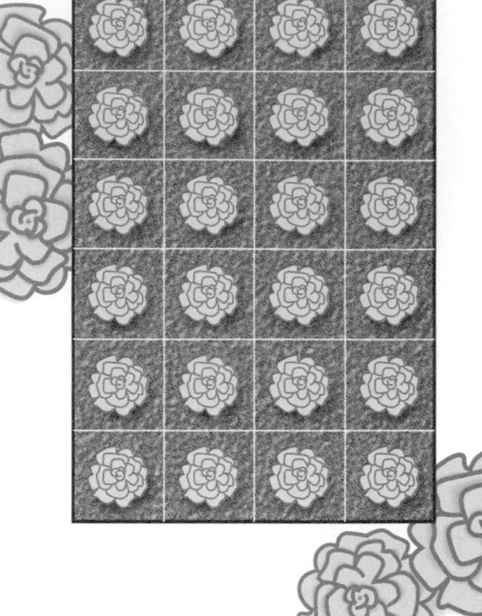

This garden is the size of 24 squares. Each square has sides that are 1 foot long. The garden has an area of 24 **square units,** or 24 square feet.

You can find the area of other rectangles by using graph paper.

Work Together

Work with a partner. Draw different-size rectangles on graph paper. Find the area of each rectangle in square inches.

Record the length, width, and area of each rectangle in a table. Organize your table from least to greatest area.

You will need
- *1-inch graph paper*

Length	Width	Area
6 inches	4 inches	24 square inches

Talk It Over

- **What if** you increase either the length or width of the rectangle. How does the area change? What if you decrease either the length or the width?
- What strategy would you use to find all the possible rectangles with an area of 24 square inches?

Check Out the Glossary
area
square unit
See page 544.

Make Connections

Cory and Elizabeth used graph paper to find all the possible rectangles with an area of 24 square inches. They wrote the lengths and widths of the rectangles in a table. They noticed that for the first rectangle, if they multiplied length times width, they got the area.

Width (in inches)	Length (in inches)	Area (in square inches)
1	24	24
2	12	24
3	8	24
4	6	24

- ▶ Would this work for the other rectangles in the table?
- ▶ Why didn't they list the rectangle with a width of 6 inches and a length of 4 inches?
- ▶ Did they list all the possible rectangles? How do you know?

Check for Understanding

CHECK

Find the area for the figure.

1
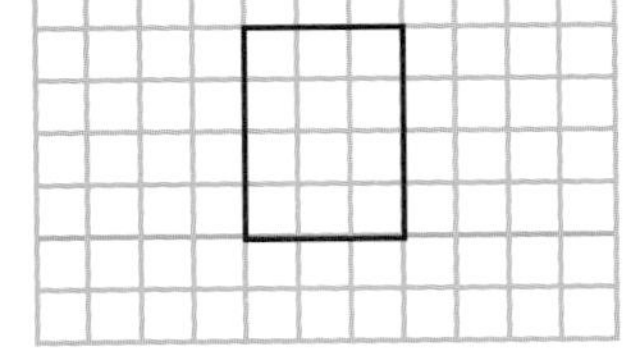

2
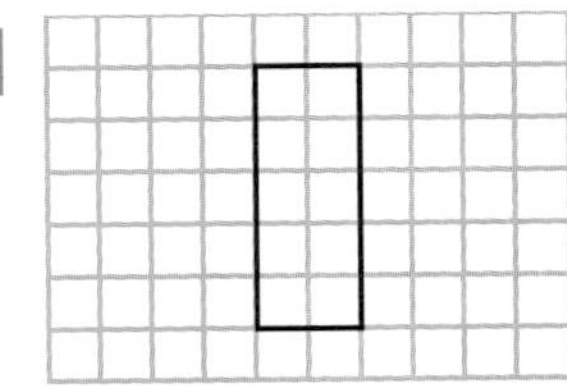

3
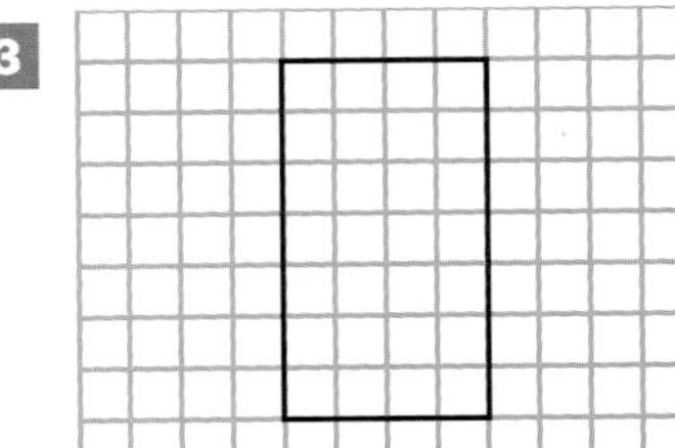

4
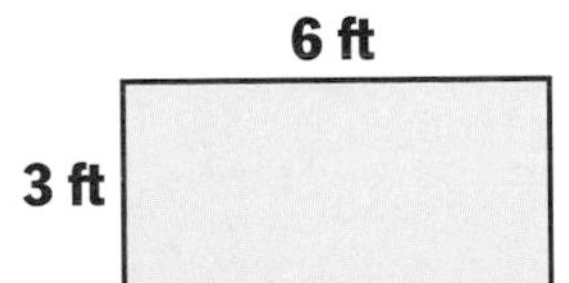

5
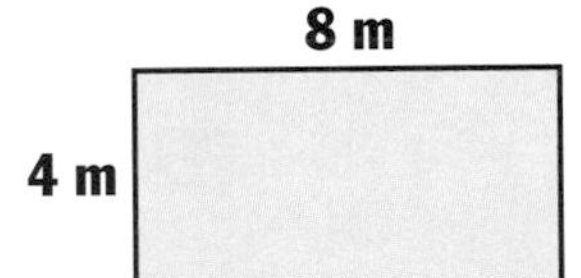

6
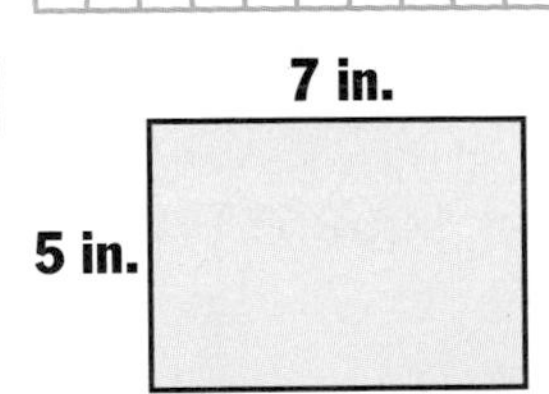

Critical Thinking: Generalize **Explain your reasoning.**

7 How many possible rectangles are there with an area of 20 square units? How can you tell without drawing them?

8 Kurt says that if you know the length of one side of a square, all you have to do to find the area is multiply the length times itself. Why is he correct?

Turn the page for Practice. ➡

PRACTICE

Practice

Find the area of the rectangle.

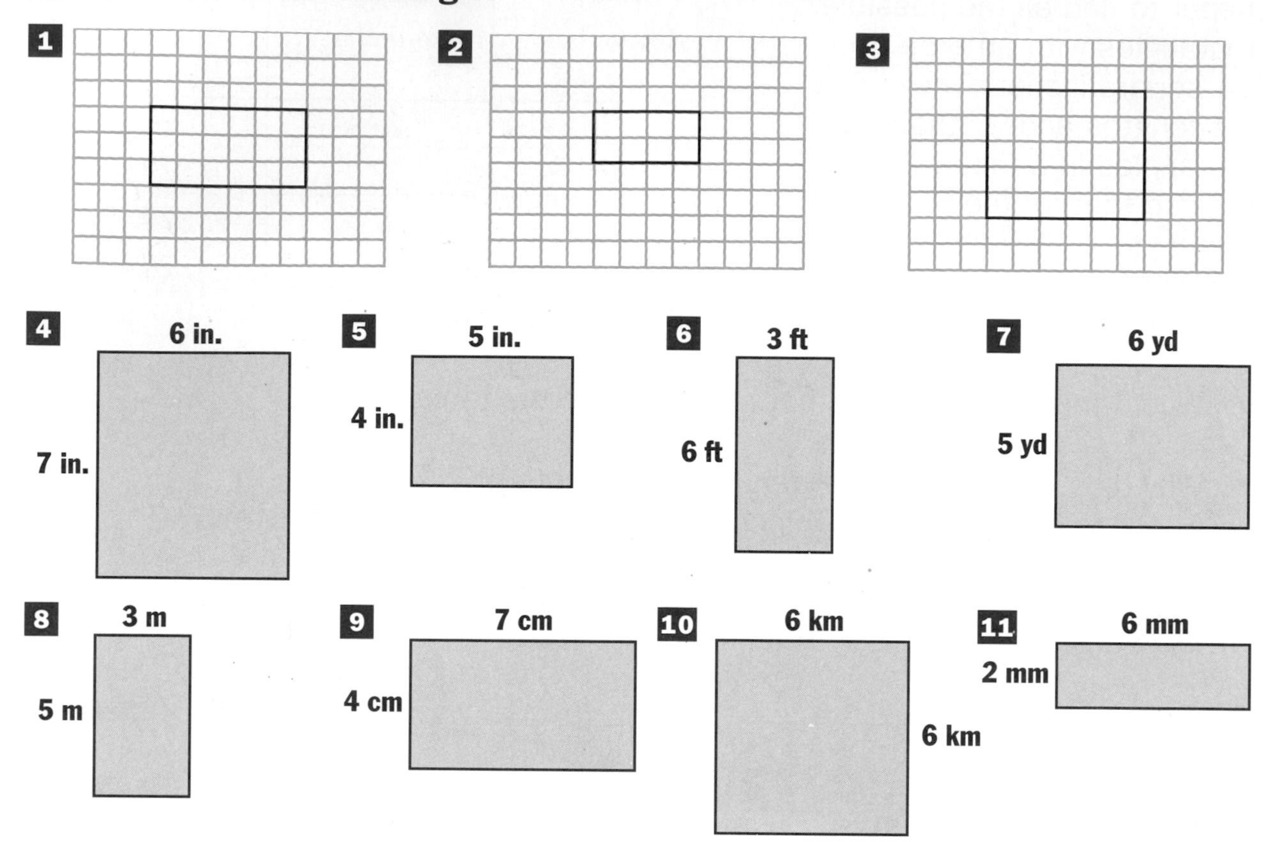

Use graph paper to draw rectangles with the following areas. How many rectangles are possible for each?

12 14 square units

13 18 square units

14 22 square units

15 25 square units

16 28 square units

17 36 square units

Solve.

18 The dimensions of four rectangles are given below. Use graph paper to draw each rectangle. Then list the areas from least to greatest.

a. length, 4 cm; width, 5 cm

b. length, 8 cm; width, 2 cm

c. length, 5 cm; width, 6 cm

d. length, 3 cm; width, 9 cm

19 Which figure has the greater area?

a. a rectangle with a length of 8 in. and a width of 2 in.

b. a square with 4-in. sides

MIXED APPLICATIONS
Problem Solving

20 **Make a decision** Mr. East wants to fence in 48 square feet of his yard for a garden. The fencing he wants costs $1.00 for each foot. He has $50.00 to spend. What dimensions should he choose? Why?

INFOBIT
The omnitheater screen at the Liberty Science Center in New Jersey is about 88 feet across and 65 feet high.

21 Mrs. McCarthy's class is going on a field trip. She thinks that her 32 students and 6 parents will each drink about 1 pint of lemonade at lunch. How many gallons of lemonade should she plan to bring?

22 Tamika is comparing two trees. One tree measures 42 ft and another measures 14 yds. Which is taller? How do you know?

23 About how large is the area of the omnitheater screen at the Liberty Science Center? **SEE INFOBIT.**

more to explore

Estimating the Area of Irregular Shapes

You can estimate the area of irregular shapes by using a grid. Count the whole squares first. Then look at the parts of squares and estimate how many whole squares they would make. Add to estimate the total area.

Scientists measure paw prints to learn more about animals. This badger paw print is about 38 square units.

Set a sheet of graph paper on the floor. Trace around your shoe on the graph paper. Estimate how many square units your shoe fits. Then count.

Extra Practice, page 518

Volume

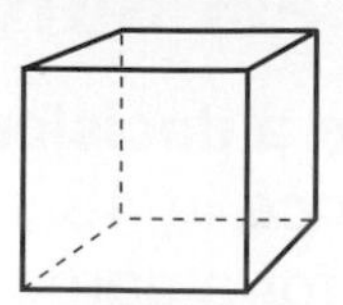
1 cubic unit

Volume is the amount of space that a 3-dimensional figure takes up. The volume of an object is measured in **cubic units.**

These rectangular prisms are made up of centimeter cubes. Each one has a volume of 24 cubic centimeters (cubic cm).

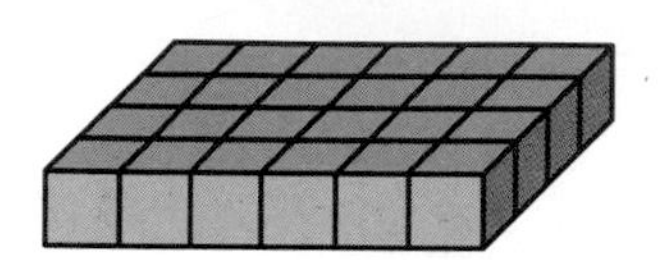

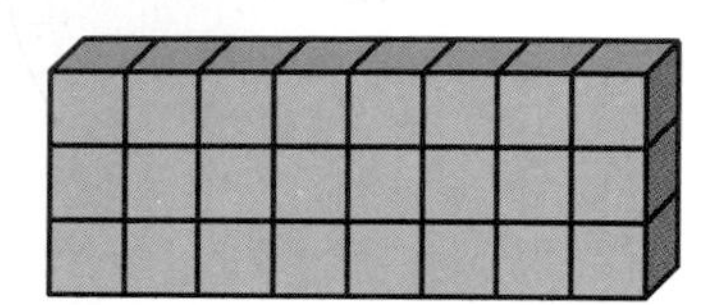

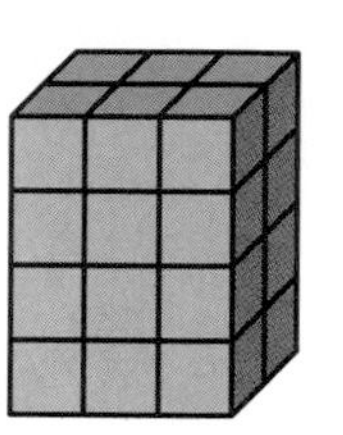

How many different rectangular prisms with a volume of 36 cubic cm can you make?

Work Together

Work with a partner. Use centimeter cubes to create rectangular prisms with a volume of 36 cubic cm.

You will need
- *centimeter cubes*

Start with a prism that is one layer high. Then build a prism that is two layers high, three layers high, and so on.

Record the height, length, width, and volume of each rectangular prism in a table like the one shown.

Height	Width	Length	Volume
1 cm	3 cm	12 cm	36 cubic cm

Check Out the Glossary
volume
cubic unit
See page 544.

Make Connections

Stacey and Gary made rectangular prisms that were 1 layer, 2 layers, 3 layers, and 4 layers high. They recorded their work in a table.

- ▶ What happens when you multiply the height by the area?
- ▶ How can you find the volume of a rectangular prism without counting the cubes?

Height (cm)	Width (cm)	Length (cm)	Area (square cm)	Volume (cubic cm)
1	2	18	36	36
2	2	9	18	36
3	2	6	12	36
4	3	3	9	36

Check for Understanding

Find the volume for the rectangular prism.

1

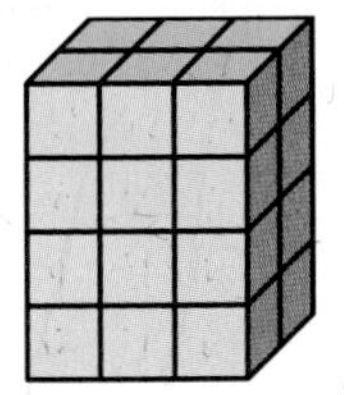

2

3

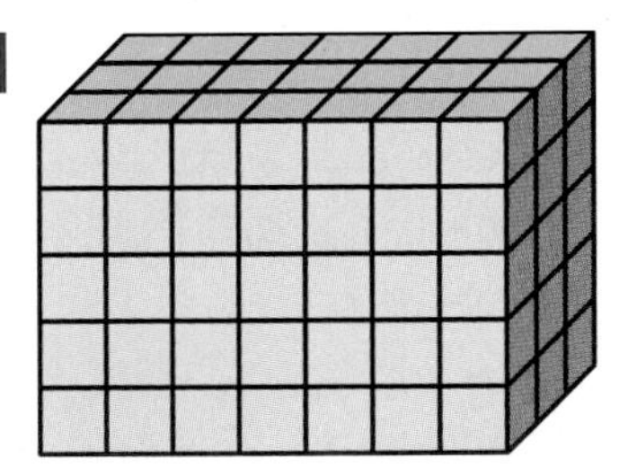

Critical Thinking: Generalize **Explain your reasoning.**

4 For what objects might it be important to know the volume?

CHECK

Practice

Find the volume for the rectangular prism.

1

2

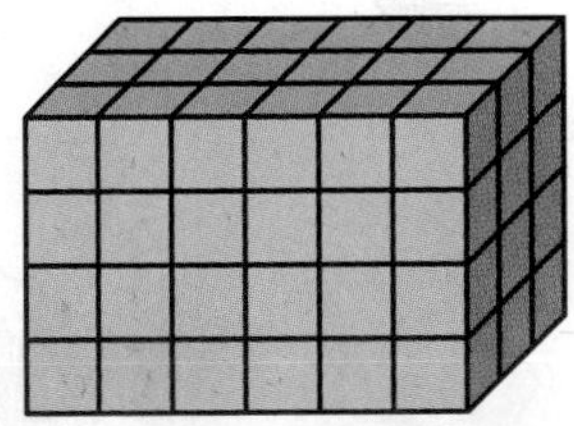

3

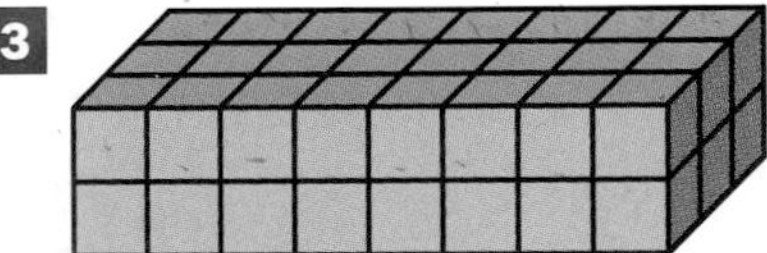

4 length–6 in.
width–5 in.
height–4 in.

5 length–8 in.
width–6 in.
height–5 in.

6 length–3 in.
width–12 in.
height–5 in.

7 length–7 in.
width–11 in.
height–6 in.

8 List the volumes in ex. 4–7 from least to greatest.

PRACTICE

Extra Practice, page 519

Problem Solvers at Work

Read
Plan
Solve
Look Back

Part 1 Use Diagrams

LEARN

Some scientists use Venn diagrams like the one on the right to help them organize and sort their work.

How can a Venn diagram help you organize your work?

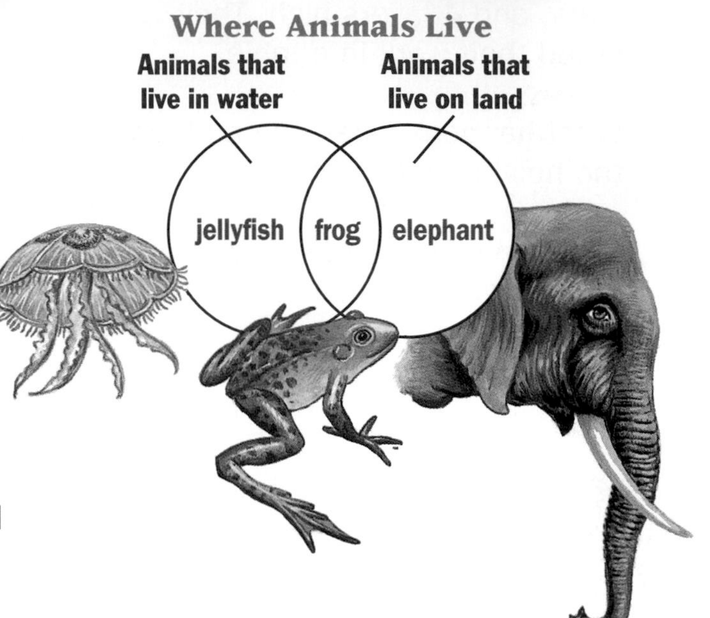

Work Together

Categories Copy the Venn diagram below. Use it to help you organize the shapes.

1. Why was the parallelogram placed in the area where the two circles overlap?

2. Where would you place a rectangle? Why?

3. Where would you place a figure that is not a polygon? Why?

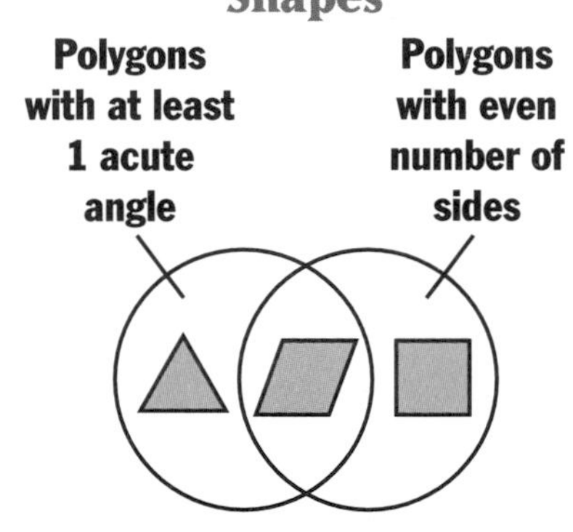

4. **What if** the category labeled *Polygons with at least 1 acute angle* was changed to *Polygons with at least 1 right angle*? Show how the Venn diagram you drew would change.

5. **Make a decision** Choose other categories that can be used to sort the shapes. Create a Venn diagram to organize your work.

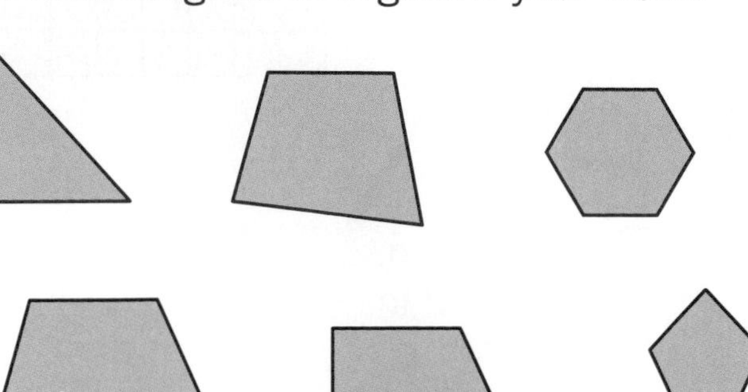

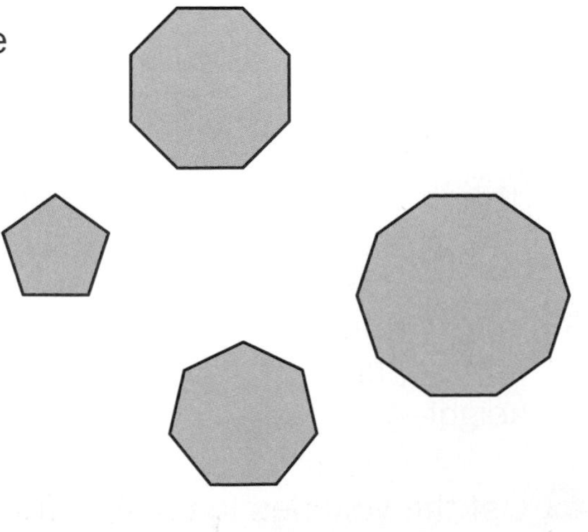

Part 2 Write and Share Problems

Students in Mrs. Jule's Class

Students with brothers | **Students with sisters**

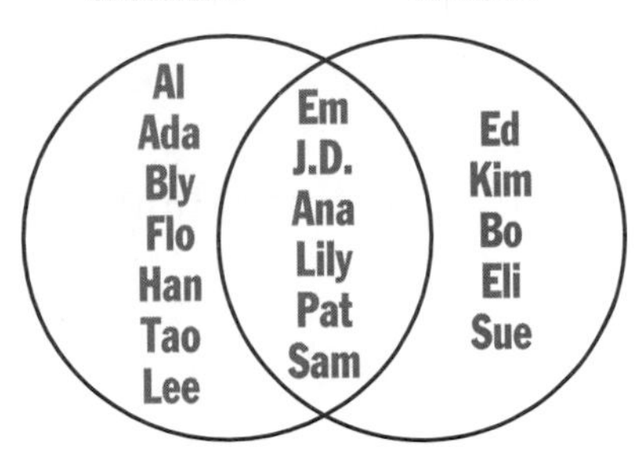

Ann, Cam, Sal, Amy

Erin wrote this problem about the information in the Venn diagram.

6 Solve Erin's problem.

7 **Write a problem** where you need to sort words or numbers using a Venn diagram.

8 Solve the new problem.

9 Trade problems. Solve at least three problems written by your classmates.

10 What was the most interesting problem you solved? Why?

CHECK

Lee's mom just had a baby girl. Where should you move Lee's name on the diagram?

Erin McCormick
St. Gabriel School
Riverdale, NY

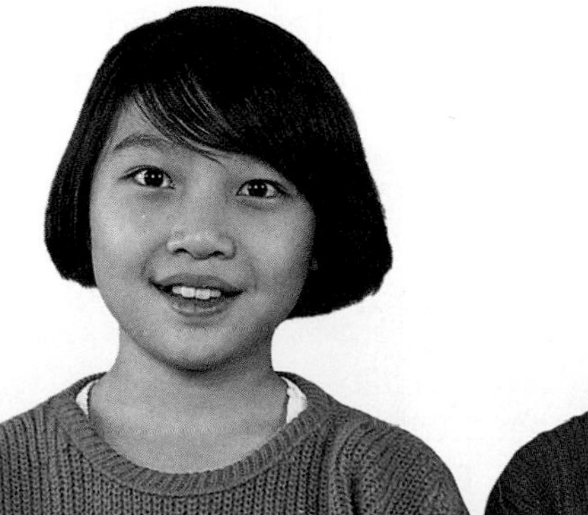

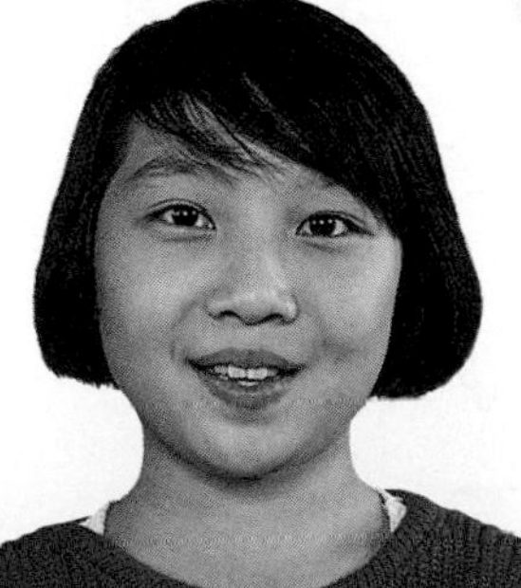

Turn the page for Practice Strategies.

Part 3 Practice Strategies

PRACTICE

Menu

Choose five problems and solve them. Explain your methods.

1 How many cubes are in this rectangular prism?

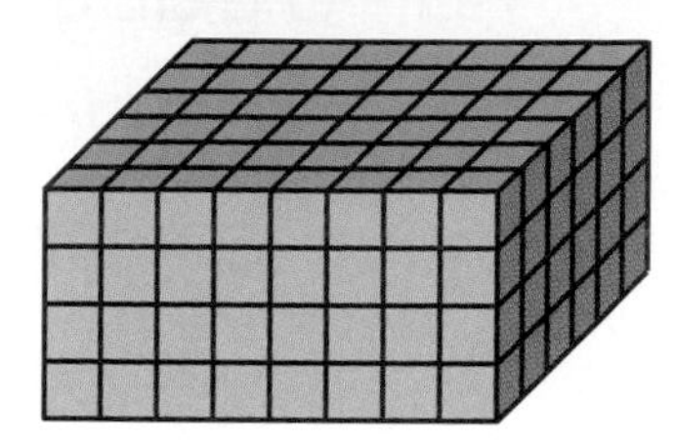

2 Helen used a calculator to create a number pattern. Find the pattern. Then give the next three numbers.
2, 3, 5, 9, 17, ■, ■, ■

3 Gordy's family ate 6 granola bars on Wednesday. That was twice as many as they ate on Monday but 3 less than they ate on Tuesday. How many granola bars are left from a box of 24?

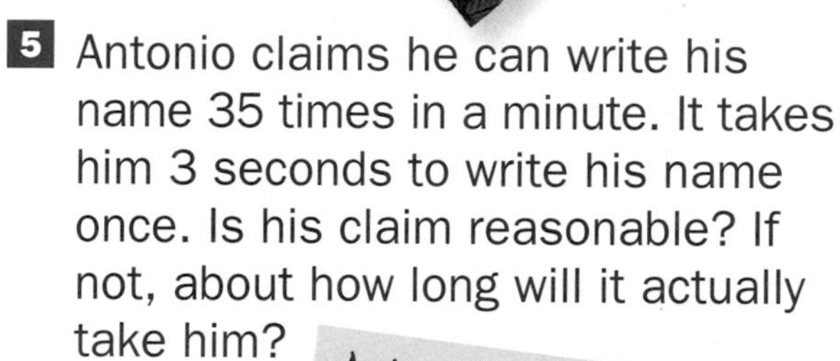

4 Jeffrey wants to buy a paintbrush set for $8.95 and as many bottles of paint as he can. Each bottle of paint costs $1.25. If he has $20.00, how many bottles of paint can he buy and still get the paintbrush set?

5 Antonio claims he can write his name 35 times in a minute. It takes him 3 seconds to write his name once. Is his claim reasonable? If not, about how long will it actually take him?

Antonio Antonio Antonio
Antonio Antonio Antonio

6 Stacy, Phil, Millie, and Theresa went to a movie. Millie sat next to Stacy, who was on Phil's left. If Theresa sat next to Phil, who was sitting in the middle?

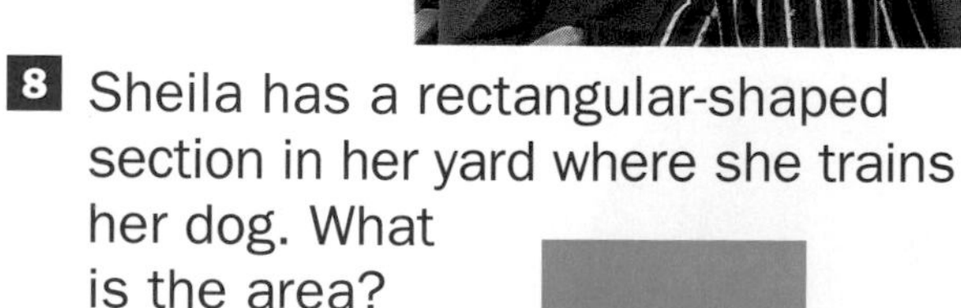

7 Use the figure to give an example of a line segment, line, ray, parallel lines, perpendicular lines, and intersecting lines.

8 Sheila has a rectangular-shaped section in her yard where she trains her dog. What is the area?

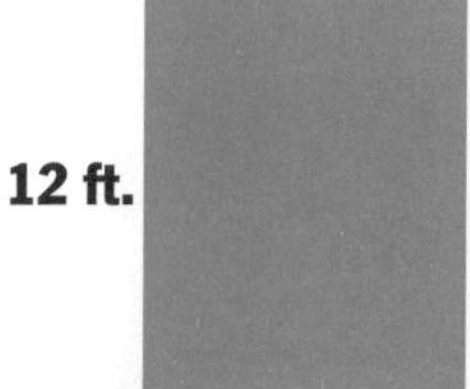

Choose two problems and solve them. Explain your methods.

9 Draw the bottom and sides of a transparent box that would hold 500 wooden cubes with 1-inch sides. Your box should waste as little space as possible. Label the length and width of the bottom and sides.

Explain how you know that your box would hold all 500 cubes.

10 **Spatial reasoning** What is the volume of this figure? How many cubes would you need to add to make this a rectangular prism?

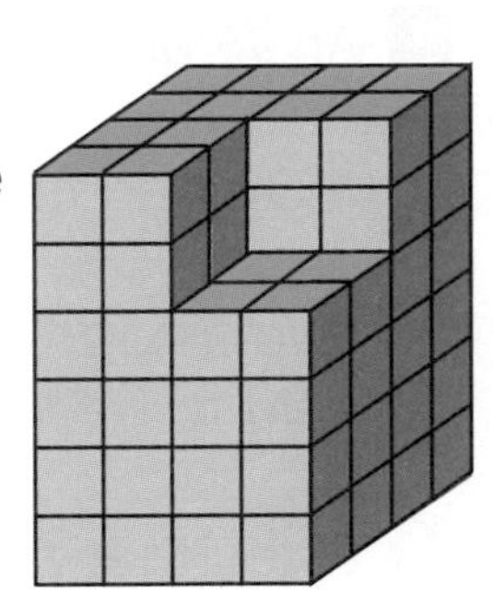

11 Choose different categories that you can use to sort the toys, games, or puzzles that you know or have heard about. Make Venn diagrams to show your work.

12 A **palindrome** is a word or number that is the same if you read it forward or backward. The word *noon* is a palindrome. Find as many other palindromes as you can.

13 **At the Computer** You can make cubes out of smaller cubes as shown on the right.

Use a drawing program to create the next three larger cubes.

Record the measurements of the cubes in a table like the one below. Write about any patterns you see.

Length	Width	Height	Number of Cubic Units
1	1	1	1
2	2	2	8
3	3	3	27
■	■	■	■
■	■	■	■
■	■	■	■

Language and Mathematics

Complete the sentence. Use a word in the chart. (pages 320–353)

1 An ■ angle is one that measures less than a right angle.

2 A 6-sided ■ is called a hexagon.

3 Two figures that are exactly the same shape and size are said to be ■.

4 Lines that are ■ never intersect.

5 The ■ of a rectangle can be found if you know its length and width.

Vocabulary
parallel
area
polygon
acute
congruent
perpendicular

Concepts and Skills

Name the figure. (pages 320, 324)

6 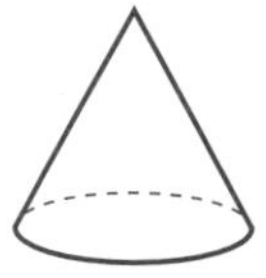

7 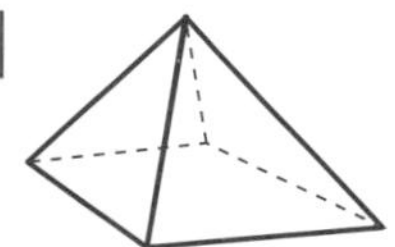

8 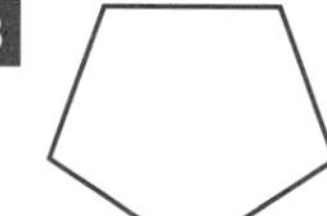

Describe the angle. (page 332)

9 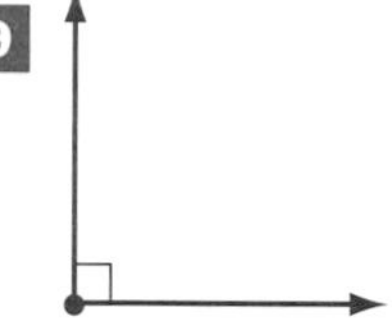

10 

11 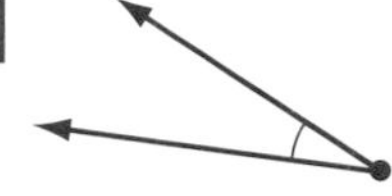

Are the figures congruent? If not, are they similar? (page 340)

12 

13 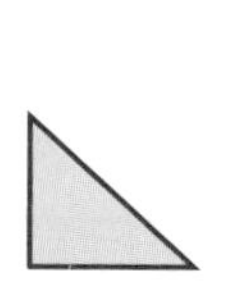 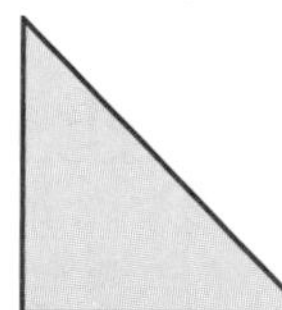

Tell whether or not the dotted line is a line of symmetry. (page 344)

14 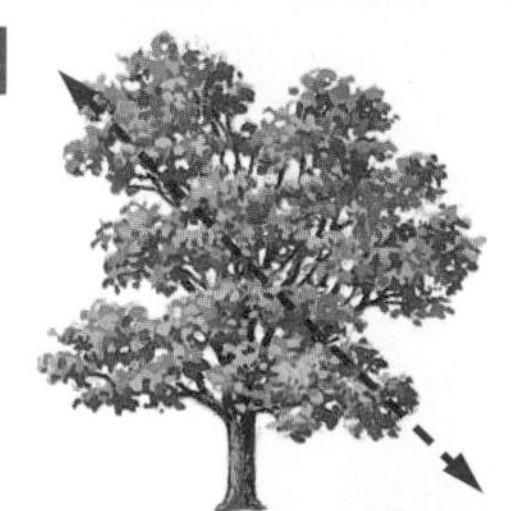

15 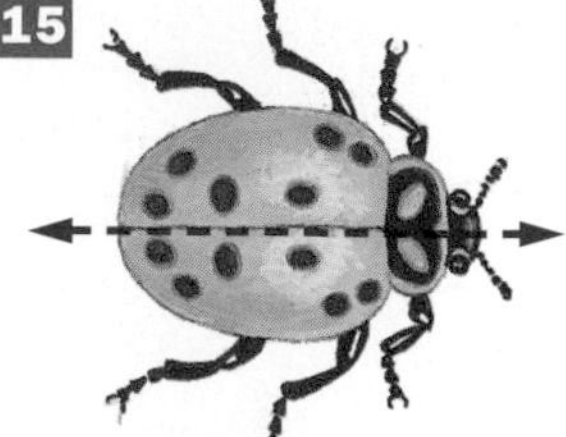

16 

Think critically. (pages 328, 346)

17 Analyze. Explain what the mistakes are, then correct them.

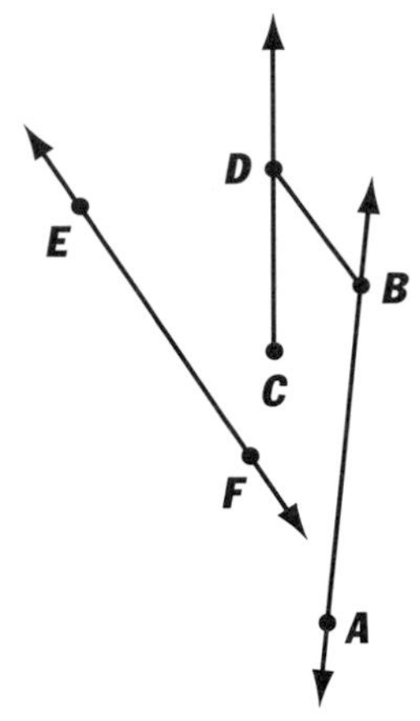

Lines: AB, BD, CD, EF

18 Analyze. Use the terms *flip, slide,* and *turn* to describe how the square has been moved.

a. 

b. 

c.

(pages 330, 354)

19 The perimeter of a rectangle measures 40 meters. The width measures 6 meters. What is the length of the rectangle?

20 What is the volume of each rectangular prism on the right? What pattern do you see? What would be the volume of the sixth rectangular prism in this pattern?

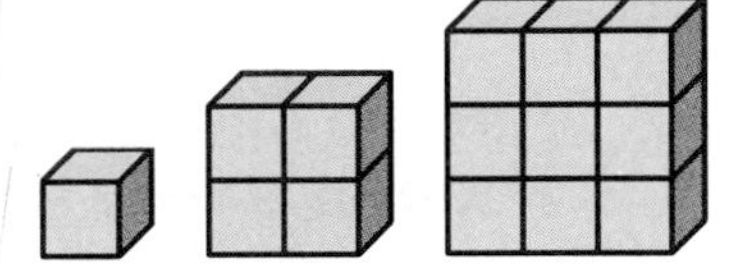

21 Draw a figure that has 3 or more lines of symmetry. Label each vertex with a letter. Use five different mathematical terms you learned in this chapter to describe your figure.

22 **Make a decision** An art store sells sheets of drawing paper for 15¢. You can buy a package of 200 sheets of the same type of paper for $17.00. What would you buy if you needed 300 sheets of paper? Make an organized list to help find the answer.

Use the figure for problems 23–25.

23 Name one pair of parallel line segments.

24 Find the area of the largest face.

25 Name one pair of perpendicular line segments.

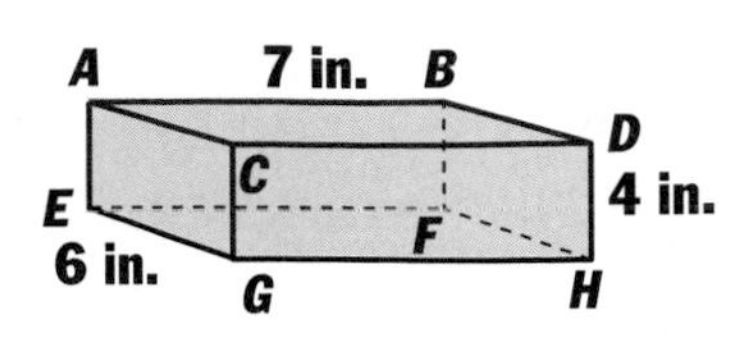

Are the figures intersecting, parallel, or perpendicular?

1 

2 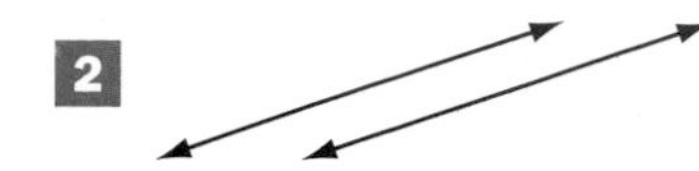

Tell if the angle is obtuse, acute, or a right angle.

3  

4 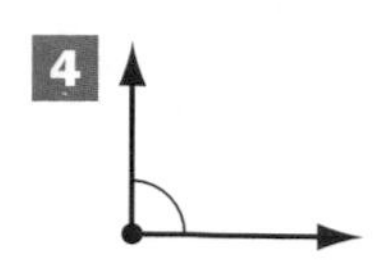

Name the figure.

5 

6 

7 

8 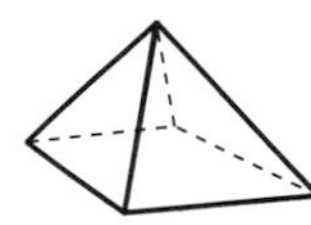

9 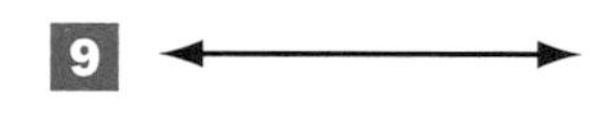

10 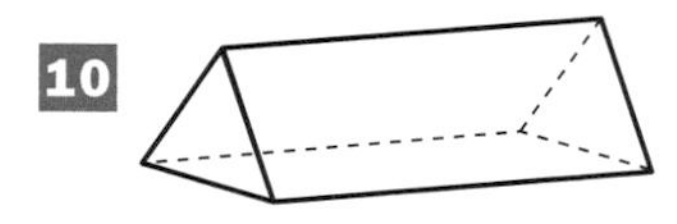

Tell if the figure is symmetric.

11 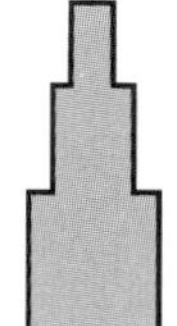

12 

Find the area.

13 

14 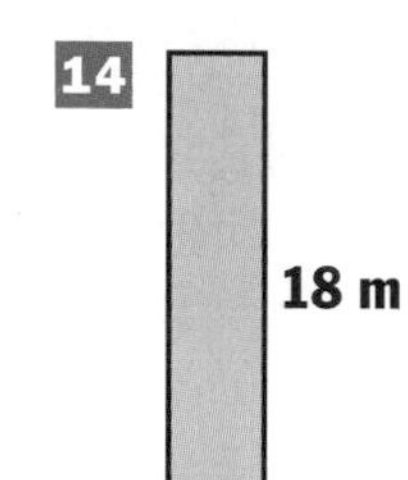

Are the figures congruent? If not, are they similar?

15 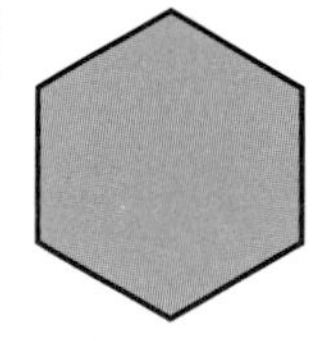 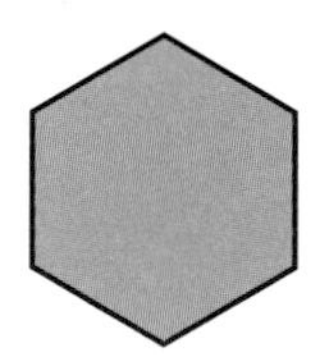

Find the volume.

16 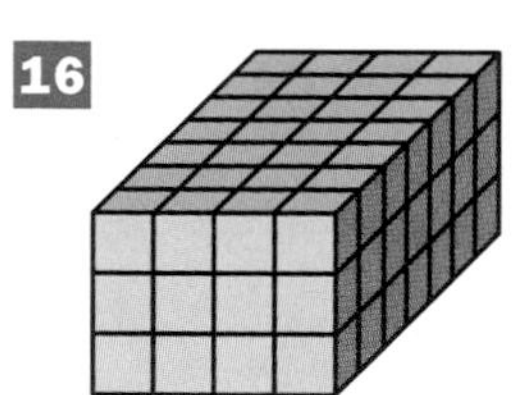

Solve.

17 An artist is arranging four paintings of Tampa, Los Angeles, Houston, and Dayton along a wall. How many different ways can she arrange the four paintings?

18 Classrooms 101, 102, 103, 104, and 105 have one computer each. All 5 computers are connected by wires to all of the others. How many connections are there in all?

19 Can you get the same position by flipping a kite as you can by turning it? sliding it? Give examples.

20 Is it possible to have parallel lines that intersect? perpendicular lines that intersect? Explain.

What Did You Learn?

Fernand Leger is a famous artist. Look at this photograph of one of his paintings.

"LE REMORGUEUR," FERNAND LEGER, GRENOBLE MUSEE

- ▶ Describe three geometric figures you see in the painting. Describe each figure in as many ways as you can.
- ▶ Draw and label three geometric figures that are not in the painting. Use the three figures you drew to create a picture of your own. The picture should show examples of slides, flips, and turns.

You will need
- *dot paper*
- *colored pencil*

A Good Answer
- accurately describes figures in the paintings using geometric terms
- identifies geometric figures that are not in the painting and correctly shows slides, flips, and turns

You may want to place your work in your portfolio.

What Do You Think

1. Do you look at the parts of a figure if you are unsure about how to name it? Why or why not?
2. List all the ways you might use to help you name a 3-dimensional figure:
   - Use the shape of the faces.
   - Use the angles in the figure.
   - Think of real-life objects.
   - Other. Explain.
3. Are you always able to determine how a figure has been moved? What method do you use?

math science technology
CONNECTION

From Bits and Pieces

1

Mosaics are made of small pieces of stone, glass, or tile cemented onto objects.

2 Some mosaics are geometric designs. Others are pictures.

3 Some of the oldest mosaics were made with colored stones cemented into walkways to make pictures and designs.

4

A volcano buried the ancient Roman city of Pompeii with ash in A.D. 79. Scientists learned about what life was like in Pompeii from mosaics, and other things they uncovered.

- What geometric shapes are found in the mosaics that are shown?

- What other things do you think scientists uncovered in Pompeii? What might still be there after centuries have passed?

Studying Mosaics

Study the mosaics on pages 362–363 and then answer the questions.

1 What different figures do you see in the mosaics?

2 How many different kinds of polygons can you find in the mosaics?

3 Use the centimeter grid to estimate the area of each section in the mosaic below.

4 What colors are used in the mosaic below? How much area of the mosaic does each color cover? Use the centimeter grid to estimate the area of each color.

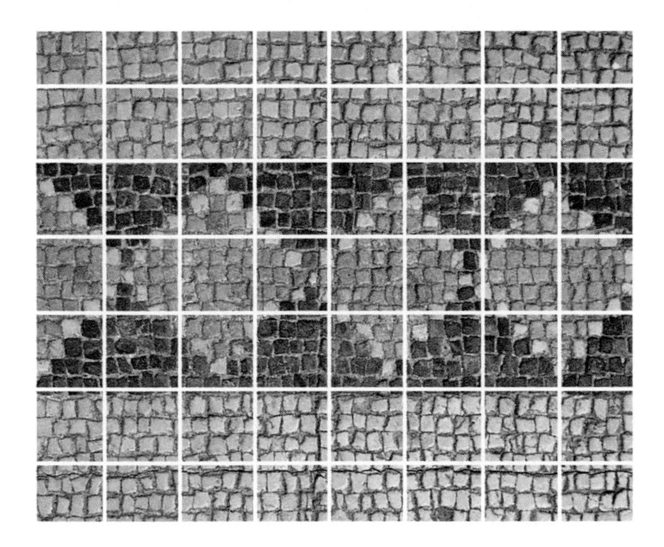

At the Computer

5 Use a drawing program to make your own mosaic. Decide whether you want to make a mosaic that is a geometric design, one that pictures something, or one that combines a picture and a geometric design. Use your computer program to repeat different shapes and use different colors.

6 Write a description of your mosaic using geometric terms.

Cultural Note

People in different cultures have created mosaic art pieces. The Greeks taught the Romans how to make them. Muslims in India made mosaics. The Mayan and Aztec Indians of Central America also made them.

Choose the letter of the correct answer.

1 If you have \$2.60, what is the greatest number of quarters you could have?

A 8
B 10
C 12
D 15

2 Jerry's class recycled 69 pounds of aluminum cans. They got \$0.33 for each pound of cans. Which is the best estimate of how much money they got?

F \$2.10
G \$18.00
H \$21.00
J \$210.00

3 What do the letters A, H, M, T, and X have in common? They are all ■.

A closed figures
B symmetrical
C similar
D congruent

4 What number when added to 28,863 gives a number between 34,000 and 35,000?

F 8,000
G 6,000
H 5,000
J 4,000

5 Complete the sentence.
$48 \times 6 = 6 \times ■ = 288$

A 6
B 48
C 54
D 288

6 The perimeter of a square is 36 cm. What is the length of one side?

F 6 cm
G 7 cm
H 8 cm
J 9 cm

7 Albert has \$1.53 worth of coins in his pocket. Which combination of coins can he *not* have?

A 6 quarters, 3 pennies
B 3 quarters, 5 dimes, 4 nickels, 8 pennies
C 4 quarters, 3 dimes, 5 nickels, 3 pennies
D 5 quarters, 2 dimes, 1 nickel, 3 pennies

8 Ada bought 4 licorice ropes that each measured 36 inches in length. If she gives a 5-inch piece to each of her 28 classmates, how much will she have left?

F 8 in.
G 5 in.
H 4 in.
J 3 in.

9

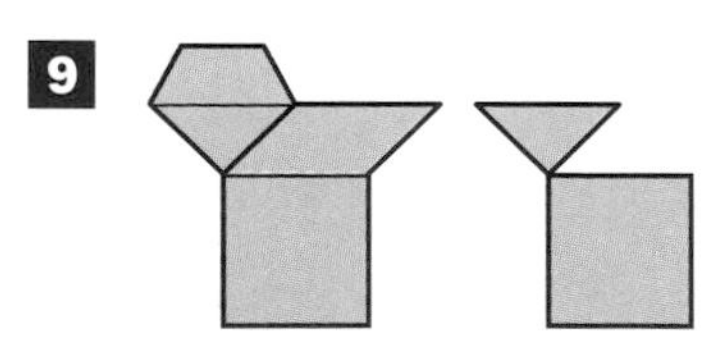

Which pattern blocks would you need to complete the design on the right to match the one on the left?

A 1 square and 1 trapezoid
B 1 trapezoid and 1 parallelogram
C 2 trapezoids
D 1 triangle and 1 square

10 A parent must accompany each group of 6 students on a trip. If 32 students are going, how many parents will be needed?

F 3
G 6
H 7
J 32

11 What is the value of the digit 3 in the number 134,927?

A 300
B 3,000
C 30,000
D 300,000

12 Marcus fills a pitcher with 3 qt of water. This is the same as ■.

F 1 gal
G 8 pt
H 12 c
J 84 c

13 What are the next three numbers in this pattern: 1, 3, 6, 10?

A 13, 16, 20
B 20, 30, 40
C 15, 20, 25
D 15, 21, 28

14 Carol has $10.00. She buys 2 notebooks for $0.99 each and a box of crayons for $2.25. The tax is $0.31. How much change will she get?

F $3.55
G $4.23
H $4.54
J $5.46

15 43×16

A 59
B 201
C 658
D 688

16 $547 \div 6$

F 9 R11
G 91
H 90 R1
J 91 R1

17 Which is the most reasonable answer? During the day, Hector drank a ■ of water.

A liter
B meter
C milliliter
D gram

18 Of the following data, 84 is the ■.
97, 59, 65, 97, 102.

F range
G median
H mode
J average

19 What is *not* shown in the diagram?

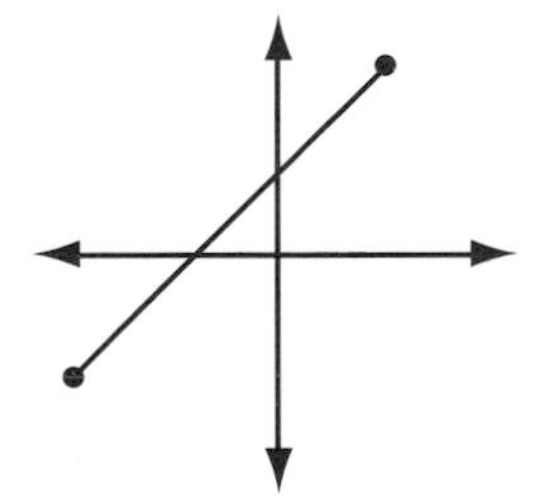

A perpendicular lines
B intersecting lines
C line segment
D parallel lines

TEST PREPARATION

CHAPTER

10 FRACTIONS AND PROBABILITY

THEME

Fun for Rainy Days

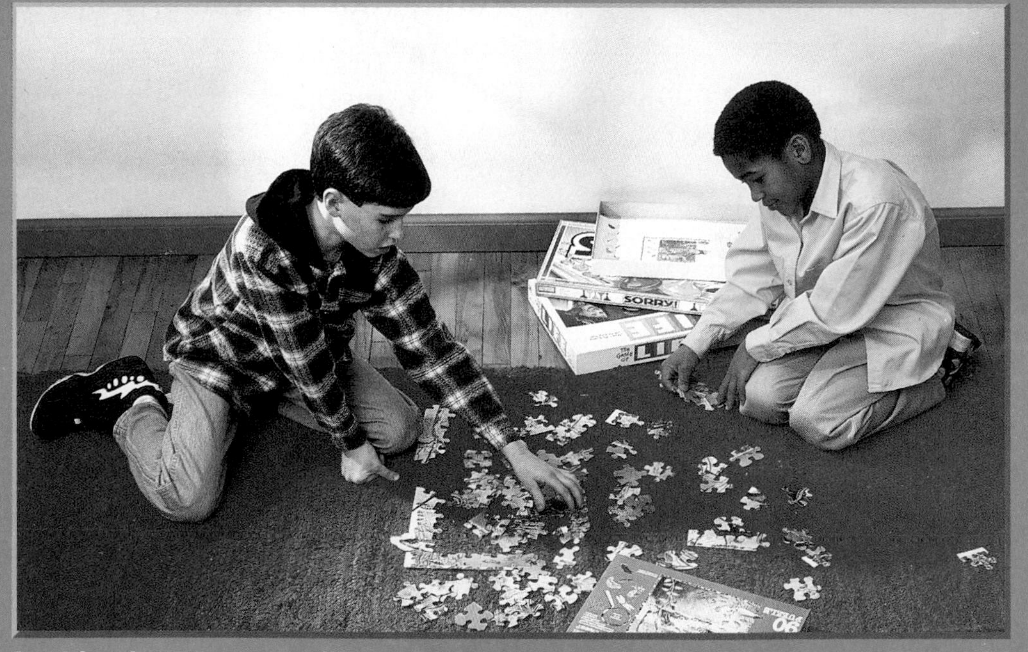

In this chapter, you will learn about activities that are great fun to do on rainy days. You will use mathematics to identify fractions and find probability to help you when it is your turn to play.

What Do You Know?

1. Which spinners show fourths? How do you know?

2. Which spinner shows eighths? How do you know?

3. Which spinner would give you the best chance of winning a game by spinning red? Explain your reasoning.

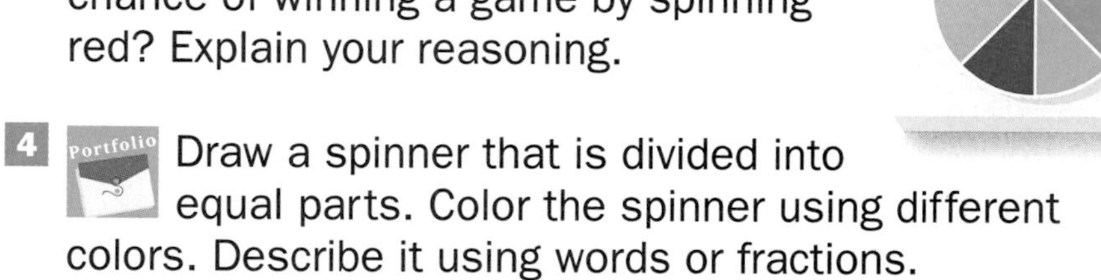

4. Portfolio Draw a spinner that is divided into equal parts. Color the spinner using different colors. Describe it using words or fractions.

Make Predictions

Each player in a basketball game has taken 10 shots. Mia has made 8 baskets, James 4, and Toshi 6. To win, a player must get 20 baskets. Predict who will be the winner.

When you make a prediction, you use what you already know to make a guess about what comes next.

1. Which player did you predict to win? to come in last? Why?

2. What fraction of the 20 baskets has each player made so far?

Vocabulary

fraction, p. 368
numerator, p. 368
denominator, p. 368
equivalent fraction, p. 376
simplest form, p. 380
improper fraction, p. 384
mixed number, p. 384
possible outcome, p. 390
probability, p.391
favorable outcome, p. 392

Part of a Whole

LEARN

This bracelet was designed on a loom from a Native American craft kit. What part of the whole bracelet is red?

You can describe the part of the bracelet that is red using words: 2 red parts out of eight equal parts.

You can also write a **fraction.**

Write: $\frac{2}{8}$ ← **numerator** / ← **denominator**

Read: Two eighths of the bracelet is red.

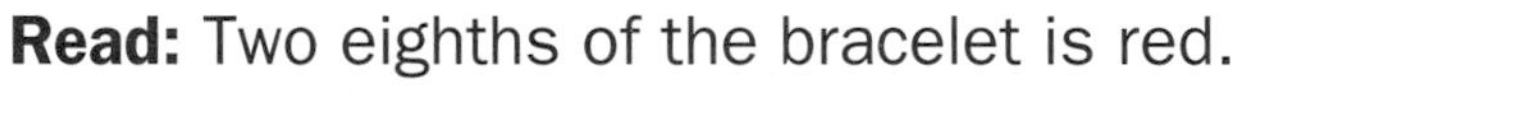

What fraction of the whole bracelet is blue? white?

$\frac{3}{8}$ ← blue parts / ← parts in all

Three eighths of the bracelet is blue.

$\frac{1}{8}$ ← white part / ← parts in all

One eighth of the bracelet is white.

Talk It Over

- ▶ **What if** you rearrange the color pattern on the bracelet. Will the fraction that is red change? the fraction that is blue? white? yellow? Explain.

- ▶ How would you describe the fraction of the bracelet that is *either* red, blue, white, *or* yellow? Explain.

Check Out the Glossary
fraction
numerator
denominator
See page 544.

You can also measure fractions of an inch using a ruler.

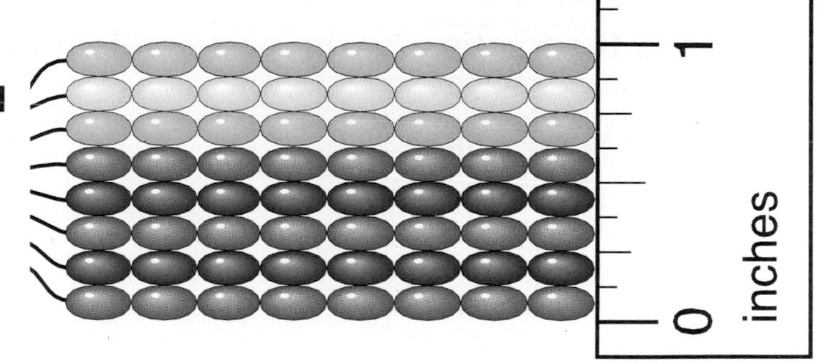

The bracelet is 1 in. wide. What part of an inch do the rows of blue and red beads make up?

Use a ruler that shows eighths to measure.

The rows of red and blue beads measure $\frac{5}{8}$ in.

More Examples

A

$\frac{3}{7}$ of the bracelet is yellow.

B

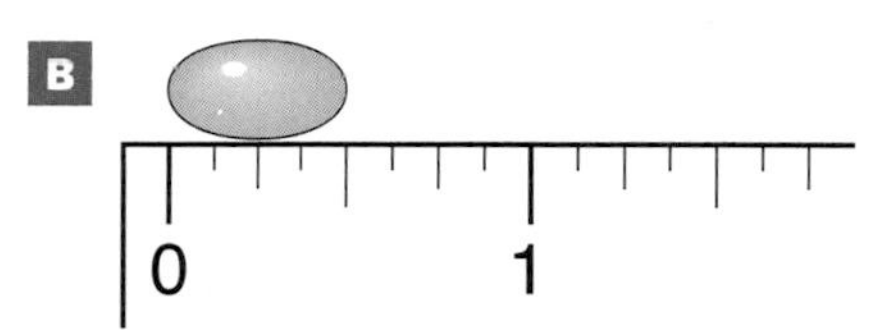

The bead is $\frac{4}{8}$ in., $\frac{2}{4}$ in., or $\frac{1}{2}$ in. long.

Check for Understanding

Write a fraction for the part that is shaded.

1 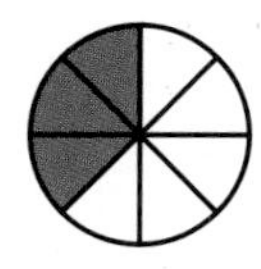

2 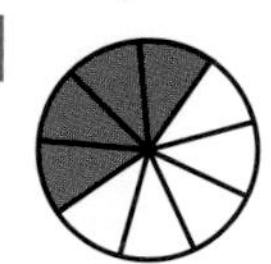

3 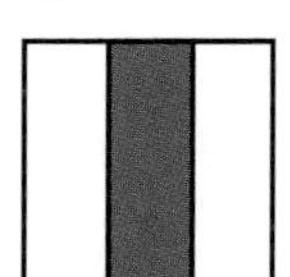

4 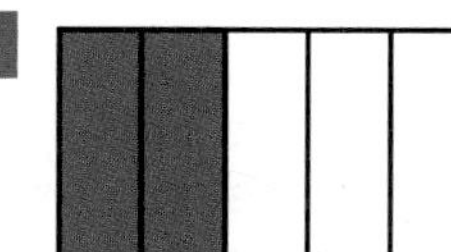

Measure the segment to the nearest $\frac{1}{8}$, $\frac{1}{4}$, and $\frac{1}{2}$ inch.

5

6

7

8 

Critical Thinking: Analyze **Explain your reasoning.**

9 Which of these squares does *not* show $\frac{1}{4}$ red?

a. 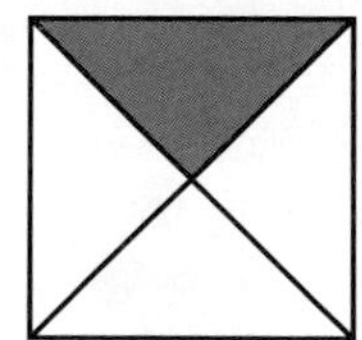

b. 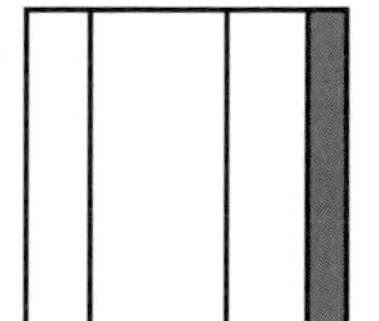

c. 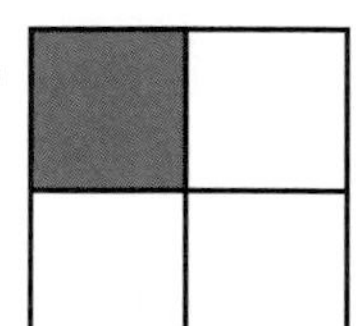

d. 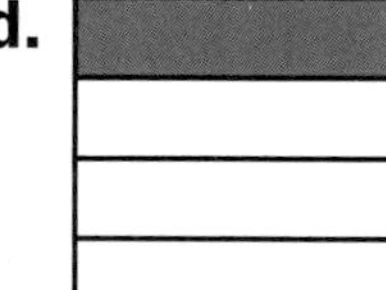

10 **What if** $\frac{1}{3}$ of an object is blue. What part is *not* blue? How can you tell?

Turn the page for Practice.

Practice

Write the fraction for the part that is white.

1 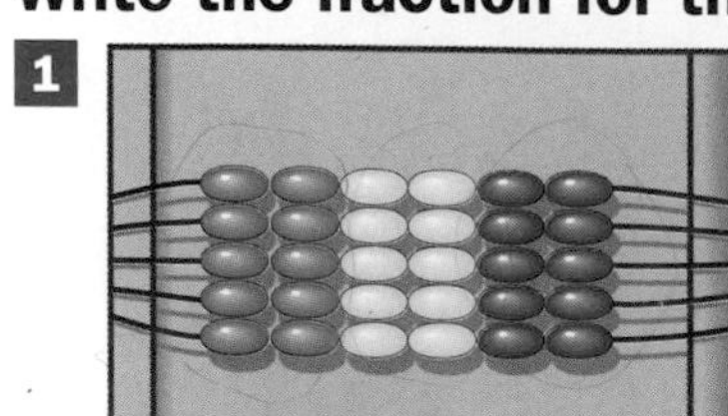

2 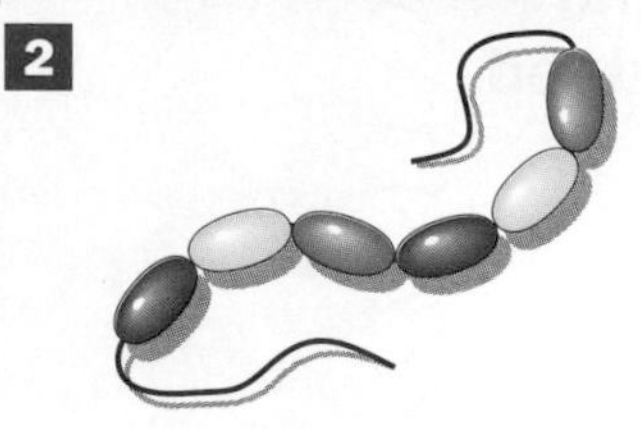

3 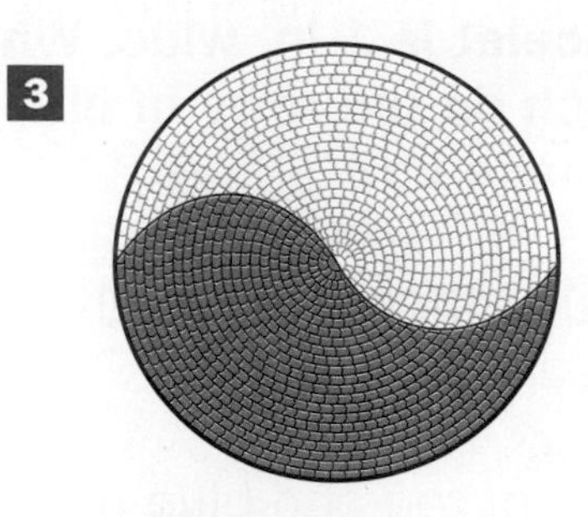

4 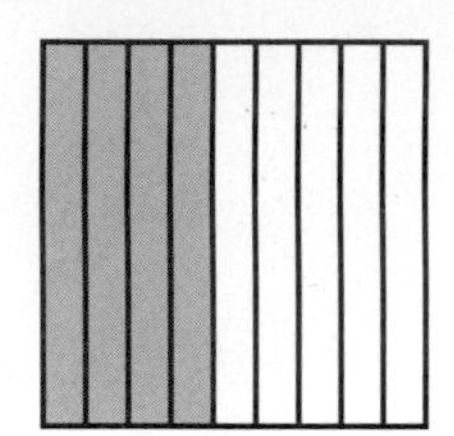

5 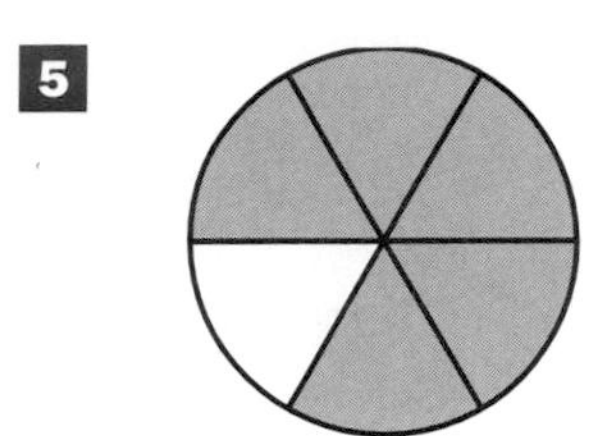

6 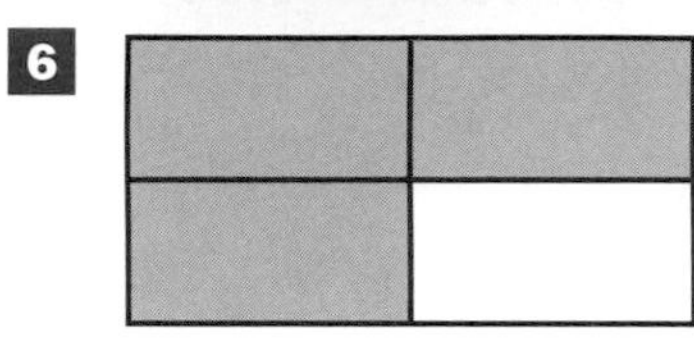

Draw a rectangle with the fraction shaded.

7 $\frac{2}{3}$ **8** $\frac{1}{4}$ **9** $\frac{5}{8}$ **10** $\frac{3}{5}$ **11** $\frac{1}{2}$ **12** $\frac{5}{9}$

13 $\frac{1}{5}$ **14** $\frac{1}{6}$ **15** $\frac{4}{7}$ **16** $\frac{4}{5}$ **17** $\frac{7}{8}$ **18** $\frac{2}{9}$

Measure the segment to the nearest $\frac{1}{8}$, $\frac{1}{4}$, and $\frac{1}{2}$ inch.

19  **20** **21**

MIXED APPLICATIONS

Problem Solving

Pencil & Paper | Calculator | Mental Math

Use the picture of the checkers and checkerboard for problems 22–24.

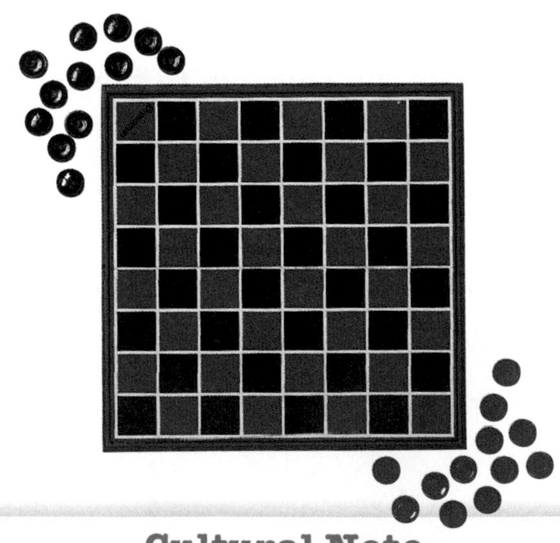

22 What fraction of the board is black squares?

23 What fraction of the board is red squares?

24 What fraction of the board is squares that touch the edges?

25 Uri is playing a card game with 3 other players. She deals out all 52 cards. Everyone gets the same number of cards. How many cards does each player get?

Cultural Note

In Britain, checkers is called *draughts.* Records show it was played during the 1500s.

Fraction Colors Game!

First, write each of the following fractions on an index card:

$\frac{1}{2}, \frac{2}{3}, \frac{1}{4}, \frac{3}{4}, \frac{3}{5}, \frac{1}{6}, \frac{5}{8}, \frac{7}{8}, \frac{3}{10}, \frac{11}{12}$.

Next, write each of the following colors on an index card: blue, green, yellow, red.

You will need
- *pencils or crayons*
- *graph paper*

Play the Game

Play with a partner.

- ▶ Place the cards in separate stacks. Mix up each stack.
- ▶ Choose a card from each stack. Each player draws a square or rectangle on graph paper and shows the fraction in the color chosen.
- ▶ If both players correctly show the same design, they each score 2 points. If both players correctly show different designs, they each score 4 points. If only one player shows the correct design, the player scores 4 points.
- ▶ Continue until a player has more than 20 points.

	Round 1	Round 2
Katie	4	
Lolita	4	

mixed review • test preparation

1 2,658 + 562 **2** 7,543 − 921 **3** 3,000 × 25 **4** 89 ÷ 3

Find the area.

5 10 cm, 14 cm

6 32 in., 24 in.

7 7 ft

8 42 cm

Extra Practice, page 520

Part of a Group

LEARN

What fraction of the chess pieces are *either* a king *or* a pawn?

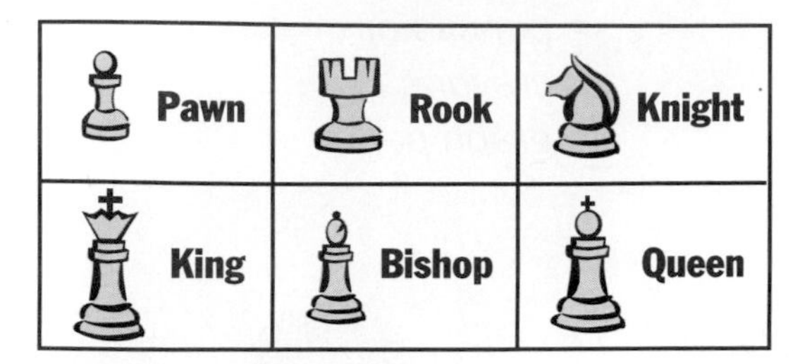

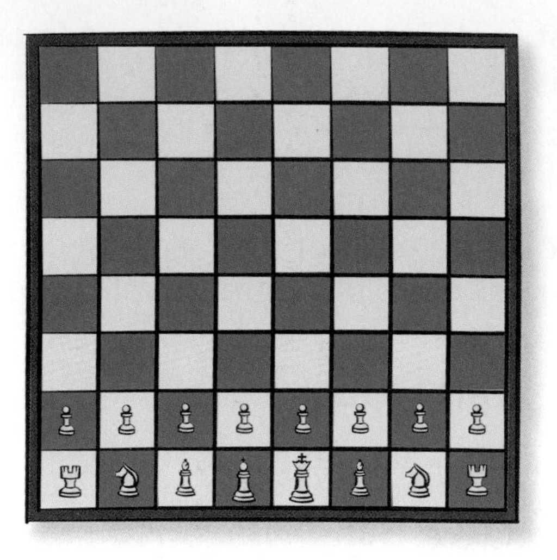

Using words or a fraction, you can describe the part of the chess pieces that are either a king or a pawn.

There are 9 king or pawn pieces out of 16 total pieces.

$$\begin{array}{r} \text{king or pawn pieces} \rightarrow \\ \text{total pieces} \rightarrow \end{array} \frac{9}{16} \begin{array}{l} \leftarrow \text{numerator} \\ \leftarrow \text{denominator} \end{array}$$

$\frac{9}{16}$ of the pieces are either a king or a pawn.

Cultural Note
Chess was first played in India. It was called *chaturanga* (CHUH-tuh-ran-gah).

More Examples

A $\frac{8}{10}$ are black.

B $\frac{5}{9}$ are blue or red.

C $\frac{4}{6}$ are not white or yellow.

CHECK

Check for Understanding

Write a fraction that names the part.

1. red
2. white
3. blue or black
4. not black

Critical Thinking: Analyze **Explain your reasoning.**

5. If you know that 9 out of 16 pieces are either a king or a pawn, how can you use this to find the fraction of the pieces that are neither a king nor a pawn?

Practice

Choose the fraction that tells which part is blue.

1

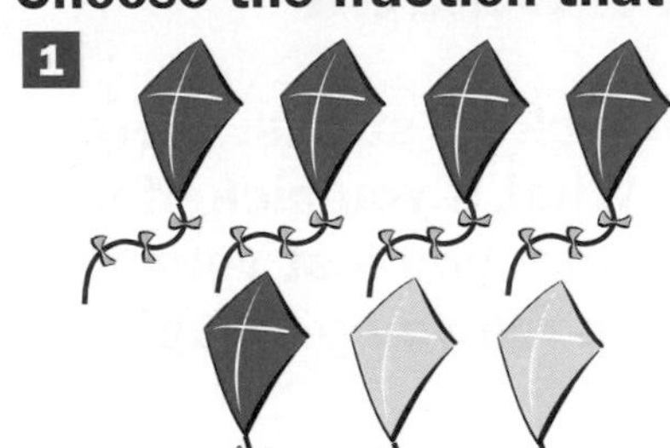

a. $\frac{5}{7}$ b. $\frac{2}{7}$ c. $\frac{7}{2}$

2

a. $\frac{3}{11}$ b. $\frac{3}{7}$ c. $\frac{3}{14}$

3

a. $\frac{2}{9}$ b. $\frac{2}{7}$ c. $\frac{2}{4}$

Use the number cubes for ex. 4–6.

3 6

3 3 4

1 1 1 2 5

4 What part of the green cubes show 1?

5 What part of all the cubes show 3?

6 What part of the blue cubes show 5?

Draw a picture, then write the fraction.

7 Five out of seven students are not wearing hats.

8 One out of four people are smiling.

9 Three out of eleven pets are dogs.

10 All of the six beanbags are red.

11 In ex. 7–10, how did the descriptions help you write the fractions?

MIXED APPLICATIONS
Problem Solving

12 The time allowed for each move during the Internet chess game was 7 min. What is the longest time this game could have taken? **SEE INFOBIT.**

13 **Write a problem** in which you name a fractional part of a group. Solve your problem. Have others solve it. Share and compare problems with other classmates.

INFOBIT
On August 26, 1996, Anatoly Karpov played the first open chess game on the Internet. The game ended after 65 moves.

Extra Practice, page 520

LEARN

Find a Fraction of a Number

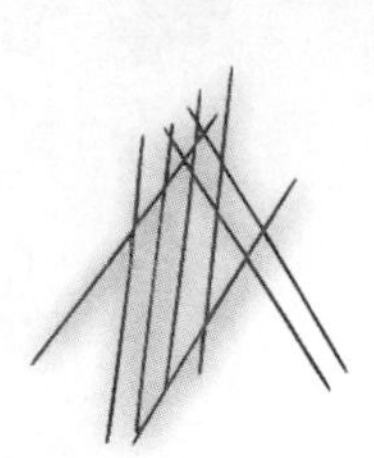

In the game of pick-up sticks, there are 24 sticks. What if you picked up $\frac{1}{3}$ of the sticks before you lost your turn. How many sticks did you pick up?

You can use counters to find the fraction of a number.

Find $\frac{1}{3}$ of 24.

First, make 3 equal groups using 24 counters.
Then, find the number of counters in 1 group.

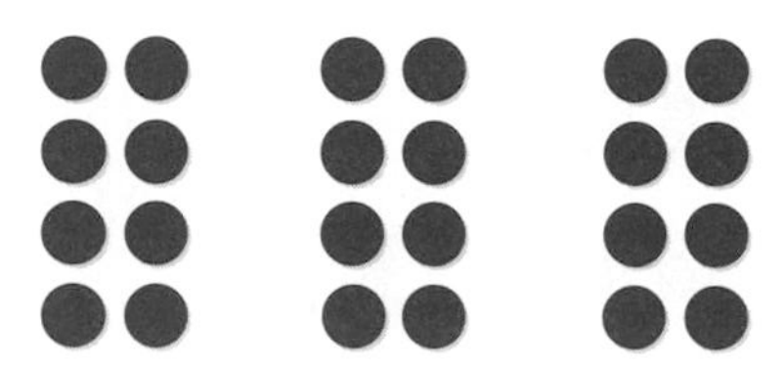

Think: $24 \div 3 = 8$

So $\frac{1}{3}$ of $24 = 8$.

You picked up 8 sticks.

Work Together

Work with a partner. Use counters to find the fraction of the number.

You will need
- *counters*

a. $\frac{1}{2}$ of 18 $\quad \frac{2}{2}$ of 18

b. $\frac{1}{3}$ of 33 $\quad \frac{2}{3}$ of 33 $\quad \frac{3}{3}$ of 33

c. $\frac{1}{5}$ of 20 $\quad \frac{2}{5}$ of 20 $\quad \frac{3}{5}$ of 20 $\quad \frac{4}{5}$ of 20 $\quad \frac{5}{5}$ of 20

Make Connections

You can use multiplication and division to find fractions of a number.

Find $\frac{2}{5}$ of 20.

Step 1	Step 2
Use the denominator. **Divide the total into that many groups.**	**Use the numerator.** **Multiply the quotient by that number.**
$20 \div 5 = 4$ **Think:** The denominator is 5.	$2 \times 4 = 8$ **Think:** The numerator is 2.

So $\frac{2}{5}$ of 20 is 8.

▶ How is the method you used with counters the same as dividing and multiplying?

▶ What do you notice about the fraction of the number when the numerator and denominator of the fraction are the same? when the numerator is less than the denominator?

Check for Understanding

Find the answer. Use any method.

1 $\frac{1}{4}$ of 8	**2** $\frac{1}{2}$ of 16	**3** $\frac{1}{8}$ of 16	**4** $\frac{1}{5}$ of 15
5 $\frac{2}{3}$ of 12	**6** $\frac{3}{4}$ of 16	**7** $\frac{1}{4}$ of 20	**8** $\frac{1}{12}$ of 12
9 $\frac{2}{5}$ of 10	**10** $\frac{1}{10}$ of 30	**11** $\frac{5}{6}$ of 12	**12** $\frac{3}{3}$ of 3

Critical Thinking: Analyze **Explain your reasoning.**

13 You know that $\frac{3}{5}$ of 15 counters are blue. How can you tell how many counters are not blue?

14 

Explain how to find $\frac{1}{7}$ of 21. How can you use the answer to find $\frac{2}{7}$ of 21? $\frac{3}{7}$ of 21?

Practice

Use the picture to help you find the missing number or fraction.

1 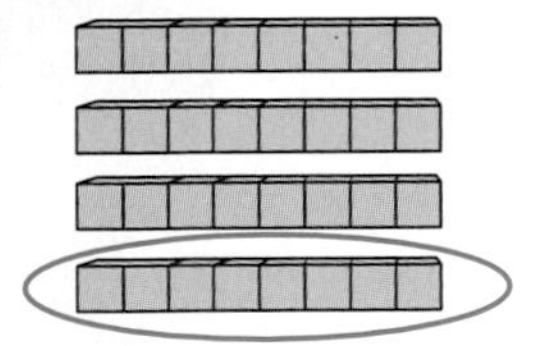

$\frac{1}{4}$ of 32 = ■

2  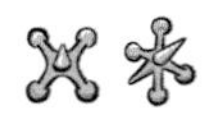 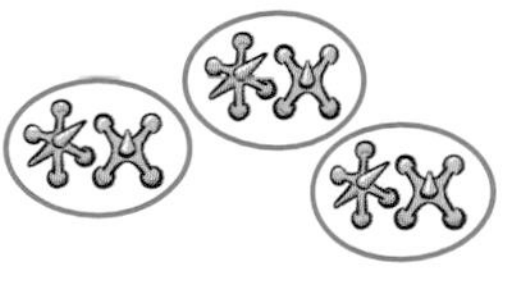

$\frac{3}{5}$ of 10 = ■

3 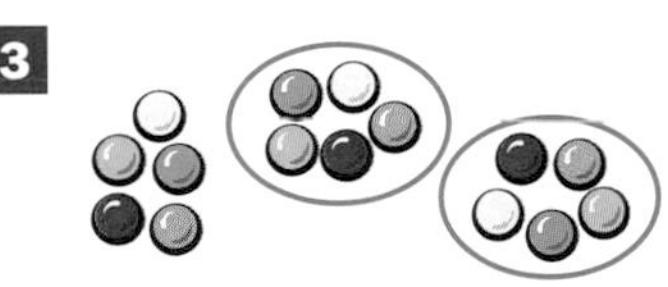

■ of 15 = 10

Find the answer. Use any method.

4 $\frac{1}{8}$ of 32	**5** $\frac{1}{5}$ of 30	**6** $\frac{3}{5}$ of 30	**7** $\frac{1}{3}$ of 27
8 $\frac{2}{3}$ of 27	**9** $\frac{4}{7}$ of 21	**10** $\frac{5}{8}$ of 16	**11** $\frac{2}{9}$ of 27

Extra Practice, page 520

Equivalent Fractions

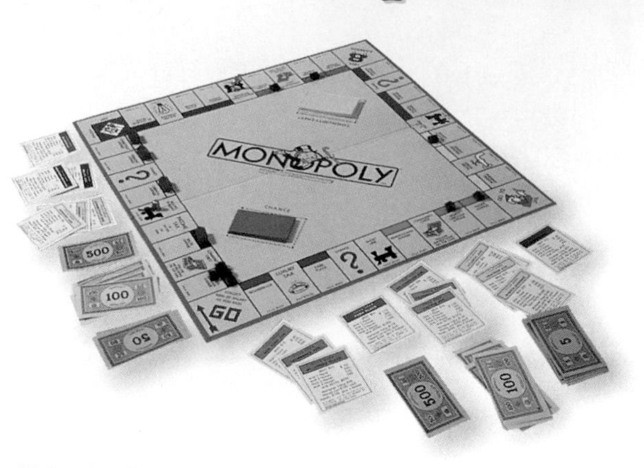

Jason is playing Monopoly. He has 4 houses and 4 hotels, so he says that $\frac{4}{8}$ of his properties are hotels. What other fractions could Jason have used to tell what part are hotels?

You can use fraction strips to find other fractions that name the same number as $\frac{4}{8}$. These are called **equivalent fractions.**

Work Together

Work in a group. Use fraction strips.

You will need
- *fraction strips*

Use eighths to show $\frac{4}{8}$.

Check each type of fraction strip. Which ones can you use to show an amount that is the same as $\frac{4}{8}$?

How many sections of each type of strip do you need to use?

Copy the table and record your work.

Equivalent Fraction for $\frac{4}{8}$

Fraction Strip Used	Number of Sections	Equivalent Fractions
$\frac{1}{8}$	4	$\frac{4}{8}$
$\frac{1}{2}$		

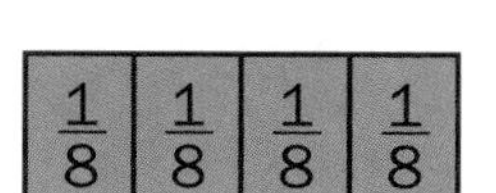

Use the same method to find equivalent fractions for $\frac{2}{3}$, $\frac{2}{8}$, $\frac{5}{10}$, and $\frac{2}{12}$.

Talk It Over

- How is the denominator in $\frac{4}{8}$ related to the numerator?
- How are the denominators in all of the fractions that are equivalent to $\frac{4}{8}$ related to their numerators?

Make Connections

You used fraction models to find equivalent fractions.

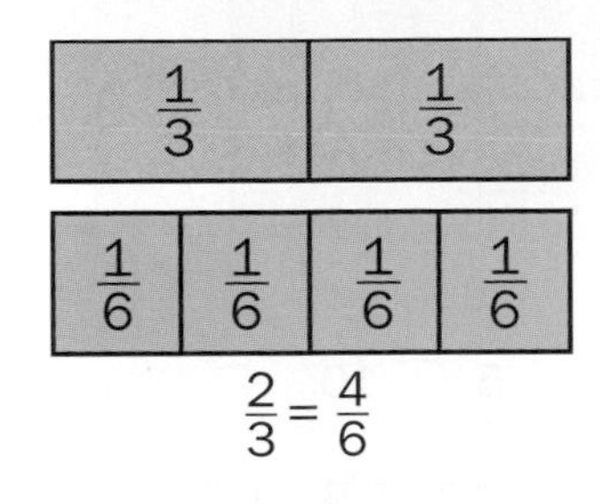

$\frac{2}{3} = \frac{4}{6}$

You can also multiply to find equivalent fractions.

To find an equivalent fraction for $\frac{2}{3}$, multiply the numerator and denominator by the same number.

$$\frac{2 \times 2}{3 \times 2} = \frac{4}{6}$$

So $\frac{2}{3}$ and $\frac{4}{6}$ are equivalent fractions.

- **What if** you divide the numerator and denominator of $\frac{4}{6}$ by the same number, 2. Does this give you an equivalent fraction? How do you know?
- Multiply or divide to check that the fractions you found are equivalent.

Check Out the Glossary
equivalent fractions
See page 544.

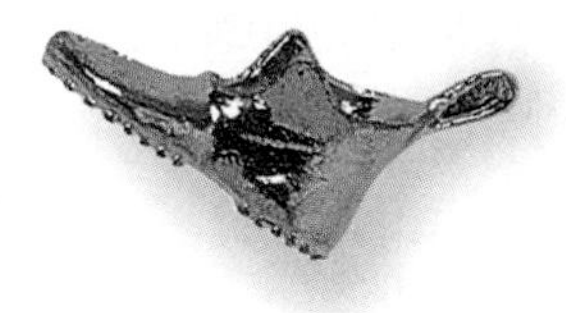

Check for Understanding

Name three equivalent fractions for the fraction.

1 $\frac{2}{3}$ **2** $\frac{1}{4}$ **3** $\frac{1}{5}$ **4** $\frac{4}{5}$ **5** $\frac{1}{2}$ **6** $\frac{1}{3}$

Complete to name the equivalent fractions.

7 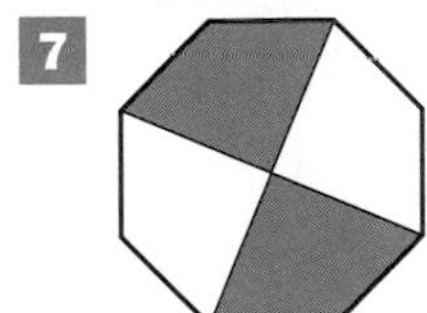

$\frac{2}{4} = \frac{■}{8}$

8 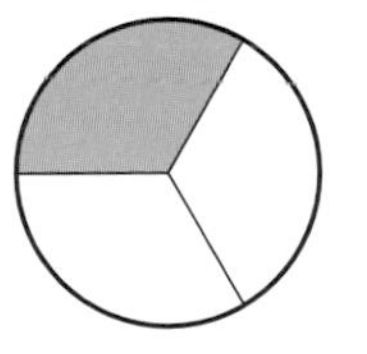

$\frac{1}{3} = \frac{■}{6}$

9 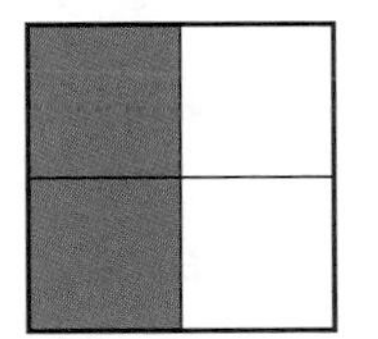 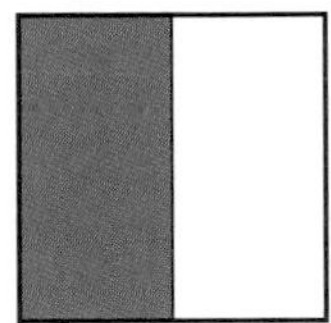

$\frac{■}{4} = \frac{1}{2}$

Critical Thinking: Analyze **Explain your reasoning.**

10 John says that no unit fraction (a fraction with a numerator of 1) is equivalent to a different unit fraction. Do you agree or disagree?

CHECK

Turn the page for Practice.

Practice

Complete to name the equivalent fraction.

1

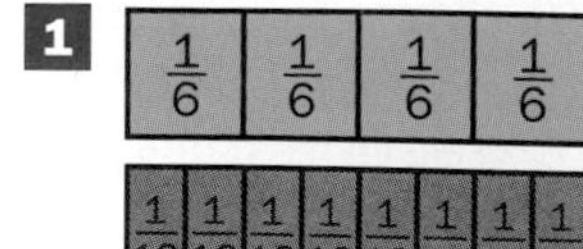

$\frac{4}{6} = \frac{\blacksquare}{12}$

2

$\frac{3}{4} = \frac{\blacksquare}{12}$

3

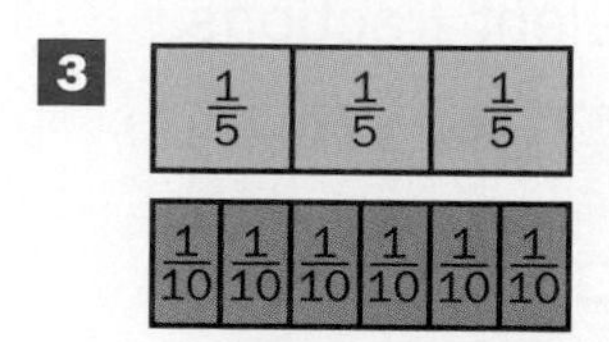

$\frac{\blacksquare}{5} = \frac{6}{10}$

4

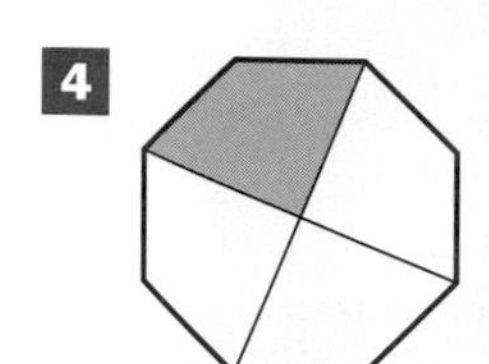

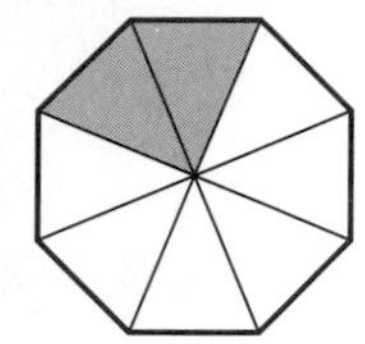

$\frac{1}{4} = \frac{\blacksquare}{8}$

5

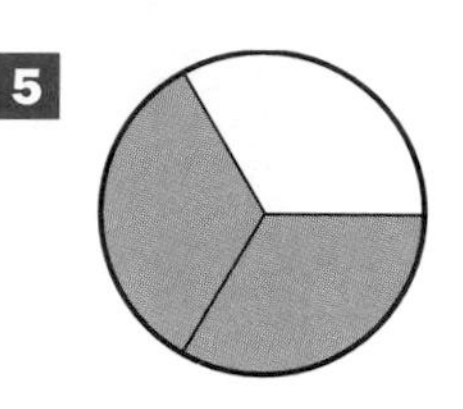

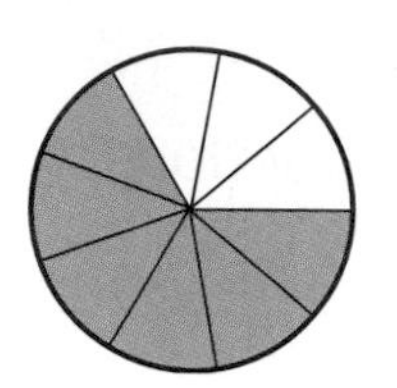

$\frac{2}{3} = \frac{\blacksquare}{9}$

6

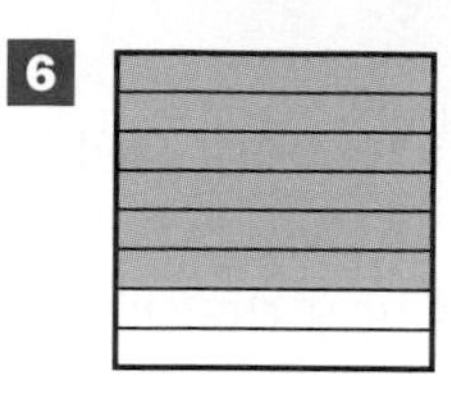

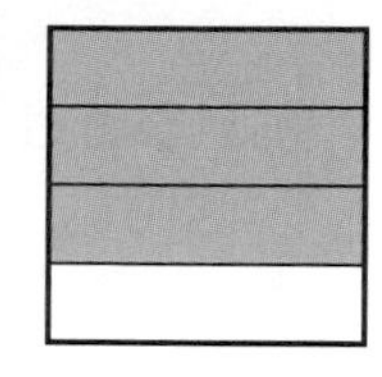

$\frac{\blacksquare}{8} = \frac{3}{4}$

7 $\frac{1}{2} = \frac{\blacksquare}{14}$ **8** $\frac{8}{24} = \frac{\blacksquare}{3}$ **9** $\frac{1}{5} = \frac{3}{\blacksquare}$ **10** $\frac{2}{3} = \frac{8}{\blacksquare}$

11 $\frac{10}{35} = \frac{\blacksquare}{7}$ **12** $\frac{3}{8} = \frac{\blacksquare}{16}$ **13** $\frac{6}{18} = \frac{2}{\blacksquare}$ **14** $\frac{9}{27} = \frac{3}{\blacksquare}$

Complete.

15 $\frac{1 \times \blacksquare}{2 \times \blacksquare} = \frac{5}{10}$ **16** $\frac{3 \times \blacksquare}{5 \times \blacksquare} = \frac{9}{15}$ **17** $\frac{1 \times \blacksquare}{4 \times \blacksquare} = \frac{3}{12}$

ALGEBRA: PATTERNS **Complete the equivalent fractions.**

18 $\frac{1}{2} = \frac{\blacksquare}{4} = \frac{\blacksquare}{6} = \frac{4}{\blacksquare} = \frac{5}{\blacksquare} = \frac{\blacksquare}{12}$

19 $\frac{1}{5} = \frac{\blacksquare}{10} = \frac{\blacksquare}{15} = \frac{4}{\blacksquare} = \frac{5}{\blacksquare} = \frac{\blacksquare}{30}$

20 $\frac{3}{4} = \frac{\blacksquare}{8} = \frac{\blacksquare}{12} = \frac{12}{\blacksquare} = \frac{15}{\blacksquare} = \frac{\blacksquare}{24}$

21 $\frac{2}{3} = \frac{\blacksquare}{6} = \frac{\blacksquare}{9} = \frac{8}{\blacksquare} = \frac{10}{\blacksquare} = \frac{\blacksquare}{18}$

22 $\frac{2}{5} = \frac{\blacksquare}{10} = \frac{6}{\blacksquare} = \frac{8}{\blacksquare} = \frac{\blacksquare}{25}$

23 $\frac{3}{7} = \frac{\blacksquare}{14} = \frac{\blacksquare}{21} = \frac{12}{\blacksquare} = \frac{15}{\blacksquare}$

Choose the letter of the equivalent fraction.

24 $\frac{5}{8}$ **a.** $\frac{5}{12}$ **b.** $\frac{10}{15}$ **c.** $\frac{10}{16}$ **d.** $\frac{8}{16}$

25 $\frac{1}{5}$ **a.** $\frac{5}{10}$ **b.** $\frac{5}{25}$ **c.** $\frac{2}{20}$ **d.** $\frac{3}{12}$

26 $\frac{16}{20}$ **a.** $\frac{10}{15}$ **b.** $\frac{4}{8}$ **c.** $\frac{8}{12}$ **d.** $\frac{4}{5}$

Make It Right

27 Here is how Miki found equivalent fractions. Find the mistake and correct it.

$\frac{12 \div 3}{20 \div 4} = \frac{4}{5}$ *$\frac{12}{20}$ and $\frac{4}{5}$ are equivalent fractions.*

MIXED APPLICATIONS
Problem Solving

28 Ricki and Mort are playing jacks. Ricki has picked up $\frac{2}{10}$ of the jacks. Mort has picked up $\frac{3}{5}$ of the jacks. Have they picked up the same number of jacks? Explain.

29 What fraction of each is made up of vowels? consonants?
a. your first name
b. your last name
c. your whole name

Use the graph for problems 30–32.

30 What fraction of the class picked chess as their favorite game?

31 How many more votes did Clue get than Monopoly?

32 What game got $\frac{1}{3}$ of the class votes?

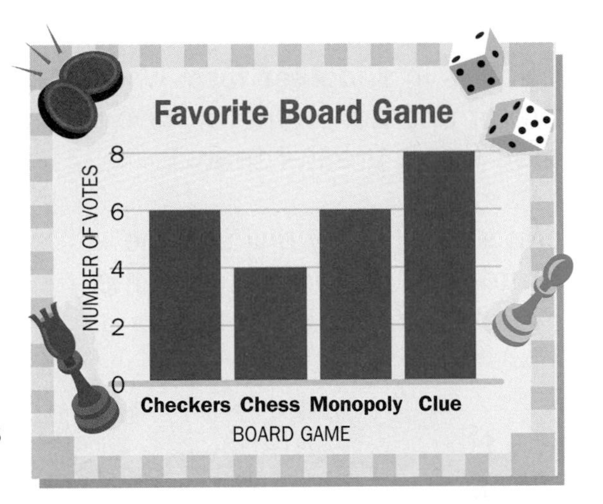

33 **Data Point** Survey your classmates about their favorite games. Graph your findings. Write a summary of the results.

Cultural Connection Pieces of Eight

Hundreds of years ago, a Spanish silver coin worth 8 *reales* (ray-AH-lez) was used. If people needed a smaller unit of money, they cut the coin.

Often, they cut the coin into 4 equal parts. How much was each part worth?

Think: The coin was worth 8 reales.
$\frac{1}{4}$ of 8 reales is 2 reales.

Each part was worth 2 reales.

Name the United States coin that is worth the fraction of a dollar.

Note: 1 United States dollar = 100 cents

1 $\frac{10}{100}$ **2** $\frac{25}{100}$ **3** $\frac{1}{100}$ **4** $\frac{5}{100}$

Simplify Fractions

LEARN

Hallie is playing Lu-lu with her friends. On her first toss, 2 out of the 4 game pieces land blank side up. Write the fraction of the pieces that land blank side up in simplest form.

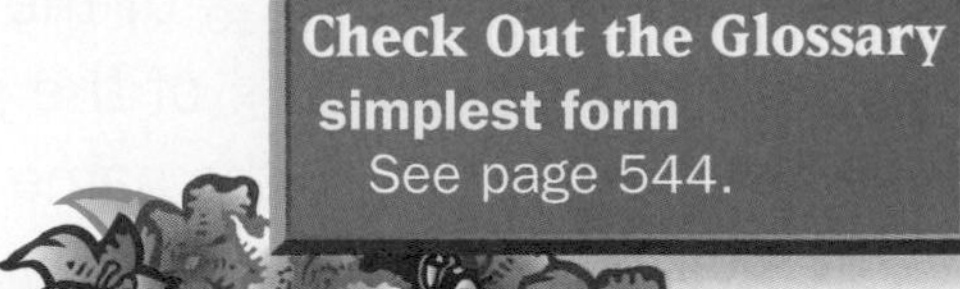

A fraction is in **simplest form** when the numerator and denominator have no common factor greater than 1.

You can simplify $\frac{2}{4}$ by dividing the numerator and denominator by the common factor that is greatest.

Step 1

Find the common factors.

Factors of 2: 1, 2
Factors of 4: 1, 2, 4
Common factors: 1 and 2

The simplest form of $\frac{2}{4}$ is $\frac{1}{2}$.

Step 2

Divide the numerator and denominator by the common factor that is greatest.

$$\frac{2}{4} = \frac{2 \div 2}{4 \div 2} = \frac{1}{2}$$

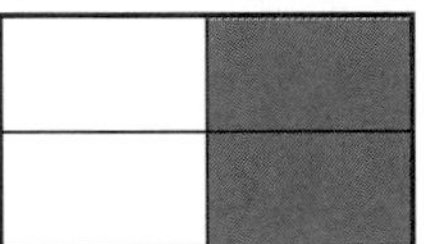

CHECK

Check for Understanding

Write the fraction in simplest form. Show your method.

1 $\frac{6}{15}$ **2** $\frac{2}{26}$ **3** $\frac{12}{18}$ **4** $\frac{10}{20}$ **5** $\frac{12}{15}$ **6** $\frac{64}{88}$

Critical Thinking: Generalize

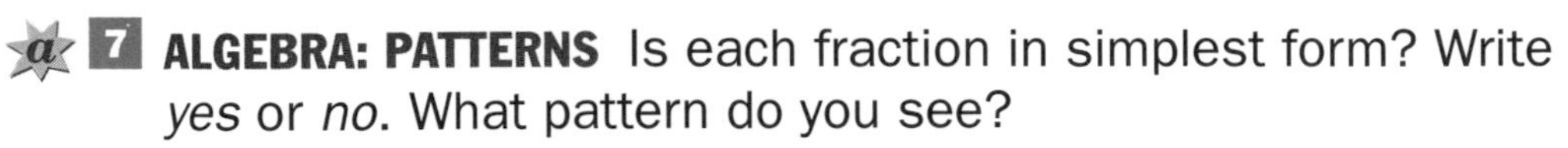

7 **ALGEBRA: PATTERNS** Is each fraction in simplest form? Write *yes* or *no*. What pattern do you see?

a. $\frac{1}{2}$ **b.** $\frac{1}{3}$ **c.** $\frac{1}{4}$ **d.** $\frac{1}{5}$ **e.** $\frac{1}{7}$ **f.** $\frac{1}{9}$

Practice

Complete to show the simplest form.

1

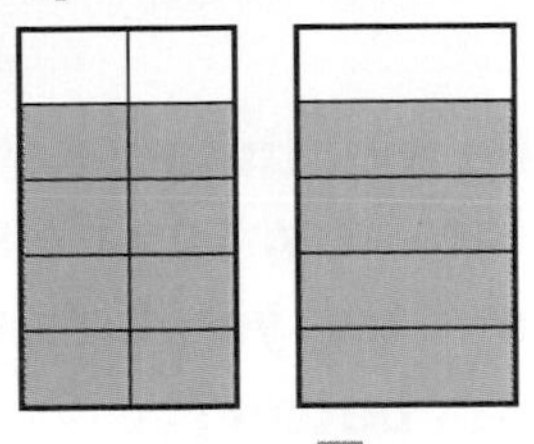

$\frac{8}{10} = \frac{\blacksquare}{\blacksquare}$

2

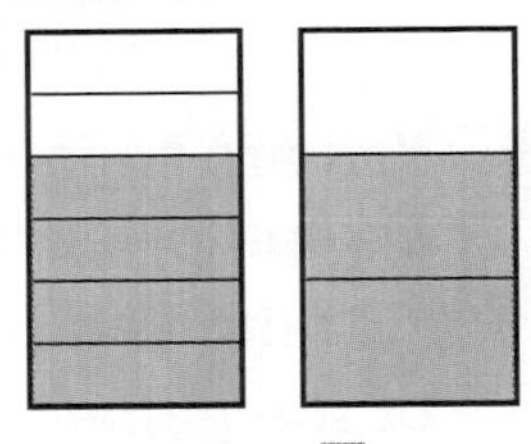

$\frac{4}{6} = \frac{\blacksquare}{\blacksquare}$

3

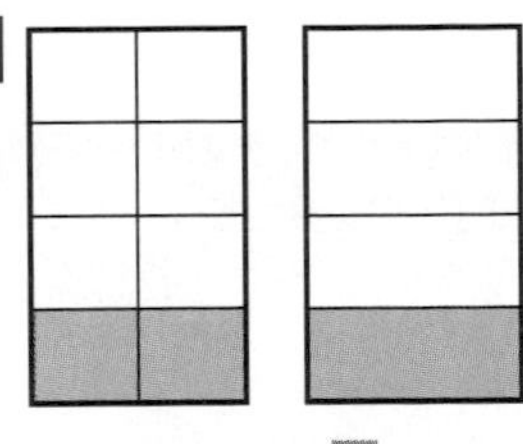

$\frac{2}{8} = \frac{\blacksquare}{\blacksquare}$

4 $\frac{6 \div 2}{10 \div \blacksquare} = \frac{3}{\blacksquare}$

5 $\frac{9 \div \blacksquare}{15 \div 3} = \frac{\blacksquare}{5}$

6 $\frac{8 \div \blacksquare}{12 \div 4} = \frac{\blacksquare}{3}$

Is the fraction in simplest form? Write *yes* or *no*.

7 $\frac{4}{10}$ **8** $\frac{1}{8}$ **9** $\frac{9}{15}$ **10** $\frac{8}{9}$ **11** $\frac{6}{21}$ **12** $\frac{6}{10}$

Write the fraction in simplest form. Show your method.

13 $\frac{7}{21}$ **14** $\frac{2}{10}$ **15** $\frac{6}{8}$ **16** $\frac{4}{16}$ **17** $\frac{9}{18}$ **18** $\frac{3}{27}$

19 $\frac{14}{16}$ **20** $\frac{15}{18}$ **21** $\frac{10}{35}$ **22** $\frac{24}{28}$ **23** $\frac{15}{24}$ **24** $\frac{12}{40}$

MIXED APPLICATIONS
Problem Solving

Use the game board for problems 25–28.

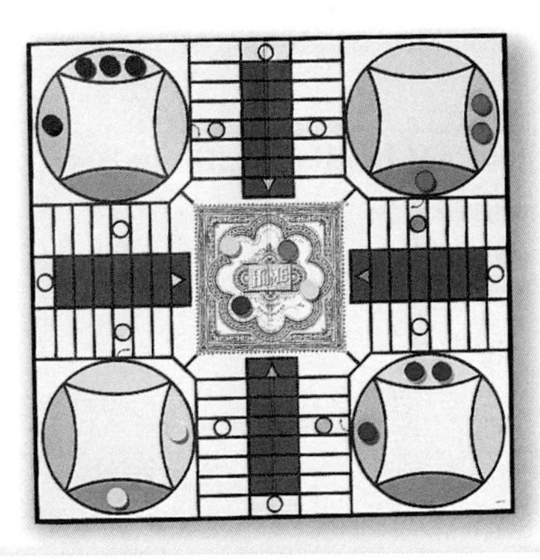

25 Sari was playing pachisi. In one toss, she scored 12 points. If the highest possible score on each toss is 25, how many more points could she have scored?

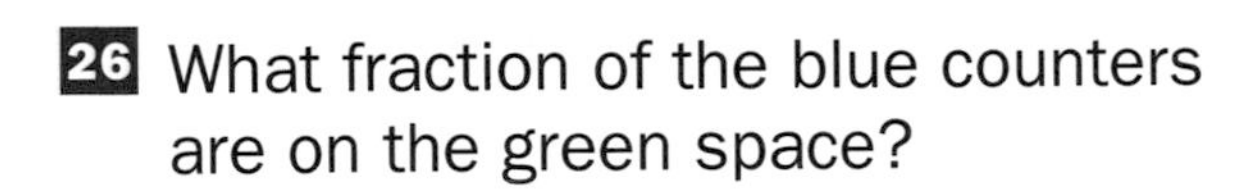

26 What fraction of the blue counters are on the green space?

27 What fraction of the red counters are on the yellow space?

28 What fraction of all the counters are in the center square?

Cultural Note

Pachisi (puh-CHEE-zee), which is considered the national game of India, has been played there for over 1,000 years. The game was played originally with cowrie shells.

mixed review • test preparation

1 258 + 625 **2** 847 − 290 **3** 4,582 × 3 **4** 8)804 **5** 10)775

Extra Practice, page 521

Compare Fractions

LEARN

You are trying to reach the eighth level of a computer game. On your first try, you finish $\frac{3}{4}$ of the game. On your next try, you finish $\frac{5}{8}$ of the game. Which try was better?

You can use a number line or fraction strips to compare and order fractions.

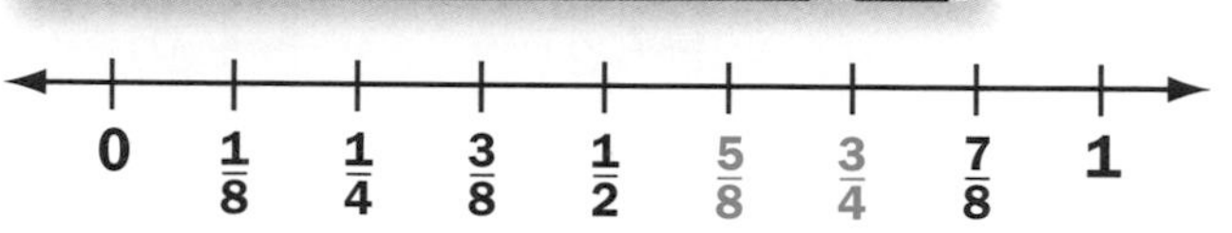

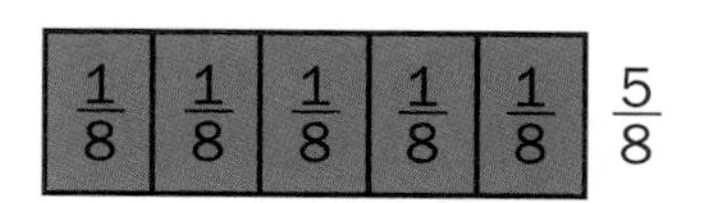

Since $\frac{3}{4} > \frac{5}{8}$, you did better on your first try.

You can also use equivalent fractions to order.

Order $\frac{1}{2}$, $\frac{1}{4}$, and $\frac{3}{8}$ from least to greatest.

Step 1	Step 2
Write equivalent fractions with the same denominator.	**Compare the numerators.**
$\frac{1}{2} = \frac{1 \times 4}{2 \times 4} = \frac{4}{8}$	$\frac{4}{8}$ **Think:** 2 is the least. 4 is the greatest.
$\frac{1}{4} = \frac{1 \times 2}{4 \times 2} = \frac{2}{8}$	$\frac{2}{8}$
$\frac{3}{8}$	$\frac{3}{8}$ From least to greatest: $\frac{1}{4}, \frac{3}{8}, \frac{1}{2}$

CHECK

Check for Understanding

Write the fractions in order from least to greatest.

1 $\frac{5}{12}, \frac{7}{12}, \frac{1}{12}$ **2** $\frac{2}{5}, \frac{4}{5}, \frac{1}{5}$ **3** $\frac{1}{2}, \frac{3}{4}, \frac{3}{8}$ **4** $\frac{1}{10}, \frac{1}{8}, \frac{1}{6}, \frac{1}{4}$

Critical Thinking: Compare

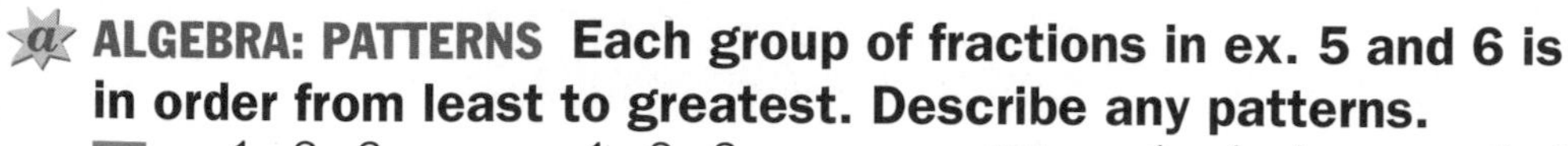

a **ALGEBRA: PATTERNS Each group of fractions in ex. 5 and 6 is in order from least to greatest. Describe any patterns.**

5 **a.** $\frac{1}{6}, \frac{2}{6}, \frac{3}{6}$ **b.** $\frac{1}{7}, \frac{2}{7}, \frac{3}{7}$ **6** **a.** $\frac{1}{10}, \frac{1}{8}, \frac{1}{6}$ **b.** $\frac{2}{9}, \frac{2}{5}, \frac{2}{4}$

Practice

Write $>$, $<$, or $=$. Use mental math when you can.

1 $\frac{1}{4}$ ● $\frac{3}{4}$ **2** $\frac{7}{8}$ ● $\frac{3}{4}$ **3** $\frac{2}{3}$ ● $\frac{5}{6}$ **4** $\frac{1}{2}$ ● $\frac{3}{8}$ **5** $\frac{2}{5}$ ● $\frac{2}{3}$

Write in order from least to greatest.

6 $\frac{3}{8}, \frac{7}{8}, \frac{1}{8}$ **7** $\frac{4}{5}, \frac{1}{5}, \frac{3}{5}$ **8** $\frac{1}{4}, \frac{1}{8}, \frac{1}{2}$ **9** $\frac{1}{16}, \frac{1}{8}, \frac{1}{4}$

10 $\frac{5}{8}, \frac{3}{8}, \frac{3}{4}$ **11** $\frac{1}{3}, \frac{5}{6}, \frac{1}{6}$ **12** $\frac{4}{5}, \frac{3}{10}, \frac{1}{2}$ **13** $\frac{1}{3}, \frac{5}{7}, \frac{4}{21}$

Write in order from greatest to least.

14 $\frac{4}{7}, \frac{2}{7}, \frac{6}{7}$ **15** $\frac{1}{3}, \frac{1}{2}, \frac{1}{6}$ **16** $\frac{3}{4}, \frac{1}{3}, \frac{5}{6}$ **17** $\frac{3}{8}, \frac{1}{2}, \frac{1}{4}$

18 $\frac{5}{7}, \frac{3}{7}, \frac{12}{21}$ **19** $\frac{1}{10}, \frac{2}{5}, \frac{1}{2}$ **20** $\frac{7}{8}, \frac{13}{16}, \frac{1}{4}$ **21** $\frac{1}{9}, \frac{5}{18}, \frac{11}{36}$

MIXED APPLICATIONS

Problem Solving

22 Three sisters each want to save \$30 to buy board games. Jen has $\frac{1}{5}$ of the amount, Jackie has $\frac{3}{10}$, and Jill has $\frac{4}{10}$. Who has the most? the least?

23 **Data Point** Use the Databank on page 541. Make a graph to show what fourth-grade students like to do. Write two statements that describe the data in your graph.

more to explore

Circle Graph

A circle graph was used to show the data from a survey.

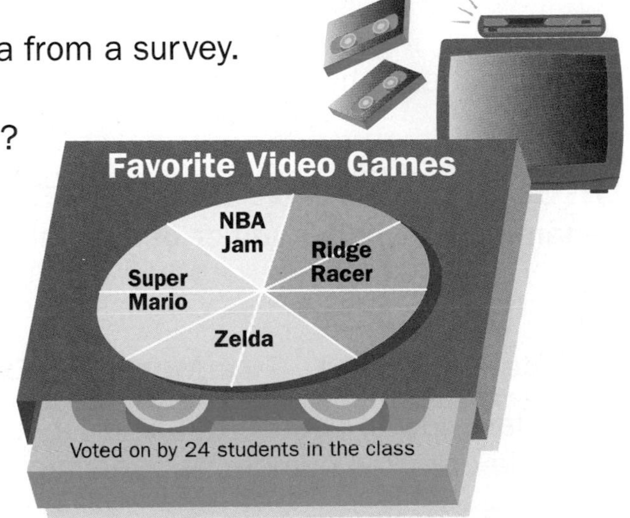

1 Did more students choose Super Mario or NBA Jam? How do you know?

2 Which video is the favorite of about half the class? How do you know?

3 The fraction of the graph showing Zelda is $\frac{2}{8}$. How many students chose Zelda? Explain.

Mixed Numbers

LEARN

Dish It Up is a game where you fill in guest checks with 4 pictures to earn tips. The diagram below represents 3 guest checks. Describe how many guest checks have been filled.

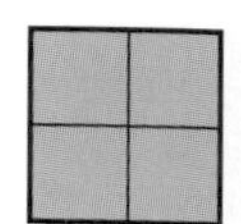
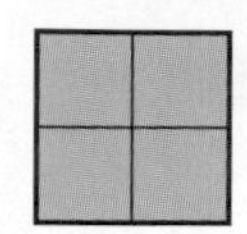
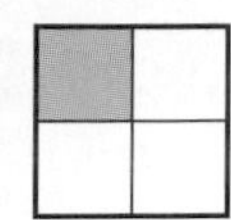

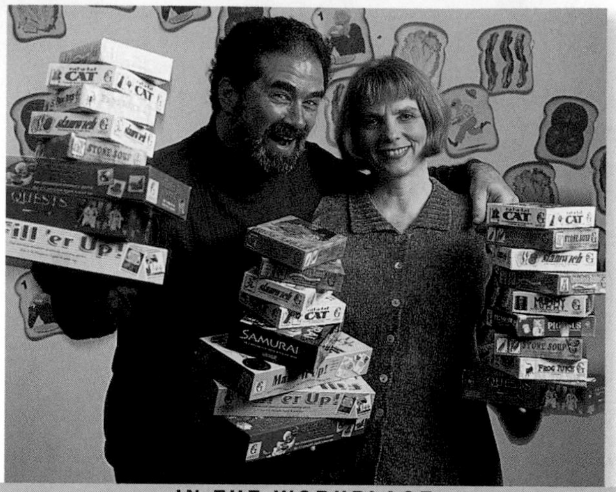

IN THE WORKPLACE

Monty and Anne Stambler are the game inventors who created Dish It Up.

You can use an improper fraction to tell how many guest checks have been filled. An **improper fraction** has a numerator that is equal to or greater than the denominator.

Think: Each check has 4 sections.

$\frac{9}{4}$ guest checks have been filled.

You can rename an improper fraction as a whole number or a **mixed number,** which is a whole number with a fraction.

Check Out the Glossary
improper fraction
mixed number
See page 544.

Step 1

Divide the numerator by the denominator.

$$\begin{array}{r} 2 \text{ R1} \\ 4\overline{)9} \\ -8 \\ \hline 1 \end{array}$$

Step 2

Write the quotient as the whole number part. Write the remainder over the divisor as the fraction part.

$2\frac{1}{4}$ **Think:** $\frac{1}{4}$ is in simplest form.

$2\frac{1}{4}$ guest checks have been filled.

CHECK

Check for Understanding

Rename as a whole number or as a mixed number in simplest form.

1 $\frac{8}{8}$ **2** $\frac{12}{6}$ **3** $\frac{15}{2}$ **4** $\frac{11}{4}$ **5** $\frac{13}{6}$ **6** $\frac{11}{3}$

Critical Thinking: Summarize **Explain your reasoning.**

7 Explain how to change a fraction that is greater than 1 to a whole or mixed number. Give examples.

Practice

Describe the colored or filled part as a fraction and as a whole or a mixed number.

1 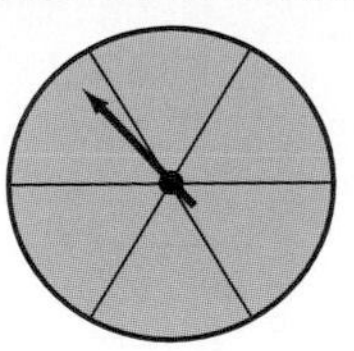 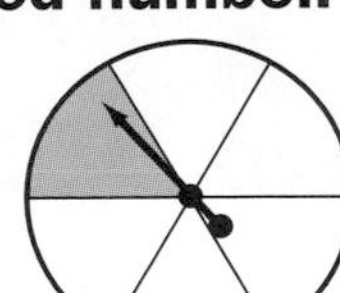

2 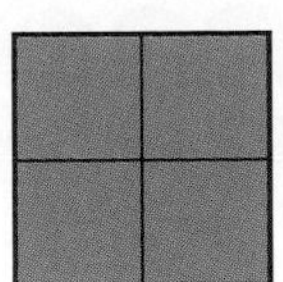 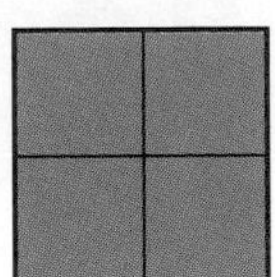

3 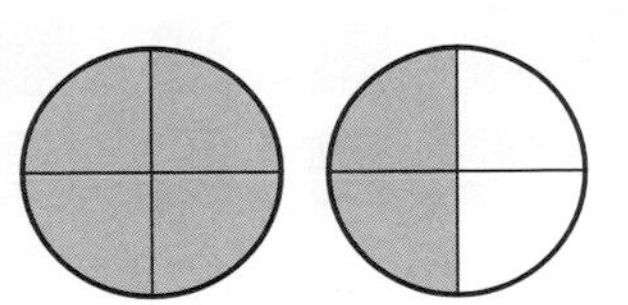

Rename as a whole number or as a mixed number in simplest form.

4 $\frac{13}{4}$ **5** $\frac{12}{12}$ **6** $\frac{15}{5}$ **7** $\frac{16}{6}$ **8** $\frac{14}{10}$ **9** $\frac{68}{8}$

10 $\frac{11}{7}$ **11** $\frac{29}{3}$ **12** $\frac{48}{9}$ **13** $\frac{21}{14}$ **14** $\frac{45}{11}$ **15** $\frac{85}{2}$

MIXED APPLICATIONS

Problem Solving

16 You watched $1\frac{1}{2}$ hours of TV. If all were $\frac{1}{2}$-hour shows, how many shows did you watch?

17 **Spatial reasoning** Suppose you have $2\frac{1}{2}$ circles. What do you need to make 3 whole circles?

more to explore

Rounding Mixed Numbers

You can use a number line to help you round a mixed number to the nearest whole number.

- If the fraction part is less than $\frac{1}{2}$, round down.
- If the fraction is $\frac{1}{2}$ or greater, round up.

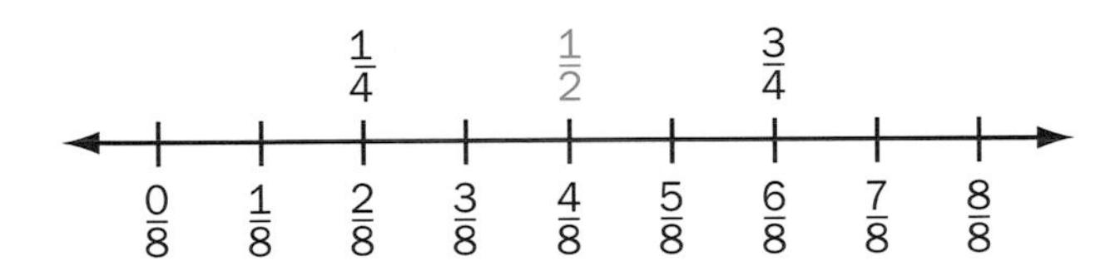

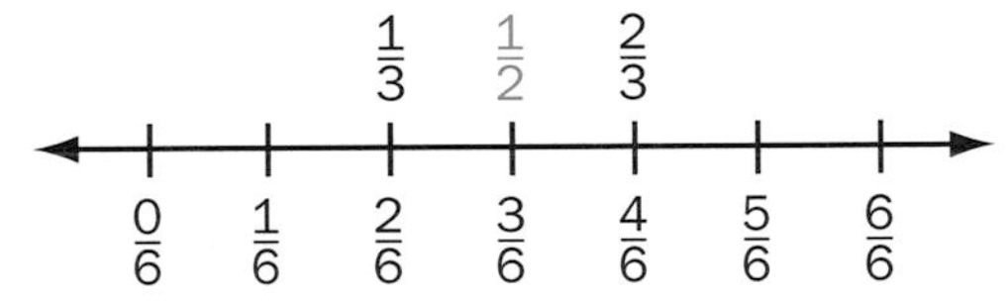

Round $2\frac{1}{4}$.

Think: $\frac{1}{4}$ is less than $\frac{1}{2}$.
$2\frac{1}{4}$ rounds down to 2.

Round $4\frac{5}{6}$.

Think: $\frac{5}{6}$ is more than $\frac{1}{2}$.
$4\frac{5}{6}$ rounds up to 5.

Estimate by rounding to the nearest whole number.

1 $1\frac{1}{8}$ **2** $3\frac{3}{4}$ **3** $1\frac{7}{8}$ **4** $1\frac{1}{6}$ **5** $4\frac{2}{3}$ **6** $5\frac{1}{2}$

Extra Practice, page 522

Write the fraction for the part that is green.

1
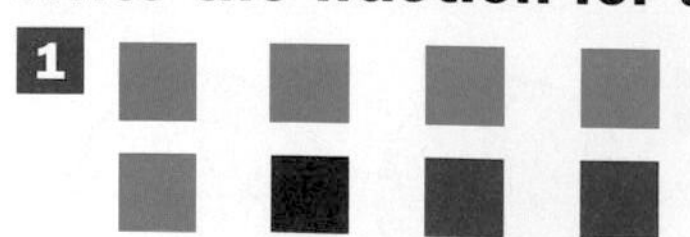

2

Complete.

3 $\frac{1}{4}$ of 12 = ■

4 $\frac{1}{2}$ of 10 = ■

Find the answer.

5 $\frac{1}{3}$ of 9

6 $\frac{2}{5}$ of 15

7 $\frac{1}{6}$ of 18

Write in simplest form.

8 $\frac{10}{16}$

9 $\frac{7}{14}$

10 $\frac{12}{19}$

11 $\frac{10}{18}$

12 $\frac{2}{14}$

Write the fractions in order from least to greatest.

13 $\frac{7}{9}, \frac{2}{9}, \frac{5}{9}$

14 $\frac{5}{6}, \frac{2}{3}, \frac{1}{3}$

15 $\frac{3}{4}, \frac{11}{12}, \frac{2}{3}$

Complete to name the equivalent fraction.

16 $\frac{1}{2} = \frac{9}{■}$

17 $\frac{4}{5} = \frac{■}{25}$

18 $\frac{3}{■} = \frac{1}{8}$

Rewrite as a whole number or as a mixed number in simplest form.

19 $\frac{4}{3}$

20 $\frac{15}{5}$

21 $\frac{12}{8}$

22 $\frac{26}{7}$

Solve. Use any method.

23 Jan, Hank, and Phil were in a long-distance race. Jan finished in $\frac{3}{4}$ of an hour. Hank took $\frac{1}{2}$ an hour to finish, and Phil took $\frac{7}{8}$ of an hour. Who finished first? last?

24 Of the 24 students in Class 4A, $\frac{3}{4}$ are 10 years old. Of the 24 students in Class 4B, $\frac{6}{8}$ are 10 years old. Do both classes have the same number of 10-year-olds? Explain.

25 

Order the following fractions from least to greatest. Explain your method.

$\frac{1}{4}, \frac{5}{6}, \frac{2}{3}, \frac{5}{12}$

developing technology sense

MATH CONNECTION

Use Fraction Tools

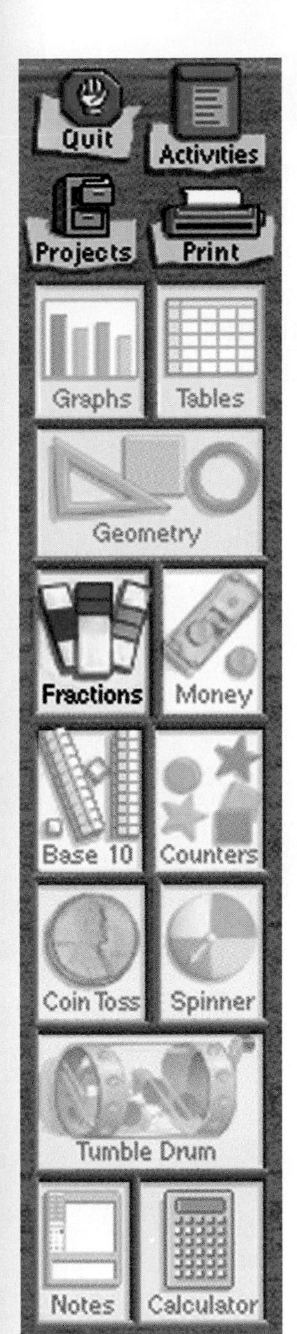

Computer-based fraction strips let you model problems the same way you do with the actual fraction strips.

- You can use the stamp tool to stamp out the fraction strips that you want.
- You can compare fraction strips and find equivalent fractions.
- You can then save your work and print it out when you want.

Work with a partner. Use actual fraction strips if you can. Then use the computer fraction strips to solve.

1 Which fraction is greater, $\frac{5}{6}$ or $\frac{3}{8}$?

2 Order the fractions from greatest to least: $\frac{3}{4}$, $\frac{9}{16}$, $\frac{7}{8}$.

3 Are $\frac{8}{12}$ and $\frac{6}{9}$ equivalent fractions?

4 What is $\frac{12}{8}$ as a mixed number in simplest form?

5 Choose three fractions. Use the computer to help you find equivalent fractions.

Critical Thinking: Generalize

6 What are the advantages of using computer fraction strips rather than actual strips?

Fraction Crossword Puzzle

Solving word puzzles is a wonderful way to spend a rainy day. In fact, crossword puzzles are so much fun, you may want to do them when it is sunny outside!

In this investigation, you and your group will write the clues for your own fraction crossword puzzle.

Here is an example to help you when you create your own puzzle.

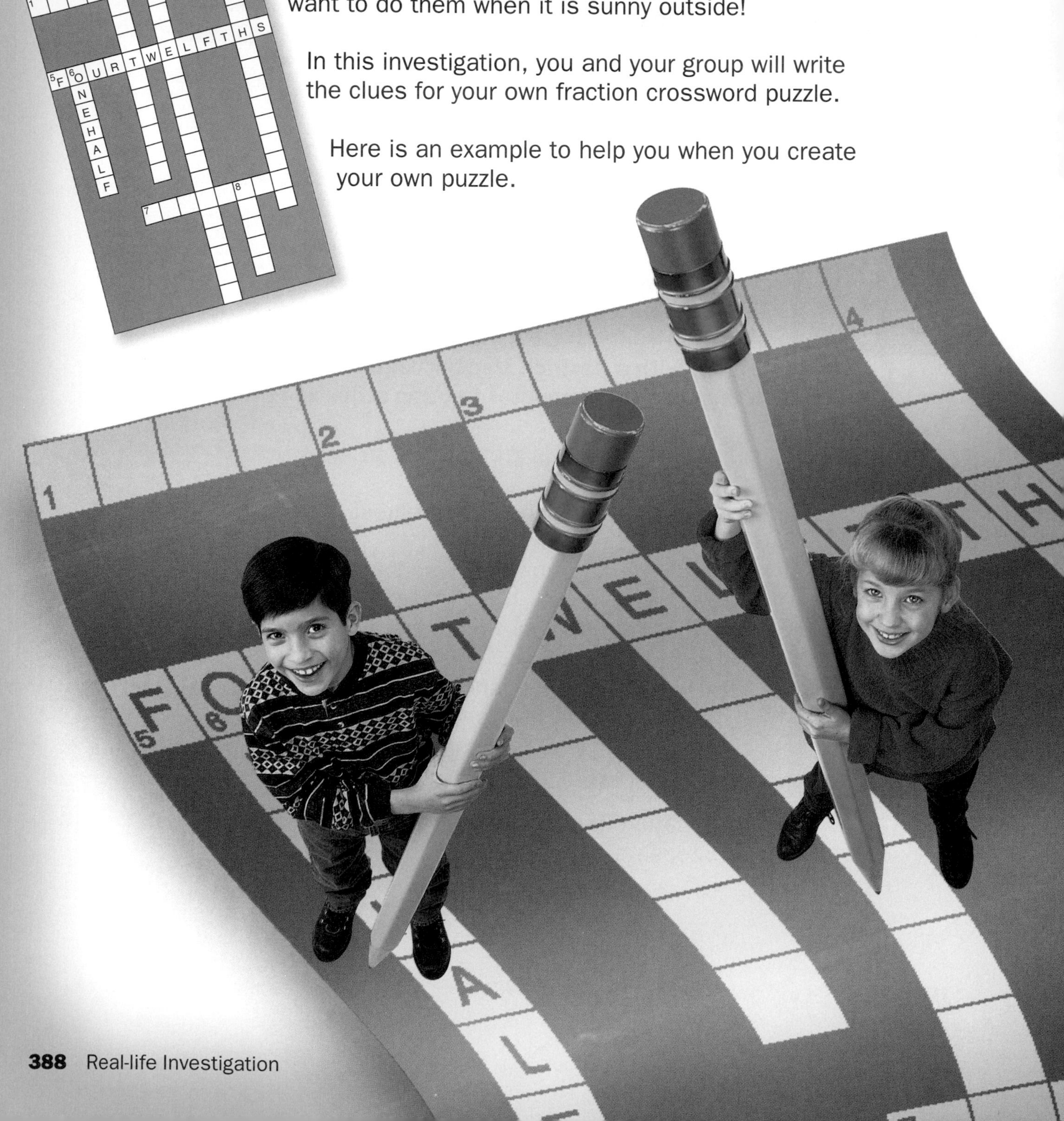

DECISION MAKING

Making Fraction Puzzles

1 Work in a group. Create the puzzle solution that contains all the fractions you will use.

2 Decide what hints you will use for each clue and the types of drawings or descriptions that will describe the fractions.

3 Include clues that use parts of a whole, parts of a group, and equivalent fractions.

4 Create a copy of the puzzle that other students can complete. Exchange your puzzles with other groups.

Reporting Your Findings

5 Portfolio Prepare a report on what you learned. Include the following:

- a copy of your puzzle and its solution
- a list of your clues and their answers
- a description of how you decided on the clues to use and what kinds of clues are the hardest to solve

6 Compare your report with the reports of other groups.

Revise your work.

- Are the puzzle, clues, and diagrams clear?
- Did you check your solutions to make sure that they are accurate?
- Check to see if your work is neat.

MORE TO INVESTIGATE

PREDICT which clues are the most difficult, then survey your classmates to check your prediction. Explain how to improve the clues so that they are easily understood.

EXPLORE other types of crossword puzzles or number puzzles in newspapers and magazines.

FIND when and where the earliest crossword puzzle was printed.

Probability

What if you and three friends are playing a game. Each time you spin the spinner and your color appears, you score a point. Each time your color does not appear, another player takes a turn. What type of spinner would you want?

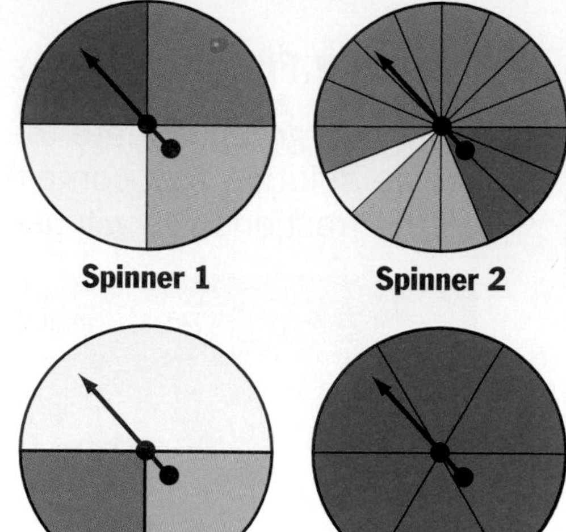

Work Together

Work with a group to play the game using different spinners.

Each member of the group chooses a different color: red, blue, green, or yellow.

Make Predictions Start with Spinner 1. Predict which player will score the most points. Then take turns spinning until a player gets 5 points.

You will need
- *spinners*

Play other rounds using a different spinner each time. Record the results in a table.

	Red	Blue	Green	Yellow
Spinner 1				
Spinner 2				

- ▶ Explain the results of each round. How did your prediction compare to the actual results?
- ▶ Did each player have the same chance of spinning his or her color in each round? Explain.

Check Out the Glossary
possible outcomes
probability
See page 544.

Make Connections

Raman's group used the **possible outcomes** to help them predict the chance of spinning each color.

Spinner 1 is divided into 4 equal parts with all four colors shown. The possible outcomes are red, blue, green, or yellow and the chance of spinning each color is the same.

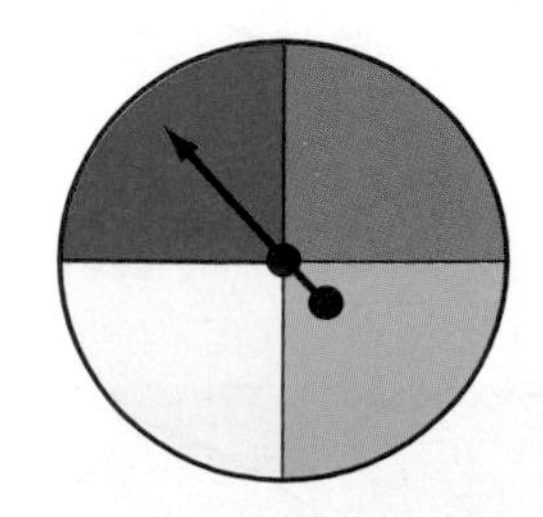

The chance, or likeliness, that something happens is called its **probability.** You can describe the probability using special words.

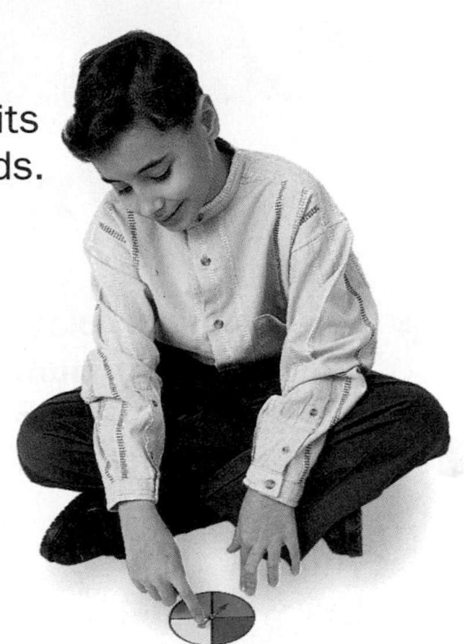

- In Spinner 1, there is an *equally likely* probability of spinning each color.
- In Spinner 2, it is *more likely* to spin blue than any other color. It is *less likely* to spin red or green. It is *unlikely* that yellow will appear.
- In Spinner 3, it is *impossible* to spin red.
- In Spinner 4, it is *certain* that red will appear.

How does knowing the probability of spinning each color help you to choose the spinner you would like to play with?

Check for Understanding

CHECK

Use the words *likely, not likely, impossible,* or *certain* to describe the probability of spinning:

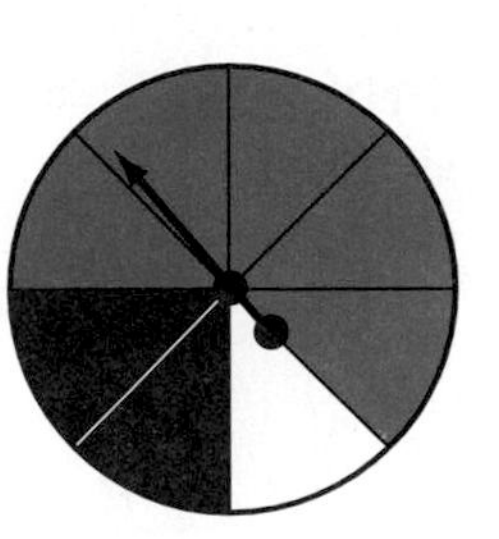

1 red.

2 white.

3 pink.

4 blue.

5 black.

6 purple.

Critical Thinking: Generalize **Give examples of how you would use the statement to help you make a decision.**

7 It is very likely that it will rain today.

8 It is impossible to get to the museum in less than an hour.

9 It is certain that the hurricane will reach this town by tomorrow.

Practice

PRACTICE

Use the words *more likely*, *less likely*, *equally likely*, *certain*, or *impossible* to describe the probability.

1 picking a red ball

2 picking a yellow ball

3 picking a striped ball

4 spinning white

5 compare spinning black to *not* spinning black

Extra Practice, page 522

Fractions and Probability

LEARN

Have you ever played a game where you need a certain number to win? What if you need to roll a 5 or 6 to win. What is the probability that you will win on your next roll?

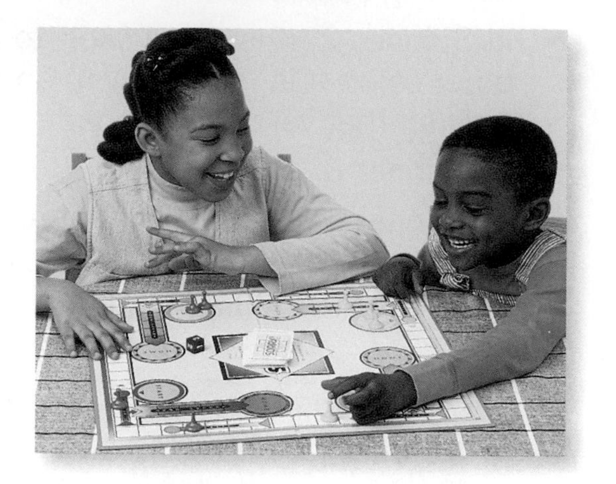

Use a fraction to show the probability.

Think: Favorable outcomes: 5, 6
Possible outcomes: 1, 2, 3, 4, 5, 6

$$\textbf{Probability} = \frac{\text{number of favorable outcomes}}{\text{number of possible outcomes}} = \frac{2}{6}$$

The probability of winning is $\frac{2}{6}$.

Check Out the Glossary
favorable outcomes
See page 544.

CHECK

Check for Understanding

What is the probability of:

1 tossing a head?

2 picking a white cube? a black cube?

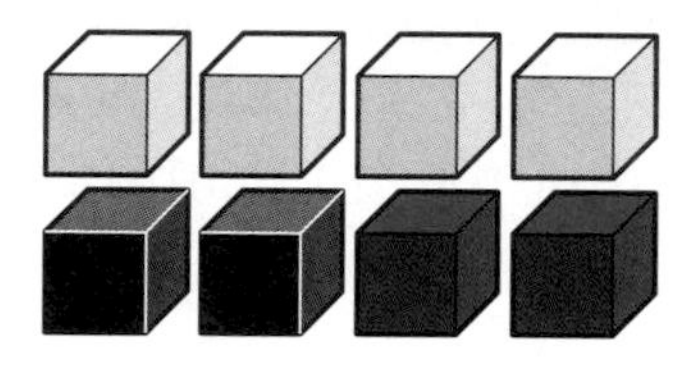

3 picking *W*? *E*?

E	S	E	E	P	W	K	S	E

4 picking a number other than 11?

0	1	5	9	11	16	22	9

Critical Thinking: Synthesize **Explain your reasoning.**

5 Journal Tell if you agree or disagree with the following statements:

a. The probability of flipping a coin and getting a head or a tail is $\frac{2}{2}$, or 1.

b. The probability of tossing a 7 on a 1–6 number cube is $\frac{0}{6}$, or 0.

Practice

Find the probability of picking the color from the bag.

1 black marker

2 white marker

3 blue marker

4 *not* a green marker

5 purple marker

6 *not* a white marker

Find the probability of picking the color from the drawer.

7 one blue sock

8 one green sock

9 *not* picking a red sock

10 *not* picking a yellow sock

11 one white sock

12 one purple sock

Find the probability of picking the color or shape.

13 small shape

14 yellow shape

15 large blue triangle

16 small yellow circle

17 *not* picking a small shape

18 *not* picking a yellow shape

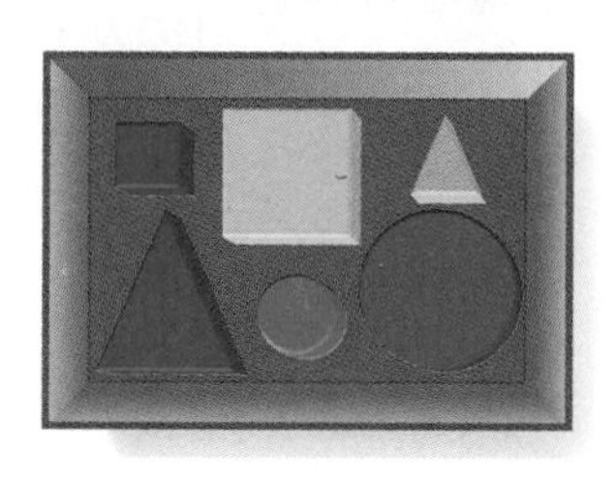

MIXED APPLICATIONS

Problem Solving

19 Earth is about $\frac{3}{10}$ land and $\frac{7}{10}$ water. If a globe ball is thrown to you, what is the probability that your right thumb will land on water?

20 **Make a decision** You need a number greater than 5 to win a game. You can spin either a 1–6 spinner or a 1–10 spinner. Which would you choose? Why?

mixed review • test preparation

Find the average.

1 Number of hot dogs sold in a game: 125, 176, 146, 163, 135

2 Number of students in a fourth grade class: 29, 30, 28, 31, 32

Tell how many vertices, faces, and edges the figure has.

3

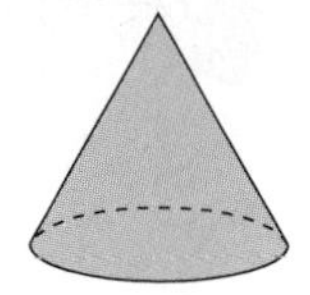

4

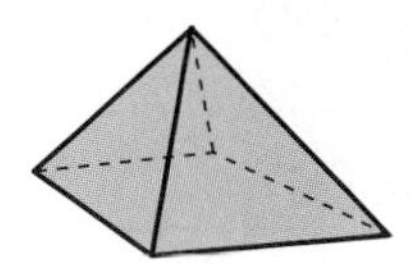

5

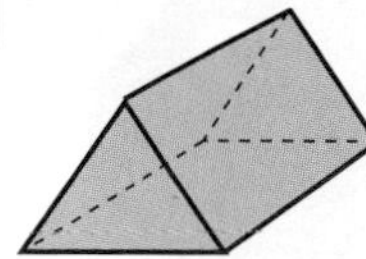

6

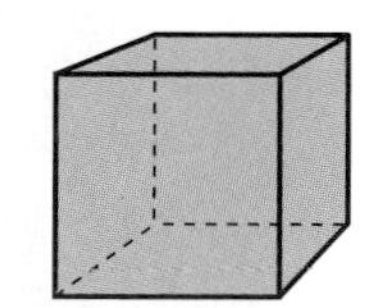

Extra Practice, page 522

Conduct an Experiment

LEARN

Read **What if you are playing a game where you choose a number. Then you toss two number cubes 60 times. On each toss, you add the numbers that appear. If the sum is the same as the number you chose, you score a point. What number should you choose?**

Sum	Tally	Times Sum Appears
2		0
3	||||	4
4	||||	4
5	~~||||~~	5
6	~~||||~~ ||	7
7	~~||||~~ ~~||||~~	10
8	~~||||~~ ~~||||~~	10
9	~~||||~~ |||	8
10	~~||||~~ |||	8
11	|||	3
12	|	1

Plan To solve the problem, you can conduct an experiment.

Solve Act it out. Toss the cubes and record how often the sums appear. A table with the results of one experiment is shown.

In this experiment, you should choose either 7 or 8.

Look Back How can you use probability to help you decide which number to choose?

CHECK

Check for Understanding

1 **What if** you score a point each time your number does not appear. What number would you choose? Why?

Critical Thinking: Synthesize **Explain your reasoning.**

2 Kaitlin and Adam are playing a board game where they toss a 1–6 number cube. To get home first, Kaitlin needs a number greater than 2 to appear six times. Adam needs a number less than 5 to appear six times. What experiment could you do to predict who will win?

PRACTICE

MIXED APPLICATIONS

Problem Solving

Tell whether or not you need to do an experiment. Then solve.

1. **ALGEBRA: PATTERNS** Kaya's quilt has a star in the first row, 3 stars in the second row, 4 stars in the third, 7 stars in the fourth, and 11 stars in the fifth. How many stars will she have in the eighth row? Explain.

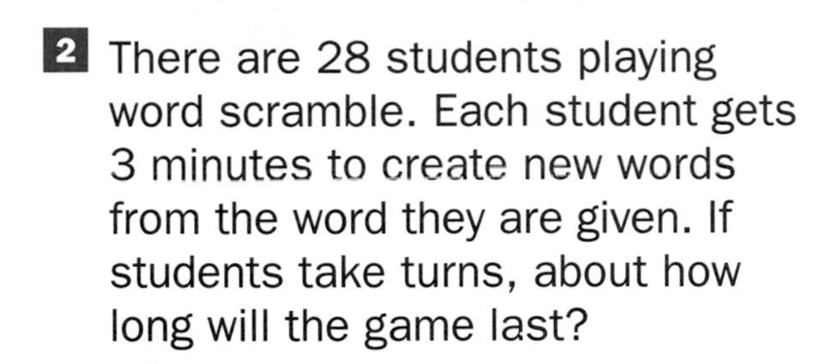

2. There are 28 students playing word scramble. Each student gets 3 minutes to create new words from the word they are given. If students take turns, about how long will the game last?

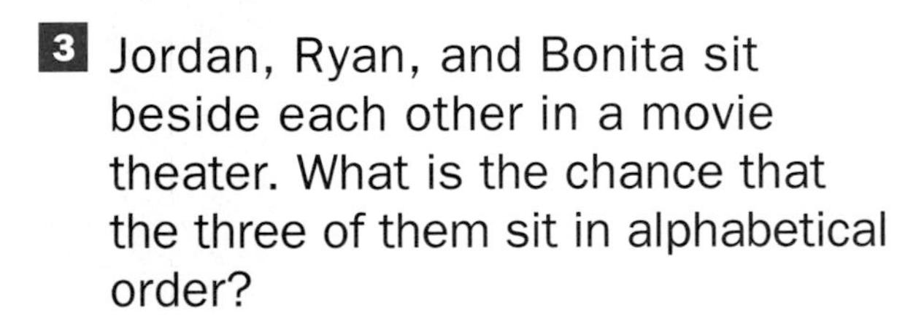

3. Jordan, Ryan, and Bonita sit beside each other in a movie theater. What is the chance that the three of them sit in alphabetical order?

4. **Make a decision** Mr. Tallchief must send his best math student to the state math contest. If you were Mr. Tallchief, how would you determine who to send?

5. Randy puts 162 basketball trading cards into plastic holders. Some holders hold 10 cards and some hold 9 cards. If all the holders are filled, how many of each size holder does he have?

6. Each of two spinners is divided into six equal parts with 3 on each part. If you spin both spinners and add the results 20 times, what sum will appear most often?

7. **What if** you flip two coins 10 times. How often will you get two heads?

8. Each number from 1 to 25 is written on index cards and mixed up. Which is most likely—picking a 1-digit number, a card with the digit 1 on it, or a card with the digit 2 on it?

9. There are 17 red cubes, 54 blue cubes, and 13 yellow cubes. Use mental math to tell how many cubes there are in all. Explain your thinking.

Extra Practice, page 523

Predict and Experiment

LEARN

How many times will you get a head if you flip a coin 30 times?

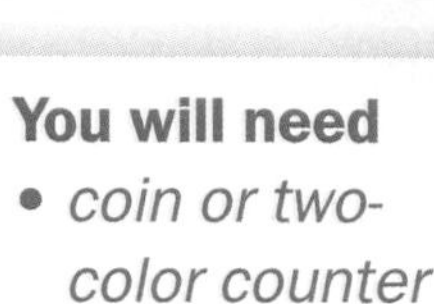

You can solve this problem in two ways: by finding the probability of getting a head or by conducting an experiment.

You will need
- *coin or two-color counter*

Work Together

Work with a partner.

First, predict the number of heads in 30 flips.

Next, experiment by flipping a coin 30 times. Record your results in a table (use *H* for heads and *T* for tails). Compare your prediction with your results.

Continue until you have flipped the coin 60 times. Compare your prediction with the new results.

► How many times do you predict you will get heads after 80 flips? Why?

Make Connections

You can use probability to predict the outcome.

Think: Favorable outcome: H
Possible outcomes: H, T

$$\textbf{Probability} = \frac{\textbf{number of favorable outcomes}}{\textbf{number of possible outcomes}} = \frac{1}{2}$$

The probability of getting a head on each flip is $\frac{1}{2}$. Since the probability is $\frac{1}{2}$, predict that 1 out of every 2 flips will be a head.

If the coin is flipped 30 times, then you would predict that a head will appear 15 times.

► A coin is tossed 100 times. Is it reasonable to predict that heads will appear 89 times? Why or why not?

Check for Understanding

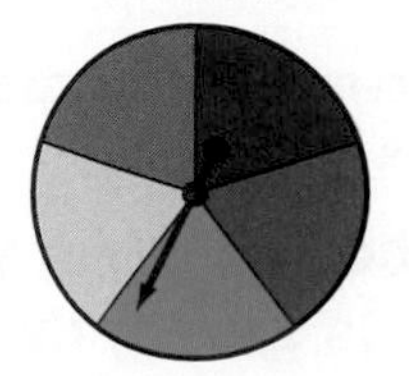

1 If you spin the spinner 50 times, is it reasonable to predict that you will spin red 10 times? Explain.

Critical Thinking: Generalize **Explain your reasoning.**

2 Suppose you have a bag with 7 green, 1 blue, and 2 yellow cubes. Each player chooses a different color and tries to pick that color cube from the bag to score a point. Each player has 10 tries. Is this game fair?

Practice

Use the spinner for problems 1–4. Write *true* or *false*. Explain your reasoning.

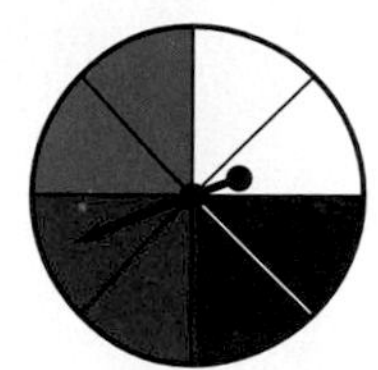

1 It is reasonable to predict that the spinner will land on red 6 out of 24 times.

2 The probability of *not* landing on white is $\frac{1}{4}$.

3 You can never expect the spinner to land on green.

4 You always expect the spinner to land on red, blue, black, or white.

5 A paper bag contains 25 cubes of one color, 10 cubes of another color, and 5 cubes of a third color. This chart shows the outcomes of 40 picks. Predict what is in the paper bag.

Red	𝍸 II
Blue	𝍸 𝍸 𝍸 𝍸 I
Green	𝍸 𝍸 II

6 Suppose you toss a quarter 50 times. Is it reasonable to predict that it will land on heads at least 25 times? Explain.

7 Suppose you roll a 1–6 number cube 30 times. Is it reasonable to predict that you will roll the number 2 at least 15 times? Explain.

8 Suppose you roll a 1–6 number cube 30 more times. Is it reasonable to predict that you will roll a number greater than 3 at least 15 times? Explain.

9 Pete has predicted that heads will appear 40 times if he flips a coin 80 times. How can he use this to predict the number of tails that will appear? Explain.

Read
Plan
Solve
Look Back

Part 1 Solve Multistep Problems

Rainy-day recess can be fun when you play games you make up yourself. In this game, there are 18 beads in a bag. Of the beads in the bag, $\frac{1}{2}$ are red, $\frac{1}{3}$ are blue, and $\frac{1}{6}$ are yellow. What is the probability of choosing a red bead in a draw?

Rules:
red = 1 point
blue = 2 points
yellow = 3 points

Pick 5 beads one at a time without looking. After each bead is chosen, return it to the bag. Find your total points. Take turns.

Work Together

1. What do you need to know to solve the problem?

2. How many beads are red? blue? yellow?

3. What is the probability of choosing a red bead in a draw?

4. What 5 beads do you have to pick to get the greatest possible score?

5. **What if** you earned 12 points on your turn. What 5 beads could you have chosen?

6. READING ARITHMETIC WRITING **Make Predictions** Predict what 5 beads you are most likely to pick. Explain your thinking.

Part 2 Write and Share Problems

Gary used the game on page 398 to write his own problem.

7 Describe the steps that you would use to solve Gary's problem.

8 Solve Gary's problem.

9 **Write a problem** about the game on page 398 that can be solved in more than one step.

10 Explain the steps you would use to solve your problem, then solve it.

11 Trade problems. Solve at least three problems written by your classmates.

12 What was the most interesting problem that you solved? Why?

CHECK

What color would you most likely not get? Why?

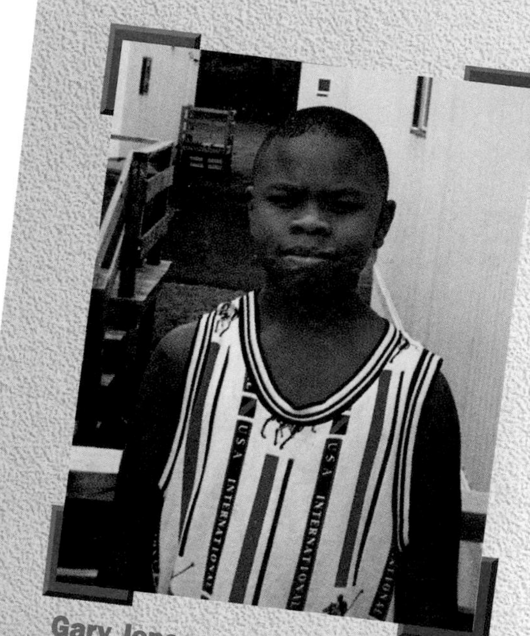

Gary Jones
Elephant's Fork Elementary School
Suffolk, VA

Turn the page for Practice Strategies.

PRACTICE

Menu

Choose five problems and solve them. Explain your methods.

1 Of the 20 students in Mr. Mario's class, $\frac{3}{10}$ got an A in math, $\frac{6}{10}$ got a B, and the rest got a C. How many students got a C in math?

2 Hector wrote his first and last name as many times as he could in 2 minutes. Predict how many times he wrote it.

Hector Gomez
Hector Gomez
Hector Gomez
Hector Gomez
Hector

3 The Gordon twins rode their bikes for 45 minutes. Then, they played Clue for $2\frac{1}{2}$ hours. They finished playing at 7:30 P.M. What time did they begin riding?

4 Paco works at the Fresh Fruit Packaging Company. He must put 146 oranges into crates that hold 15 oranges each. What is the least number of crates Paco will need to pack all of these oranges?

5 Mandy wants to call her cousin in London, England, at 7:00 P.M. London is 8 hours ahead. If the time on her watch is 8:30 A.M., how much longer should she wait before making the call?

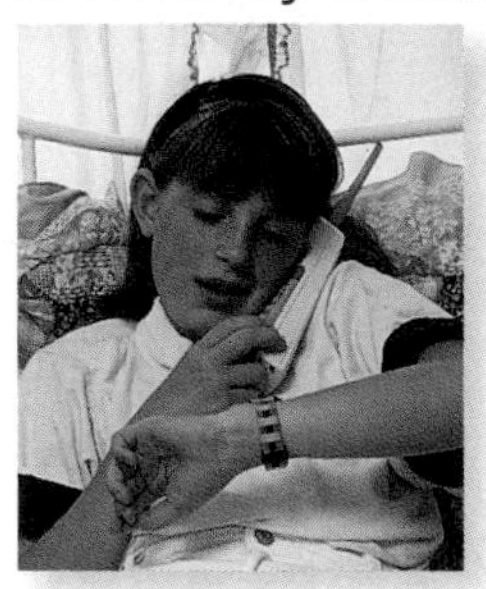

6 Taylor has $7.00. A hamburger costs $3.95, a milk shake costs $2.75, and an apple pie costs $2.50. Explain how she can use estimation to find out if she has enough money to buy all three.

7 **ALGEBRA: PATTERNS** Gail's goal is to jog for 45 min each day. She jogs 10 min on the first day, 15 min on the second day, and 20 min on the third day. In how many more days will she reach her goal if she continues this pattern? How many days will it take in all?

8 **Logical reasoning** In a survey, 147 people saw Disney's *The Hunchback of Notre Dame* and 167 people saw *Harriet the Spy.* If 23 people saw both movies, how many people saw only *Harriet the Spy*? How many saw only *The Hunchback*?

Choose two problems and solve them. Explain your methods.

9 **Make a decision** You have $20 to buy food to make a lunch for four. Use the chart to make a shopping list for your lunch.

10 **Make a decision** You and three friends are planning a dinner. You will each contribute $7 to pay for dinner. Plan a dinner and estimate the cost. Do you have enough money?

Food	Price
Chicken	$1.29 each pound
Hamburgers	$2.39 each pound
Potatoes	$0.69 each pound
Apples	$1.29 each pound
Lettuce (head)	$1.39 each
Peas	$0.99 each 10-oz bag
Milk (1 qt)	$1.49
Frozen yogurt ($\frac{1}{2}$ gal)	$3.99

11 **a.** Make a table or chart showing all the possible products that you can get from multiplying the results of rolling two 1–6 number cubes.

b. Experiment to find the product of two cubes in each of 40 tosses. Record your results.

c. Use a graph to show the results.

d. Compare your graph to your table.

12 **At the Computer** You can use graphing and probability programs to conduct experiments that would otherwise take a very long time to do.

- Predict how many times you will get a head if you toss a coin 200 times.
- Run the program so that the computer does the experiment. How close were your predictions?
- Predict how many times you will get a head if you toss a coin 500 times. What probability did you use?
- Run the program again for 500 tosses. How close was your prediction this time?

Language and Mathematics

Complete the sentence. Use a word in the chart. (pages 368–397)

1 You can write $\frac{4}{6}$ in ■ as $\frac{2}{3}$.

2 $\frac{4}{8}$ and $\frac{3}{6}$ are ■ fractions.

3 5 is the ■ in the fraction $\frac{5}{10}$.

4 $3\frac{1}{2}$ is a ■.

Vocabulary
simplest form
numerator
equivalent
mixed number
denominator

Concepts and Skills

Name the part that is white. (page 368)

5 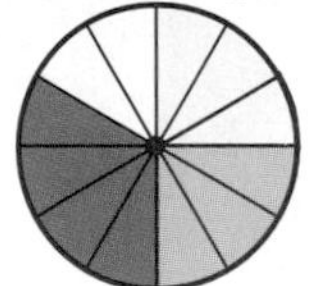

6  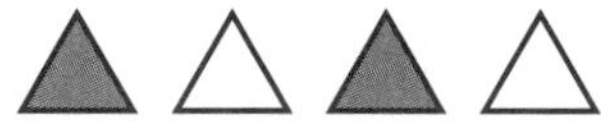

7 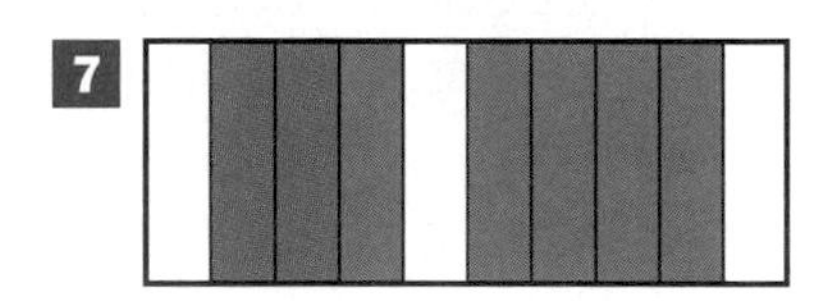

Write the letter of the equivalent fraction. (page 376)

8 $\frac{4}{5}$ **a.** $\frac{4}{10}$ **b.** $\frac{8}{10}$ **c.** $\frac{12}{16}$ **d.** $\frac{2}{10}$

9 $\frac{9}{12}$ **a.** $\frac{4}{8}$ **b.** $\frac{3}{9}$ **c.** $\frac{1}{4}$ **d.** $\frac{6}{8}$

Write in order from least to greatest. (page 382)

10 $\frac{1}{2}, \frac{1}{4}, \frac{1}{3}$ **11** $\frac{3}{8}, \frac{1}{8}, \frac{5}{8}$ **12** $\frac{3}{4}, \frac{1}{3}, \frac{5}{6}$ **13** $\frac{1}{2}, \frac{7}{8}, \frac{3}{4}$

14 $\frac{1}{12}, \frac{5}{6}, \frac{2}{3}$ **15** $\frac{3}{8}, \frac{1}{4}, \frac{3}{4}$ **16** $\frac{1}{2}, \frac{3}{10}, \frac{1}{5}$ **17** $\frac{1}{4}, \frac{2}{3}, \frac{11}{12}$

Write in order from greatest to least. (page 382)

18 $\frac{1}{3}, \frac{5}{8}, \frac{1}{2}$ **19** $\frac{3}{10}, \frac{1}{5}, \frac{3}{5}$ **20** $\frac{7}{8}, \frac{1}{3}, \frac{3}{4}$ **21** $\frac{1}{3}, \frac{3}{6}, \frac{2}{3}$

Write the fraction as a whole number or mixed number. (page 384)

22 $\frac{17}{2}$ **23** $\frac{15}{5}$ **24** $\frac{56}{8}$ **25** $\frac{26}{4}$ **26** $\frac{14}{3}$

27 $\frac{55}{6}$ **28** $\frac{39}{9}$ **29** $\frac{40}{7}$ **30** $\frac{13}{2}$ **31** $\frac{48}{5}$

32 If you flip a coin 50 times, how many times would you predict that you will get heads? (page 396)

33 If you roll a 1–6 number cube 30 times, how many times would you predict that you will roll a 3 or 4? (page 396)

34 What is the probability of landing on red? (page 392)

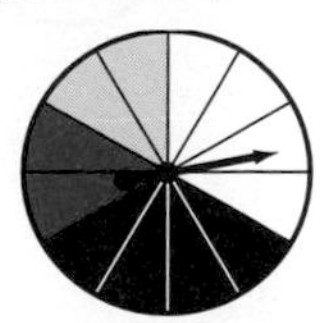

35 What is the probability of picking a white cube? (page 392)

Think critically. (page 374)

36 Analyze. Explain what mistake was made, and then correct it.

$\frac{2}{4}$ of 32 $32 \div 4 = 8$

$8 \div 2 = 4$

So $\frac{2}{4}$ of $32 = 4$.

MIXED APPLICATIONS
Problem Solving

Pencil & Paper | Calculator | Mental Math

(pages 394, 398)

37 Dave put together $\frac{1}{4}$ of his 16 model airplanes. Bryan put together $\frac{1}{3}$ of his 15 model airplanes. Who needs to put together more model airplanes? How many?

38 How long will it take you to write the name of your city and state 25 times? Make an estimate. Conduct an experiment to find the actual time.

39 Gus and Peter are almost finished playing Scrabble. There are only 8 tiles left to pick from. Gus needs the letter *E* to make a word. He sees that 10 *E*s have been used on the board. If there are 12 *E*s in all and Peter does not have any, what is the probability that Gus will pick an *E* next from the pile? Explain your reasoning.

40 Jesse visits people in a nursing home for 30 minutes each week. She spends $\frac{1}{3}$ of the time with her favorite resident, Mr. Washington. If she visits the nursing home 40 weeks each year, how many minutes does she spend with Mr. Washington each year?

chapter test

Name the fraction or mixed number for the part that is white.

1 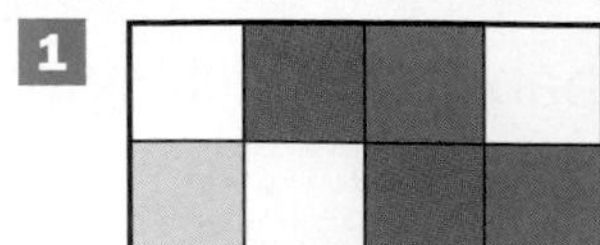

2 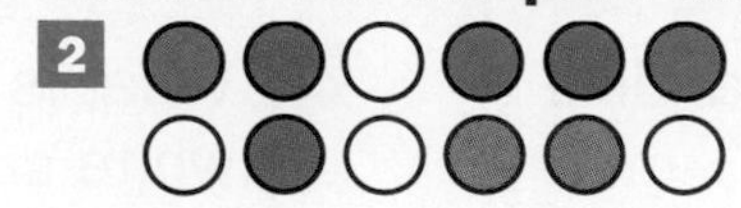

3 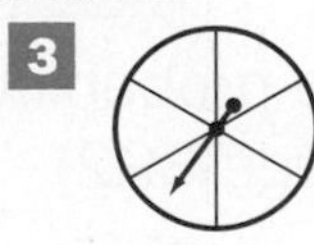  

Write an equivalent fraction.

4 $\frac{3}{8}$

5 $\frac{12}{18}$

6 $\frac{1}{3}$

Write in order from least to greatest.

7 $\frac{3}{7}, \frac{1}{7}, \frac{4}{7}$

8 $\frac{5}{6}, \frac{3}{4}, \frac{2}{3}$

9 $\frac{2}{3}, \frac{1}{4}, \frac{1}{8}$

Find the fraction of the number.

10 $\frac{1}{8}$ of 24

11 $\frac{3}{5}$ of 45

12 $\frac{2}{3}$ of 36

13 $\frac{3}{7}$ of 21

Write the fraction as a whole number or a mixed number.

14 $\frac{15}{7}$

15 $\frac{32}{4}$

16 $\frac{71}{8}$

17 $\frac{62}{9}$

18 $\frac{42}{6}$

Use the spinner for ex. 19–21.

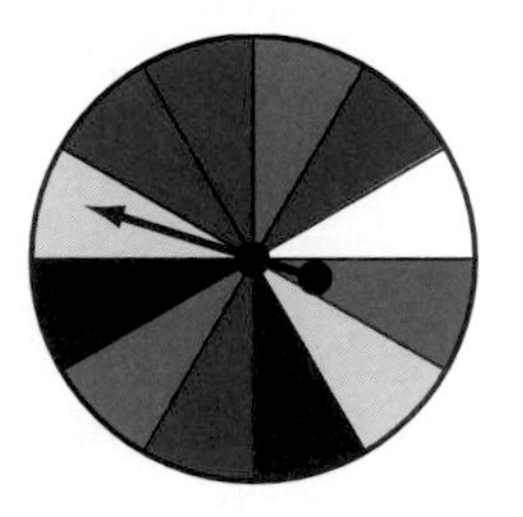

19 What is the probability of landing on blue?

20 Which color do you have the greatest probability of landing on? What is the probability?

21 What is the probability of landing on either black or white?

Solve. Conduct an experiment to solve if necessary.

22 Toyland sells 8 different brands of in-line skates. Last week, the store had 12 pairs of each brand. If the store sells 8 pairs, how many pairs will be left?

23 Of the 20 fourth graders, $\frac{1}{4}$ walk to school. Of the 24 third graders, $\frac{1}{6}$ walk to school. How many children walk to school from both grades combined?

24 If you close your eyes and point to a different word in a textbook 20 times, about how many times will you point to a word with the letter *e* in it?

25 Brad said the alphabet backwards. About how long did it take him?

What Did You Learn?

Tony used the counters to model that $\frac{4}{6}$ of the counters were red. Then, he drew this diagram to prove that the fraction $\frac{2}{3}$ is equivalent to $\frac{4}{6}$.

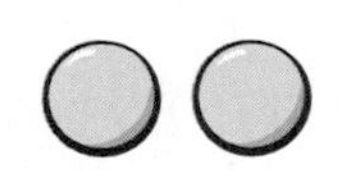

$\frac{4}{6} = \frac{2}{3}$

- Use counters to model two different fractions that are equivalent to $\frac{2}{3}$.
- Draw a diagram of your models to prove that they are equivalent to $\frac{2}{3}$. You may need to explain your diagrams.
- Prove that $\frac{1}{2} = \frac{4}{8}$ is true using a method other than a model or diagram.

A Good Answer

- names two different fractions that are equivalent to $\frac{2}{3}$
- shows a diagram with words or labels to prove that the fractions are equivalent
- uses a method other than a model or diagram to prove that $\frac{1}{2} = \frac{4}{8}$

You may want to place your work in your portfolio.

What Do You Think

1. Do you understand what a probability is and how to predict the probability of an event? If you are unsure, what do you do?
2. List all the ways you might use to compare or order fractions.
   - Use fraction models.
   - Use a grid.
   - Use a number line.
   - Other. Explain.

math science technology CONNECTION

That's the way the Ball Bounces

Referees drop a ball from overhead to check if the ball has the correct amount of air. The ball should bounce back somewhere between their waist and shoulder.

Try It Out

Work with a partner.

- Collect different basketballs to test. Make sure the amounts of air in the balls are different.
- Place a yardstick against the wall as shown.
- One partner holds the ball so that the bottom of the ball is at the same level as the top of the yardstick.
- The other partner watches how high the bottom of the ball reaches on the first bounce.
- Bounce each ball three times. Record your data in a table.

Do you think the height that the ball is dropped from also affects how it bounces? Check your answer by bouncing each ball from different heights. Record the data in a table.

What else may affect the way the ball bounces?

Cultural Note

One of the most intricately decorated balls is the *temari* (te-MAH-ree), from Japan. In the 1800s, they were used to play *mari* games. Today, they are used mainly for decoration.

Comparing Bounces

Use the data you collected to answer these questions.

1. What is the average height each ball bounced?

2. What can you say about the way the air in the ball affects the way it bounces?

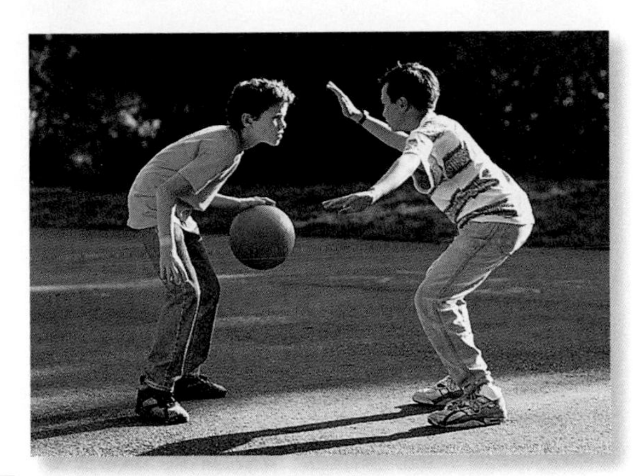

3. The balls were all dropped from 36 in. On average, what fraction of that height did each ball reach on the first bounce?

4. What if a ball that is dropped from 36 in. bounces back 12 in. Would you predict that the ball had too little air or too much air? Explain.

At the Computer

- Use graphing software to make a table and a bar graph of your data.
- Write a short report that explains your findings about how well different basketballs bounce.
- Combine all of the class's data together and present the project to one of the school's physical education teachers.

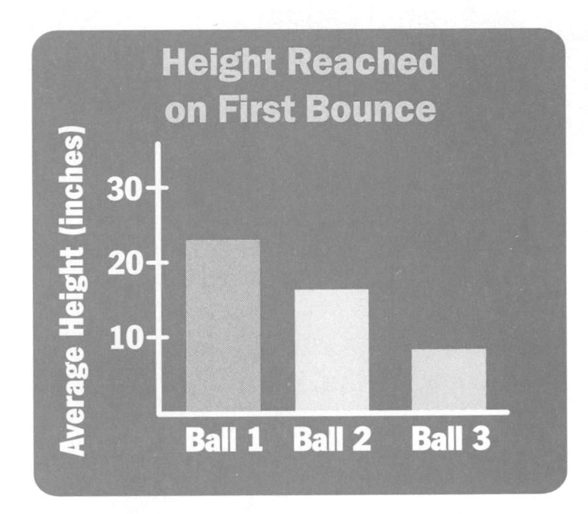

CHAPTER

11

USING FRACTIONS

THEME

Food for Thought

In this chapter, you will look at dishes and interesting recipes from various cultures. You will see how fractions are used to measure and compare foods that you might like to eat.

What Do You Know

Use the recipe for problems 1–3.

1 Draw rectangles to represent fractions. Show the amounts of sugar and salt.

2 A teaspoon is $\frac{1}{3}$ of a tablespoon. What fraction of a tablespoon is 2 teaspoons?

3  Double the recipe for the pie filling. Explain how you found the amounts for each ingredient. Use a drawing to help explain.

Write a Report

Choose a favorite vegetable and write a short report about it. Describe it, tell how it tastes, what size it is, and how much it weighs. Find out how many classmates also like this vegetable.

A report gives information about a subject. When you write a report, you usually do research first to learn more about the subject.

1 What fraction of students in your class like this vegetable?

2 What research did you do to answer question 1?

Vocabulary

fraction, p. 410
simplest form, p. 411
common denominator, p. 414
numerator, p. 414
mixed number, p. 434

Add Fractions

You can mix $\frac{1}{3}$ cup of pineapple juice with $\frac{1}{3}$ cup of orange juice to make pineapple-orange drink. How much drink will this make?

To solve this problem, you need to add **fractions.**

Work Together

Work with a partner to explore adding fractions.

Use fraction strips to model the problem. Find $\frac{1}{3} + \frac{1}{3}$. Write a number sentence to show your work.

Continue using fraction strips to add these fractions:

a. $\frac{1}{2} + \frac{1}{2}$ **b.** $\frac{1}{5} + \frac{3}{5}$ **c.** $\frac{5}{8} + \frac{1}{8}$

d. $\frac{3}{4} + \frac{1}{4}$ **e.** $\frac{3}{4} + \frac{3}{4}$ **f.** $\frac{3}{10} + \frac{7}{10}$

g. $\frac{5}{6} + \frac{5}{6}$ **h.** $\frac{7}{10} + \frac{9}{10}$ **i.** $\frac{7}{12} + \frac{11}{12}$

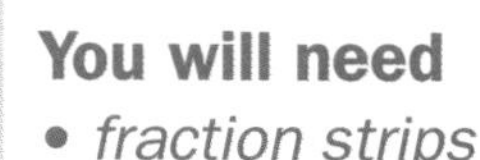

You will need
- *fraction strips*

Check Out the Glossary
fraction
simplest form
See page 544.

Talk It Over

- What method did you use to find the sums?
- How much drink will you make if you mix $\frac{1}{3}$ cup of pineapple juice with $\frac{1}{3}$ cup of orange juice?
- How would you show $\frac{10}{6}$ in another way?
- How would you show $\frac{10}{10}$ in another way?

Make Connections

Clive and Lisa added these fractions. They wrote number sentences to show what they did.

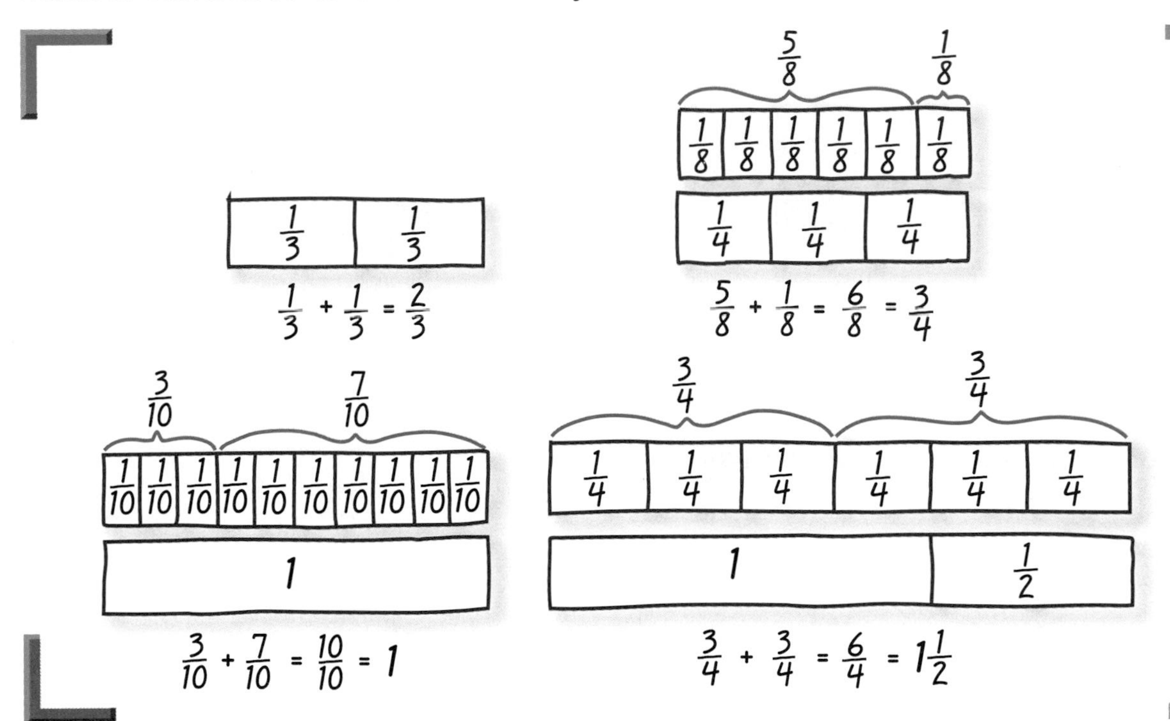

▶ Check if the sums in your number sentences are in **simplest form.** If not, write them in simplest form.

Check for Understanding

Use the models to complete the number sentence.

1

$\frac{1}{12}$	$\frac{1}{12}$	$\frac{1}{12}$	$\frac{1}{12}$	$\frac{1}{12}$	$\frac{1}{12}$	$\frac{1}{12}$	$\frac{1}{12}$	$\frac{1}{12}$	$\frac{1}{12}$

$\frac{1}{6}$	$\frac{1}{6}$	$\frac{1}{6}$	$\frac{1}{6}$	$\frac{1}{6}$

$\frac{7}{12} + \frac{3}{12} = \frac{\blacksquare}{12} = \frac{\blacksquare}{6}$

2

$\frac{1}{4}$	$\frac{1}{4}$	$\frac{1}{4}$	$\frac{1}{4}$	$\frac{1}{4}$

1	$\frac{1}{4}$

$\frac{3}{4} + \frac{2}{4} = \frac{\blacksquare}{4} = 1\frac{\blacksquare}{4}$

Add. Write the sum in simplest form.

3 $\frac{2}{3} + \frac{2}{3}$

4 $\frac{3}{5} + \frac{4}{5}$

5 $\frac{8}{10} + \frac{6}{10}$

6 $\frac{9}{12} + \frac{11}{12}$

7 $\frac{5}{8} + \frac{5}{8}$

Critical Thinking: Generalize

8 Do you always need to simplify the sum after you add fractions? Why or why not? Give an example.

CHECK

Turn the page for Practice.

PRACTICE

Practice

Use models to complete the numbers sentences.

1 $\frac{1}{5}$ $\frac{1}{5}$

$\frac{1}{5} + \frac{1}{5} = \frac{■}{5}$

2 $\frac{1}{6}$ $\frac{1}{6}$ $\frac{1}{6}$ $\frac{1}{6}$ $\frac{1}{6}$

$\frac{2}{6} + \frac{3}{6} = \frac{■}{6}$

3

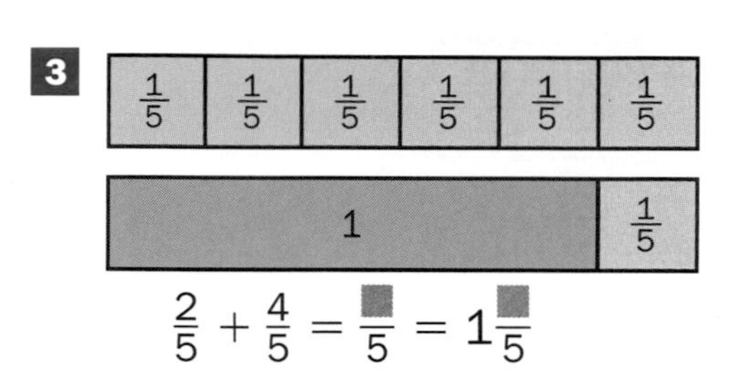

$\frac{2}{5} + \frac{4}{5} = \frac{■}{5} = 1\frac{■}{5}$

4

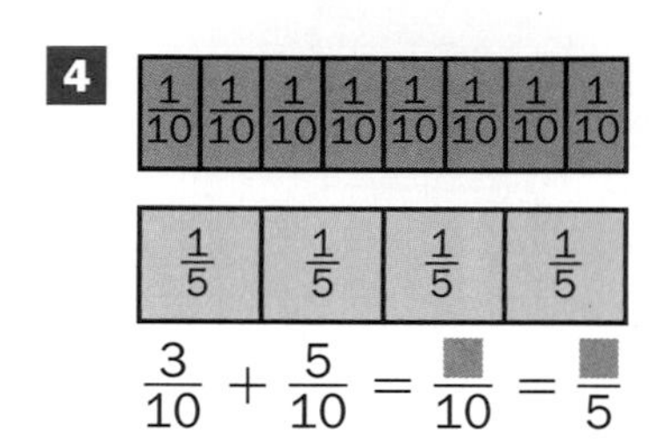

$\frac{3}{10} + \frac{5}{10} = \frac{■}{10} = \frac{■}{5}$

Add. You may use models if you wish.

5 $\frac{3}{10} + \frac{8}{10}$

6 $\frac{5}{8} + \frac{7}{8}$

7 $\frac{1}{3} + \frac{2}{3}$

8 $\frac{1}{8} + \frac{2}{8}$

9 $\frac{4}{5} + \frac{4}{5}$

10 $\frac{3}{6} + \frac{4}{6}$

11 $\frac{2}{6} + \frac{3}{6}$

12 $\frac{9}{10} + \frac{2}{10}$

13 $\frac{7}{12} + \frac{9}{12}$

14 $\frac{2}{4} + \frac{2}{4}$

15 $\frac{8}{12} + \frac{11}{12}$

16 $\frac{1}{8} + \frac{6}{8}$

17 $\frac{1}{5} + \frac{3}{5}$ **18** $\frac{1}{6} + \frac{4}{6}$ **19** $\frac{3}{10} + \frac{4}{10}$ **20** $\frac{4}{5} + \frac{1}{5}$ **21** $\frac{1}{4} + \frac{3}{4}$

22 $\frac{3}{8} + \frac{1}{8}$ **23** $\frac{1}{4} + \frac{1}{4}$ **24** $\frac{2}{6} + \frac{2}{6}$ **25** $\frac{5}{8} + \frac{3}{8}$ **26** $\frac{1}{12} + \frac{2}{12}$

ALGEBRA Complete the table.

27

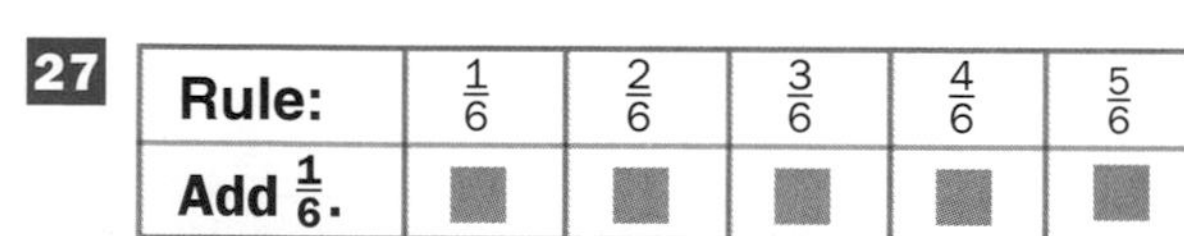

Rule:	$\frac{1}{6}$	$\frac{2}{6}$	$\frac{3}{6}$	$\frac{4}{6}$	$\frac{5}{6}$
Add $\frac{1}{6}$.	■	■	■	■	■

28

Rule:	$\frac{8}{12}$	$\frac{9}{12}$	$\frac{10}{12}$	$\frac{11}{12}$
Add $\frac{3}{12}$.	■	■	■	■

ALGEBRA Write >, <, or =.

29 $\frac{1}{4} + \frac{2}{4}$ ● 1 **30** $\frac{6}{12} + \frac{8}{12}$ ● 1 **31** $\frac{3}{6} + \frac{3}{6}$ ● 1 **32** $\frac{7}{8} + \frac{3}{8}$ ● 1

33 $\frac{4}{12} + \frac{8}{12}$ ● 1 **34** $\frac{5}{10} + \frac{4}{10}$ ● 1 **35** $\frac{3}{5} + \frac{3}{5}$ ● 1 **36** $\frac{2}{6} + \frac{3}{6}$ ● 1

Make It Right

37 Megan used fraction strips to add $\frac{3}{4} + \frac{1}{4}$. Find the mistake, then correct it.

$\frac{3}{4} + \frac{1}{4} = \frac{4}{16}$, or $\frac{1}{4}$

Hexagon Roll Game!

Write the following numbers on each side of two number cubes: 0, $\frac{1}{6}$, $\frac{2}{6}$, $\frac{3}{6}$, $\frac{4}{6}$, $\frac{5}{6}$.

You will need
- *number cubes*
- *triangle and hexagon pattern blocks*

Play the Game

Play with a partner to form hexagons. Let each triangle pattern block stand for $\frac{1}{6}$ and each hexagon stand for 1.

- ► The first player rolls the two number cubes, then shows the fractions with the triangle pattern blocks.
- ► Find the sum by combining the pattern blocks. Write a number sentence to show what you did.
- ► If the triangle pattern blocks form a whole hexagon, call "Hexagon!" to score 1 point.
- ► Take turns and continue playing for 5 rounds. The player with the most points is the winner.

Which fractions would you like to roll each time? Explain.

mixed review • test preparation

Write in order from least to greatest.

1 $\frac{2}{4}, \frac{2}{8}, \frac{2}{16}$

2 $\frac{5}{7}, \frac{2}{7}, \frac{6}{7}$

3 $\frac{1}{3}, \frac{11}{13}, \frac{1}{7}$

4 $\frac{2}{3}, \frac{5}{8}, \frac{5}{6}$

Write *certain, likely, unlikely,* or *impossible*.

5 spinning a number less than 8 on a 1–9 spinner

6 rolling a number greater than 9 on a 1–6 number cube

7 choosing a blue cube from a bag that has seven blue cubes

8 flipping a head each time in 100 tries

Extra Practice, page 524

Add Fractions

LEARN

Your class is creating a 12-month recipe calendar. You plan to have 4 main dish recipes, 3 appetizer recipes, and 5 dessert recipes. What fraction of the calendar contains recipes for appetizers and desserts?

Add: $\frac{3}{12} + \frac{5}{12}$ **Think:** $\frac{3}{12}$ are appetizers, $\frac{5}{12}$ are desserts.

Check Out the Glossary
common denominator
numerator
See page 544.

In the last lesson, you used fraction strips to add fractions with a **common denominator.**

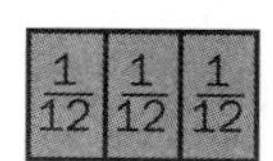
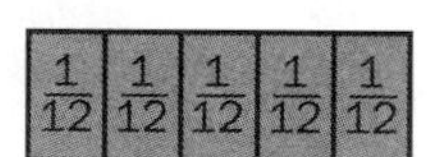
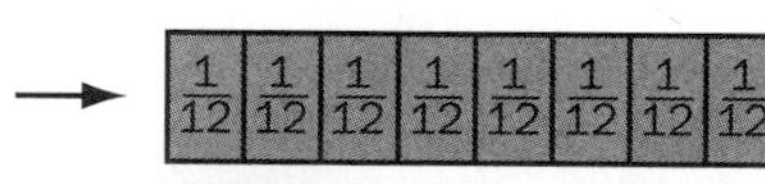

$\frac{3}{12} + \frac{5}{12} = \frac{8}{12}$, or $\frac{2}{3}$

Here is another method.

Step 1	Step 2	Step 3
Add the numerators.	**Use the common denominator.**	**Write the sum in simplest form.**
$\frac{3}{12} + \frac{5}{12} = \frac{8}{}$	$\frac{3}{12} + \frac{5}{12} = \frac{8}{12}$	$\frac{8}{12} = \frac{8 \div 4}{12 \div 4} = \frac{2}{3}$

$\frac{2}{3}$ of the recipes are for appetizers and desserts.

More Examples

A $\frac{4}{16} + \frac{6}{16} + \frac{1}{16} = \frac{11}{16}$

B $\frac{3}{4} + \frac{2}{4} = \frac{5}{4}$

$\frac{5}{4} = 1\frac{1}{4}$

C $\frac{5}{6} + \frac{1}{6} = \frac{6}{6}$

$\frac{6}{6} = 1$

CHECK

Check for Understanding

Add. Write the sum in simplest form.

1 $\frac{1}{6} + \frac{4}{6}$

2 $\frac{1}{9} + \frac{2}{9}$

3 $\frac{4}{12} + \frac{5}{12}$

4 $\frac{3}{10} + \frac{1}{10} + \frac{8}{10}$

Critical Thinking: Analyze **Explain your reasoning.**

5 Does changing the order in which you add fractions change the sum? Why or why not? Give examples to support your answer.

Practice

Add. Write the sum in simplest form.

1 $\frac{1}{10} + \frac{2}{10}$

2 $\frac{2}{5} + \frac{2}{5}$

3 $\frac{2}{12} + \frac{3}{12}$

4 $\frac{1}{5} + \frac{2}{5}$

5 $\frac{2}{5} + \frac{3}{5}$

6 $\frac{5}{10} + \frac{5}{10}$

7 $\frac{1}{10} + \frac{3}{10}$

8 $\frac{5}{12} + \frac{1}{12}$

9 $\frac{1}{8} + \frac{5}{8}$

10 $\frac{7}{8} + \frac{3}{8}$

11 $\frac{1}{6} + \frac{1}{6} + \frac{3}{6}$

12 $\frac{1}{9} + \frac{4}{9} + \frac{3}{9}$

13 $\frac{1}{3} + \frac{2}{3} + \frac{2}{3}$

14 $\frac{2}{11} + \frac{5}{11} + \frac{6}{11}$

MIXED APPLICATIONS
Problem Solving

15 A basic muffin recipe for 12 muffins calls for $\frac{1}{4}$ cup of sugar and a $\frac{1}{2}$ cup of raisins. If you make 2 batches, how much sugar will you need? how many cups of raisins?

16 **What if** you added $\frac{1}{8}$ cup of filberts and $\frac{1}{8}$ cup of hazelnuts to the recipe. How many cups of nuts is that?

17 **Spatial reasoning** Is it possible to draw a triangle on dot paper that touches 8 dots but has no dots on the inside? Show an example.

18 There are 12 muffins in each batch. How many muffins are there in 6 batches?

Cultural Connection
Kpelle Rice Measurement

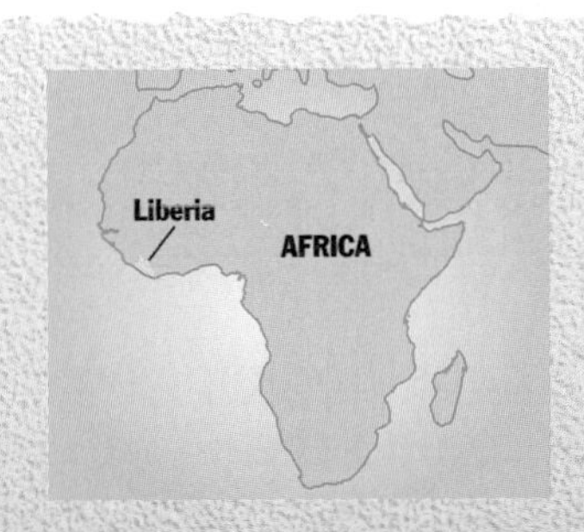

The Kpelle (PE-lay) people of Liberia, Africa, measure rice by using a *sâmo-ko* (sah-MOH-koh), or "salmon cup." A sâmo-ko holds about 2 cups of rice.

Larger amounts of rice may be measured with a *bôke* (BAH-kee), or "bucket." A bôke is about 12 sâmo-ko or $\frac{1}{2}$ of a *tin.* The largest Kpelle measure for rice is a *boro* (BAH-roh), or "bag." A bôke is $\frac{1}{4}$ of a boro.

▶ How are the units the Kpelle people use to measure rice related to each other?

Extra Practice, page 524

Subtract Fractions

Miko is making *bobotie* (BOH-boh-tī). She has a box of raisins in her cupboard that is $\frac{5}{8}$ filled. She will use $\frac{3}{8}$ of the box for her meal. What part of the box will be left?

To solve the problem, you need to subtract fractions.

Cultural Note

Bobotie is a spicy dish made in southern Africa. It is usually made in a casserole dish with onions, bread, ground beef, curry powder, almonds, and raisins.

Work Together

Work with a partner to explore subtracting fractions.

Use fraction strips to model the problem. Write a number sentence to show your work.

You will need
- *fraction strips*

Continue using fraction strips to subtract these fractions:

a. $\frac{2}{3} - \frac{1}{3}$ **b.** $\frac{4}{8} - \frac{3}{8}$ **c.** $\frac{3}{5} - \frac{1}{5}$

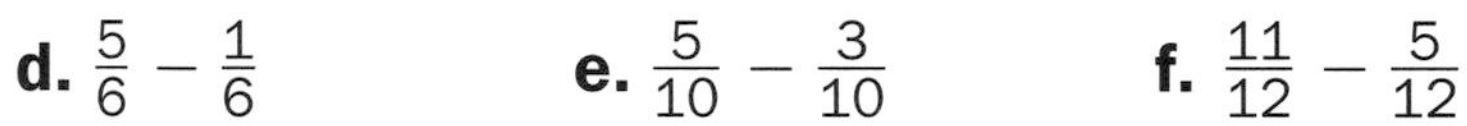

d. $\frac{5}{6} - \frac{1}{6}$ **e.** $\frac{5}{10} - \frac{3}{10}$ **f.** $\frac{11}{12} - \frac{5}{12}$

Talk It Over

- ▶ What part of the box of raisins is left?
- ▶ What method did you use to find the differences?

Make Connections

Kendall and Lori drew fraction strips to show their work.

$\frac{5}{8} - \frac{3}{8} = \frac{2}{8}$, or $\frac{1}{4}$ or

$$\begin{array}{r} \frac{5}{8} \\ -\frac{3}{8} \\ \hline \frac{2}{8}, \text{ or } \frac{1}{4} \end{array}$$

So $\frac{1}{4}$ of the box of raisins will be left over.

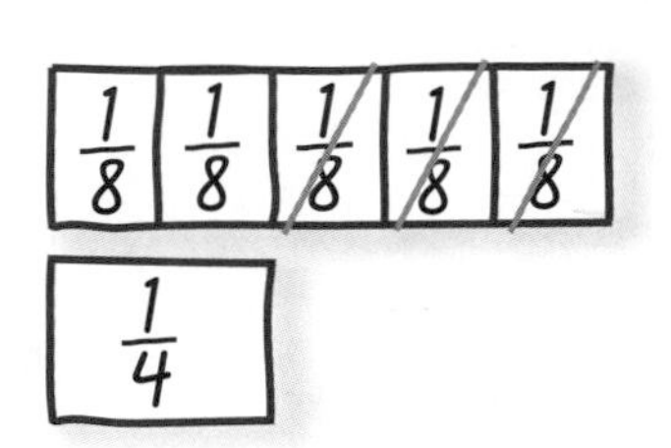

▶ Check if the differences in your number sentences are in simplest form. If not, write them in simplest form.

CHECK

Check for Understanding

Use models to complete the number sentence.

1 $\frac{7}{8} - \frac{2}{8} = \frac{■}{8}$

2

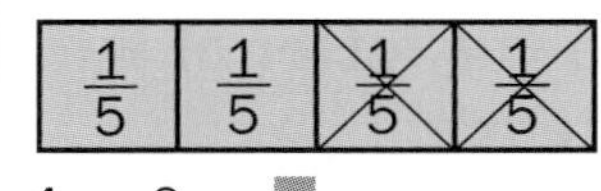

$\frac{4}{5} - \frac{2}{5} = \frac{■}{5}$

3 $\frac{9}{10} - \frac{5}{10} = \frac{■}{10} = \frac{■}{5}$

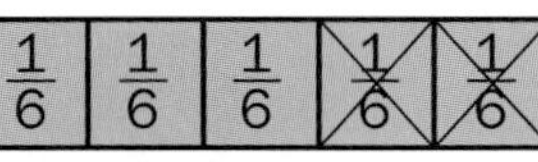

$\frac{5}{6} - \frac{2}{6} = \frac{■}{6} = \frac{■}{2}$

Subtract using models. Write the difference in simplest form.

5 $\frac{7}{8} - \frac{5}{8}$ **6** $\frac{2}{4} - \frac{1}{4}$ **7** $\frac{4}{5} - \frac{1}{5}$ **8** $\frac{3}{4} - \frac{2}{4}$ **9** $\frac{5}{6} - \frac{3}{6}$ **10** $\frac{7}{8} - \frac{1}{8}$

11 $\frac{4}{8} - \frac{2}{8}$ **12** $\frac{5}{8} - \frac{1}{8}$ **13** $\frac{1}{2} - \frac{1}{2}$ **14** $\frac{7}{10} - \frac{1}{10}$ **15** $\frac{7}{12} - \frac{4}{12}$

Critical Thinking: Summarize

16 

Describe how you would use fraction strips to subtract $\frac{7}{10} - \frac{5}{10}$. Tell how you would simplify the answer.

Turn the page for Practice.

Practice

Use models to complete the number sentence.

1 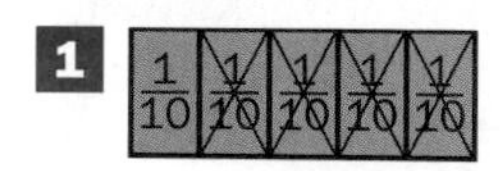

$\frac{5}{10} - \frac{4}{10} = \frac{■}{10}$

2 $\frac{6}{8} - \frac{3}{8} = \frac{■}{8}$

3 $\frac{4}{5} - \frac{3}{5} = \frac{■}{5}$

4 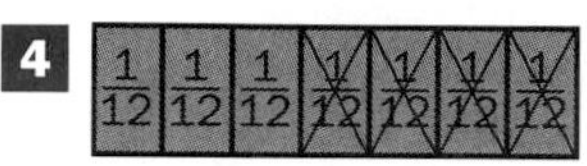

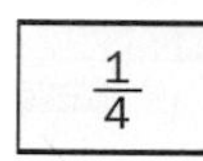

$\frac{7}{12} - \frac{4}{12} = \frac{■}{12} = \frac{■}{4}$

5 $\frac{3}{4} - \frac{1}{4} = \frac{■}{4} = \frac{■}{2}$

6 $\frac{4}{6} - \frac{1}{6} = \frac{■}{6} = \frac{■}{2}$

Subtract. Write the difference in simplest form.

7 $\frac{8}{12} - \frac{2}{12}$ **8** $\frac{4}{6} - \frac{3}{6}$ **9** $\frac{7}{10} - \frac{4}{10}$ **10** $\frac{5}{8} - \frac{2}{8}$ **11** $\frac{9}{12} - \frac{2}{12}$

12 $\frac{4}{6} - \frac{2}{6}$ **13** $\frac{9}{12} - \frac{6}{12}$ **14** $\frac{8}{10} - \frac{2}{10}$ **15** $\frac{8}{12} - \frac{5}{12}$ **16** $\frac{3}{6} - \frac{1}{6}$

ALGEBRA Choose the letter of the number that completes the number sentence.

17 $\frac{3}{6} - \frac{2}{6} = ■$ **a.** $\frac{8}{12}$ **b.** $\frac{1}{6}$ **c.** $\frac{5}{8}$ **d.** $\frac{1}{3}$

18 $\frac{8}{12} - \frac{■}{12} = \frac{1}{12}$ **a.** 9 **b.** 1 **c.** 12 **d.** 7

19 $\frac{3}{8} - \frac{2}{■} = \frac{1}{8}$ **a.** 3 **b.** 2 **c.** 8 **d.** 4

20 $\frac{■}{10} - \frac{3}{10} = \frac{3}{10}$ **a.** 1 **b.** 2 **c.** 5 **d.** 6

21 **ALGEBRA: PATTERNS** Find the difference. For each row, describe any patterns you see.

a. $\frac{2}{5} - \frac{2}{5}$ $\frac{5}{6} - \frac{5}{6}$ $\frac{1}{8} - \frac{1}{8}$ $\frac{3}{10} - \frac{3}{10}$

b. $\frac{1}{2} - 0$ $\frac{2}{3} - 0$ $\frac{3}{4} - 0$ $\frac{7}{12} - 0$

Make It Right

22 Darren drew fraction strips to find $\frac{5}{8} - \frac{1}{8}$. Tell what the mistake is and correct it.

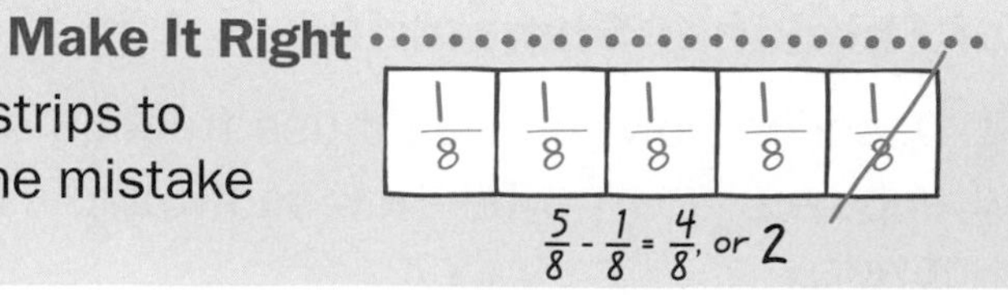

MIXED APPLICATIONS
Problem Solving

Use the pictograph for problem 23.

23 **a.** How many students ate eggs for breakfast?

b. How many more students ate cold cereal than hot cereal?

c. Can you tell how many students were surveyed? Why or why not?

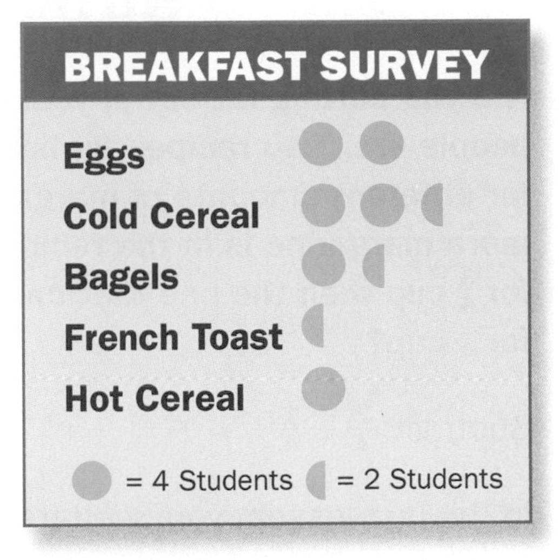

24 Ada needs $\frac{7}{8}$ cup of almonds for her bobotie. If she has $\frac{3}{8}$ cup, how many more cups does she need?

25 Mr. Hodges bought $\frac{3}{4}$ lb of roast beef. He used $\frac{1}{4}$ lb of roast beef for each of his two roast beef sandwiches. How much roast beef does he have left?

26 **Write a problem** that can be solved by subtracting fractions with like denominators. Solve it. Trade with a classmate and solve each other's problems.

more to explore

Fractions in Circle Graphs

You can add and subtract fractions to solve problems using information from a circle graph.

The circle graph shows what part of the total amount of money was spent on different lunch items.

To find the amount spent on a sandwich and fries, add the fractions.

Think: $\frac{3}{8} + \frac{1}{8} = \frac{4}{8}$, or $\frac{1}{2}$

So $\frac{1}{2}$ the money was spent on a sandwich and fries.

Money Spent for Lunch

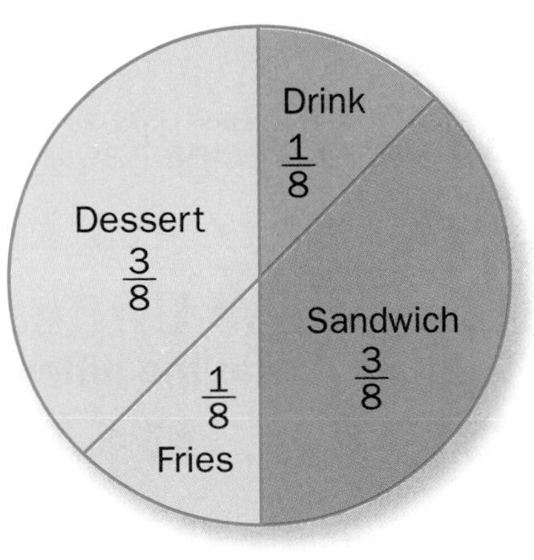

Solve. Use the circle graph.

1 Was a greater fraction of the money spent for dessert or for fries? Explain.

2 **What if** the drink was bought first. What part of the money was left after paying for it?

Extra Practice, page 524

Subtract Fractions

LEARN

Are you cutting fat out of your diet? Many young people are. Two recipes for banana bread call for different amounts of margarine. How much more margarine is in the recipe that calls for $\frac{3}{4}$ cup than the one that calls for $\frac{1}{4}$ cup?

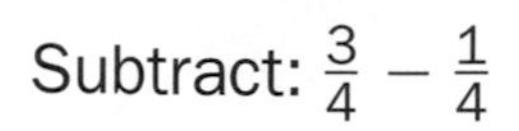

Subtract: $\frac{3}{4} - \frac{1}{4}$

In the last lesson, you used fraction strips to subtract fractions with a common denominator.

$\frac{1}{4}$	$\frac{1}{4}$	~~$\frac{1}{4}$~~

$\frac{1}{2}$

$$\frac{3}{4} - \frac{1}{4} = \frac{2}{4} = \frac{1}{2}$$

Cultural Note

In the Caribbean, bananas are used for many foods, such as banana bread. Even the leaves have uses. Sometimes, they are woven together to create shelter from the sun.

Here is another method.

Step 1	Step 2	Step 3
Subtract the numerators.	**Use the common denominator.**	**Write the difference in simplest form.**
$\frac{3}{4} - \frac{1}{4} = \frac{2}{}$	$\frac{3}{4} - \frac{1}{4} = \frac{2}{4}$	$\frac{2}{4} = \frac{2 \div 2}{4 \div 2} = \frac{1}{2}$

The second recipe has $\frac{1}{2}$ cup more margarine.

CHECK

Check for Understanding

Subtract. Write the difference in simplest form.

1 $\frac{5}{7} - \frac{4}{7}$ **2** $\frac{9}{11} - \frac{5}{11}$ **3** $\frac{8}{9} - \frac{2}{9}$ **4** $\frac{12}{12} - \frac{4}{12}$ **5** $\frac{10}{16} - \frac{2}{16}$

6 $\frac{3}{7} - \frac{2}{7}$ **7** $\frac{7}{16} - \frac{5}{16}$ **8** $\frac{10}{14} - \frac{3}{14}$ **9** $\frac{8}{10} - \frac{4}{10}$ **10** $\frac{7}{8} - \frac{1}{8}$

Critical Thinking: Analyze **Explain your reasoning.**

11 Does changing the order in which you subtract fractions change the answer? Why or why not? Give examples to support your answer.

Practice

Subtract. Write the difference in simplest form.

1 $\frac{7}{11} - \frac{4}{11}$ **2** $\frac{6}{7} - \frac{2}{7}$ **3** $\frac{9}{10} - \frac{9}{10}$ **4** $\frac{4}{7} - \frac{1}{7}$ **5** $\frac{6}{9} - \frac{4}{9}$ **6** $\frac{8}{11} - \frac{5}{11}$

7 $\frac{8}{8} - \frac{5}{8}$ **8** $\frac{7}{12} - \frac{5}{12}$ **9** $\frac{8}{9} - \frac{5}{9}$ **10** $\frac{6}{12} - \frac{3}{12}$ **11** $\frac{5}{10} - \frac{3}{10}$

12 $\frac{5}{16} - \frac{1}{16}$ **13** $\frac{11}{12} - \frac{7}{12}$ **14** $\frac{7}{10} - \frac{2}{10}$ **15** $\frac{5}{8} - \frac{3}{8}$ **16** $\frac{5}{12} - \frac{1}{12}$

ALGEBRA Write >, <, or =.

17 $\frac{7}{11} - \frac{4}{11}$ ● $\frac{5}{11} - \frac{2}{11}$ **18** $\frac{10}{16} - \frac{5}{16}$ ● $\frac{1}{4}$ **19** $\frac{5}{8} - \frac{1}{8}$ ● $\frac{5}{6} - \frac{1}{6}$

20 $\frac{3}{4} - \frac{1}{4}$ ● $\frac{3}{4} - \frac{2}{4}$ **21** $\frac{2}{3} - \frac{1}{3}$ ● $\frac{7}{9} - \frac{1}{9}$ **22** $\frac{4}{5} - \frac{2}{5}$ ● $\frac{9}{10} - \frac{7}{10}$

MIXED APPLICATIONS
Problem Solving

23 Rashid pours some milk from a full pint container. There are 12 ounces of milk left. What fraction of the container did he pour out? (Hint: 16 ounces = 1 pint)

24 The school cafeteria sold 1,212 chicken fingers at lunchtime. Each student who bought lunch got 3 chicken fingers. How many students bought lunch?

25 Mel walks $\frac{7}{10}$ mi to school. He walks $\frac{3}{10}$ mi in 5 minutes. Shelley walks $\frac{4}{5}$ mi to school. She walks $\frac{2}{5}$ mi in 5 minutes. Who is closer to school after walking for 5 minutes? Explain your thinking.

26 **Data Point** Survey ten or more family members or friends to find out what their favorite food is. Display the data using a graph. Write a paragraph explaining what your graph shows.

mixed review • test preparation

1 876 + 364 **2** 627 − 429 **3** 583×24 **4** $563 \div 3$

Write the number in standard form.

5 four thousand, sixty-two

6 one hundred ten thousand, six

7 thirty-two thousand, fourteen

8 ninety-nine thousand, ninety-nine

Add. Write the sum in simplest form.

1 $\frac{2}{6} + \frac{4}{6}$

2 $\frac{3}{12} + \frac{6}{12}$

3 $\frac{5}{7} + \frac{5}{7}$

4 $\frac{1}{6} + \frac{3}{6} + \frac{4}{6}$

5 $\frac{7}{16} + \frac{4}{16}$

6 $\frac{3}{8} + \frac{3}{8}$

7 $\frac{3}{4} + \frac{2}{4}$

8 $\frac{5}{10} + \frac{7}{10}$

Subtract. Write the difference in simplest form.

9 $\frac{8}{12} - \frac{1}{12}$

10 $\frac{7}{11} - \frac{7}{11}$

11 $\frac{5}{6} - \frac{2}{6}$

12 $\frac{9}{10} - \frac{7}{10}$

13 $\frac{15}{16} - \frac{3}{16}$

14 $\frac{8}{8} - \frac{2}{8}$

Write >, <, or =.

15 $\frac{1}{2} + \frac{1}{2}$ ● 1

16 $\frac{9}{12} - \frac{3}{12}$ ● $\frac{5}{12} - \frac{1}{12}$

17 $\frac{4}{7} + \frac{6}{7}$ ● 1

18 $\frac{6}{10} - \frac{4}{10}$ ● $\frac{4}{10}$

19 $\frac{3}{4} + \frac{1}{4}$ ● $\frac{5}{8} + \frac{2}{8}$

20 $\frac{7}{8} - \frac{3}{8}$ ● $\frac{15}{16} - \frac{6}{16}$

Solve.

21 A pizza is cut into 12 equal pieces. Paco eats 3 of the pieces, Len and Ben eat 2 pieces each. What fraction of the pizza is left?

22 Karen baked two batches of cookies. She used $\frac{3}{4}$ pound of dough for each batch. How much dough did she use altogether?

23 Molly lives $\frac{7}{8}$ mi from the mall. She walked $\frac{3}{8}$ mi toward the mall before meeting a friend. She walked another $\frac{2}{8}$ mi before meeting another friend. How far away from the mall is Molly when she meets the second friend?

24 Dylan started cooking dinner at 3:00 P.M. He took $\frac{3}{4}$ h to make the tuna casserole, $\frac{1}{4}$ h to make biscuits, $\frac{1}{4}$ h to make soup, and $\frac{3}{4}$ h to make dessert. What time did Dylan finish cooking dinner? Explain your methods.

25 Journal Explain how subtracting fractions less than one with like denominators and subtracting whole numbers are alike and are different.

Find Area Using Fractions of Shapes

Sometimes, you can find the area of a shape if it is a part of another shape you know.

The side of the square is 8 units long. So the area of the square is 64 square units.

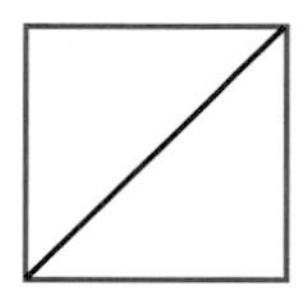

Each triangle is $\frac{1}{2}$ of the square.

So the area of each triangle is $\frac{1}{2}$ of 64, or 32 square units.

"ON THE POINTS," WASSILY KANDINSKY, MUSEE NATIONAL D'ART MODERNE, PARIS, FRANCE

The square above is outlined in each figure.
Use fractions to help you find the area of the entire shape.

1 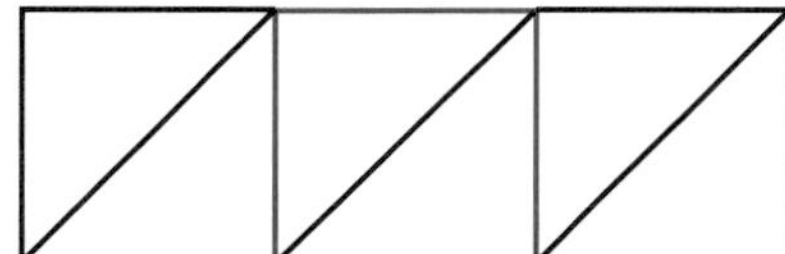

2 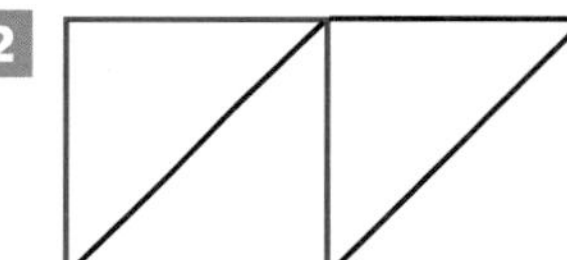

3 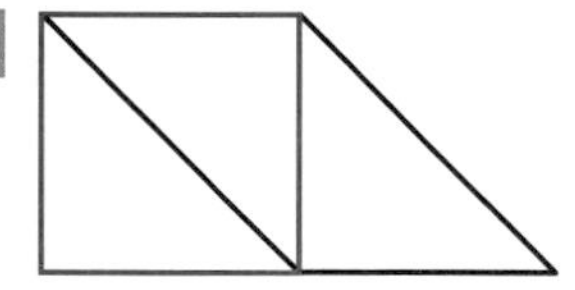

4 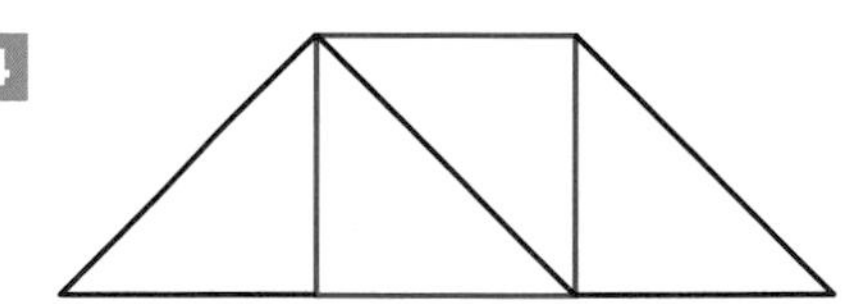

5 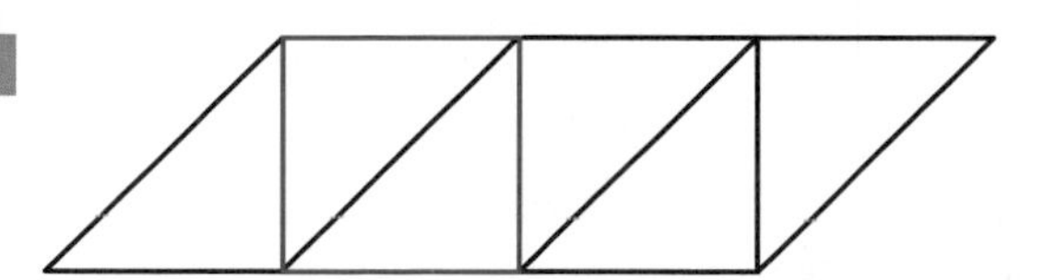

6

Plan a Class Lunch

Suppose you are in charge of preparing lunch for the students in your class. What kinds of decisions would you have to make? Here are just a few.

- What kinds of foods would you serve?
- How much of each food would you prepare?
- How would you decide if you needed to make more of one kind of food?

READING ARITHMETIC WRITING

Write a Report In this activity, you will collect and combine information to prepare an imaginary lunch. You will use what you have learned about adding and subtracting fractions to help you collect and present the data in a report.

DECISION MAKING

Planning Lunch

1 Work in a group to plan the class lunch. Survey your classmates to find out what they like to eat.

2 Use your data to decide:
- ▶ what food you will serve.
- ▶ what fraction of a pound or a serving tray you will use as a serving portion.
- ▶ the total amount of each type of food you will need to serve lunch.

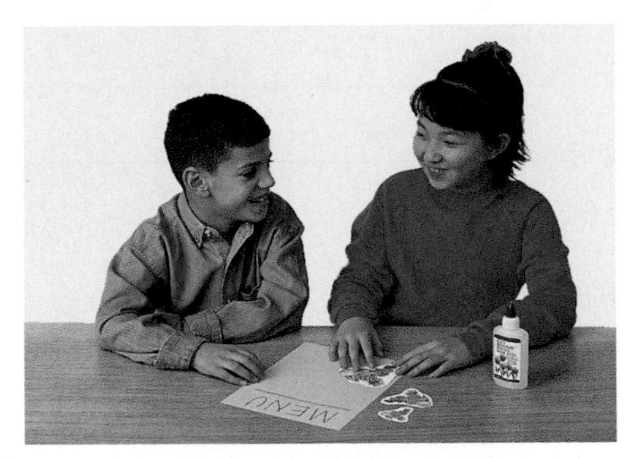

3 Create a menu for your classmates to choose from. Have them order lunch. Keep track of how much food is served.

Reporting Your Findings

4 Portfolio Prepare a report on what you learned. Include the following:

- ▶ a graph that shows the results of your survey
- ▶ statements about the results of the survey and how you used the results to decide what your menu would contain
- ▶ a table that shows the total amounts for each type of food, the amounts that were ordered, and the amounts that were left

5 Tell what changes you would make if you planned another lunch.

Revise your work.
- ▶ Did you include all of the data required?
- ▶ Did you add or subtract the fractions correctly?
- ▶ Is the table organized and accurate?

MORE TO INVESTIGATE

PREDICT how the menu and the amounts served would change if you prepared lunch for kindergarten students.

EXPLORE whether your menu contains healthy, balanced meals. Explain how you can change it to be more healthy.

FIND what foods are most popular in your own school cafeteria. Interview the cafeteria staff or survey other grades. Report your findings to the class.

Find a Common Denominator

LEARN

You can use fraction strips to help you find equivalent fractions with a common denominator.

Work Together

Work with a partner to compare fraction strips.

You will need
- *fraction strips*
- *number cube*

Write the following numbers on the faces of a number cube: $\frac{1}{2}$, $\frac{2}{3}$, $\frac{3}{4}$, $\frac{5}{6}$, $\frac{7}{12}$, $\frac{11}{12}$.

Toss the number cube twice to get two fractions. Show both fractions using fraction strips.

Find equivalent fractions for both fractions using only one type of fraction strip. Record the equivalent fractions you found.

Fractions	Equivalent Fractions
$\frac{2}{3}$	
$\frac{3}{4}$	

Continue tossing the number cube for other pairs of fractions.

▶ Explain how you decided which common fraction strips to use.

Making Connections

Tom and Anne showed equivalent fractions for both $\frac{2}{3}$ and $\frac{3}{4}$ using twelfth strips.

$\frac{1}{3}$	$\frac{1}{3}$
$\frac{1}{12}$ $\frac{1}{12}$ $\frac{1}{12}$ $\frac{1}{12}$	$\frac{1}{12}$ $\frac{1}{12}$ $\frac{1}{12}$ $\frac{1}{12}$

$\frac{2}{3}$ is equivalent to $\frac{8}{12}$.

$\frac{1}{4}$	$\frac{1}{4}$	$\frac{1}{4}$
$\frac{1}{12}$ $\frac{1}{12}$ $\frac{1}{12}$	$\frac{1}{12}$ $\frac{1}{12}$ $\frac{1}{12}$	$\frac{1}{12}$ $\frac{1}{12}$ $\frac{1}{12}$

$\frac{3}{4}$ is equivalent to $\frac{9}{12}$.

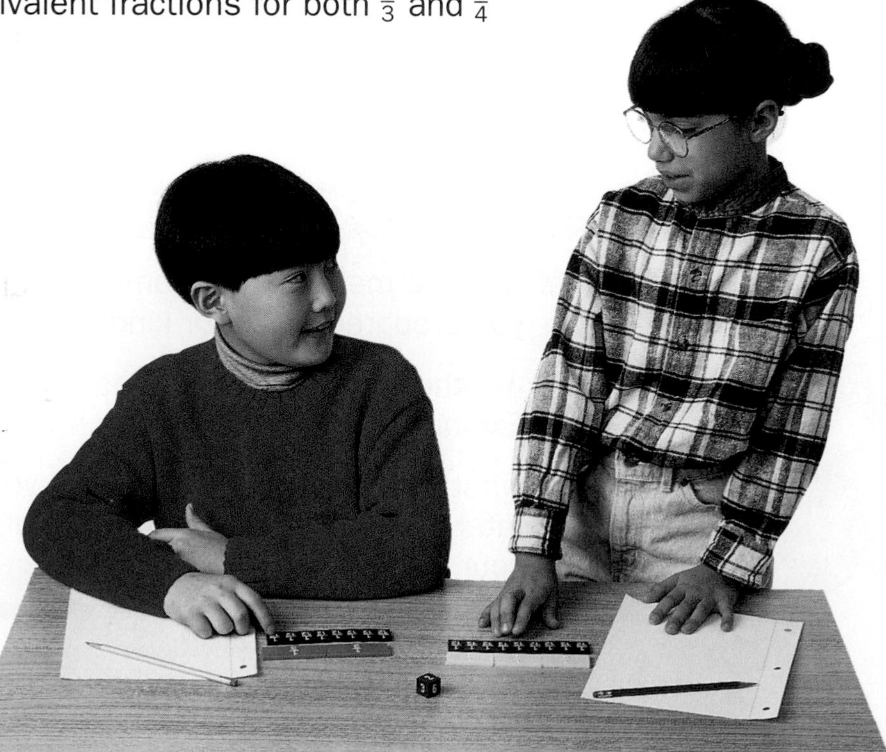

You can find a common denominator by using multiples.

Compare the multiples of the uncommon denominators.

$\frac{2}{3}$ has a denominator of 3. Multiples of 3: 3, 6, 9, **12**

$\frac{3}{4}$ has a denominator of 4. Multiples of 4: 4, 8, **12**

$\frac{2}{3}$ and $\frac{3}{4}$ can be written as equivalent fractions with 12 as the common denominator.

▶ Are there different common denominators that you can use for fractions? Explain why or why not.

Check for Understanding

Write as equivalent fractions with common denominators.

1 $\frac{1}{4}$ and $\frac{4}{12}$

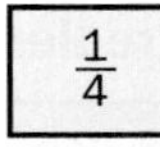

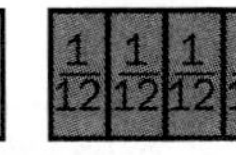

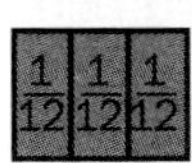

2 $\frac{2}{5}$ and $\frac{3}{10}$

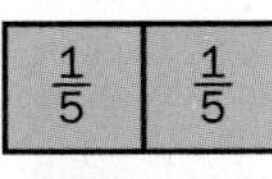

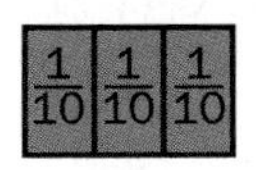

1/10 1/10 1/10 1/10

3 $\frac{1}{4}$ and $\frac{3}{8}$ **4** $\frac{2}{3}$ and $\frac{3}{6}$ **5** $\frac{1}{6}$ and $\frac{1}{4}$ **6** $\frac{1}{2}$ and $\frac{1}{3}$ **7** $\frac{3}{4}$ and $\frac{5}{6}$

Critical Thinking: Analyze **Explain your reasoning.**

8 

How would you use multiples to find the common denominator for these three fractions: $\frac{1}{3}$, $\frac{1}{2}$, $\frac{1}{6}$?

Practice

Name the common denominator.

1 $\frac{1}{3}$ and $\frac{5}{6}$ **2** $\frac{3}{4}$ and $\frac{5}{8}$ **3** $\frac{1}{4}$ and $\frac{5}{12}$ **4** $\frac{1}{2}$ and $\frac{5}{12}$ **5** $\frac{2}{3}$ and $\frac{7}{9}$

6 $\frac{2}{5}$ and $\frac{2}{3}$ **7** $\frac{5}{6}$ and $\frac{1}{4}$ **8** $\frac{1}{2}$ and $\frac{3}{7}$ **9** $\frac{3}{4}$ and $\frac{1}{6}$ **10** $\frac{6}{7}$ and $\frac{1}{14}$

Write as equivalent fractions with common denominators.

11 $\frac{2}{3}$ and $\frac{1}{6}$ **12** $\frac{1}{4}$ and $\frac{5}{8}$ **13** $\frac{5}{6}$ and $\frac{5}{12}$ **14** $\frac{9}{10}$ and $\frac{2}{5}$ **15** $\frac{1}{2}$ and $\frac{3}{5}$

16 $\frac{1}{3}$ and $\frac{4}{5}$ **17** $\frac{5}{6}$ and $\frac{1}{2}$ **18** $\frac{1}{2}$ and $\frac{2}{5}$ **19** $\frac{1}{4}$ and $\frac{2}{3}$ **20** $\frac{1}{2}$ and $\frac{5}{8}$

CHECK

PRACTICE

Extra Practice, page 525

LEARN

Add and Subtract Fractions with Unlike Denominators

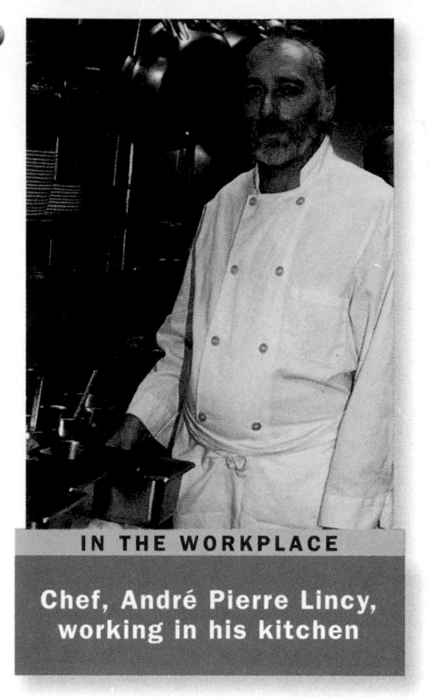

IN THE WORKPLACE

Chef, André Pierre Lincy, working in his kitchen

Have you ever thought about what it is like to cook for over 1,000 people a day? Part of a head chef's job is to create new recipes. You start by testing small amounts.

To work with recipes, you need to know how to add and subtract fractions with unlike denominators.

How much broth is used in this recipe?

How much more wild rice is used than orzo pasta?

Recipe for Wild Rice	
Liquids	**Dry Ingredients**
$\frac{3}{8}$ cup cider	$\frac{1}{3}$ cup wild rice
$\frac{1}{2}$ cup coconut milk	$\frac{1}{4}$ cup orzo pasta
$\frac{3}{4}$ cup chicken broth	$\frac{5}{12}$ cup basmati rice
$\frac{3}{8}$ cup beef broth	

Work Together

Work with a partner. Use fraction strips to model the problems.

You will need
- *fraction strips*

Add to find the total amount of broth.

Subtract to find the difference in the amounts of rice and pasta.

Use fraction strips to find the sum or difference:

a. $\frac{3}{8} + \frac{1}{2}$ **b.** $\frac{1}{2} + \frac{3}{4}$ **c.** $\frac{1}{3} + \frac{1}{4}$ **d.** $\frac{1}{4} + \frac{5}{12}$

e. $\frac{5}{12} - \frac{1}{6}$ **f.** $\frac{3}{4} - \frac{3}{8}$ **g.** $\frac{5}{12} - \frac{1}{4}$ **h.** $\frac{5}{12} - \frac{1}{3}$

Talk It Over

► How did you use fraction strips to add and subtract fractions when their denominators were different?

Make Connections

You can use what you know about adding and subtracting fractions with like denominators to add and subtract fractions with unlike denominators.

Add: $\frac{3}{4} + \frac{3}{8}$

Step 1

Find equivalent fractions with a common denominator.

$$\frac{3}{4} = \frac{3 \times 2}{4 \times 2} = \frac{6}{8}$$
$$+\frac{3}{8} \qquad\qquad +\frac{3}{8}$$

Think:
Multiples of 4: 4, 8
Multiples of 8: 8
8 is the common denominator.

$\frac{1}{4}$	$\frac{1}{4}$	$\frac{1}{4}$	$\frac{1}{8}$	$\frac{1}{8}$	$\frac{1}{8}$

$\frac{1}{8}$	$\frac{1}{8}$	$\frac{1}{8}$	$\frac{1}{8}$	$\frac{1}{8}$	$\frac{1}{8}$	$\frac{1}{8}$	$\frac{1}{8}$	$\frac{1}{8}$

Step 2

Add the numerators.
Use the common denominator.

$$\frac{6}{8} + \frac{3}{8} = \frac{9}{8} = 1\frac{1}{8}$$

$\frac{1}{8}$	$\frac{1}{8}$	$\frac{1}{8}$	$\frac{1}{8}$	$\frac{1}{8}$	$\frac{1}{8}$	$\frac{1}{8}$	$\frac{1}{8}$	$\frac{1}{8}$

1	$\frac{1}{8}$

Subtract: $\frac{1}{3} - \frac{1}{4}$

Step 1

Find equivalent fractions with a common denominator.

$$\frac{1}{3} = \frac{1 \times 4}{3 \times 4} = \frac{4}{12}$$
$$-\frac{1}{4} = \frac{1 \times 3}{4 \times 3} = -\frac{3}{12}$$

Think:
Multiples of 3: 3, 6, 9, 12
Multiples of 4: 4, 8, 12
12 is the common denominator.

$\frac{1}{3}$			
$\frac{1}{12}$	$\frac{1}{12}$	$\frac{1}{12}$	$\frac{1}{12}$

Step 2

Subtract the numerators.
Use the common denominator.

$$\frac{4}{12} - \frac{3}{12} = \frac{1}{12}$$

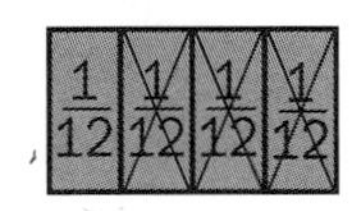

Check for Understanding

Add or subtract using any method.

1 $\frac{3}{8} + \frac{1}{4}$ **2** $\frac{2}{12} + \frac{4}{6}$ **3** $\frac{9}{10} - \frac{7}{10}$ **4** $\frac{5}{6} - \frac{1}{3}$

Critical Thinking: Analyze **Explain your reasoning.**

5 Find $\frac{5}{6} - \frac{1}{3}$ in two ways, using 6, then 12, as the common denominator. What do you notice about the answers?

CHECK

Turn the page for Practice.

Practice

Use the models to complete the number sentence.

1

$\frac{1}{4}$		$\frac{1}{4}$		$\frac{1}{4}$		$\frac{1}{8}$
$\frac{1}{8}$	$\frac{1}{8}$	$\frac{1}{8}$	$\frac{1}{8}$	$\frac{1}{8}$	$\frac{1}{8}$	$\frac{1}{8}$

$\frac{3}{4} + \frac{1}{8} = \blacksquare$

2

$\frac{1}{10}$	$\frac{1}{10}$	$\frac{1}{10}$	$\frac{1}{10}$	$\frac{1}{10}$	$\frac{1}{10}$	$\frac{1}{10}$	$\frac{1}{5}$	
$\frac{1}{10}$	$\frac{1}{10}$	$\frac{1}{10}$	$\frac{1}{10}$	$\frac{1}{10}$	$\frac{1}{10}$	$\frac{1}{10}$	$\frac{1}{10}$	$\frac{1}{10}$

$\frac{7}{10} + \frac{1}{5} = \blacksquare$

3

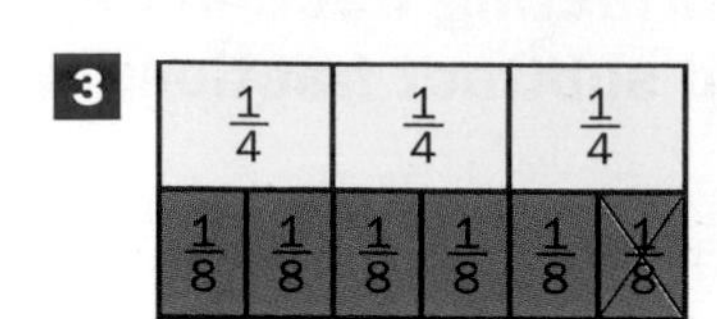

$\frac{3}{4} - \frac{1}{8} = \blacksquare$

Find the equivalent fraction. Then add or subtract.

4 $\frac{2}{9} = \frac{\blacksquare}{9}$, $+\ \frac{2}{3} = \frac{\blacksquare}{9}$, $\frac{\blacksquare}{9}$

5 $\frac{1}{5} = \frac{\blacksquare}{10}$, $+\ \frac{4}{10} = \frac{\blacksquare}{10}$, $\frac{\blacksquare}{10}$

6 $\frac{3}{5} = \frac{\blacksquare}{15}$, $+\ \frac{1}{3} = \frac{\blacksquare}{15}$, $\frac{\blacksquare}{15}$

7 $\frac{2}{3} = \frac{\blacksquare}{9}$, $+\ \frac{5}{9} = \frac{\blacksquare}{9}$, $\frac{\blacksquare}{9}$

8 $\frac{4}{5} = \frac{\blacksquare}{10}$, $-\ \frac{5}{10} = \frac{\blacksquare}{10}$, $\frac{\blacksquare}{10}$

9 $\frac{5}{8} = \frac{\blacksquare}{8}$, $-\ \frac{1}{2} = \frac{\blacksquare}{8}$, $\frac{\blacksquare}{8}$

10 $\frac{5}{9} = \frac{\blacksquare}{9}$, $-\ \frac{1}{3} = \frac{\blacksquare}{9}$, $\frac{\blacksquare}{9}$

11 $\frac{1}{2} = \frac{\blacksquare}{6}$, $-\ \frac{1}{6} = \frac{\blacksquare}{6}$, $\frac{\blacksquare}{6}$

Add or subtract using any method.

12 $\frac{9}{10} + \frac{3}{5}$

13 $\frac{7}{8} - \frac{1}{2}$

14 $\frac{7}{12} - \frac{2}{6}$

15 $\frac{1}{3} + \frac{5}{6}$

16 $\frac{1}{5} + \frac{3}{10}$

17 $\frac{1}{3} + \frac{5}{12}$

18 $\frac{3}{10} + \frac{8}{10}$

19 $\frac{3}{4} + \frac{2}{8}$

20 $\frac{7}{8} - \frac{1}{8}$

21 $\frac{11}{12} - \frac{2}{3}$

22 $\frac{5}{6} - \frac{1}{2}$

23 $\frac{1}{2} - \frac{1}{6}$

24 $\frac{2}{3} - \frac{1}{6}$

25 $\frac{2}{3} - \frac{5}{12}$

26 $\frac{4}{5} - \frac{3}{10}$

27 $\frac{7}{8} - \frac{3}{4}$

Make It Right

28 This is how Tamika found $\frac{5}{12} - \frac{1}{3}$. Tell what mistake was made, then correct it.

$\frac{5}{12} - \frac{1}{3} = \frac{4}{9}$

MIXED APPLICATIONS
Problem Solving

Solve. Use any method you choose.

29 Orville gave his little brother $\frac{7}{8}$ cup of raisins. His brother ate $\frac{1}{4}$ cup for a snack and $\frac{1}{2}$ cup for lunch. He brought the rest of the raisins home. How much did he bring home? Explain your thinking.

30 **Make a decision** Jan has $\frac{11}{12}$ yd of lace. She needs $\frac{1}{6}$ yd for her shirt collar, $\frac{1}{2}$ yd for her skirt, $\frac{3}{4}$ yd for her dress, and $\frac{1}{4}$ yd for her pants. She wants to add the lace to as many clothes as she can. Which should she add lace to? Explain.

31 **Write a problem** that has $\frac{7}{12}$ as the answer. Compare your problem with those of other classmates. How are they similar? different?

32 **Data Point** Use the Databank on page 542. Draw a graph to show the data. Explain how you decided what graph to choose.

more to explore

Estimating Sums and Differences of Fractions

You can use a number line to estimate the sum or difference of two fractions.

Decide if the fraction is close to 0, $\frac{1}{2}$, or 1. Then add or subtract your estimates.

Estimate: $\frac{3}{4} + \frac{1}{16}$

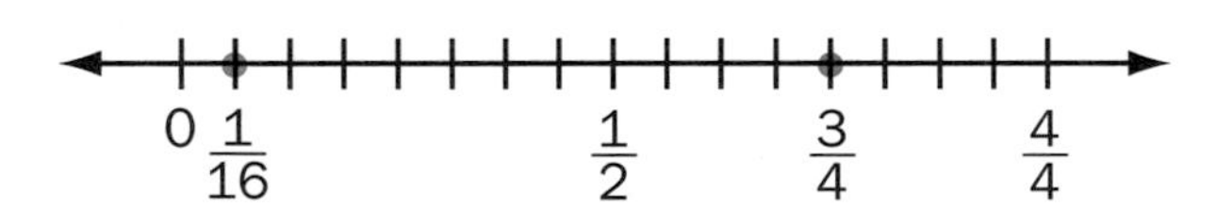

Think: $\frac{3}{4}$ is closer to 1.
$\frac{1}{16}$ is closer to 0.
$1 + 0 = 1$

$\frac{3}{4} + \frac{1}{16}$ is about 1.

Estimate: $\frac{11}{12} - \frac{4}{6}$

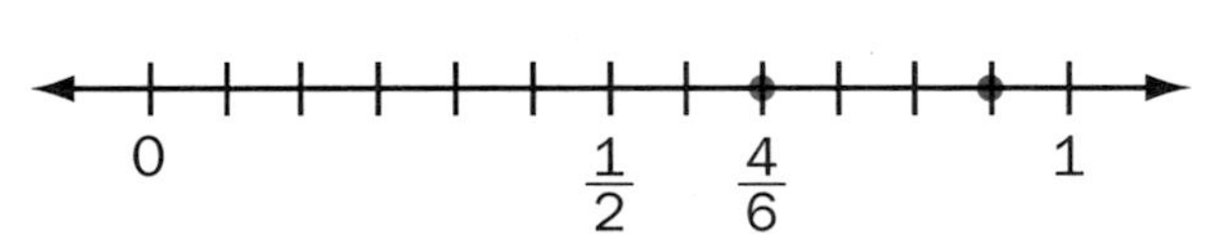

Think: $\frac{11}{12}$ is about 1.
$\frac{4}{6}$ is close to $\frac{1}{2}$.
$1 - \frac{1}{2} = \frac{1}{2}$

$\frac{11}{12} - \frac{4}{6}$ is about $\frac{1}{2}$.

Estimate the sum or difference.

1 $\frac{14}{16} + \frac{7}{8}$

2 $\frac{3}{5} + \frac{4}{7}$

3 $\frac{4}{5} - \frac{3}{4}$

4 $\frac{9}{10} - \frac{1}{16}$

Extra Practice, page 526

Problem-Solving Strategy

Read
Plan
Solve
Look Back

Draw a Picture

LEARN

Read **Suppose you are designing packages for a new line of frozen dinners. The rectangular trays are divided into sections. Plan $\frac{1}{3}$ of the tray for potatoes, $\frac{3}{8}$ for a main course, $\frac{1}{12}$ for green beans, and $\frac{1}{12}$ for squash. How much of the tray is available for a dessert?**

Plan To solve the problem, you can draw a picture.

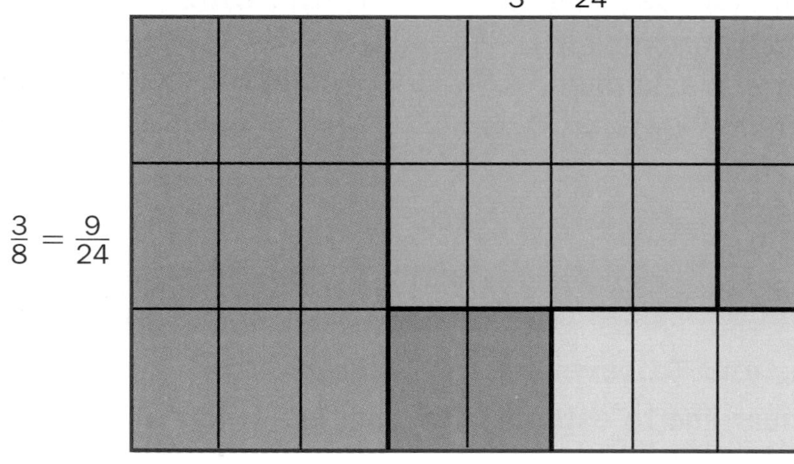

Solve A common denominator is 24, so divide the tray into 24 sections. When you fill in the portions, $\frac{3}{24}$, or $\frac{1}{8}$, is left empty.

$\frac{1}{8}$ of the tray is available for dessert.

Look Back What other way could you have solved this problem?

CHECK

Check for Understanding

1 **What if** the space for potatoes changes to $\frac{1}{4}$ of the tray. Now, how much will be left for the dessert?

Critical Thinking: Generalize

2 Which kinds of problems will you most likely solve by using the strategy of drawing a picture?

MIXED APPLICATIONS

Problem Solving

PRACTICE

1 There are 175 students who want to see a play. The theater has rows of seats with 6 seats in each row. About how many rows will the students need to reserve to see the play? Explain your thinking.

2 The Turners framed their family portrait. The framed portrait is 12 inches wide and 16 inches long. The frame is 1 inch wide all around. What is the area of the picture without the frame?

3 Jacy writes the first ten letters of the alphabet on separate index cards. He turns them over and chooses one card. Which did he most likely choose: a vowel, a consonant that comes after *D*, or a consonant that comes before *D*? Explain your method.

4 **What if** the principal arranges to buy popcorn for the 175 students. There are 30 boxes with 6 bags of popcorn in each box. Use mental math to determine if there is enough popcorn for each student to get a bag. Explain your methods.

5 Ms. Hopkins is storing 205 baking pans in two cupboards. Each section of one cupboard holds 20 pans, and each section of the other cupboard holds 25 pans. She wants to fill each section that is used. How many sections in each cupboard will she use to store the pans?

6 The school fair has a game in which you must throw beanbags into different-color boxes. A bag in the red box is worth 8 points, blue is 12, and yellow is 24. To win, you must score exactly 48 points. You get up to 5 throws to win. How many different ways can you win? Explain your thinking.

7 There are three food trays for a breakfast buffet. Each tray contains either eggs, potatoes, or fruit. The trays are placed in a row on the table. How many different ways can the trays be arranged on the table?

8 Cora, Kurt, Nathan, and Toni are sitting beside each other at a play. Nathan is between Cora and Kurt. Kurt is at the end of the row. In what order are they sitting?

9 Suppose a serving bowl is divided into 12 equal sections. The first $\frac{1}{4}$ of the bowl is for olives, $\frac{1}{3}$ is for carrot sticks, $\frac{1}{6}$ for celery sticks, and $\frac{1}{12}$ is for radishes. How much of the bowl is left for the dip?

Extra Practice, page 526

Add and Subtract Mixed Numbers

Sometimes, when you add and subtract with fractions, the numbers you have to use are mixed numbers.

Work Together

Work with a partner to add and subtract mixed numbers.

You will need
- *fraction strips*
- *index cards*

Write these numbers on index cards:
$1\frac{1}{12}$, $1\frac{3}{12}$, $1\frac{5}{12}$, $1\frac{7}{12}$, $1\frac{11}{12}$, $2\frac{1}{12}$,

$2\frac{5}{12}$, $2\frac{7}{12}$, $2\frac{11}{12}$, $3\frac{5}{12}$, $3\frac{7}{12}$, $3\frac{11}{12}$.

Mix up the cards. Choose two cards. Use fraction strips to find the total. Record your work in a table.

Choose another pair of cards. Use fraction strips to find the difference. Record your work in another table.

Repeat the activity five times.

KEEP IN MIND
- You may have to find equivalent fractions.
- You may need to regroup whole numbers or fractions.

Card 1	Card 2	Total
$2\frac{5}{12}$	$2\frac{11}{12}$	

Card 1	Card 2	Difference
$2\frac{7}{12}$	$1\frac{5}{12}$	

Talk It Over

- What methods did you use to add and subtract mixed numbers?
- How did you know when to regroup the fraction part of a mixed number?

mixed number A number that has a whole number and a fraction.

Make Connections

You can use fraction strips to add and subtract mixed numbers.

Add: $1\frac{7}{12} + 1\frac{3}{12}$

Step 1

Model each mixed number.

$$\begin{array}{r} 1\frac{7}{12} \\ +\ 1\frac{3}{12} \\ \hline \end{array}$$

1

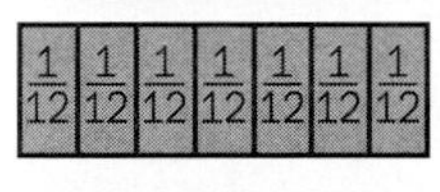

1

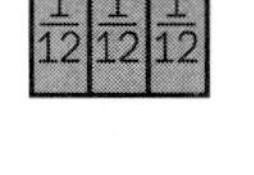

Step 2

Add. Simplify the sum if necessary.

$$\begin{array}{r} 1\frac{7}{12} \\ +\ 1\frac{3}{12} \\ \hline 2\frac{10}{12} \end{array} = 2\frac{5}{6}$$

1

1

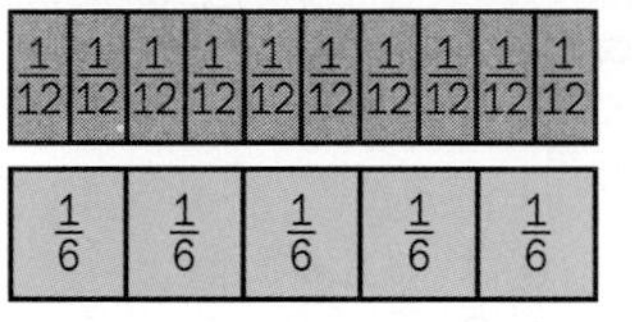

Subtract: $2\frac{7}{12} - 1\frac{5}{12}$

Step 1

Model the greater mixed number.

$$\begin{array}{r} 2\frac{7}{12} \\ -\ 1\frac{5}{12} \\ \hline \end{array}$$

1

1

1/12 1/12 1/12 1/12 1/12 1/12 1/12

Step 2

Subtract. Simplify the sum if necessary.

$$\begin{array}{r} 2\frac{7}{12} \\ -\ 1\frac{5}{12} \\ \hline 1\frac{2}{12} \end{array} = 1\frac{1}{6}$$

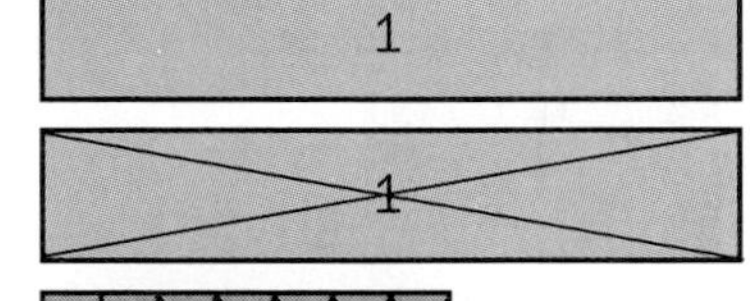

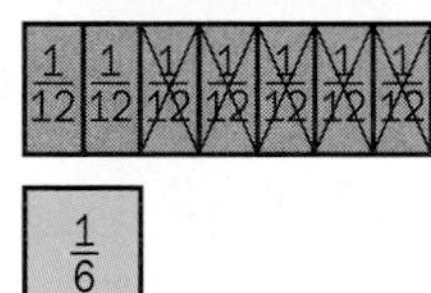

Check for Understanding

Add or subtract.

1 $3\frac{3}{8} + 2\frac{1}{8}$ **2** $1\frac{5}{12} + 2\frac{3}{12}$ **3** $1\frac{4}{12} + 1\frac{10}{12}$ **4** $2\frac{4}{6} + 1\frac{2}{6}$

5 $5\frac{9}{10} - 3\frac{6}{10}$ **6** $3\frac{5}{6} - 2\frac{2}{6}$ **7** $3\frac{5}{8} - 1\frac{1}{8}$ **8** $2\frac{3}{4} - 2\frac{1}{4}$

CHECK

Critical Thinking: Analyze **Explain your reasoning.**

9 Do you always need to rename when you add mixed numbers? Show an example.

Turn the page for Practice.

Practice

Complete the number sentence.

1 $2\frac{3}{5} + 1\frac{1}{5} = 3\frac{■}{5}$

1

$\frac{1}{5}$ $\frac{1}{5}$ $\frac{1}{5}$

1

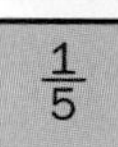

2 $3\frac{4}{10} - 3\frac{2}{10} = \frac{■}{10} = \frac{■}{5}$

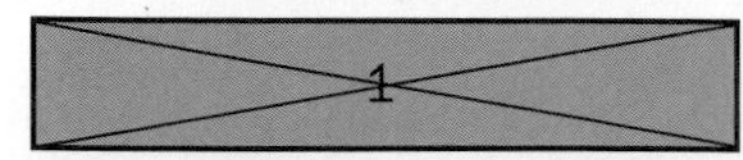

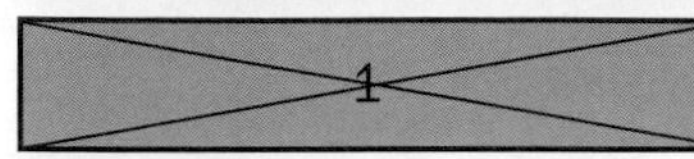

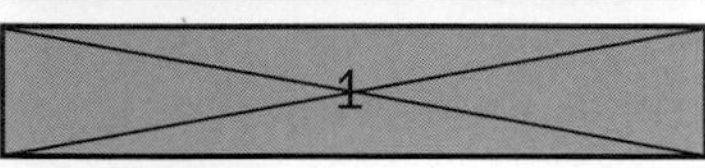

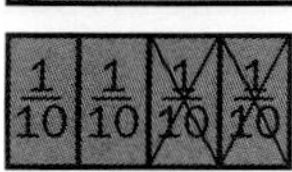

$\frac{1}{5}$

3 $1\frac{3}{8} + 1\frac{4}{8} = 2\frac{■}{8}$

$\frac{1}{8}$ $\frac{1}{8}$ $\frac{1}{8}$

1

$\frac{1}{8}$ $\frac{1}{8}$ $\frac{1}{8}$ $\frac{1}{8}$

4 $3\frac{2}{3} - 1\frac{1}{3} = 2\frac{■}{3}$

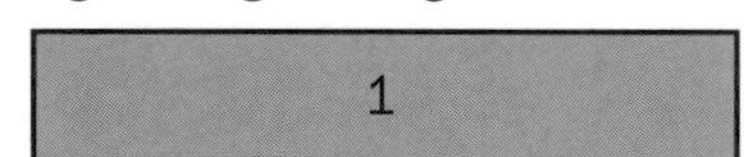

1

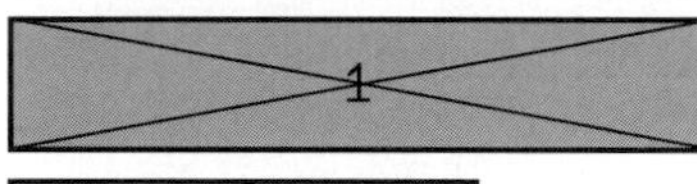

$\frac{1}{3}$ $\frac{1}{3}$

ALGEBRA **Complete the table. Write the answers in simplest form.**

5

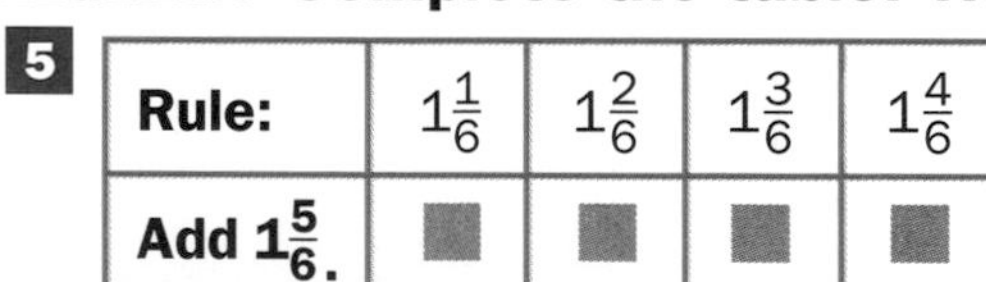

Rule:	$1\frac{1}{6}$	$1\frac{2}{6}$	$1\frac{3}{6}$	$1\frac{4}{6}$
Add $1\frac{5}{6}$.	■	■	■	■

6

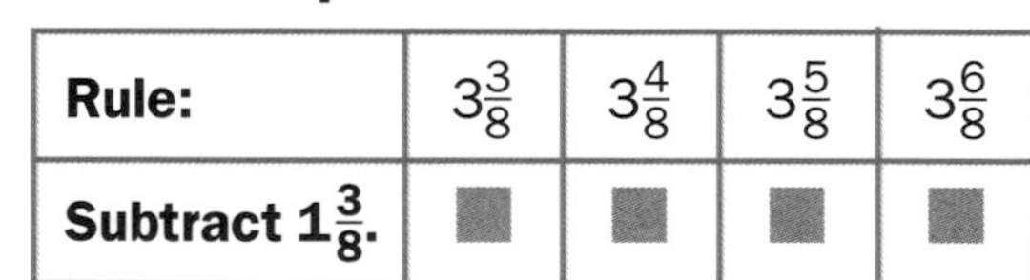

Rule:	$3\frac{3}{8}$	$3\frac{4}{8}$	$3\frac{5}{8}$	$3\frac{6}{8}$
Subtract $1\frac{3}{8}$.	■	■	■	■

Add or subtract. Rename and simplify when necessary.

7 $2\frac{1}{5} + 1\frac{3}{5}$

8 $2\frac{4}{12} + 1\frac{5}{12}$

9 $1\frac{3}{10} + 2\frac{8}{10}$

10 $2\frac{7}{8} + 1\frac{7}{8}$

11 $4\frac{6}{8} - 3\frac{1}{8}$

12 $4\frac{11}{12} - 2\frac{2}{12}$

13 $1\frac{5}{6} - 1\frac{3}{6}$

14 $4\frac{3}{4} - 2\frac{1}{4}$

Make It Right

15 Sid used fraction strips to add $3\frac{3}{4} + 2\frac{2}{4}$.
Explain what the mistake is, then correct it. $3\frac{3}{4} + 2\frac{2}{4} = 5\frac{1}{4}$

MIXED APPLICATIONS
Problem Solving

16 Pablo made $2\frac{3}{4}$ cups of guacamole and $3\frac{1}{2}$ pounds of ground beef. After his family ate, there were $1\frac{1}{4}$ cups of guacamole left and $2\frac{1}{2}$ pounds of ground beef. How much of each did his family eat?

17 **Make a decision** George spent $1\frac{1}{2}$ h doing math homework and $1\frac{1}{2}$ h reading. Meg spent $1\frac{1}{4}$ h doing math and $\frac{3}{4}$ h reading. Who spent more time? Explain.

18 Suppose vowels are worth 9¢, consonants are worth 4¢, and words beginning with *R* are worth an extra 7¢. Find three words that are worth more than 60¢.

19 **Write a problem** in which you must either add or subtract mixed numbers with like denominators in order to solve it. Use fraction strips to solve it.

20 Rosalita is making a tropical fruit punch for a birthday party. She uses a recipe that mixes $6\frac{3}{4}$ cups of pineapple juice with $5\frac{1}{4}$ cups of guava juice. How many cups of tropical punch will there be for the party?

21 Michael is making a bowl of mixed nuts as a snack for family and friends. He has $3\frac{7}{8}$ pounds of peanuts and $2\frac{3}{8}$ pounds of Brazil nuts. Which type of nut does he have more of? How many more pounds of it does he have?

more to explore

Estimate Sums and Differences of Mixed Numbers

You can round to estimate the sum or difference of two mixed numbers.

Estimate: $3\frac{3}{4} + 2\frac{1}{7}$

Think: $\frac{3}{4} > \frac{1}{2}$ Round $3\frac{3}{4}$ up to 4.

$\frac{1}{7} < \frac{1}{2}$ Round $2\frac{1}{7}$ down to 2.

$3\frac{3}{4} + 2\frac{1}{7}$ is about 4 + 2, or 6.

Estimate: $5\frac{3}{8} - 2\frac{1}{16}$

Think: $\frac{3}{8} < \frac{1}{2}$ Round $5\frac{3}{8}$ down to 5.

$\frac{1}{16} < \frac{1}{2}$ Round $2\frac{1}{16}$ down to 2.

$5\frac{3}{8} - 2\frac{1}{16}$ is about 5 − 2, or 3.

Estimate the sum or difference.

1 $4\frac{12}{13} + 3\frac{5}{6}$

2 $1\frac{1}{15} + 2\frac{1}{7}$

3 $\frac{15}{16} + 3\frac{1}{12}$

4 $2\frac{8}{9} + \frac{1}{10}$

5 $5\frac{4}{5} - \frac{6}{7}$

6 $4\frac{2}{3} - 1\frac{7}{8}$

7 $1\frac{8}{9} - 1\frac{1}{16}$

8 $3\frac{3}{4} - 1\frac{1}{16}$

Extra Practice, page 527

Problem Solvers at Work

Read
Plan
Solve
Look Back

Part 1 Choose the Operation

LEARN

This table shows some of the foods that were brought to a party to celebrate United Nations Day. Most of the food was eaten, but there were some leftovers.

Kind of Food/Country	Amount at Beginning of Party	Amount at End of Party
Taco salad/Mexico	2 bowls	$\frac{2}{3}$ bowl
Banana bread/Caribbean countries	3 loaves	$\frac{1}{2}$ loaf
Baked bananas/African countries	2 trays	$\frac{5}{8}$ tray
Paella/Spain	3 trays	$\frac{1}{8}$ tray
Vegetable rice/Japan	2 bowls	$\frac{1}{3}$ bowl

Work Together

Solve. Tell which operation you used.

1 How many bowls of taco salad and vegetable rice were eaten altogether?

2 Next time, the party organizers will bring in twice as many loaves of banana bread. How many loaves will they bring in?

3 How much more baked bananas than paella was left at the end of the party?

4 **What if** the banana bread was cut up into a total of 72 slices and there were 24 people. What is the greatest number of slices each person could eat if they all got the same amount?

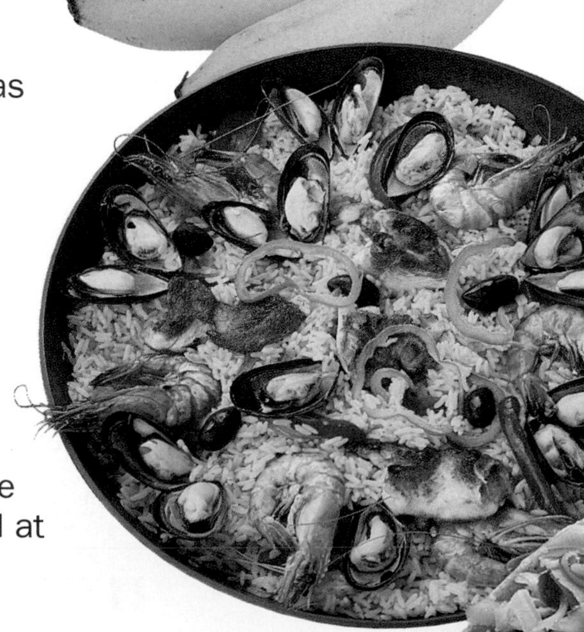

5 READING ARITHMETIC WRITING **Write a Report** Write a report about a food on the table. Tell what the food is like. Tell how much was served at the party and how much was eaten.

Part 2 Write and Share Problems

Morgan used the information in the table to write a problem.

Food	Preparation Time	Cooking Time
Meat loaf	$\frac{1}{2}$ hour	1 hour
Chili	$\frac{1}{2}$ hour	$2\frac{1}{2}$ hours
Pizza	$\frac{1}{4}$ hour	$\frac{1}{2}$ hour
Tomato soup	$\frac{1}{4}$ hour	$1\frac{1}{4}$ hours

CHECK

How much longer does it take chili to cook than pizza?

Morgan McLuen
Piney Grove Elementary School
Charlotte, NC

6 Solve Morgan's problem.

7 Change Morgan's problem so that you must use a different operation to solve it.

8 Solve your new problem. Explain why you must use a different operation to solve your problem than to solve Morgan's problem.

9 **Write a problem** of your own about foods from the table. Use foods other than those Morgan used. Solve your problem. Explain why you chose the operation you did to solve it.

10 Trade problems with another classmate. Solve each other's problems. Compare your solutions. Talk about why you chose the operations.

Turn the page for Practice Strategies.

Menu

Choose five problems and solve them. Explain your methods.

1 **Make a decision** You have 3 h to do any activities you want. Create a schedule to show which activities you will do and how long they will take.

ACTIVITY	TIME
Homework	
Baseball	

2 Mía is carrying $2\frac{3}{8}$ lb of apples and $1\frac{3}{8}$ lb of pears. Al is carrying $2\frac{1}{4}$ lb of nuts and 1 lb of grapes. Whose bag is heavier? How much heavier?

3 **Spatial reasoning** How many different ways can you get three postage stamps from the post office so that they are all attached?

4 Brendan said the alphabet as quickly as he could. How many seconds did it take him to say it?

A B C D E F G H I
J K L M N O P Q R
S T U V W X Y Z

5 Nathan woke up, took a shower for 15 min, got dressed in 15 min, ate breakfast in $\frac{1}{2}$ h, and then took $\frac{1}{2}$ h to walk to school. He got to school at 8:30 A.M. What time did he wake up?

6 A large pizza costs $8.95 for 8 slices. About how much will 4 large pies cost? If 16 students share them equally, how many slices will each student get?

7 **Logical reasoning** Lena bought $2\frac{3}{4}$ lb of cold cuts. She bought $\frac{1}{4}$ lb bologna, $\frac{1}{2}$ lb ham, $\frac{1}{2}$ lb roast beef, and some turkey. How much turkey did she buy?

8 **ALGEBRA: PATTERNS** In 1994, a school population was 300 students. The population was 305 in 1995, 315 in 1996, 330 in 1997, and 350 in 1998. If the same pattern continues, how many students will there be in this school in the year 2000?

Choose two problems and solve them. Explain your methods.

9 You have to fill a large container with exactly 2 gallons of water. The picture shows the containers you can use to measure. How will you use the containers to pour exactly 2 gallons of water? (Hint: 4 quarts = 1 gallon)

10 **What if** you have to buy paint to paint your classroom walls and ceiling. The paint you plan to buy covers 200 sq ft per gallon. How many gallons should you buy?

11 **Data Point** Survey your class or another class to find out their opinion about a specific television program. Have them rate the show by telling you whether they like the show, neither like nor dislike the show, or dislike the show. Draw a graph or make a chart displaying the results of your survey. Then write a statement about what your graph or table shows.

12 **At the Computer** You can use a graphing program to make circle graphs to represent data.

Study your school lunch menu for 20 school days. Estimate how often each meal is served. Then estimate the fraction of the total number of meals for each type of meal.

Use a graphing program to draw a circle graph showing the fraction of the total number of meals each type of meal represents.

Compare your estimate with the graph. How close was your estimate?

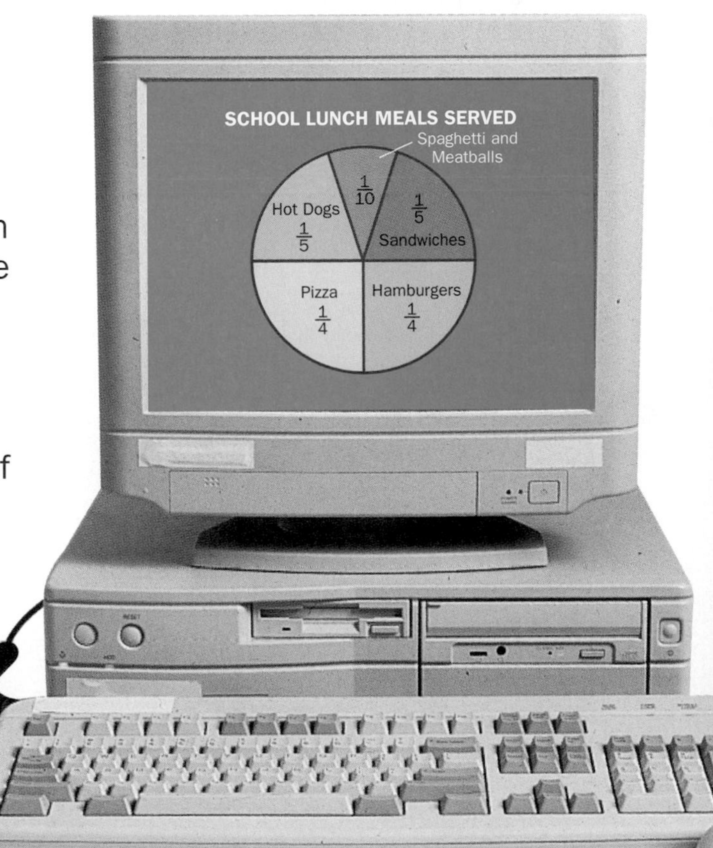

Extra Practice, page 527

Language and Mathematics

Complete the sentence. Use a word in the chart. (pages 410–437)

1 To find $\frac{3}{4} - \frac{1}{8}$, you change $\frac{3}{4}$ and $\frac{1}{8}$ into fractions with ■ denominators.

2 $\frac{2}{5}$ and $\frac{4}{10}$ are ■ fractions.

3 $1\frac{7}{9}$ is a ■.

4 $\frac{3}{5}$ and $\frac{3}{7}$ have ■ denominators.

Vocabulary
- unlike
- mixed number
- improper
- common
- equivalent

Concepts and Skills

Complete the number sentence. (page 410)

5 $2\frac{2}{5} + 2\frac{1}{5} = 4\frac{■}{5}$

6 $1\frac{3}{8} + 1\frac{4}{8} = 2\frac{■}{8}$

7 $2\frac{2}{3} - 1\frac{1}{3} = 1\frac{■}{3}$

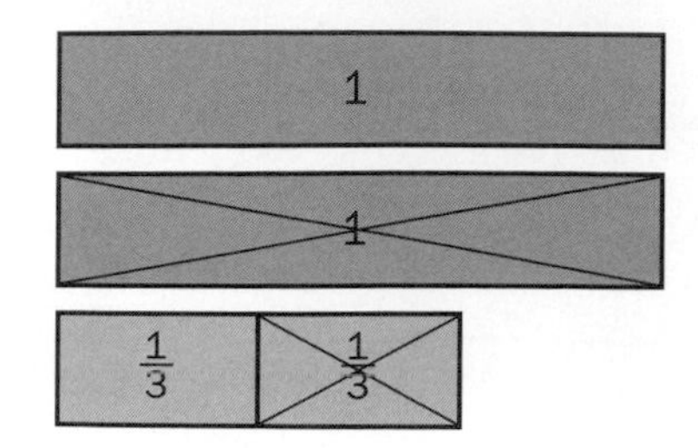

8 $2\frac{4}{5} - 1\frac{2}{5} = 1\frac{■}{5}$

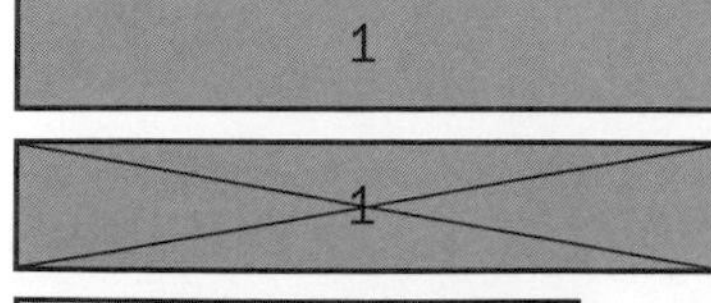

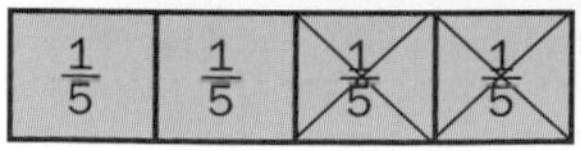

Write as equivalent fractions with common denominators. (page 426)

9 $\frac{1}{3}$ and $\frac{1}{4}$ **10** $\frac{2}{5}$ and $\frac{3}{10}$ **11** $\frac{1}{6}$ and $\frac{3}{4}$ **12** $\frac{2}{3}$ and $\frac{7}{12}$

Find the sum or difference. Write the answer in simplest form. (pages 410, 416)

13 $\frac{3}{8} + \frac{4}{8}$ **14** $\frac{1}{6} + \frac{3}{6}$ **15** $\frac{1}{12} + \frac{3}{12} + \frac{2}{12}$ **16** $\frac{2}{3} + \frac{2}{3}$

17 $\frac{5}{6} - \frac{4}{6}$ **18** $\frac{10}{16} - \frac{5}{16}$ **19** $\frac{9}{10} - \frac{4}{10}$ **20** $\frac{11}{12} - \frac{3}{12}$

Think critically. (page 410)

21 Analyze. Explain what the mistake is, then correct it.

$\frac{2}{3} + \frac{2}{3} = \frac{4}{3}$

Simplify: $\frac{4}{3} = 3\frac{1}{3}$

MIXED APPLICATIONS

Problem Solving

(pages 432, 438)

22 Rewrite the table so that the months are listed in order from least to greatest number of inches of precipitation.

Average Monthly Precipitation for Albany, New York

Selected Months	Inches
January	$2\frac{2}{5}$
February	$2\frac{3}{10}$
March	3
April	$2\frac{9}{10}$
May	$3\frac{3}{10}$

23 **Logical reasoning** Eli surveyed his class to find out their favorite foods. He reported these results to his teacher: $\frac{1}{4}$ like pizza best, $\frac{2}{4}$ like hot dogs best, and $\frac{3}{4}$ like cheeseburgers best. His teacher told him that his results could not be correct. How could his teacher know this?

24 Lonato must put a drape around the edge of a table for the science fair. The table is $3\frac{1}{4}$ ft long and $2\frac{3}{4}$ ft wide. How long a drape will Lonato need?

25 **What if** you start at the hiking station and hike 3 mi north, 5 mi east, 6 mi south, and 5 mi west. How far and in what direction would you have to hike to return to the station?

Find the sum or difference. Write the answer in simplest form.

1 $\frac{1}{8} + \frac{3}{8}$

2 $\frac{1}{6} + \frac{4}{6}$

3 $\frac{2}{10} + \frac{1}{10} + \frac{2}{10}$

4 $\frac{3}{4} + \frac{3}{4}$

5 $\frac{10}{12} - \frac{8}{12}$

6 $\frac{5}{6} - \frac{2}{6}$

7 $\frac{7}{8} - \frac{1}{8}$

8 $\frac{11}{16} - \frac{1}{16}$

9 $\frac{3}{4} + \frac{5}{8}$

10 $\frac{7}{12} + \frac{1}{3}$

11 $\frac{4}{5} - \frac{1}{10}$

12 $\frac{1}{2} - \frac{1}{6}$

Complete the number sentence.

13 $1\frac{5}{8} + 1\frac{2}{8} = 2\frac{■}{8}$

14 $1\frac{3}{12} + 1\frac{4}{12} = 2\frac{■}{12}$

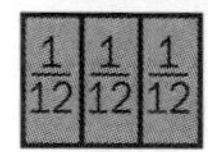

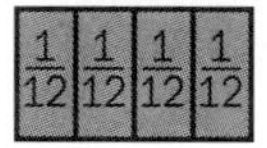

15 $2\frac{3}{4} - 1\frac{2}{4} = 1\frac{■}{4}$

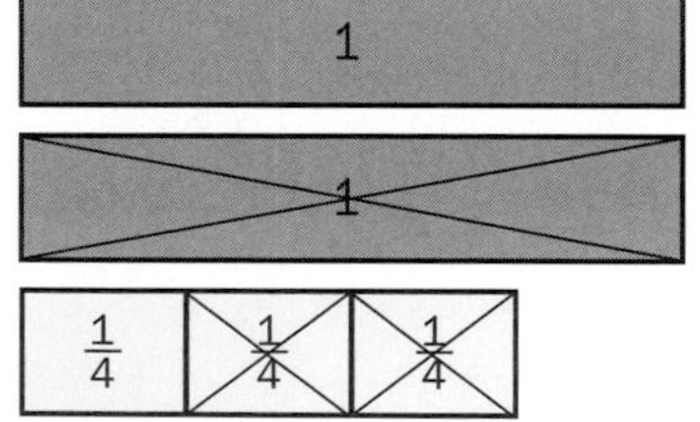

16 $2\frac{3}{5} - 1\frac{1}{5} = 1\frac{■}{5}$

Solve.

17 The store has 6 more bags of unsalted corn chips than bags of salted corn chips. If there are 30 bags altogether, how many contain unsalted corn chips?

18 Mrs. Gomez bought 5 lb of potatoes. She cooked $1\frac{1}{2}$ lb on Monday for dinner and $\frac{1}{2}$ lb on Tuesday for lunch. How many pounds of potatoes are left?

19 The Wongs ordered 2 pizzas. Each of the 4 family members ate $\frac{3}{8}$ of a pizza. How much pizza was left over?

20 Bo draws a rectangle that is 10 in. long and 8 in. wide. She draws 2 lines to divide it into 4 rectangles. What are the length and width of each of the smaller rectangles?

What Did You Learn?

Use the chart to plan the meals for a day.

▶ Find the total amount for each type of food. Use models or drawings to show your work.

▶ Write two statements that compare the amounts of vegetables and fruits in your meals to the amount of cereal, rice, and pasta. Use fractions in your answer.

The meals should contain:

- ❑ 2-3 servings of milk or yogurt
- ❑ 2-3 servings of meat or beans
- ❑ 3-5 servings of vegetables
- ❑ 2-4 servings of fruit
- ❑ 6-11 servings of bread, cereal, rice, or pasta

Serving Size of Various Foods			
Milk or yogurt	1 cup	Bread	1 slice
Cooked beans	$1\frac{1}{2}$ cups	Chopped vegetables	$\frac{1}{2}$ cup
Meat, poultry, fish	2 ounces	Fruits	$\frac{1}{3}$ cup
Cereal, rice, pasta	$\frac{1}{2}$ cup	Fruit or vegetable juice	$\frac{3}{4}$ cup

U.S. Department of Agriculture

A Good Answer

- provides meals with the servings requested
- uses models or drawings to show the totals
- correctly compares the amounts

You may want to place your work in your portfolio.

What Do You Think 

1 Can you add and subtract fractions as well as you add and subtract whole numbers? Explain.

2 What methods have you used to add and subtract fractions?
- Models
- Drawings
- Mental math
- Other. Explain.

3 Do you know when an answer can be simplified? Give examples.

math science technology
CONNECTION

Eating Right

Cultural Note

In 1757, a British doctor, James Lind, discovered that eating fruits and vegetables rich in vitamin C could prevent the disease known as scurvy. The British navy then required sailors to eat lemons and limes each day.

Nutrients are substances in foods that are used by your cells to create energy and to help your cells grow and repair themselves. There are six groups of nutrients—proteins, carbohydrates, fats, vitamins, minerals, and water.

Fats give the body long-lasting energy and help the body store vitamins and build tissue.

Proteins are used by the body for growth and to build and repair cells. They keep muscles, skin, hair, and nails healthy.

Vitamins are needed by the body to grow and function.

Minerals are used by the body to build new cells and control important body processes.

Carbohydrates are used by the body as the main source of energy.

Water helps to keep the body temperature normal, dissolves some vitamins, and helps bring other nutrients to the cells in the body.

► Describe a healthy lunch. How is each group represented?

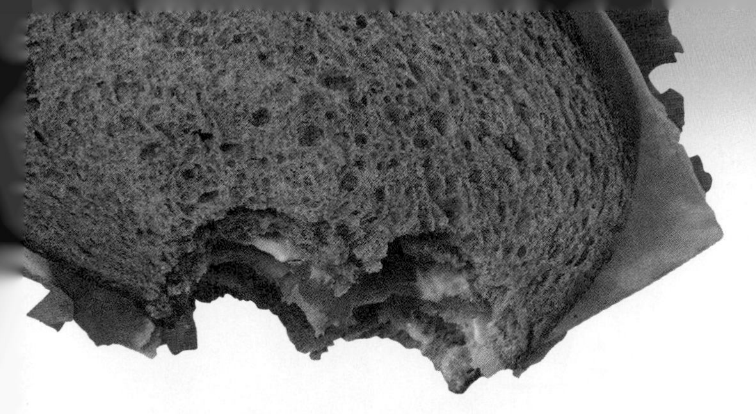

A Balanced Meal

The table below shows the fraction of the suggested amounts of iron, calcium, vitamin A, and vitamin C found in various foods.

Use the table for problems 1–3.

1 Which of the foods supplies the greatest amount of iron in each serving?

2 Which of the foods supplies the least amount of calcium?

3 Order from greatest to least by the amounts shown the vitamins and minerals in green beans.

At the Computer

4 Use a spreadsheet to keep track of the number of servings of each of the foods in the table you eat each day for a week. Also show the fraction of each type of vitamin and mineral.

5 Use a word processing program to write a short report that describes what the data shows.

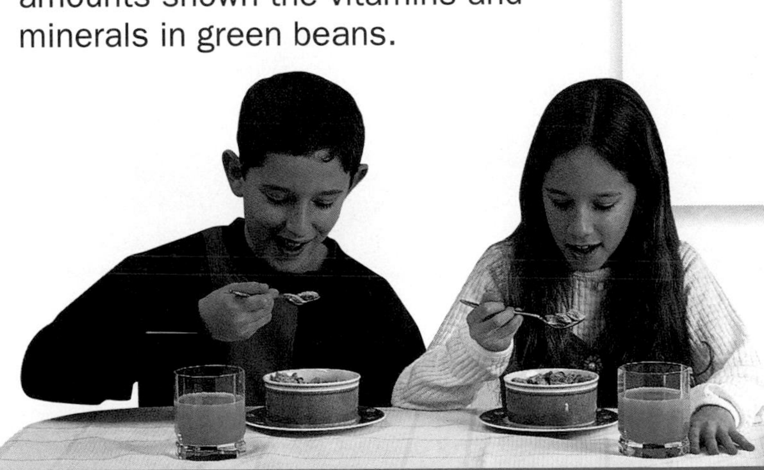

Vitamins and Minerals in One Serving (fraction of suggested amount)

	Cereal	Spaghetti	Green Beans	Tomato Soup	Skimmed Milk
Iron	$\frac{2}{5}$	$\frac{1}{10}$	$\frac{1}{25}$	$\frac{1}{25}$	0
Calcium	$\frac{2}{25}$	0	$\frac{1}{50}$	$\frac{1}{50}$	$\frac{3}{10}$
Vitamin A	0	0	$\frac{2}{25}$	$\frac{1}{10}$	$\frac{1}{10}$
Vitamin C	0	0	$\frac{1}{10}$	$\frac{3}{10}$	$\frac{1}{25}$

CHAPTER

12 DECIMALS

THEME

Communication

How many different forms of communication can you think of? In this chapter, you will discover how telephones, computers, and news reporters all use mathematics.

What Do You Know?

There are many unique ways to provide information about products or services. However, advertising by mail is still very popular.

Use the mail advertisements for problems 1–3.

1 Does Wheat All give more or less than 1 g of extra cereal in each box? Explain.

2 Wheat Chex gives $\frac{3}{10}$ g extra cereal in each box. How can you write $\frac{3}{10}$ as a decimal? Use a diagram to show your work.

3 Portfolio Choose at least two fractions from the advertisements. Then rename the fractions as decimals. Explain your work.

Write an Essay

Think about all the numbers in your life—your age, address, telephone number, and so on. Write an essay telling how these numbers give information about you.

An essay presents ideas about a topic.

1 What are some things these numbers tell about you?

2 Which of your numbers can you show as a decimal?

Vocabulary

decimal, p. 450 **decimal point,** p. 451 **mixed number,** p. 454

Decimals Less Than 1

LEARN

Have you ever painted on a computer screen? Painting programs let you pick a style and a pattern.

What part of this 10-by-10 square is painted with stripes? with dots?

You can use a fraction or a **decimal** to name the parts that are striped and dotted.

Part	Fraction	Decimal	Read
Stripes	$\frac{3}{10}$	0.3	three tenths
Dots	$\frac{45}{100}$	0.45	forty-five hundredths

Check Out the Glossary
decimal
decimal point
See page 544.

Work Together

Work with a partner to model these decimals. Use graph paper to create your own decimal squares.

a. two tenths
b. fifty-two hundredths
c. nine tenths
d. seventy-eight hundredths
e. nine hundredths
f. ninety hundredths

You will need
- *graph paper*
- *crayons, colored pencils, or markers*

Talk It Over

- How did you model tenths? hundredths?
- How many hundredths are in a tenth?
- How do your models for nine tenths and ninety hundredths compare? Why?

Make Connections

You can use a model or a place-value chart to help you understand and write a decimal.

Note: The **decimal point** separates the ones from the tenths.

Ones		Tenths	Hundredths
0	•	9	

Read: nine tenths
Write: 0.9

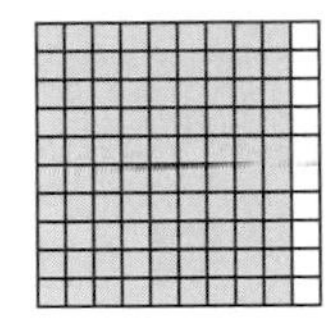

Ones		Tenths	Hundredths
0	•	7	8

Read: seventy-eight hundredths
Write: 0.78

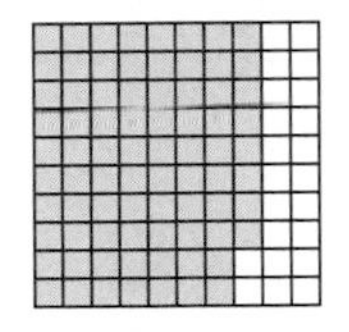

Ones		Tenths	Hundredths
0	•	0	9

Read: nine hundredths
Write: 0.09

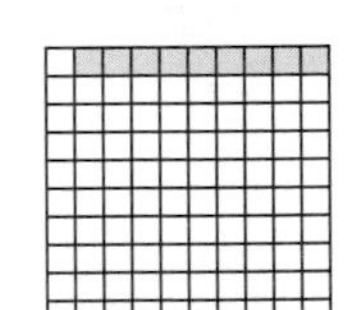

Ones		Tenths	Hundredths
0	•	9	0

Read: ninety hundredths
Write: 0.90

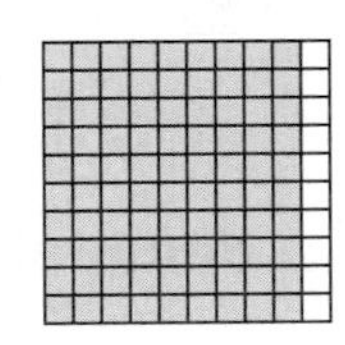

▶ When reading a decimal, which place name do you use?

▶ What do the zeros mean in the decimal 0.09?

Check for Understanding

Write a fraction and a decimal for the part that is shaded.

1 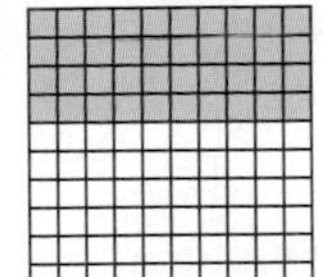 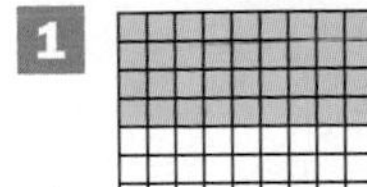

2 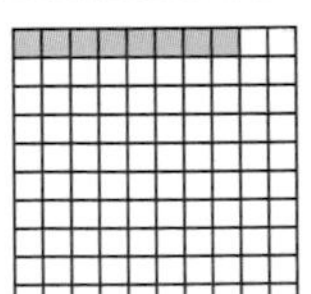 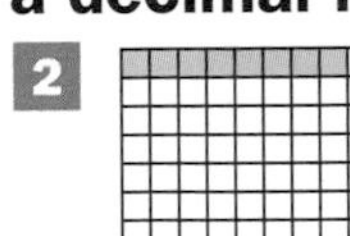

3 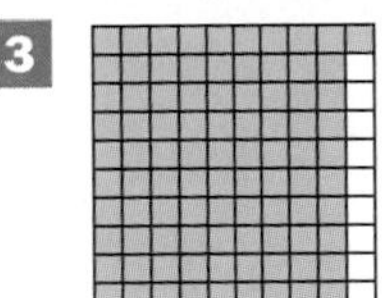 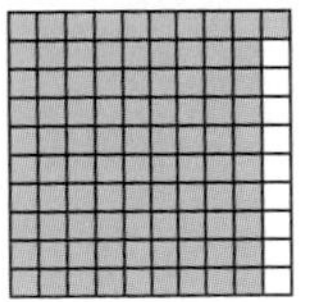

4 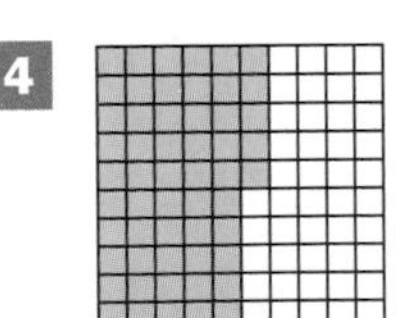

Write a decimal.

5 $\frac{1}{10}$ **6** $\frac{6}{10}$ **7** $\frac{60}{100}$ **8** $\frac{57}{100}$ **9** $\frac{98}{100}$ **10** $\frac{4}{100}$

11 seven tenths **12** twelve hundredths **13** five hundredths

Critical Thinking: Analyze **Explain your reasoning.**

14 **What if** a decimal square stands for one dollar. Describe the models that would show one dime and one penny.

CHECK

Turn the page for Practice.

Practice

Write the word name.

1

Ones		Tenths	Hundredths
0	•	2	8

2

Ones		Tenths	Hundredths
0	•	3	3

3

Ones		Tenths	Hundredths
0	•	7	

4

Ones		Tenths	Hundredths
0	•	0	5

5 0.5 **6** 0.4 **7** 0.08 **8** 0.37 **9** 0.17 **10** 0.93

Write a decimal for the part that is shaded.

11 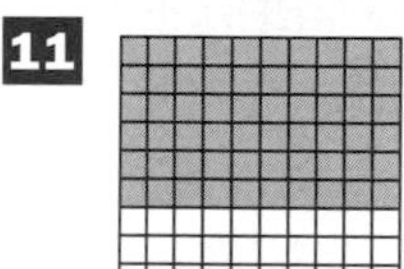

12 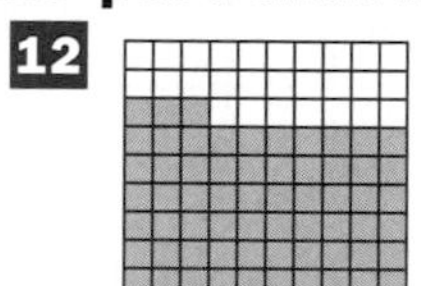

13 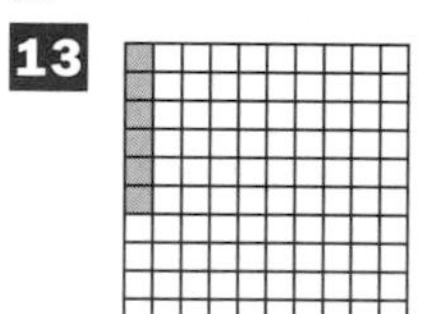

14 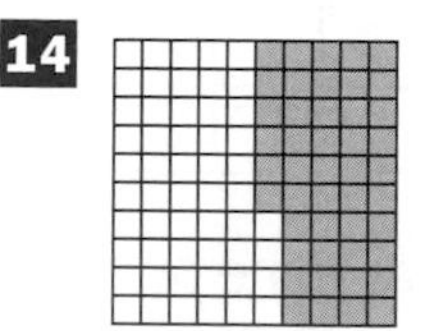

Write a decimal.

15 $\frac{7}{10}$ **16** $\frac{8}{10}$ **17** $\frac{49}{100}$ **18** $\frac{2}{100}$ **19** $\frac{93}{100}$ **20** $\frac{68}{100}$

21 $\frac{2}{10}$ **22** $\frac{9}{100}$ **23** $\frac{77}{100}$ **24** $\frac{5}{10}$ **25** $\frac{40}{100}$ **26** $\frac{7}{100}$

27 eight tenths **28** one tenth **29** twenty-one hundredths

30 seven hundredths **31** three tenths **32** ninety-nine hundredths

Match a money amount to its word name.

33 one tenth of a dollar **a.** \$0.45

34 forty-five hundredths of a dollar **b.** \$0.05

35 five tenths of a dollar **c.** \$0.10

36 five hundredths of a dollar **d.** \$0.25

37 one fourth of a dollar **e.** \$0.50

Make It Right

38 Marcel wrote a decimal for three hundedths. Explain the error and then correct it.

0.3

MIXED APPLICATIONS
Problem Solving

39 Susan painted this design on her computer screen. What decimal tells how much is painted with wavy lines?

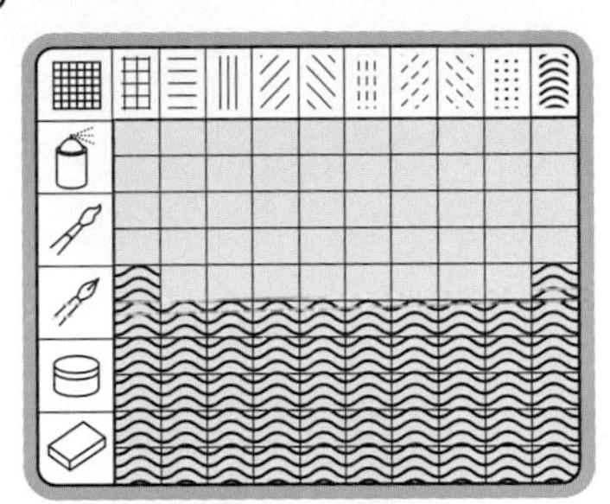

40 Calvin painted this design on his computer screen. What decimal tells how much he painted altogether?

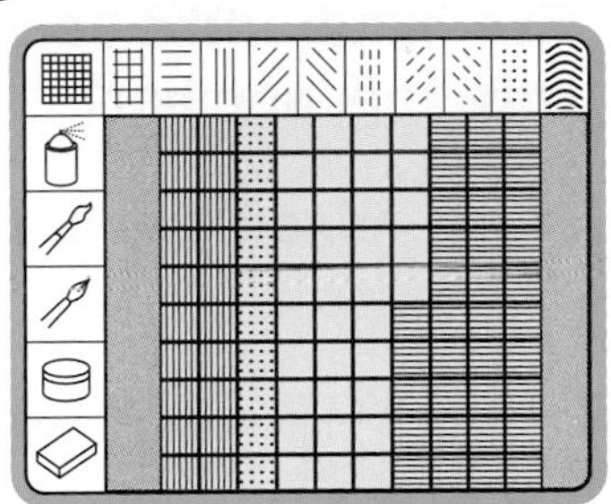

41 Marguerite is 3 years older than her sister, Janice. Janice is 6 years younger than her brother, Tommy, who is 1 year older than his best friend, Lee. If Lee is 14 years old, how old is Marguerite?

42 Rosa bought a painting program and a morphing program for her computer. Each cost the same amount. The total bill was $159. How much did each software program cost?

43 **Make a decision** A 3-hour tour of two museums costs $19 and starts in 1 hour. A 2-hour tour on space communication costs $10. Which tour would you take? Why?

44 **Write a problem** that is a set of directions for painting a 10-by-10 computer screen. Trade directions with a partner and draw each other's design.

more to explore

Thousandths

The first decimal square is divided into hundredths. Think of dividing each hundredth into 10 equal parts. So the second square shows thousandths.

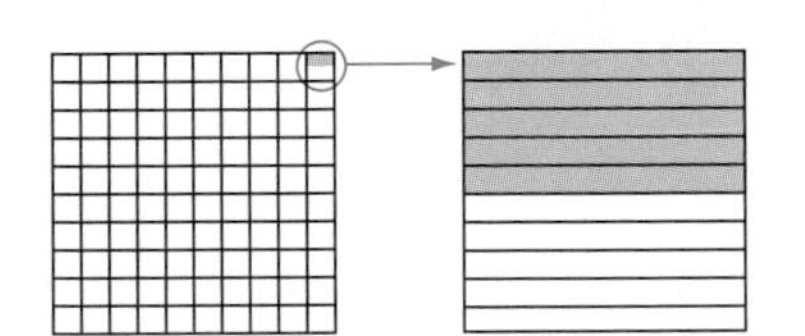

Ones		Tenths	Hundredths	Thousandths
0	.	0	0	5

Read: five thousandths

Fraction: $\frac{5}{1,000}$

Decimal: 0.005

▶ What do the zeros mean in 0.005?

Write a decimal.

1 $\frac{7}{1,000}$ **2** $\frac{9}{1,000}$ **3** $\frac{10}{1,000}$ **4** $\frac{25}{1,000}$ **5** $\frac{115}{1,000}$ **6** $\frac{502}{1,000}$

Extra Practice, page 528

Decimals Greater Than 1

LEARN

Dolphins make clicking noises underwater that bounce back if the sound hits an object. Suppose an echo returns in one and forty-five hundredths seconds. How can you write this number?

You can write a **mixed number** as a decimal.

Model

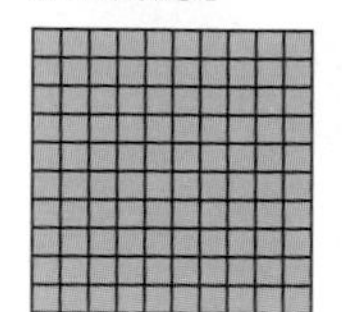
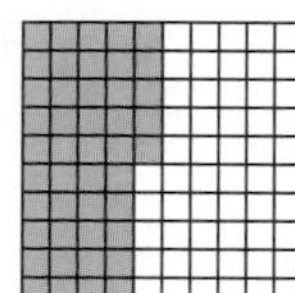

Decimal			
Ones		Tenths	Hundredths
1	•	4	5

mixed number A number that has a whole number and a fraction.

Mixed Number $1\frac{45}{100}$

Read: one *and* forty-five hundredths
Write: 1.45

More Examples

A $4\frac{7}{10}$

Decimal			
Ones		Tenths	Hundredths
4	•	7	

Read: four *and* seven tenths
Write: 4.7

B $72\frac{4}{100}$

Tens	Ones		Tenths	Hundredths
7	2	•	0	4

Read: seventy-two *and* four hundredths
Write: 72.04

CHECK

Check for Understanding

Write a mixed number and a decimal to tell how much is shaded.

1

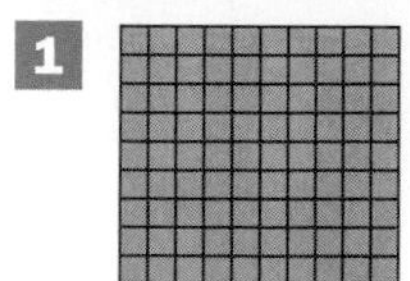
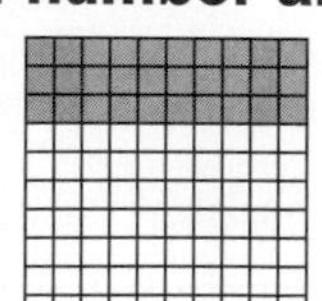

2

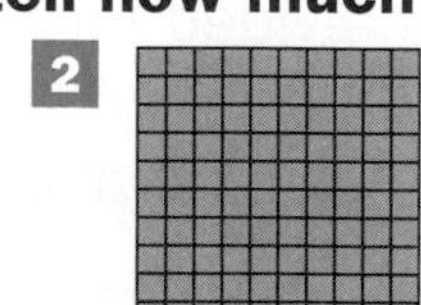
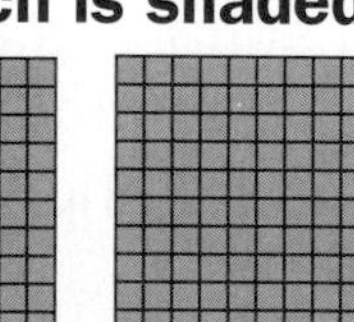
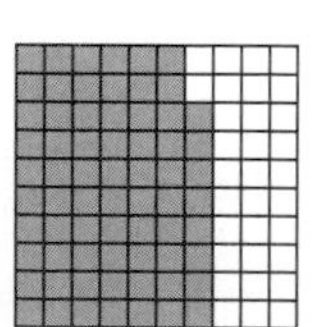

Write a decimal.

3 $8\frac{5}{10}$ **4** $4\frac{18}{100}$ **5** $1\frac{3}{100}$ **6** $7\frac{8}{10}$ **7** $12\frac{9}{100}$

Critical Thinking: Analyze **Explain your reasoning.**

8 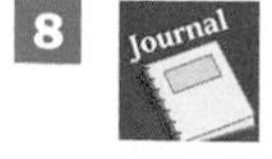

Do the two decimals name the same number? Include drawings of models with your explanations.

a. 3.5 and 3.05 **b.** 3.5 and 3.50

Practice

Write a mixed number and a decimal to tell how much is shaded.

1 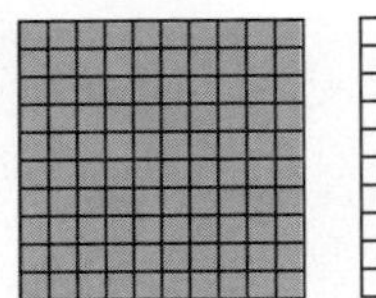 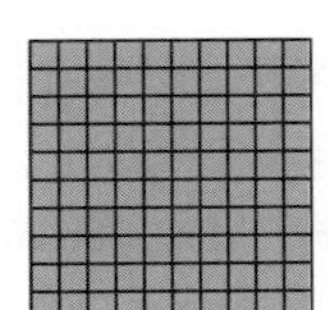

2  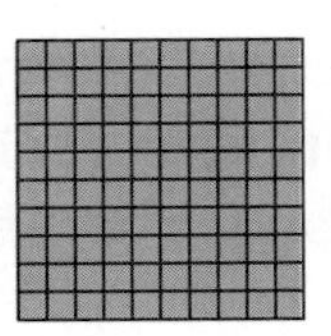 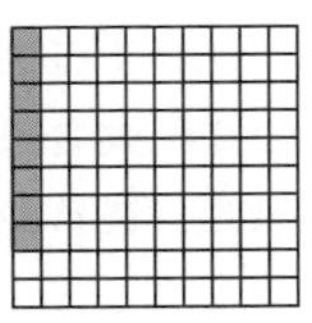

Write a decimal.

3 $9\frac{6}{10}$ 4 $6\frac{3}{100}$ 5 $42\frac{78}{100}$ 6 $6\frac{7}{10}$ 7 $15\frac{2}{10}$

8 $\frac{24}{100}$ 9 $2\frac{50}{100}$ 10 $1\frac{8}{10}$ 11 $7\frac{4}{100}$ 12 $28\frac{5}{10}$

13 $3\frac{18}{100}$ 14 $12\frac{1}{10}$ 15 $100\frac{3}{100}$ 16 $75\frac{99}{100}$ 17 $31\frac{43}{100}$

18 nineteen and 9 tenths

19 sixty-one and three tenths

20 fifty-two and eight hundredths

21 two and seventeen hundredths

Write the word name.

22 7.2 23 9.07 24 12.32 25 4.6 26 10.01 27 75.8

MIXED APPLICATIONS
Problem Solving

28 When you used the computer spell-check for your 100-word paragraph on dolphins, you found 14 misspelled words. What decimal shows the part of the words that were misspelled?

29 Nicholas wants to buy a baseball for \$3.95 and a catcher's mitt for \$99.95. He has saved \$84.50. How much more money does he need?

30 Fay watched a movie that lasted two and a half hours. What decimal shows how long the movie was?

31 **Write a problem** that uses one or more decimals greater than 1. Solve it and ask others to solve it.

mixed review • test preparation

1 $75 + 99$ 2 $175 + 325$ 3 $1,741 + 29$ 4 $748 - 329$ 5 $4,326 - 2,968$

6 3×924 7 $7 \times 1,045$ 8 29×87 9 $64 \div 8$ 10 $1,800 \div 20$

Extra Practice, page 528

Compare and Order Decimals

LEARN

The San Francisco earthquake in 1994 had a Richter-scale rating of 6.8. Ratings give us information about the strengths of earthquakes.

Compare and order the ratings of these big earthquakes.

Richter-Scale Ratings	
Portugal in 1775	8.6
California in 1906	8.3
China in 1920	8.5

You can use a number line to compare and order decimal

8.6 > 8.5, and 8.5 > 8.3

From greatest to least: 8.6, 8.5, 8.3
From least to greatest: 8.3, 8.5, 8.6

You can also compare and order decimals without a number line.

Compare and order 6.29, 7.27, and 6.25.

Step 1	Step 2	Step 3
Line up the decimal points.	**Compare the other two decimals.**	**Order the decimals.**
6.29 7.27 6.25	6.29 6.25	From greatest to least: 7.27, 6.29, 6.25
Think: 7 > 6, so 7.27 is the greatest decimal.	**Think:** 0.2 = 0.2 Compare hundredths. 0.09 > 0.05, so 6.29 > 6.25.	From least to greatest: 6.25, 6.29, 7.27

CHECK

Check for Understanding

Model each decimal. Then write in order from greatest to least.

1. 17.9, 18.6, 18.8
2. 4.64, 4.74, 4.7
3. \$2.36, \$2.28, \$2.39

Critical Thinking: Analyze **Explain your reasoning.**

4. If you compare two decimals, will the greater number always have more digits?

Practice

Compare. Write >, <, or =.

1 0.5 ● 0.50 **2** 3.2 ● 3.5 **3** 0.67 ● 0.68 **4** 3.9 ● 3.90

5 7.3 ● 7.37 **6** $9.63 ● $9.36 **7** 2.04 ● 2.4 **8** 54.18 ● 45.81

Write in order from greatest to least.

9 0.43, 0.34, 0.40 **10** 0.77, 0.70, 0.07 **11** 13.8, 13.3, 18.3

Write in order from least to greatest.

12 3.61, 3.09, 3.9 **13** 0.9, 0.93, 0.85 **14** 5.53, 15.05, 5.3

Problem Solving

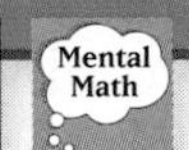

Use the table for problems 15–17.

15 Which earthquake had a rating of six and nine tenths?

16 Which earthquake was weaker than Flores Island, Indonesia, but stronger than California, United States?

17 Write the locations of the three strongest earthquakes in order from greatest to least.

Richter-Scale Ratings	
Kobe, Japan, in 1995	7.2
California, United States, in 1994	7.0
Flores Island, Indonesia, in 1992	7.5
Cabanatuan, Philippines, in 1990	7.5
Armenia in 1988	6.9
Mexico City, Mexico, in 1985	8.1

more to explore

Fractions on a Calculator

One way to compare fractions is to rename them as decimals. Divide the numerator by the denominator to find each decimal.

Compare $\frac{3}{10}$ and $\frac{6}{8}$.

$3 \div 10 =$ *0.3* $6 \div 8 =$ *0.75*

Since $0.75 > 0.3$, $\frac{6}{8} > \frac{3}{10}$.

Rename the fractions as decimals and then compare.

1 $\frac{6}{10}$ ● $\frac{12}{100}$ **2** $\frac{80}{100}$ ● $\frac{4}{5}$ **3** $\frac{9}{12}$ ● $\frac{7}{10}$ **4** $\frac{1}{4}$ ● $\frac{3}{10}$

Extra Practice, page 528

midchapter review

Write a decimal.

1 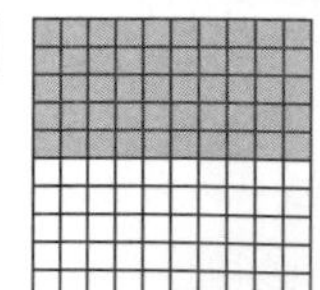

2 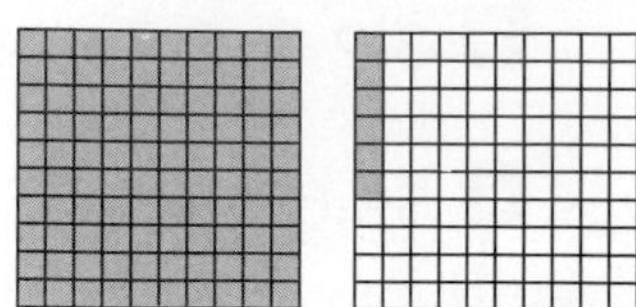

3 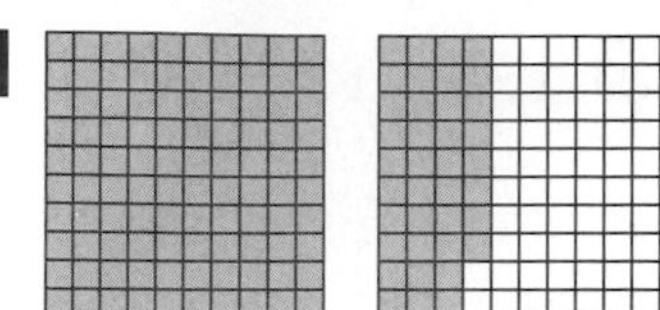

4 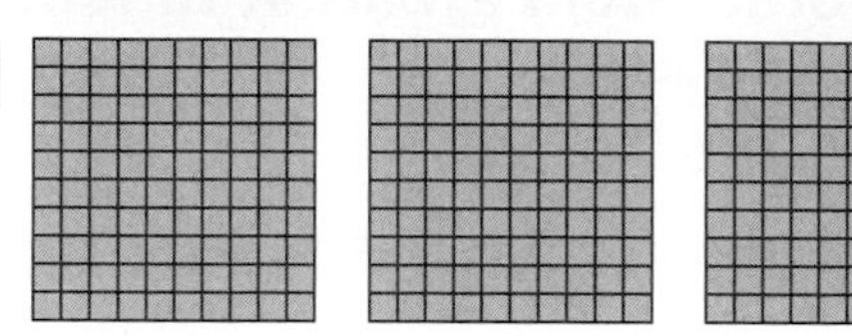

5 $\frac{46}{100}$

6 $2\frac{7}{10}$

7 $14\frac{93}{100}$

8 four tenths

9 six and ninety-three hundredths

Write the word name.

10 0.83

11 2.6

12 5.72

13 9.04

Compare. Write >, <, or =.

14 4.52 ● 4.25

15 3.08 ● 3.80

16 5.70 ● 5.7

17 2.23 ● 2.32

18 14.26 ● 4.62

19 9.2 ● 9.02

Write in order from greatest to least.

20 0.86, 0.83, 0.94

21 1.39, 1.93, 1.43

22 2.46, 2.44, 2.4

Solve.

23 Suppose your family phone bill shows that your brother's most expensive call cost $15.73. Your sister's most expensive call cost $14.92. Whose most expensive call cost more?

24 Six schools compare the average number of snow days they take each year: 3.3, 3.2, 3.5, 3.3, 3.6, and 3.4 days. List the averages from least to greatest.

25 Journal Write three different decimals using the digits 5, 0, and 8. Explain how you would order the decimals you wrote.

developing algebra sense

MATH CONNECTION

Graph a Function

Have you watched cable TV? If so, you probably know that your family can order "Pay-per-View" movies.

Suppose each movie costs $2. How much would it cost a family to order 4 movies? 5 movies?

ALGEBRA You can multiply. You can also use a function table and a graph to "see" what the function looks like.

Rule: Multiply by 2. (Cost of Pay-per-View Movies)	
Input (Number of Movies)	**Output (Cost)**
1	$2
2	$4
3	$6
4	■
5	■

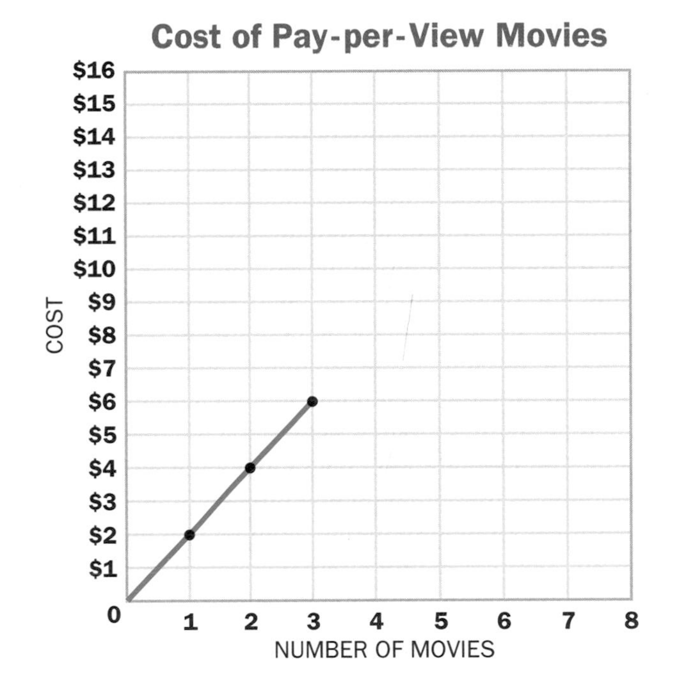

Graph the ordered pairs from the function table. Then draw a line through and beyond the points. Use the line to find 1×2, 2×2, 3×2, 4×2, and 5×2.

Four movies cost $8, and 5 movies cost $10.

What if the cost of each movie increases to $3. Make a function table and graph to show what the function looks like.

- How does your graph compare to the graph above?
- Predict what the graph of movies that each cost $1 would look like. Explain your reasoning.

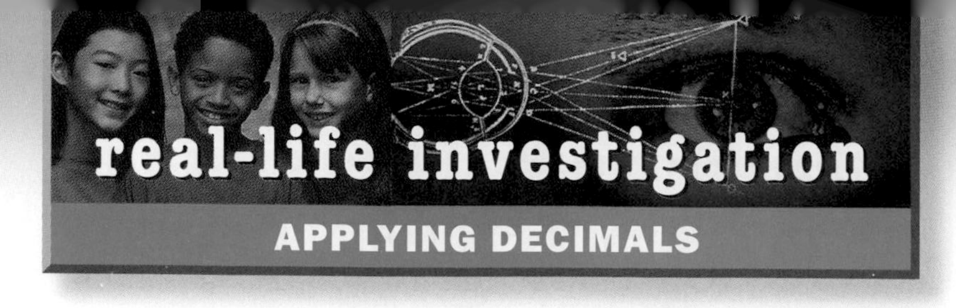

Newsletter Layouts

A monthly newsletter is a great way to let people know about your classroom. Stories can cover what has happened as well as upcoming events. It is fun to put in cartoons, photos, and puzzles about your classmates.

You will need
- *paper*
- *glue*
- *scissors*
- *centimeter ruler*

You can plan the space needed for the columns, photos, and drawings by making a layout.

A layout shows:
- the length and width of the entire page.
- the length and width of each column, photo, or drawing on the page, and the spaces between them.
- the top, bottom, left, and right margins.

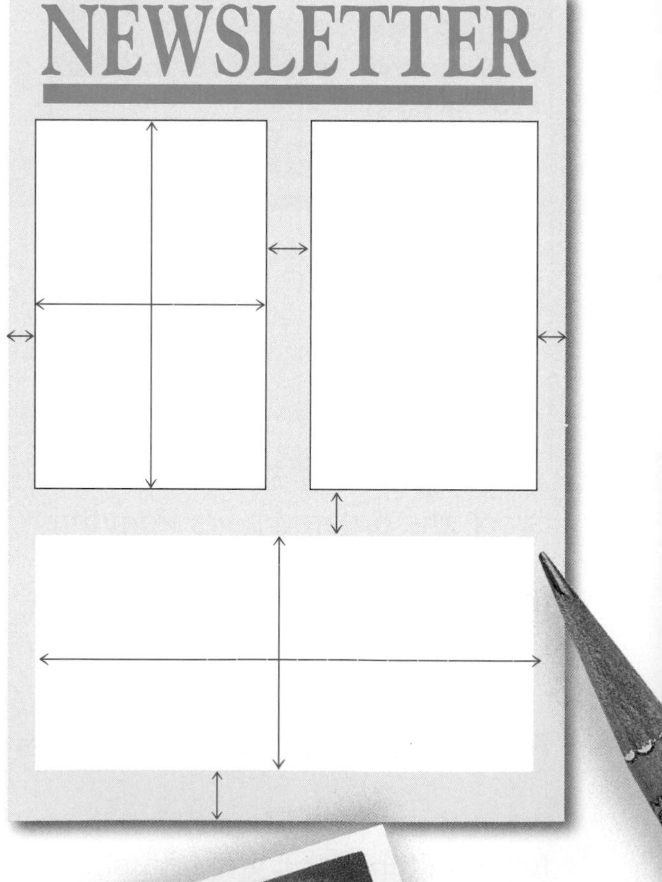

The Grand Canyon
Our trip to the Grand Canyon was terrific! We rode on the burros down the side of the canyon. Some of us were

This column is 4.2 cm wide.

You need to think about:
- the title for the newsletter.
- what stories and other features you want in your paper.
- what size they can be so that everything fits on the page.

DECISION MAKING

Creating a Newsletter

1 Work with a group. Choose a page size for your newsletter. Measure its length and width to the nearest tenth of a centimeter. Then agree on the measurement of the columns, the space between the columns, and the margins.

2 Agree on the part of the newsletter each of you will contribute. Type or print your news items to fit the columns.

3 Measure and mark boxes on the page for any photos, drawings, or other items, such as a puzzle or cartoon.

4 When you have all your text and pictures, cut them out and paste them into your layout.

Reporting Your Findings

5 Portfolio Prepare to present your page or section. Include the following:

- How you gathered your information and came up with your ideas.
- What measurements you used to do your layouts.
- Give a brief description of how you made all the items on the page fit together and why you put them in certain places on the page.

6 

Write an Essay Write an essay telling why you believe your newsletter is important.

Revise your work.
- Are your measurements correct?
- Is your newsletter clear and organized? Is it easy to read?
- Did you proofread your work?

MORE TO INVESTIGATE

PREDICT some different uses for newsletters.

EXPLORE the lengths and widths of columns in your local newspaper.

FIND how your local newspaper lays out its pages.

Problem-Solving Strategy

Read
Plan
Solve
Look Back

Solve a Simpler Problem

LEARN

Read **Suppose that you and your brother work at a local newspaper. You earn $6.40 an hour, and your brother earns $12.15 an hour. You both work an 8-hour day. How much do both of you earn each day altogether?**

Plan You can sometimes see how to solve a problem by first solving a simpler problem.

Solve Try using smaller or easier numbers instead of the numbers in the original problem.

A earns $5 an hour. B earns $10 an hour.

Now try the simpler problem.

A: $8 \times \$5 = \$40$ B: $8 \times \$10 = \$80$
Total Pay: $\$40 + \$80 = \$120$

Now solve the original problem the same way.

A: $8 \times \$6.40 = \$51.20$ B: $8 \times \$12.15 = \$97.20$
Total Pay: $\$51.20 + \$97.20 = \$148.40$

You both earn $148.40 each day altogether.

Look Back How could you solve the problem another way?

Check for Understanding

1 **What if** your sister starts to work at the paper. She earns $15.80 an hour. How much will the three of you earn each day?

Critical Thinking: Analyze

2 Explain how you could use a simpler problem to solve problem 1.

MIXED APPLICATIONS

Problem Solving

PRACTICE

1 You have grown fruit to sell. There are 12 tomatoes for $0.40 each and 15 watermelons for $1.75 each. If you sell all your fruit, how much money will you have?

2 There are 192 countries in the world. In how many countries is *Peanuts* not published?
SEE INFOBIT.

INFOBIT
The cartoon strip *Peanuts* was first published in October 1950. It currently appears in 2,600 newspapers in 75 countries and in 21 languages.

3 Marcus scores 12 baskets for his team. Three of the baskets are worth 3 points, while the rest are worth 2 points. How many more 3-point baskets does he need to score 40 points?

4 **Data Point** Use the Databank on page 543. Show the story of the ballpoint pen by making and labeling a time line.

5 Helen rolls two 1–6 number cubes to get a factor. She rolls the cubes again to get another factor. What is the greatest product she can get? Explain your reasoning.

Use the tax table for problems 6–10.

6 What is the tax for an item that costs $6.90?

7 What is the tax for an item that costs $16.24?

8 What is the tax for an item that costs $54.96?

9 The price tag on an electronic diary is $28.99. What is the cost of the diary, with tax?

10 **Write a problem** using the tax table. Solve it and have others solve it.

Tax Table

Amount	Tax	Amount	Tax
6.49–6.57	0.50	9.24–9.32	0.72
6.58–6.66	0.51	9.33–9.41	0.73
6.67–6.75	0.52	9.42–9.50	0.74
6.76–6.84	0.53	9.51–9.59	0.75
6.85–6.93	0.54	9.60–9.68	0.76
6.94–7.02	0.55	9.69–9.77	0.77

Extra Practice, page 529

Estimate Sums and Differences

LEARN

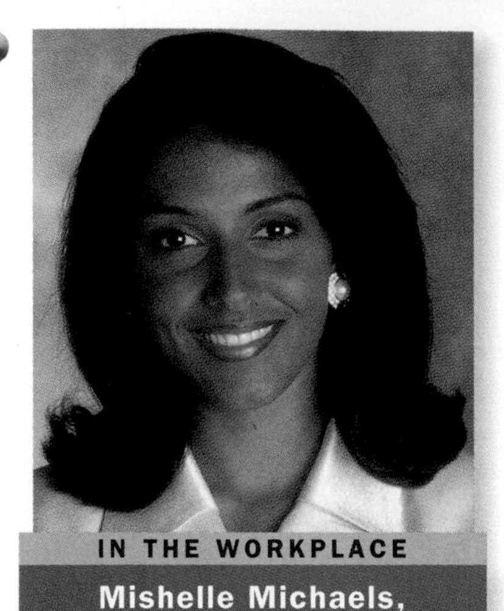

IN THE WORKPLACE

Mishelle Michaels, meteorologist, WHDH-TV, Boston, MA

How do you know what the weather is going to be like? Meteorologists, such as Mishelle Michaels, help you with their forecasts. Suppose Boston got 3.2 inches, 5.7 inches, and 4.6 inches of snow in one week. About how much snow would Mishelle say that Boston got?

You can round decimals to the nearest whole numbers to add or subtract mentally.

Estimate: $3.2 + 5.7 + 4.6$

Read each addend. Look at the tenths place. Then estimate the sum.

$$\begin{array}{ccccccc} 3.2 & + & 5.7 & + & 4.6 & & \\ \downarrow & & \downarrow & & \downarrow & & \\ 3 & + & 6 & + & 5 & = & 14 \end{array}$$

Think: $2 < 5$ Round down. **Think:** $7 > 5$ Round up. **Think:** $6 > 5$ Round up.

She would say about 14 inches of snow fell that week.

More Examples

A Estimate: $2.57 + 8.08$

$$\begin{array}{ccccc} 2.57 & + & 8.08 & & \\ \downarrow & & \downarrow & & \\ 3 & + & 8 & = & 11 \end{array}$$

B Estimate: $12.50 - 5.76$

$$\begin{array}{ccccc} 12.50 & - & 5.76 & & \\ \downarrow & & \downarrow & & \\ 13 & - & 6 & = & 7 \end{array}$$

CHECK

Check for Understanding

Estimate. Round to the nearest whole number.

1 $6.7 + 3.2$ **2** $1.50 + 6.82 + 8.46$ **3** $12.65 - 6.09$

Critical Thinking: Compare **Explain your reasoning.**

4 How is estimating sums and differences with amounts of money similar to estimating with decimals? How is it different?

Practice

Estimate. Round to the nearest whole number.

1 8.1 + 6.3
2 1.9 + 5.2
3 7.52 + 8.73
4 12.03 + 2.61
5 5.4 − 3.6
6 10.9 − 2.5
7 6.74 − 1.28
8 17.5 − 5.9
9 2.7 + 5.3 + 9.1
10 10.45 + 6.52 + 2.81
11 9.62 + 5.07 + 4.25

Estimate. Round to the nearest dollar.

12 \$5.75 − \$5.45
13 \$8.89 + \$3.75
14 \$8.79 + \$6.11
15 \$12.05 − \$7.50

Problem Solving

Pencil & Paper | Calculator | Mental Math

Use the shot list for problems 16–17.

16 Round the time for each shot in the TV sports story. About how long is the sports story?

17 Suppose there are 29.5 seconds to tell the story. The director decides to open with the sports replay. About how many seconds are left for other shots?

Cultural Connection The Russian Abacus

In Russia, many salespeople calculate a customer's bill on a *schoty* (SHOH-tee). The schoty is a kind of abacus. It can be used to show decimal numbers.

This schoty shows the number 36.05. Explain how you think the positions of the beads show this number.

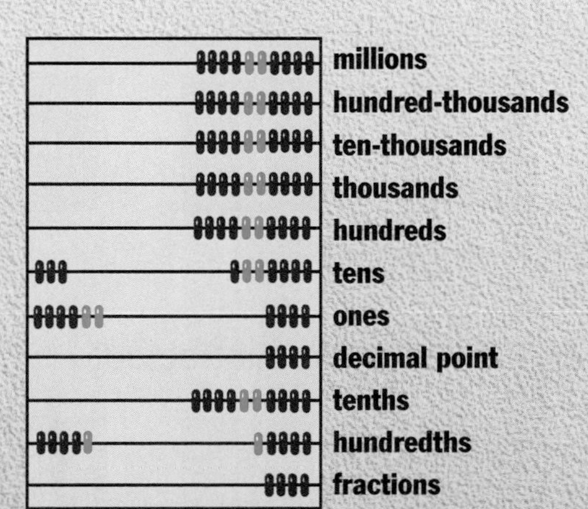

Draw a picture of a schoty to show the decimal.

1 8.14
2 5.37
3 21.03
4 58.75

Add and Subtract Decimals

You can use decimal squares to explore addition and subtraction with decimals.

Work Together

Work with a partner to add and subtract decimals.

You will need
- *decimal squares*
- *crayons, colored pencils, or markers*
- *scissors*

Choose any four digits from 0 to 9. Use the digits to complete the addition example and the subtraction example.

$$\begin{array}{r} 2.\square\square \\ +1.\square\square \\ \hline \end{array} \qquad \begin{array}{r} 2.\square\square \\ -1.\square\square \\ \hline \end{array}$$

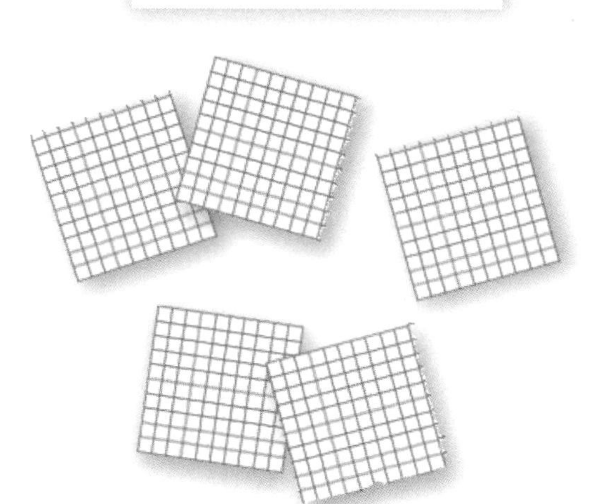

Then use the decimal squares to find the exact answers. Color and cut them to find the sum and the difference. Record your answers.

Repeat the activity two more times.

▶ How did you use the decimal squares to add? to subtract?

Make Connections

Here is how Mark and Sondra added.

$$\begin{array}{r} 2.\boxed{3}\boxed{5} \\ +1.\boxed{7}\boxed{0} \\ \hline 4.0\ 5 \end{array}$$

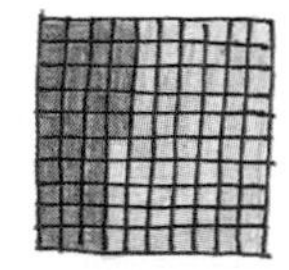
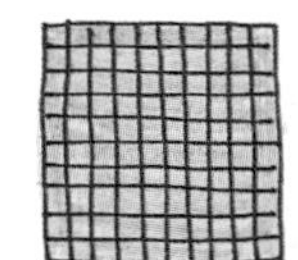
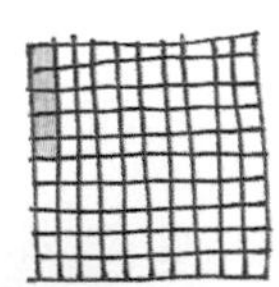

Here is how Mark and Sondra subtracted.

$$\begin{array}{r} 2.\boxed{3}\boxed{7} \\ -1.\boxed{5}\boxed{0} \\ \hline 0.8\ 7 \end{array}$$

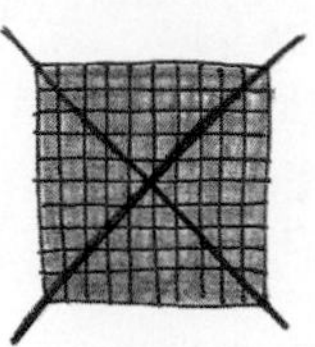
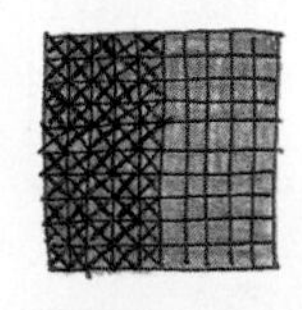
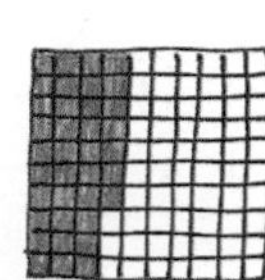

▶ Explain how Mark and Sondra added and subtracted.

▶ How would you model 1.45 + 2.07 + 1.95?

CHECK

Check for Understanding

Use the models to complete the number sentence.

1. 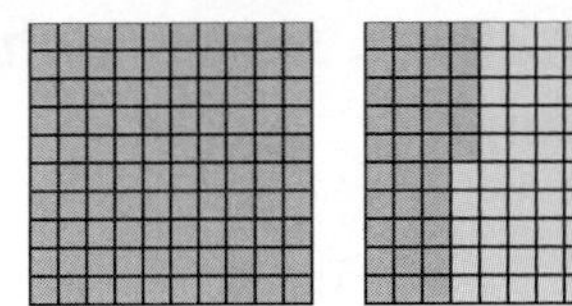 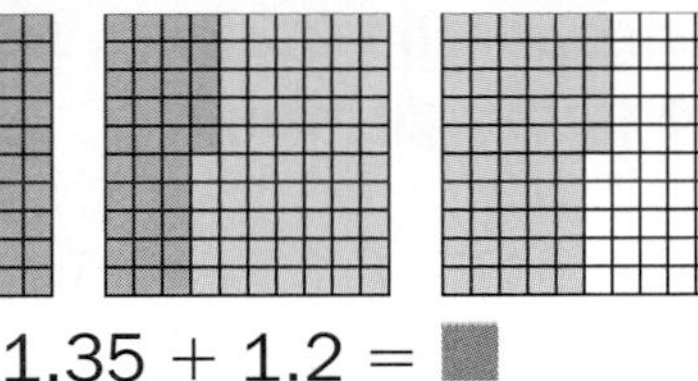

   1.35 + 1.2 = ■

2. 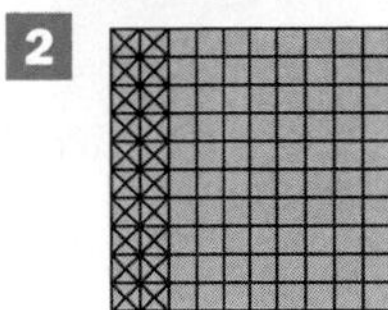 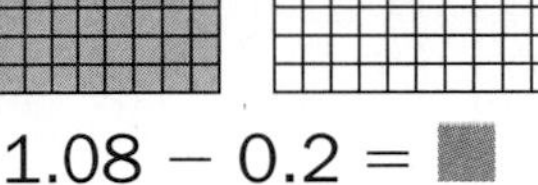

   1.08 − 0.2 = ■

Use decimal squares to add or subtract.

3. 0.8 + 0.3
4. 1.5 + 1.8
5. 1.6 − 0.2
6. 2.15 − 1.73
7. 3.06 − 0.89

Critical Thinking: Generalize **Explain your reasoning.**

8. When do you need to regroup when adding?

PRACTICE

Practice

Use the models to complete the number sentence.

1. 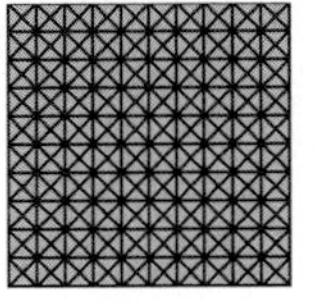 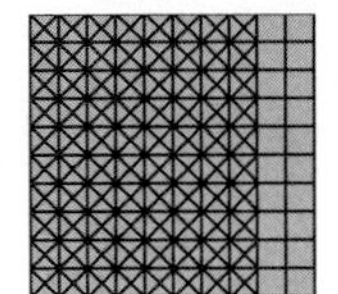 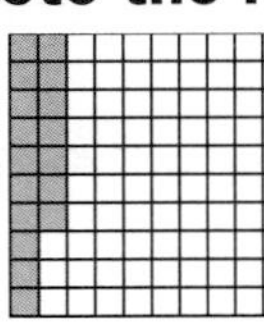

   2.17 − 1.8 = ■

2. 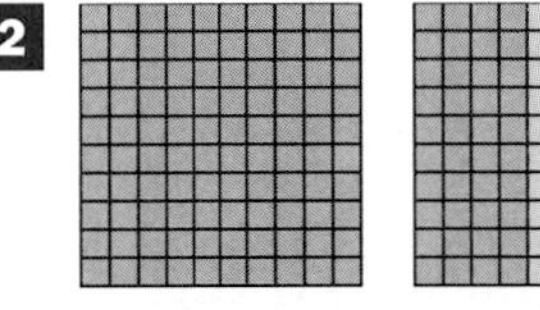

   1.4 + 0.29 = ■

Use decimal squares to add or subtract.

3. 0.4 + 0.3
4. 0.8 + 0.5
5. 1.42 + 1.85
6. 1.66 + 0.45
7. 2.09 + 1.97
8. 0.7 − 0.1
9. 1.4 − 0.5
10. 1.28 − 0.40
11. 2.07 − 1.45
12. 3.45 − 1.96
13. 1.5 + 1.7
14. 1.50 + 1.25
15. 1.8 + 0.29
16. 1.33 + 1.9
17. 2.7 − 1.9
18. 1.3 − 0.4
19. 2.11 − 1.03
20. 2.09 − 1.64
21. 0.9 + 0.07
22. 0.9 − 0.3
23. 1.85 + 0.11
24. 3 − 1.09
25. 2.2 − 0.8
26. 1.16 + 2.8
27. 2.04 − 0.38
28. 1.3 + 1.83

Extra Practice, page 530

Add Decimals

LEARN

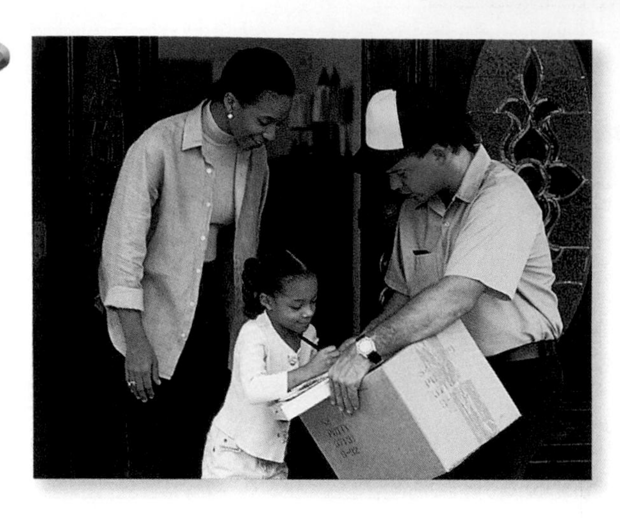

Do you love to get mail? Suppose you get a package. The gift inside the box weighs 5.85 oz and the box and wrapping weighs 2.38 oz. How much will the package weigh?

Estimate: 5.85 + 2.38 **Think:** 6 + 2 = 8

You have already used models to add decimals. You can also use pencil and paper.

Add: 5.85 + 2.38

Think: Adding decimals is like adding whole numbers.

Whole Numbers

$$\begin{array}{r} {}^{1\,1} \\ 585 \\ +\ 238 \\ \hline 823 \end{array}$$

Decimals

$$\begin{array}{r} {}^{1\;\,1} \\ 5.85 \\ +\ 2.38 \\ \hline 8.23 \end{array}$$

↑ **Write the decimal point in the sum.**

You can also use a calculator. 5.85 + 2.38 = 8.23

The package will weigh 8.23 oz.

CHECK

Check for Understanding

Add using any method. Estimate to check that your answer is reasonable.

1 $\begin{array}{r} 0.47 \\ +\ 0.59 \\ \hline \end{array}$ **2** $\begin{array}{r} 2.07 \\ +\ 3.91 \\ \hline \end{array}$ **3** $\begin{array}{r} 6.98 \\ +\ 4.50 \\ \hline \end{array}$ **4** $\begin{array}{r} 17.55 \\ 9.87 \\ +\ \ 2.04 \\ \hline \end{array}$ **5** $\begin{array}{r} \$24.50 \\ 4.95 \\ +\ \ 19.75 \\ \hline \end{array}$

6 $\begin{array}{r} 5.4 \\ +\ 3.7 \\ \hline \end{array}$ **7** $\begin{array}{r} \$3.05 \\ +\ \ 2.31 \\ \hline \end{array}$ **8** $\begin{array}{r} 26.27 \\ +\ 12.80 \\ \hline \end{array}$ **9** $\begin{array}{r} 23.25 \\ 1.60 \\ +\ \ 4.08 \\ \hline \end{array}$ **10** $\begin{array}{r} 38.50 \\ 21.36 \\ +\ \ 3.54 \\ \hline \end{array}$

Critical Thinking: Analyze **Explain your reasoning.**

11 How would you estimate and add 19.5 and 24.83?

Practice

Add using any method. Remember to estimate.

1 $\begin{array}{r} 0.82 \\ +\ 0.16 \\ \hline \end{array}$ **2** $\begin{array}{r} 0.8 \\ +\ 0.5 \\ \hline \end{array}$ **3** $\begin{array}{r} 5.7 \\ +\ 0.9 \\ \hline \end{array}$ **4** $\begin{array}{r} 6.4 \\ +\ 2.9 \\ \hline \end{array}$ **5** $\begin{array}{r} 2.7 \\ +\ 3.6 \\ \hline \end{array}$

6 $\begin{array}{r} 0.52 \\ +\ 0.98 \\ \hline \end{array}$ **7** $\begin{array}{r} 2.79 \\ +\ 3.52 \\ \hline \end{array}$ **8** $\begin{array}{r} 5.82 \\ +\ 8.34 \\ \hline \end{array}$ **9** $\begin{array}{r} 7.69 \\ +\ 0.49 \\ \hline \end{array}$ **10** $\begin{array}{r} \$6.38 \\ +\ 7.97 \\ \hline \end{array}$

11 $\begin{array}{r} 12.13 \\ 6.45 \\ +\ 9.80 \\ \hline \end{array}$ **12** $\begin{array}{r} \$24.25 \\ 9.06 \\ +\ 1.45 \\ \hline \end{array}$ **13** $\begin{array}{r} 49.72 \\ 3.74 \\ +\ 10.07 \\ \hline \end{array}$ **14** $\begin{array}{r} 19.26 \\ 9.80 \\ +\ 6.25 \\ \hline \end{array}$ **15** $\begin{array}{r} 35.30 \\ 7.74 \\ +\ 20.80 \\ \hline \end{array}$

16 4.20 + 0.73 **17** 0.26 + 5.19 **18** 27.30 + 4.84 + 30.75

19 7.85 + 4.11 **20** 6.33 + 3.90 **21** 33.14 + 12.99 + 0.70

MIXED APPLICATIONS
Problem Solving

Use the price list for problems 22-26.

22 How much did it cost to send 3 packages that weigh 1 lb each?

23 How much did it cost to mail 2 paddles weighing 1 lb each and three 2-lb life preservers?

24 How much did it cost altogether to send 1-lb, 2-lb, 3-lb, 4-lb, and 5-lb boxes of chocolates?

25 What did it cost to send three 4-lb laptop computers and two 1-lb computer games?

26 **Write a problem** that uses information from the price list. Solve it and ask others to solve it.

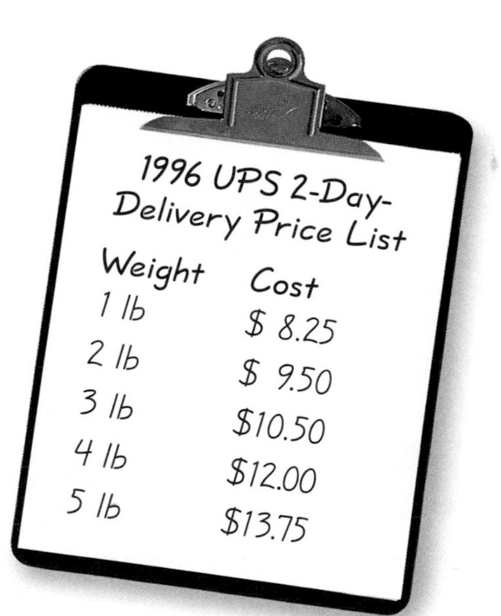

1996 UPS 2-Day-Delivery Price List

Weight	Cost
1 lb	$ 8.25
2 lb	$ 9.50
3 lb	$10.50
4 lb	$12.00
5 lb	$13.75

mixed review • test preparation

1 8 ft = ■ in. **2** 2 mi = ■ yd **3** 6 gal = ■ qt **4** 16 oz = ■ lb

5 3 cm = ■ mm **6** 7,000 m = ■ km **7** 10 kg = ■ g **8** 2,000 mL = ■ L

ALGEBRA: PATTERNS Complete.

9 $\frac{1}{3} = \frac{■}{6} = \frac{■}{12} = \frac{8}{■} = \frac{16}{■} = \frac{■}{96}$

10 $\frac{1}{6} = \frac{2}{■} = \frac{3}{■} = \frac{■}{24} = \frac{■}{30} = \frac{6}{■}$

Extra Practice, page 530

Subtract Decimals

LEARN

Telescopes allow us to see into space. The mirror of a large reflecting telescope at the Palomar Observatory in California is 5.08 m wide. The mirror on the Hubble space telescope is 4.29 m wide. Which mirror is wider? How much wider?

Estimate: 5.08 − 4.29

Think: 5 − 4 = 1

You have already used models to subtract decimals. You can also use pencil and paper.

Above: Hubble space telescope. Left: Palomar Observatory.

Subtract: 5.08 − 4.29

Think: Subtracting decimals is like subtracting whole numbers.

Whole Numbers	**Decimals**
9	9
4 10 18	4 10 18
~~5~~ ~~0~~ ~~8~~	~~5~~ . ~~0~~ ~~8~~
− 4 2 9	− 4 . 2 9
7 9	0 . 7 9

Write the decimal point in the difference.

You can also use a calculator. 5.08 − 4.29 = *0.79*

The mirror on the Palomar telescope is 0.79 m wider.

CHECK

Check for Understanding

Subtract using any method. Estimate to check that your answer is reasonable.

1 0.8 − 0.3

2 9.3 − 6.5

3 6.75 − 4.03

4 12.91 − 7.48

5 \$39.70 − 19.99

Critical Thinking: Generalize **Explain your reasoning.**

6 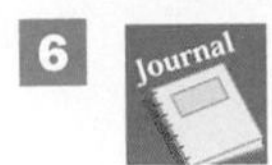 Tell how you could check decimal sums and differences.

Practice

Subtract using any method. Remember to estimate.

1. $0.8 - 0.6$	2. $4.9 - 0.3$	3. $7.1 - 2.5$	4. $9.4 - 3.8$	5. $1.2 - 0.7$
6. $0.63 - 0.12$	7. $6.39 - 2.11$	8. $2.43 - 1.36$	9. $9.57 - 7.64$	10. $4.02 - 1.67$
11. \$12.82 − 7.51	12. \$27.25 − 4.06	13. $39.70 - 8.64$	14. $15.21 - 1.50$	15. $24.90 - 3.21$

16. 9.6 − 7.2
17. 4.5 − 2.8
18. 3.50 − 1.06
19. \$5.08 − \$2.17
20. 7.8 − 4.2
21. \$8.23 − \$3.04
22. 9.31 − 4.56
23. 11.42 − 4.02
24. 6.2 − 1.5
25. 5.06 − 2.40
26. \$8.41 − \$3.02
27. 25.19 − 18.05
28. 9.26 − 1.46
29. 5.03 − 3.39
30. 6.72 − 5.56
31. \$22.74 − \$7.38

MIXED APPLICATIONS

Problem Solving

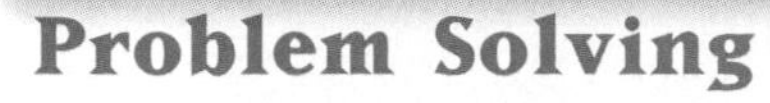

32. **Logical reasoning** We are two numbers between 0 and 1. Our difference is 0.43. When you add us together, the sum is 1.53. When you subtract one of us from 1, the difference is 0.02. What numbers are we?

33. Felipe created a secret code. The first ten letters in his code are:
A—0.1 *D*—0.13 *G*—0.22
B—0.11 *E*—0.2 *H*—0.23
C—0.12 *F*—0.21 *I*—0.3
Follow Felipe's pattern. What decimal would be used for *T*?

34. The main body of the Hubble telescope is 13.12 m long and 4.27 m wide. How much longer is it than it is wide?

35. **Write a problem** that is a message to a friend using Felipe's code in problem 33. Ask a friend to decode your message.

mixed review • test preparation

1. $\frac{2}{3} + \frac{2}{3}$
2. $\frac{3}{5} + \frac{4}{5}$
3. $\frac{11}{12} + \frac{7}{12}$
4. $\frac{1}{6} + \frac{5}{6}$
5. $\frac{5}{8} + \frac{7}{8}$
6. $\frac{4}{5} - \frac{3}{5}$
7. $\frac{9}{10} - \frac{3}{10}$
8. $\frac{14}{16} - \frac{10}{16}$
9. $\frac{11}{12} - \frac{7}{12}$
10. $\frac{3}{4} - \frac{3}{4}$

Extra Practice, page 531

Problem Solvers at Work

Read
Plan
Solve
Look Back

LEARN

Part 1 Write a Number Sentence

Do weather reports ever catch your attention while you are channel surfing? It may be because meteorologists use interesting comparisons, unusual stories, and maps to capture your attention.

Cultural Note

During the days between July 3 and August 11, the ancient Egyptians noticed a star rising with the sun. They thought the star made the sun hotter. The star, Sirius, is part of a constellation shaped like a dog. Today, we call July 3–August 11 the "dog days of summer."

Rainfall

Date	This Year	Last Year
July 22	0.02 in.	0.00 in.
July 23	1.13 in.	0.18 in.
July 24	1.08 in.	0.84 in.
July 25	0.45 in.	2.62 in.
July 26	0.97 in.	3.09 in.

Work Together

Solve the problem. Write number sentences to record your work.

1. What is the difference between the greatest and least rainfalls from July 22 to July 26 this year?

2. What was the total rainfall from July 22 to July 26 this year?

3. Did it rain more altogether this year or last year on the dates given? How much more?

4. **What if** the normal rainfall from July 22 to July 26 is 2.85 in. How much over this was the rainfall last year?

5. READING ARITHMETIC WRITING **Write an Essay** Kristie used a calculator to solve problem 3. She said that you have to use the MEMORY key to find the answer. Do you agree or disagree with her? Explain your reasoning in an essay.

Part 2 Write and Share Problems

Kenya looked in an almanac to find information on the precipitation in several major United States cities. Kenya used the data to write a problem.

Place	Annual Precipitation (in inches)
Albuquerque, NM	8.12
Mobile, AL	64.64
Charlotte, NC	43.16
Detroit, MI	30.97
San Francisco, CA	19.71
Minneapolis, MN	26.36

CHECK

What is the difference in precipitation between Charlotte, North Carolina, and Detroit, Michigan?

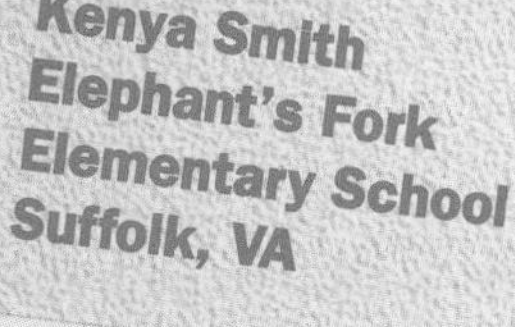

Kenya Smith
Elephant's Fork Elementary School
Suffolk, VA

6 Write a number sentence that will help you solve Kenya's problem. Then solve it.

7 Change Kenya's problem so that you would complete this number sentence to solve it:
19.71 − 8.12 = ■. Then solve it.

8 **Write a problem** of your own using the data in the table.

9 Trade problems. Solve at least three problems written by your classmates.

10 What was the most difficult problem that you solved? Why?

Turn the page for Practice Strategies.

PRACTICE

Menu

Choose five problems and solve them. Explain your methods.

1 Carol bought 4 concert tickets that cost $9.50 each. She got back $12 in change. How much did she give the cashier?

2 Calvin says that the probability of choosing his favorite color is $\frac{3}{6}$. What is his favorite color?

3 Annie has a red sweatshirt, a green sweatshirt, and a blue sweatshirt. She has white shorts and grey sweatpants. How many different combinations of outfits does she have?

4 You subscribe to 8 magazines. You donate 1 out of every 4 magazines to your classroom. How many magazines do you donate to your classroom?

5 **ALGEBRA: PATTERNS** John is watching a news program on TV. There are four parts to the program. The first three parts were shown on the 2nd, 9th, and 16th of the month. If this pattern continues, on what day will the fourth part be shown?

6 Juan is buying plant food for his vegetable garden. He needs to know the area of the rectangular garden. One side of the garden measures 9 ft and the perimeter is 40 ft. What is the area?

7 Roberto has $20.00. He wants to buy a CD for $8.99 and two cassettes for $3.00 each. How much change should he get back? Write one or more number sentences and solve them.

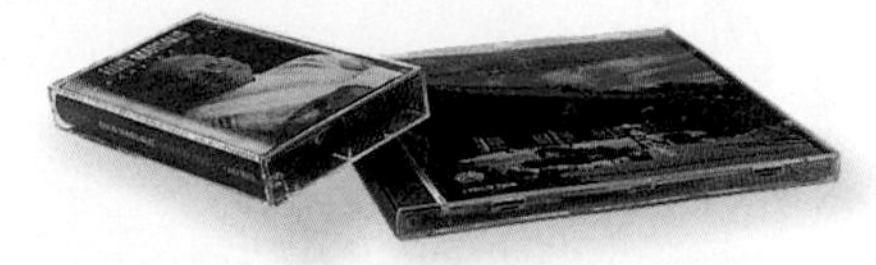

8 A radio station plays 4 commercials every hour of a 24-hour day, except for Sundays on which it plays 2 commercials every hour. How many commercials does the station play in one week?

Choose two problems and solve them. Explain your methods.

9 Use the table to plan the possible combinations of two-person swim teams for the race. What would be the combined time for the two swimmers on each team?

100 Meters Freestyle	
Swimmer 1	42.36 seconds
Swimmer 2	44.57 seconds
Swimmer 3	47.54 seconds
Swimmer 4	43.56 seconds
Swimmer 5	48.01 seconds

10 **Logical reasoning** There are three people standing in line to make purchases. Jon is not first in line. Ali is standing in front of Sandy. Sandy is not last in line. Who is first, second, and third in line?

11 Suppose you have 28 ft of picture-framing wood. Sketch and label the size of one picture you could frame. Now try using the 28 ft for two pictures.

12 **At the Computer** A computer spreadsheet can arrange data in rows and columns and make calculations.

Telephone City is keeping track of the sales for its stores in Newport and Maple. They will use the information to decide how often to order the items listed on the spreadsheet.

Use a spreadsheet program to arrange the data below into rows and columns and make calculations.

Telephone City Sales
Sales for September 3, 1997

Item	A Sales for Newport	B Sales for Maple	C[A + B] Total
Stick-on pad and pencil	$354.24	$75.75	$429.99
Shoulder rest	$108.76	$104.89	
Long jack cord	$135.21	$199.45	
Earpiece amplifier	$91.00	$367.19	
30-ft expandable cord	$723.22	$605.60	

Extra Practice, page 531

Language and Mathematics

Complete the sentence. Use a word in the chart. (pages 450–471)

1 The ■ of two decimals will always be greater than either decimal.

2 In the decimal 0.08, the 8 is in the ■ place.

3 The ■ separates whole numbers from decimals.

4 There are 10 hundredths in 1 ■.

5 You can estimate the sum of two or more decimals by ■.

Vocabulary
- decimal point
- tenth
- rounding
- sum
- hundredths
- ten

Concepts and Skills

Write a decimal. (page 450)

6 

7 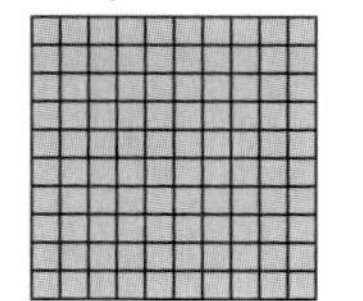

8  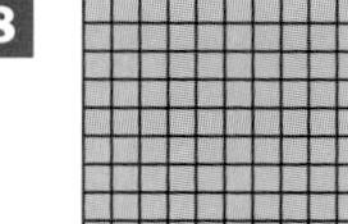

9 $\frac{7}{100}$

10 $3\frac{6}{10}$

11 $7\frac{32}{100}$

12 $\frac{54}{100}$

13 $17\frac{3}{10}$

14 $24\frac{74}{100}$

15 four and six tenths

16 eight and thirty-two hundredths

Compare. Write >, <, or =. (page 456)

17 2.76 ● 2.67

18 6.56 ● 5.66

19 9.30 ● 9.3

Write in order from least to greatest. (page 456)

20 0.82, 0.88, 0.28

21 1.93, 1.39, 1.79

22 4.14, 4.04, 4.4

Write in order from greatest to least. (page 456)

23 0.07, 0.70, 7.07

24 2.28, 2.09, 2.25

25 5.64, 5.90, 15.03

Estimate the sum or difference. (page 464)

26 2.34 + 5.85

27 12.08 + 3.50

28 6.82 − 4.37

29 13.70 − 6.93

30 9.91 + 3.02

31 20.39 + 0.77

32 8.03 − 2.74

33 38.71 − 4.01

Add. (pages 466, 468)

34 0.26 + 0.44

35 4.5 + 3.6

36 14.62 + 0.94

37 2.20 + 9.68

38 \$18.56 + 57.39

Subtract. (pages 466, 470)

39 0.43 − 0.21

40 13.2 − 9.6

41 23.15 − 6.40

42 64.35 − 36.80

43 \$10.05 − 8.78

Think critically. (pages 450, 464, 466)

44 Analyze. When Carla estimates decimal sums, she rounds each addend to the nearest whole number. What is the greatest amount she could be off by using this method to estimate the sum of two decimals? Give an example.

45 Generalize. Explain how fractions help you to understand decimals.

46 Generalize. Explain how adding and subtracting decimals is like adding and subtracting whole numbers. How is it different?

MIXED APPLICATIONS Problem Solving

Pencil & Paper · Calculator · Mental Math (pages 462, 472)

47 It rained 2.34 in. on Tuesday, 1.67 in. on Wednesday, and 0.43 in. on Thursday. What was the total rainfall for the three days?

48 Fernando's team swam the relay in 41.89 seconds. Mara's team swam it in 43.06 seconds. Who won? By how much?

49 Cal is saving for a new microphone. It costs \$39.95 plus \$3.10 tax. So far he has saved \$26.72. How much more does he need to buy the microphone?

50 Mrs. Kelly's class raised \$147.58 in April for the Children's Hospital. They raised \$23.18 the first week, \$37.82 the second week, and \$59.45 the third week. How much did the class raise during the fourth week in April?

Write the number as a decimal.

1

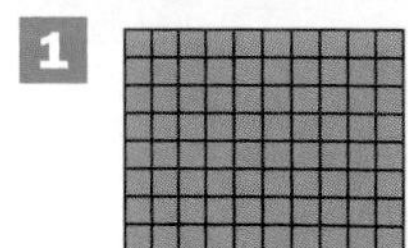

2

3 $\frac{1}{100}$

4 $2\frac{7}{10}$

5 eight and three hundredths

6 twelve and three tenths

Complete. Write >, <, or =.

7 1.13 ● 1.3

8 4.8 ● 4.80

9 1.50 ● 1.05

Write in order from least to greatest.

10 0.56, 0.65, 0.5

11 3.06, 2.99, 3.6

Estimate the sum or difference.

12 4.09 + 0.95

13 \$24.32 + \$6.75

14 7.82 − 2.09

15 5.12 − 2.23

Add.

16 $\begin{array}{r} 0.67 \\ +\,0.43 \\ \hline \end{array}$

17 $\begin{array}{r} 3.60 \\ +\,5.46 \\ \hline \end{array}$

18 $\begin{array}{r} 62.35 \\ +\;\;0.48 \\ \hline \end{array}$

Subtract.

19 $\begin{array}{r} 0.52 \\ -\,0.10 \\ \hline \end{array}$

20 $\begin{array}{r} \$10.00 \\ -\quad 0.09 \\ \hline \end{array}$

21 $\begin{array}{r} 4.42 \\ -\,1.08 \\ \hline \end{array}$

Solve.

22 Cary ran around a rectangular track one time. The track is 125.5 yd long and 85.5 yd wide. How far did she run?

23 Lane walks 1.20 mi to a bus and rides another 7.25 mi to work. Her boss drives 8.50 mi to work. Who lives farther from work? How can you tell?

24 At the end of the summer Jed was riding his bike 264 mi each week. This was 8 times as far as he rode at the beginning of the summer. How far did he ride at the beginning of the summer?

25 Walter bought a \$3.95 hamburger and \$1.75 fruit cup for lunch. He paid with a \$10 bill. The clerk gave him 4 one-dollar bills, 1 quarter, and 1 nickel for his change. Is this correct? Why or why not?

What Did You Learn?

You can communicate with people around the country or around the world by telephone.

- ▶ Use the telephone bills. Write a paragraph comparing the cost of calls to three cities.
- ▶ Compare the totals for all the calls on each bill. Use models or drawings to explain how you found the amounts.

SPEEDY TELEPHONE CARRIER

Calls to	Minutes	Total
Des Moines, Iowa	5	$0.45
Jackson, Mississippi	13	$1.17
Nashua, New Hampshire	22	$2.98
Norfolk, Virginia	5	$0.62
Kyoto, Japan	20	$10.68

RELIABLE TELEPHONE CARRIER

Calls to	Minutes	Total
Des Moines, Iowa	5	$0.34
Jackson, Mississippi	13	$1.09
Nashua, New Hampshire	22	$2.78
Norfolk, Virginia	5	$1.28
Kyoto, Japan	20	$9.68

A Good Answer

- gives accurate comparisons of the costs of calls to different cities
- uses models or drawings to find and compare the phone bills

You may want to place your work in your portfolio.

What Do You Think?

1 Does it help to think about subtracting whole numbers when you subtract decimals? Why or why not?

2 How do you know which digits to add or subtract when you solve problems that include decimals?
- Line up the decimal points.
- Use 10-by-10 grids and group entire grids, rows, or single squares.
- Use place value to add digits in the same place.
- Other. Explain.

math science technology
CONNECTION

Communication and Technology

Humans have always looked for better and faster ways to communicate. Science and technology have influenced communication greatly.

Cultural Note

The Egyptians wrote messages on a material made from a grasslike plant known as *papyrus* more than 5,000 years ago. Before papyrus, people wrote on clay tablets or stones. Paper was invented by the Chinese about 2,000 years ago.

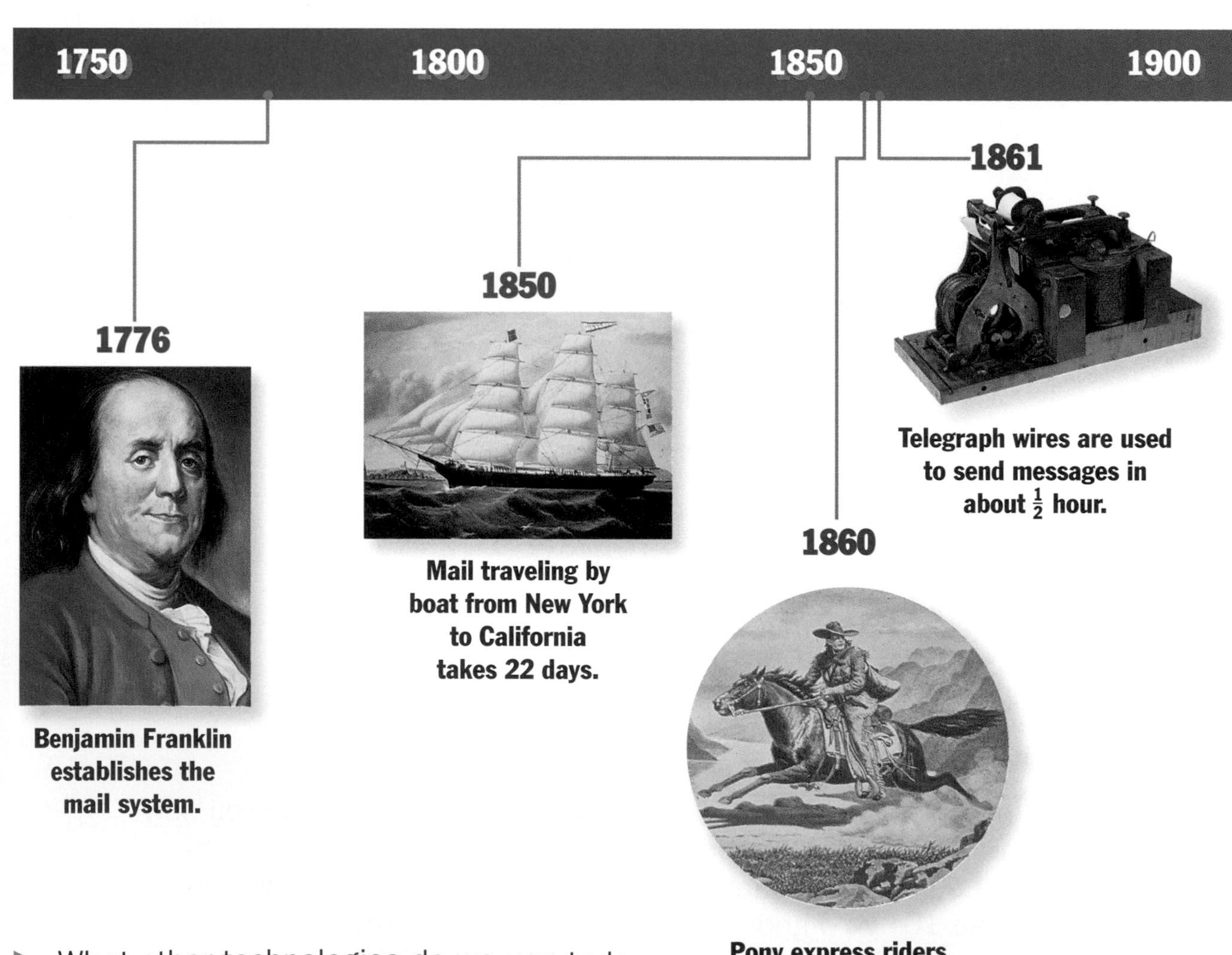

1776 Benjamin Franklin establishes the mail system.

1850 Mail traveling by boat from New York to California takes 22 days.

1860 Pony express riders carrying mail on horseback over a trail of almost 2,000 miles take 10 days.

1861 Telegraph wires are used to send messages in about $\frac{1}{2}$ hour.

- ► What other technologies do we use today to communicate?
- ► What are some of the advantages of today's communication technology?

Cost of Stamps

Use the table for problems 1–3.

1 Which increase in postage cost was greater, from 1968 to 1971 or from 1991 to 1995?

2 During which years did postage cost increase by more than 2 cents?

3 Since you were born, how much has the cost of mailing a 1-ounce letter increased?

1950 **Today**

Messages are sent almost instantaneously by transmitting information by way of communication satellites.

At the Computer

4 Use a graphing program to make a bar graph that shows the cost of mailing for the years given.

5 Write two statements that describe the data in your bar graph.

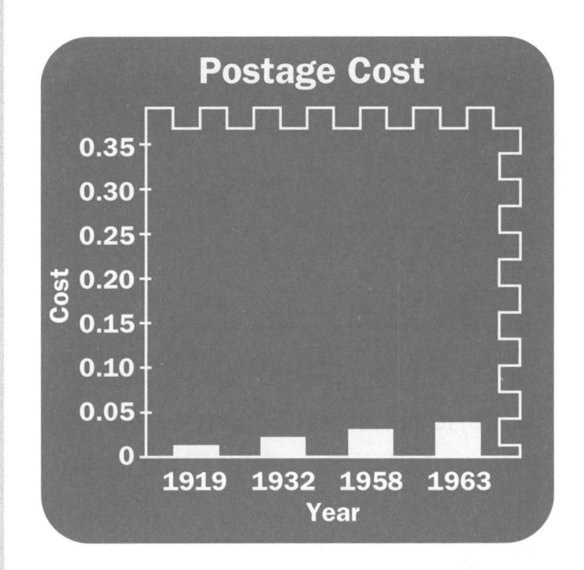

Postage Cost of Mailing a 1-ounce letter			
Year	**Cost**	**Year**	**Cost**
1919	$0.02	1978	$0.15
1932	$0.03	1981 (March)	$0.18
1958	$0.04	1981 (November)	$0.20
1963	$0.05	1985	$0.22
1968	$0.06	1988	$0.25
1971	$0.08	1991	$0.29
1974	$0.10	1995	$0.32
1975	$0.13		

Choose the letter of the best answer.

1 What is the least number possible using the digits 1, 3, 4, and 0?

A 3,104
B 1,340
C 1,304
D 1,034

2 Carter puts his baseball card collection in 14 shoe boxes with 488 cards in each box. How many cards does he have?

F 8,712
G 6,832
H 6,382
J 502

3 Kim ran around the track in 65.19 seconds. Jenny ran around it in 71.04 seconds. How much faster was Kim's time?

A 5.85 seconds
B 6.15 seconds
C 6.85 seconds
D 5.95 seconds

4 Jacob had \$25. He spent $\frac{3}{5}$ of his money on a present for his mother. How much money does he have left?

F \$15.00
G \$ 5.00
H \$10.00
J \$12.50

5 $3,864 \div 8$

A 408
B 438
C 480
D 483

6 Which numbers are ordered from least to greatest?

F 0.48, 1.23, 1.05
G 1.01, 0.92, 0.50
H 0.02, 0.20, 2.00
J 8.08, 0.80, 0.88

7 Katie plans to build a fence around her garden. The garden measures 12 ft by 8 ft. How much fencing will she need?

A 96 ft
B 48 ft
C 40 ft
D 32 ft

8 Which subtraction is shown?

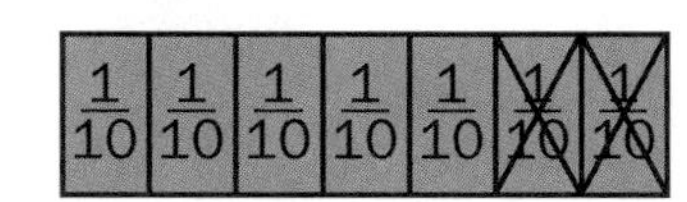

F $0.7 - 0.5$
G $7 - 2$
H $0.5 - 0.2$
J $0.7 - 0.2$

9 $7 \times 1,765$

A 13,355
B 12,553
C 12,355
D 8,925

10 Alan has $\frac{3}{4}$ cup of rice. He uses $\frac{1}{2}$ cup of rice. How much rice is left?

F $\frac{1}{4}$ cup
G $\frac{1}{2}$ cup
H $\frac{3}{4}$ cup
J 1 cup

11 Jay finishes work at 4:00 P.M. If he works for 7 h 45 min each day, what time does he start?

A 8:15 A.M
B 8:15 P.M.
C 9:15 A.M.
D 9:15 P.M.

12 The numbers 0.43, 0.65, 0.97, 0.74, and 0.31 are all ■.

F less than 1
G greater than 1
H equal to 1
J less than $\frac{8}{10}$.

13 You can write 2.04 as ■.

A two and four tenths
B twenty-four hundredths
C two and four hundredths
D not given

14 Which statement is true?

F A circle is a polygon.
G Quadrilaterals always have a right angle.
H An obtuse angle is less than a right angle.
J Perpendicular lines meet at right angles.

15 Which letter has a line of symmetry?

A F **B** G **C** R **D** T

16 Which is another way to write 4,065?

F 40 tens 65 ones
G 406 tens 5 ones
H 4 thousands 65 tens
J 40 hundreds 65 tens

17 Which fractions are equivalent?

A $\frac{1}{2}$ and $\frac{1}{12}$
B $\frac{1}{2}$ and $\frac{3}{9}$
C $\frac{1}{2}$ and $\frac{7}{14}$
D $\frac{1}{2}$ and $\frac{2}{2}$

Use the bar graph for problems 18–19.

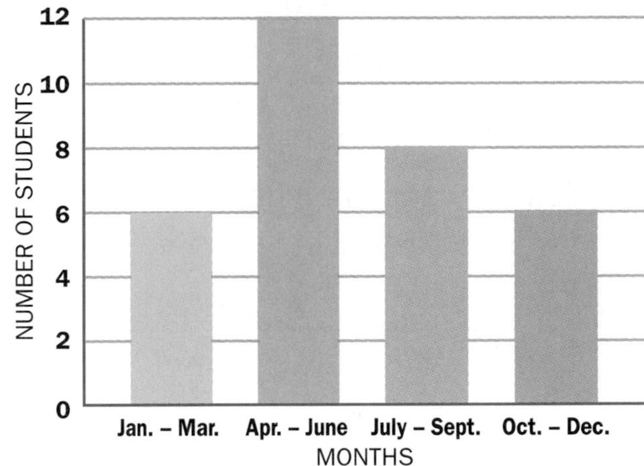

18 Which statement is *not* true about the number of students?

F More than $\frac{1}{2}$ were born in months before July.
G $\frac{5}{8}$ were born in months between April and September.
H $\frac{3}{4}$ were born in months after September.
J $\frac{1}{4}$ were born in July through September.

19 Which can you find?

A the number of students born in January
B the total number of students
C the birth date of each student
D the age of the students

extra practice

Numbers in Your World page 3

Write a sentence in which the number is used in the way given.

1 Number: 6
Used: to show order

2 Number: 8
Used: to count

3 Number: 16
Used: to show order

4 Number: 24
Used: to measure

5 Number: 31
Used: to name

6 Number: 91
Used: to measure

7 Number: 57
Used: to name

8 Number: 25
Used: to count

9 Number: 36
Used: to measure

Birthday Party
You're invited!
Sam Nelson will be 9.

September 19
1:00 P.M.
1056 First Street
RSVP: 888-8888

Solve.

10 Tell how each number on Sam's birthday invitation is used.

11 Jan's birthday is 5 days later than Sam's. What is the date of her birthday?

Building to Thousands page 7

Write the number.

1 9 hundreds 2 tens

2 7 hundreds 9 tens 8 ones

3 29 tens 8 ones

4 84 tens 3 ones

5 76 tens 4 ones

6 100 tens 9 ones

7 1 thousand 5 hundreds 6 ones

8 5 thousands 3 tens 9 ones

Solve.

9 Write the greatest 3-digit number using the cards at the right. Then, give the number that is 1,000 more.

10 Write the least 3-digit number using the cards at the right. Then, give the number that is 100 less.

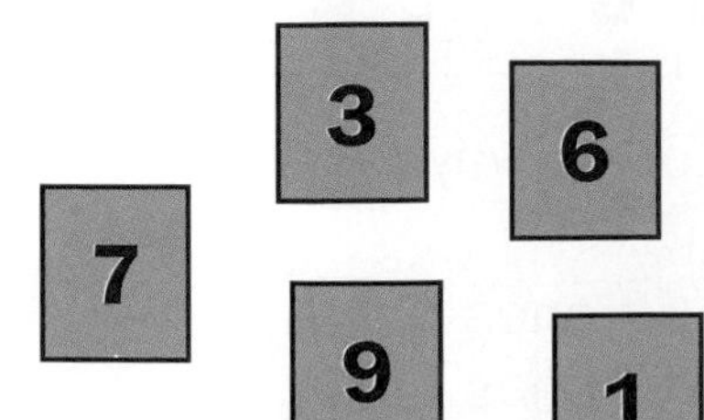

Thousands page 11

Write the number in standard form and in expanded form.

1 eight thousand, six

2 nine thousand, sixty-four

3 six thousand, two hundred ten

4 four thousand, six hundred thirty-five

5 nineteen thousand, fifty

6 two hundred thousand, four hundred

7 eighty-three thousand, six hundred seventy-two

8 two hundred six thousand, three hundred forty-five

9 eight hundred ten thousand, five hundred twenty

Solve.

10 Tell what number this model would represent if you added 4 more thousands, 2 more hundreds, and 9 more ones.

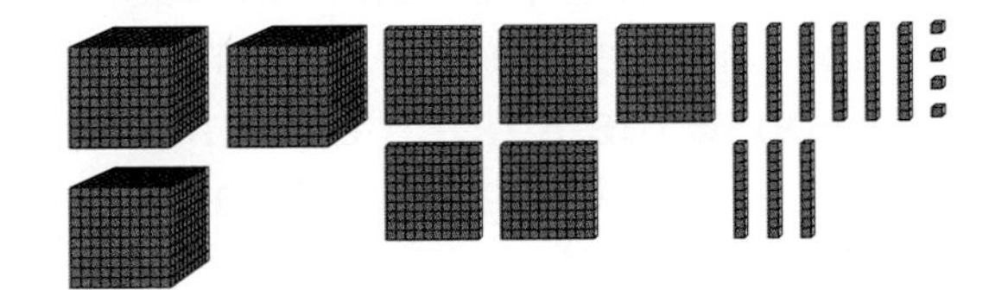

Compare and Order Numbers page 13

Compare. Write >, <, or =.

1 962 ● 926

2 3,916 ● 3,691

3 7,019 ● 7,901

4 12,934 ● 21,934

5 405,196 ● 405,196

6 801,359 ● 891,530

Order the numbers from least to greatest.

7 568; 856; 658

8 4,925; 49,250; 49,025

9 7,092; 9,207; 9,389

10 94,210; 9,421; 94,021

11 14,281; 8,365; 14,821

12 17,392; 17,923; 1,732

Solve.

13 Postage stamps come in sheets of 100. How many full sheets of stamps can you receive if you purchase 513 stamps?

14 You purchase 345 stamps and receive as many full sheets of stamps as possible. How many stamps are on the partial sheet?

15 Carolee has $35. Earlier, she bought a pair of gloves for $20 and socks for $10. How much money did she start with?

16 **Logical reasoning** Pat is creating his own secret code. The letter *A* is 1, *B* is 3, *C* is 5, and *D* is 7. What pattern is he using?

Problem-Solving Strategy: Make a Table page 15

Solve.

1 Akoni has 11 marbles. He has 6 more red marbles than blue marbles. He has twice as many blue marbles as yellow marbles. How many of each color does he have?

2 Ashley paid $8.95 for a book. Keira paid $7 more than Ashley for her book. Ken paid $2 less than Keira for his book. How much did Keira and Ken each pay for their books?

3 **Logical reasoning** A stuffed animal, a toy guitar, and a doll were prizes at a carnival. Brandon did not choose the doll. Keisha did not choose the stuffed animal. Cassie chose the guitar. What did Keisha choose?

4 Antonio works at a local grocery store. He earns $6 for each hour he works. Last week, he worked 2 hours on Tuesday, 4 hours on Friday, and 2 hours on Saturday. How much did he earn that week?

Round Numbers page 23

Round to the nearest ten.

1 47 **2** 314 **3** 4,098 **4** 6,152 **5** 79,996

Round to the nearest hundred.

6 448 **7** 9,105 **8** 45,051 **9** 38,649 **10** 508,954

Round to the nearest thousand.

11 4,099 **12** 9,601 **13** 25,741 **14** 98,389 **15** 399,563

Solve.

16 When rounding to the nearest ten, what is the greatest number that rounds to 620?

17 When rounding to the nearest ten, what is the least number that rounds to 620?

18 A special code is on facing pages in Joanne's secret code book. The total of the page numbers is 49. What are the numbers of the pages on which you will find the secret code?

19 When Onida walks 3 blocks to the park and another 4 blocks to the grocery, she is halfway to school. How many blocks will she walk from home to school and back?

Millions page 27

Write the number in standard form and in expanded form.

1. five million, six hundred three thousand, two hundred twenty-five
2. four million, nine hundred fifty-six thousand, one
3. six million, three hundred ninety-two thousand, nine hundred
4. three million, five hundred seven thousand, ninety
5. two million, one hundred nine thousand, sixty-seven
6. nine million, two hundred thousand, six hundred thirty-seven

Solve.

7. Suppose you save 100 pennies every day. If you continue at that rate, how many days will it take you to save 1,000,000 pennies? Explain your answer.
8. The highest volcano on Mars is recorded at 78,000 feet above the surface. Is this an actual or a rounded number? Explain your answer.

Problem Solvers at Work page 31

Solve.

1. May surveyed 26 students to find out their favorite color. She recorded the data in a tally table. Which color was favored by the most students?

Favorite Colors of Students

Color	Number of Students
Red	~~||||~~ ||
Blue	~~||||~~ |||
Yellow	||||
Green	~~||||~~ ||

2. Suppose the choices—both green—of two new students are added to the table. Will this change the answer to problem 1? Explain.
3. Describe what happens if two students change their choices—one from blue to yellow and one from green to yellow.
4. There are about 900 species of endangered birds. In 1978, there were about 300 species of endangered birds. By how many has the number of endangered species increased?
5. A hummingbird visits about 2,000 flowers each day to gather nectar. Use this information to write and answer a question.

Count Money and Make Change page 41

Use play money to show the amount of money in two ways.

1. 26¢
2. $1.02
3. $6.36
4. $11.45

Solve.

5. Sari bought a toy for $6.25. How much change did she get from $10?
6. List at least ten ways to show 75¢ using nickels, dimes, and quarters.

Compare, Order, and Round Money page 43

Write the amounts in order from least to greatest.

1. $1.15, $1.51, $5.51, $5.11
2. $25.83, $25.28, $5.75, $15.57

Solve.

3. An amount rounded to the nearest 10¢ is $11.50, to the nearest dollar is $12, and to the nearest $10 is $10. What could the amount be?
4. Karly paid with a ten-dollar bill. Her change was 4 one-dollar bills, 2 quarters, 1 nickel, and 3 pennies. How much was her purchase?

Addition Strategies page 45

Add mentally.

1. 200 + 568
2. $746 + $222
3. 462 + 321
4. $123 + $321
5. 816 + 183
6. 195 + 308 + 105
7. 520 + 113 + 280

Solve.

8. Banana Hut sold 215 shirts one week and 174 shirts the next week. How many shirts were sold in all?
9. Which would you rather have—1 ten-dollar bill, 11 quarters, and 10 pennies or 12 one-dollar bills, 3 quarters, and 1 dime? Explain.

Estimate Sums page 47

Estimate. Round to the nearest ten or ten cents.

1 \$0.55 + \$0.97 **2** \$0.92 + \$0.21 **3** 58 + 99 **4** 48 + 22

Estimate. Round to the nearest hundred or dollar.

5 153 + 246 **6** 637 + 152 **7** \$6.39 + \$9.67 **8** 8,469 + 743

Solve.

9 A model costs \$15.96 and a puzzle costs \$4.24. Can you purchase both items with \$20.00? Explain.

10 Bly paid with a five-dollar bill and 3 quarters. His change was 21¢. How much was his purchase?

Add Whole Numbers page 51

Add. Remember to estimate.

1 $\begin{array}{r} 43 \\ +\ 36 \\ \hline \end{array}$ **2** $\begin{array}{r} 72 \\ +\ 89 \\ \hline \end{array}$ **3** $\begin{array}{r} 385 \\ +\ 828 \\ \hline \end{array}$ **4** $\begin{array}{r} 3{,}764 \\ +\ \ \ 144 \\ \hline \end{array}$ **5** $\begin{array}{r} \$3.26 \\ +\ \ 8.75 \\ \hline \end{array}$

Solve.

6 Last year, attendance was about 95,000 people. This year, 9,500 more people are expected. What is the expected attendance this year?

7 Po, Rae, and Dee sell pads, rings, and dolls. The first letter of the item and name are not the same. Po sells dolls. What do Rae and Dee sell?

Three or More Addends page 53

Add. Remember to estimate.

1 $\begin{array}{r} 32 \\ 65 \\ +\ 19 \\ \hline \end{array}$ **2** $\begin{array}{r} \$4.28 \\ 6.53 \\ +\ \ 8.12 \\ \hline \end{array}$ **3** $\begin{array}{r} 6{,}203 \\ 475 \\ +\ \ \ 634 \\ \hline \end{array}$ **4** $\begin{array}{r} 356 \\ 482 \\ 27 \\ +\ \ 72 \\ \hline \end{array}$ **5** $\begin{array}{r} 1{,}111 \\ 3{,}333 \\ 88 \\ +\ \ \ \ 77 \\ \hline \end{array}$

Solve.

6 You have a twenty-dollar bill. Can you buy three craft items costing \$6.46, \$11.31, and \$3.55? Explain.

a **7** **ALGEBRA: PATTERNS** Show the next step.

Subtraction Strategies page 59

Subtract mentally.

1 42 − 9 **2** 53¢ − 12¢ **3** 56 − 15 **4** $805 − $795

Use the menu to solve.

5 Liza and Evan have $8.00 together. What can they have for lunch if they spend as much money as possible?

Menu			
Hot dog	$1.75	Hamburger	$3.25
Milk	.85	Lemonade	.95
Pretzel	.95	Chips	.75

Estimate Differences page 61

Estimate. Round to the nearest ten or ten cents.

1 84 − 39 **2** $4.44 − $2.52 **3** $42.28 − $16.15 **4** 567 − 296

Estimate. Round to the nearest hundred or dollar.

5 851 − 363 **6** 847 − 385 **7** 8,096 − 304 **8** $28.03 − $6.93

Estimate. Round to the nearest thousand or ten dollars.

9 3,053 − 1,843 **10** $63.95 − $8.72 **11** 8,763 − 2,933 **12** 9,393 − 279

Solve.

13 Last year the Skating Program earned $596. This year the goal of the organizers is to earn $725. About how much more do they hope to earn this year?

14 **ALGEBRA** Find the value for each symbol. Every symbol has the same value when it appears.

♣ + ♣ = 8

♦ + ♣ = 10

♦ + ♣ + ♣ = Δ

Subtract Whole Numbers page 65

Subtract. Remember to estimate.

1 48 − 25

2 84 − 37

3 419 − 62

4 $35.86 − 17.88

5 $7.24 − 2.55

Solve.

6 A sign is 18 inches wide. How much space will be on the left side if it is centered on a wall that is 36 inches wide?

7 A sketch pad costs $3.98, a brush costs $2.19, and a set of paints costs $5.68. What is the total cost?

Problem-Solving Strategy: Choose the Operation page 67

Solve.

1 Lily's family drove 1,045 miles to a family reunion. Then, they drove 485 miles to a friend's home. How far did they travel in all?

2 Park tickets cost $1.50 for each car and $2.00 for each person. How much change will 2 people in a car get from a ten-dollar bill?

3 Caleb counts 18 wheels on bicycles and cars. There are more cars than bicycles. How many cars and bicycles are there?

4 A ranger's shift is from 10:30 A.M. until 5:00 P.M. with $\frac{1}{2}$ hour off for lunch and a $\frac{1}{2}$ hour break. How long does the ranger work?

Subtract Across Zero page 69

Subtract. Remember to estimate.

1 $\begin{array}{r} 603 \\ -\ 72 \\ \hline \end{array}$

2 $\begin{array}{r} 500 \\ -\ 387 \\ \hline \end{array}$

3 $\begin{array}{r} \$50.30 \\ -\ 26.15 \\ \hline \end{array}$

4 $\begin{array}{r} 8{,}060 \\ -\ 926 \\ \hline \end{array}$

5 $\begin{array}{r} 9{,}001 \\ -\ 6{,}504 \\ \hline \end{array}$

Solve.

6 **Write a problem** that uses the numbers 3,050 and 5,008 and can be solved using subtraction.

7 Hal has $40. A new coat costs $28.73. How much change will he receive?

Problem Solvers at Work page 73

Solve.

1 Mitchell has two friends on the Internet. Akoni lives 3,045 miles away, and Juan lives 1,268 miles away. How much closer does Mitchell live to Juan than to Akoni?

2 One amusement park charges $37.75 for each adult ticket. Children under 12 can get in for $24.25. Mr. and Mrs. Hill are going to the park. How much will their tickets cost?

3 Jewel receives 35¢ in change. One coin is a quarter. List all the other coins she might receive to make up the total amount.

4 **Spatial reasoning** Use graph paper to arrange and shade six squares in as many different shapes as you can. Each square must share at least one side with another square.

Time page 83

Write the time in two different ways.

1

2

3

Solve.

4 School starts at quarter to nine. Jocelyn arrives at 9:01. Is she late or on time? How do you know?

5 Benjamin saved $22.89. Then he found $1.72 in the park. How much money does he have now?

Elapsed Time page 87

Tell what time it will be:

1 1 h 30 min after 6:15 P.M.

2 40 min after 2:10 A.M.

Solve.

3 Use the digits 2, 4, 7, and 9 to write numbers that are greater than seven thousand, five hundred.

4 Frank's soccer practice always ends at 4:00 P.M. It lasts 90 minutes. What time does practice start?

Problem-Solving Strategy: Work Backward page 89

Solve. Explain your methods.

1 School starts at 8:45 A.M. It takes Ray 15 minutes to walk to school. What is the latest time he can leave for school and not be late?

2 Kathy had $1.58 left after buying a book for $9.95. How much money did she have before her purchase?

3 January has 44,640 minutes. During leap years, February has 41,760 minutes. April has 43,200 minutes. Write the numbers in order from least to greatest.

4 Write the extra information in this problem. Then solve it. Oscar pays 25¢ for a 5-mile bus trip to Len's house. How far does he ride in all to visit Len and return home?

Range, Median, and Mode page 93

Find the median, range, and mode for the set of data.

1 Weights of Dogs (in pounds):
68 13 25 88 49 68 52

2 Amounts Saved by 7 Students:
\$95 \$60 \$3 \$36 \$86 \$36 \$57

Use the data in ex. 1–2 to tell whether the sentence is *true* or *false*. Then, explain your answer.

3 The heaviest dog weighs 52 pounds more than the lightest dog.

4 More students saved \$36 than any other amount.

Solve.

5 Yolanda buys stickers costing 35¢, 50¢, 85¢, 35¢, and 75¢. She says the mode price is 50¢. Tell why you agree or disagree with her.

6 **Spatial reasoning** List as many ways as you can to sort these shapes.

 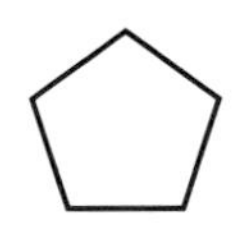  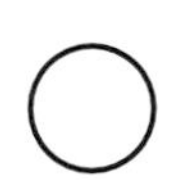 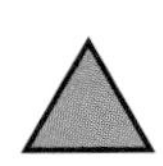 

Pictographs page 95

Use the pictograph for problems 1–6.

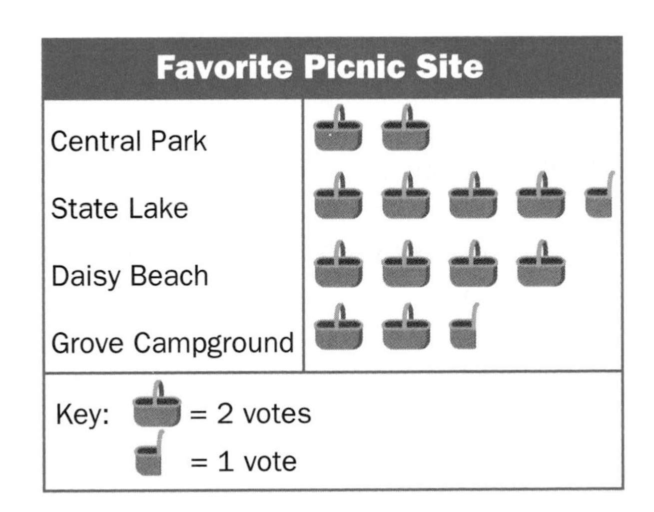

1 Which site has the most votes? How many votes did it receive?

2 How many students chose Grove Campground as the picnic site?

3 How many more students chose State Lake than chose Daisy Beach as the picnic site?

4 How many students were surveyed?

5 How would the pictograph change if each stands for 1 vote?

6 Which site has votes that can be added to the votes for Central Park to equal the sum of the votes for the other two sites?

Bar Graphs page 103

Use the bar graph for problems 1–5.

1 Which animal is the slowest? What is its speed?

2 Which two animals have the same speed? How do you know?

3 How much faster is a cheetah than a lion?

4 A human's fastest speed is about 28 miles per hour. Which animals are faster than a human?

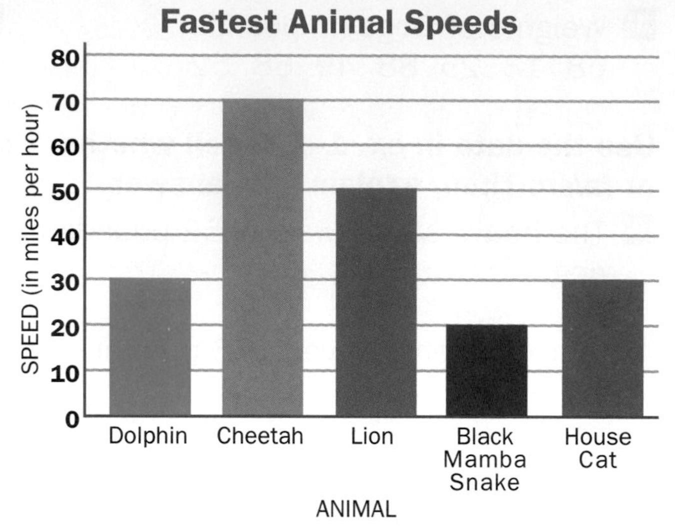

5 Can you list the animals in order from fastest to slowest by looking at the bars and not reading the scale numbers? Explain your answer.

Ordered Pairs page 105

Use the map to find the ordered pair for the location.

1 Picnic table

2 Slide

3 Wading pool

Tell what is found at the location on the map.

4 (1, 0)

5 (3, 2)

Solve.

6 The entrance to the park is above the wishing well and to the left of the park bench. What is the ordered pair that describes its location?

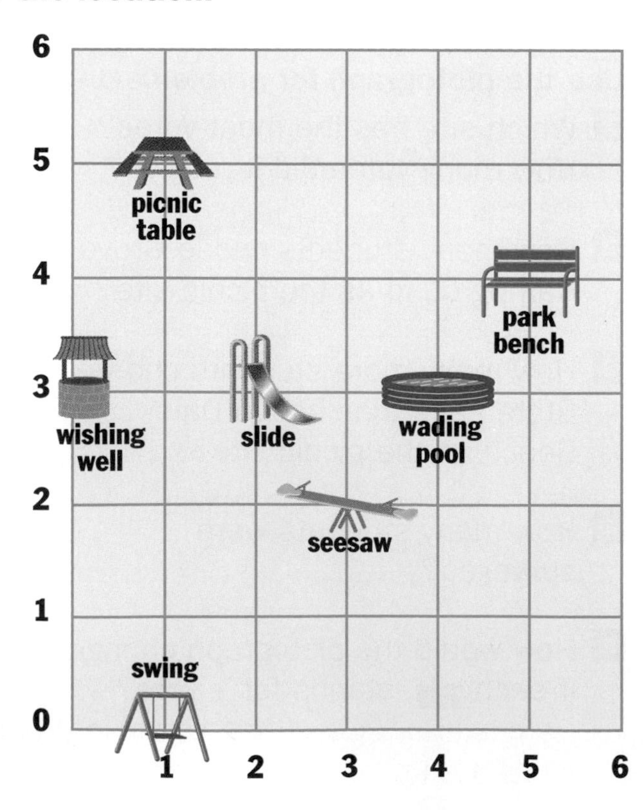

Line Graphs page 109

Use the line graph for problems 1–5.

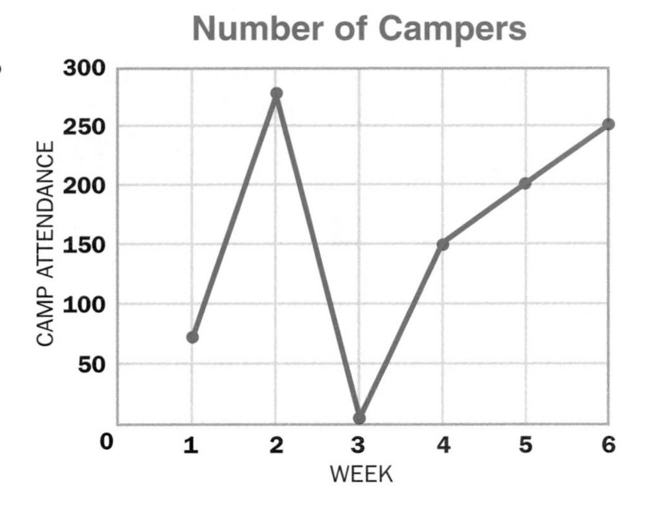

1. How many campers attended week 4?

2. During which week were there the fewest campers? Why do you think this happened?

3. Which weeks had the highest attendance?

4. Between which two weeks did the attendance increase the most?

5. What is the range in the number of campers?

6. Art class is held from 2:15 P.M. to 3:45 P.M. How long is the class?

Problem Solvers at Work page 113

Solve. Explain your methods.

1. The Backyard Club paid $15.62 for a pizza and juices. They gave a tip of $2.50. How much did they spend in all?

2. Keiko gave the clerk 2 five-dollar bills and 3 quarters. She received 2 dimes and 4 pennies in return. How much was her purchase?

3. **Logical reasoning** Jan is in front of Bob but behind Flo on the trail. Gary is in front of Flo. Who is in the lead on the trail?

4. Add a sentence so that you can solve the problem, then solve it. Cal has a flute lesson at 3:15 P.M. It takes him 20 minutes to walk home. What time will he get home?

5. In 1995, Mother's Day was on May 14 and Father's Day was on June 18. In 1996, Mother's Day was on May 12 and Father's Day was on June 16. How many days were between these special days in each of the years?

6. Use the line graph at the top of the page. When the attendance for week 7 is graphed, the line will go up. How many students could be attending that session? Explain your answer.

Meaning of Multiplication page 127

Multiply using any method.

1 3×8 **2** 8×2 **3** 4×5 **4** 7×4 **5** 9×3

6 1×5 **7** 2×6 **8** 3×4 **9** 6×6 **10** 3×5

Solve.

11 The baseball league has 6 teams. Each team has 9 positions. What is the least number of players in the league?

12 Use tiles to find two numbers that have a product of 12. Write a multiplication sentence for each way you can find.

2 Through 5 as Factors page 131

Multiply using any method.

1 3×2 **2** 3×6 **3** 4×4 **4** 6×5

5 9×2 **6** 3×7 **7** 4×7 **8** 7×5

9 2×4 **10** 3×3 **11** 4×8 **12** 5×3

Solve.

13 Leonard made four 2-point baskets and two 3-point baskets. How many points did he score altogether?

14 Latisha practices tennis for 1 hour every other day. How many hours does she practice tennis in 2 weeks?

6 and 8 as Factors page 133

Multiply using any method.

1 6×3 **2** 8×0 **3** 6×8 **4** 8×9 **5** 6×2

6 6×1 **7** 8×8 **8** 6×7 **9** 8×7 **10** 6×9

11 8×3 **12** 6×4 **13** 8×5 **14** 6×6 **15** 8×4

Solve.

16 Arthur's high school plays 8 football games each year. How many games can he attend during his 4 years of high school?

17 Kenicka bought a drink for $0.75 and a hot dog for $1.50. She received $3.00 in change. How much money did she give the clerk?

7 and 9 as Factors page 137

Multiply mentally.

1 7×2	**2** 9×1	**3** 7×7	**4** 9×4	**5** 7×9
6 9×5	**7** 7×1	**8** 9×7	**9** 7×0	**10** 9×9
11 7×3	**12** 7×6	**13** 9×6	**14** 7×8	**15** 9×8

Solve.

16 Marik competed in a kite-making contest. The judges gave him these scores: 9, 9, 8, 8, 9, 9. How many points did he score in all?

17 Each of 7 teams has 4 players. Write a multiplication and an addition sentence to show the total number of players.

Problem-Solving Strategy: Find a Pattern page 139

Solve. Explain your methods.

1 Chris alternates sprinting and jogging. She sprints for 30 seconds and then jogs for 3 minutes. She has trained 9 minutes. Will she sprint or jog for the next 30 seconds?

2 **Spatial reasoning** Copy and complete the diagram.

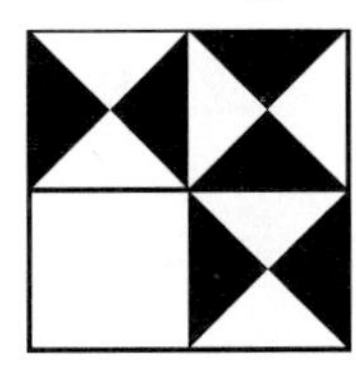

3 Jason treated himself and 2 friends to a baseball game and a juice. Tickets were $6 each and juice was $1 each. How much did he spend?

4 Write the information needed to solve the problem. Then solve it. The school has 6 jump ropes and 5 balls for each class. How many balls does the school have in all?

Three Factors page 141

Multiply.

1 $(6 \times 2) \times 4$	**2** $4 \times (8 \times 2)$	**3** $5 \times (2 \times 4)$	**4** $3 \times (9 \times 2)$	**5** $2 \times (5 \times 3)$
6 $(3 \times 7) \times 3$	**7** $(6 \times 2) \times 2$	**8** $4 \times (2 \times 7)$	**9** $2 \times (2 \times 3)$	**10** $(9 \times 2) \times 2$

Solve.

11 Casey lives 2 miles from school. He rides his bike to and from school 3 days each week. How many miles does he ride during his commute in 1 week?

12 Nathan does 15 minutes of warm-up stretches, 45 minutes of exercise, and 15 minutes of cool-off stretches. To finish at 6:45 P.M., what time should he start?

Meaning of Division page 149

Draw an array. Find the quotient.

1 $14 \div 2$ **2** $48 \div 8$ **3** $16 \div 4$ **4** $18 \div 9$ **5** $27 \div 3$

6 $21 \div 7$ **7** $36 \div 4$ **8** $12 \div 3$ **9** $24 \div 6$ **10** $3 \div 3$

Solve. Use the bar graph to answer.

11 Which game had the most points scored? Which had the least?

12 The league record is 40 points scored in 5 games. Did this team break the record? Explain.

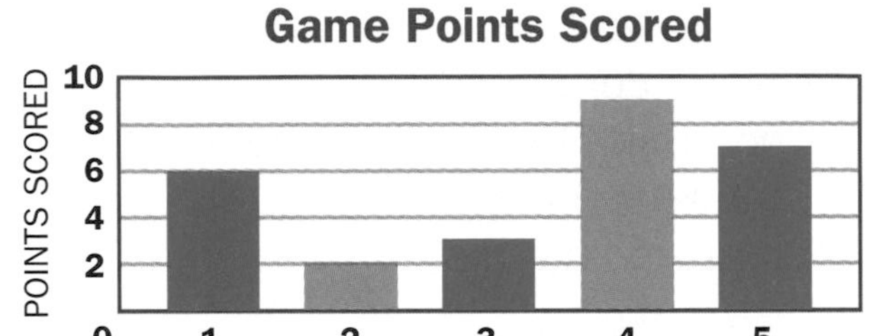

2 Through 5 as Divisors page 151

Divide.

1 $6 \div 2$ **2** $18 \div 3$ **3** $24 \div 3$ **4** $32 \div 4$ **5** $30 \div 5$

6 $2\overline{)18}$ **7** $3\overline{)15}$ **8** $4\overline{)20}$ **9** $5\overline{)25}$ **10** $5\overline{)30}$

Solve.

11 A 2-digit number can be divided evenly by 3 and 4. The sum of its digits is 6. What is the number?

12 Find the value for each symbol. The value is the same in all equations.

♣ × ♣ = 25
♦ ÷ ♣ = 3
♦ + ♣ + ♣ = ♥

6 Through 9 as Divisors page 153

Divide.

1 $18 \div 6$ **2** $49 \div 7$ **3** $72 \div 8$ **4** $63 \div 9$ **5** $54 \div 9$

6 $6\overline{)42}$ **7** $7\overline{)56}$ **8** $7\overline{)35}$ **9** $8\overline{)64}$ **10** $9\overline{)27}$

Solve.

11 The food for a picnic costs $45. The cost is to be split evenly between 9 people. How much will each person pay?

12 Each tour group consists of 5 people. A class of 30 students decides to take the tour. How many tour groups will they need for everyone?

Fact Families page 155

Find the fact family for the set of numbers.

1 2, 5, 10 **2** 4, 9, 36 **3** 6, 9, 54 **4** 9, 9, 81

Multiply or divide.

5 6×5 **6** $63 \div 7$ **7** 8×6 **8** $24 \div 4$ **9** 3×7

10 $72 \div 8$ **11** 2×7 **12** $40 \div 8$ **13** 6×7 **14** $56 \div 7$

Solve.

15 One rose costs $2. How much would it cost to buy each member of your family one rose? Use your data to write a fact family.

16 **ALGEBRA: PATTERNS** Find the rule. Then show the next set of dots.

•   • • •   • • • • •

Remainders page 157

Divide.

1 $20 \div 4$ **2** $38 \div 5$ **3** $32 \div 7$ **4** $65 \div 9$ **5** $55 \div 6$

6 $28 \div 3$ **7** $45 \div 7$ **8** $38 \div 6$ **9** $77 \div 8$ **10** $50 \div 8$

Solve.

11 A pictograph shows that 18 students like hockey and 24 students like tennis. What could each symbol stand for if only whole symbols are used in the graph?

12 Kelly is making banners to take to the basketball game. Each banner will be 3 feet long. She has a roll of paper 25 feet long. How many banners can she make?

Problem Solvers at Work page 161

Solve. Explain your methods.

1 Kiley works 2 hours each day at the golf-pro shop. He earns $24 for 2 days work. How much does he earn per hour?

2 Chane bowled two games and scored 85 and 113. Emilio scored 98 and 99. Who scored more points in all? How do you know?

3 **Spatial reasoning** How many small cubes make up the large block?

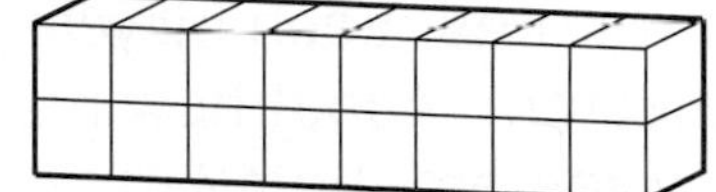

4 Felix, Marla, Hillary, and Liam are on a relay team. Marla runs before Liam. Felix follows Hillary. Hillary follows Liam. In what order do they run?

Multiplication Patterns page 171

Multiply mentally.

1 4 × 20 **2** 3 × 30 **3** 8 × 60 **4** 5 × 80 **5** 2 × 50

6 3 × 700 **7** 6 × 500 **8** 7 × 300 **9** 4 × 400 **10** 6 × 200

11 8 × 3,000 **12** 4 × 6,000 **13** 3 × 4,000 **14** 6 × 5,000 **15** 7 × 8,000

Solve.

16 It takes Karen 7 minutes to print a report from her computer. How long will it take to print 60 reports?

17 **ALGEBRA: PATTERNS** Explain how you found your answer.
5, 12, 19, 26, 33, ■, ■

18 Carlos gets paid $2 for every word in his article. If his article has 7,000 words, how much will he be paid?

19 Lee gave 6 and 800 as factors with a product of 4,800. Give another pair of factors with the same product.

Estimate Products page 173

Estimate the product.

1 6 × 39 **2** 6 × 96 **3** 7 × 43 **4** 8 × 60 **5** 5 × 86

6 7 × 568 **7** 6 × 701 **8** 4 × 652 **9** 5 × 979 **10** 6 × 337

11 4 × 6,593 **12** 5 × 7,593 **13** 4 × 9,103 **14** 6 × 5,045 **15** 3 × 8,453

16 39×5 **17** 56×6 **18** 19×7 **19** 14×8 **20** 84×6

Solve.

21 Bicycle helmets cost $9 each. Kneepads cost $3 each. About how much will 187 helmets and 308 kneepads cost?

22 **Logical reasoning** Which two numbers do *not* belong? Explain your reasoning.
20, 27, 62, 70, 328, 429

23 The bicycle race started at 9:30 A.M., and the last person finished the race at 4:45 P.M. How long did the race last?

24 Identify any extra information. There are 387 bicycles in each of four stores. Each bicycle has 5 reflectors. About how many bicycles are there?

Use Models to Multiply page 177

Find the total number of squares in the rectangle without counting all the squares.

1 

2 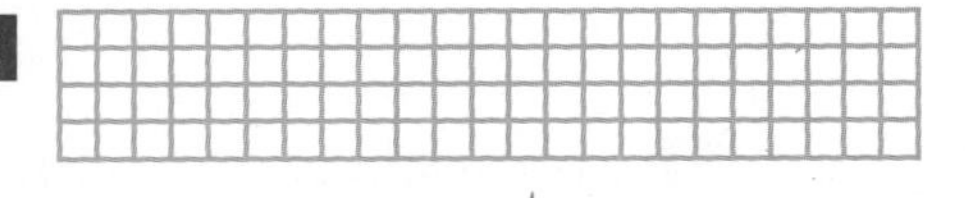

3 

4 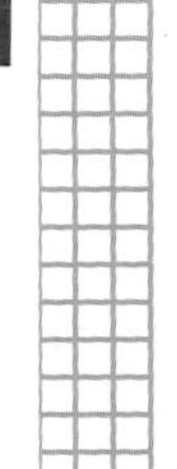

5 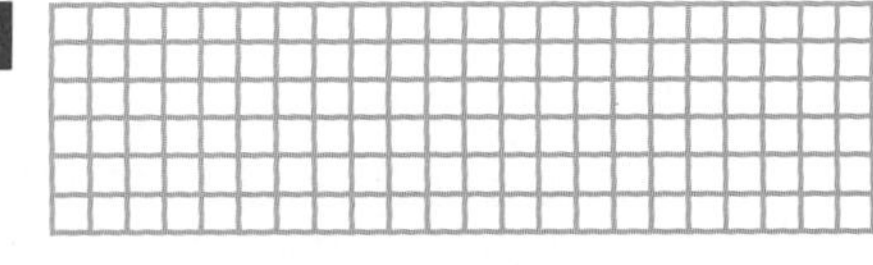

6 

Multiply using any method.

7. 39×5
8. 93×6
9. 19×4
10. 91×3
11. 72×5

Multiply 2-Digit Numbers page 181

Multiply.

1. 83×3
2. 93×6
3. 82×4
4. 91×3
5. 72×5
6. 75×8
7. 97×4
8. 43×5
9. 78×2
10. 27×6
11. 6×55
12. 8×99
13. 7×30
14. 4×76
15. 5×36
16. 4×61
17. 2×58
18. 3×42
19. 9×22
20. 7×44
21. 3×48
22. 5×47
23. 4×83
24. 3×98
25. 5×26

Solve.

26. Peter gives $3 to charity from every piñata he sells. If he sold 68 piñatas this week, how much money will he give?
27. The large piñata holds 768 candies. The small piñata holds 75 candies. How many more candies does a large piñata hold?

Problem-Solving Strategy:
Solve Multistep Problems page 183

Solve.

1. Carole is designing outfits for five action figures. Each action figure has outfits that sell for $2.39, $2.99, $2.99, and $3.99. What is the cost of buying all the outfits for the action figures?

2. Lee bakes 55 batches of cookies. Each batch has 9 pounds of cookies. If he needs to meet a target of 550 pounds of cookies, how many more pounds of cookies does Lee need to bake?

3. Karen rounds the median price to estimate the cost of rolls of paper. Her prices are $45.59, $37.50, $46.75, $39.00, and $51.99. What is Karen's estimate for 9 rolls of paper?

4. James walks 4 miles every morning and 6 miles every evening. Steve walks 7 miles every morning and 2 miles every evening. Who walks more in a week? How much more?

5. Robert leaves work at 5:00 P.M. He shops for 45 minutes and then takes 30 minutes to get home. What time does he get home?

6. A bag of fruit costs $4, and a box of fruit costs seven times as much as a bag. How much will Mei spend if she buys 7 bags and 4 boxes?

Multiply Greater Numbers page 191

Multiply.

1. 984×2
2. 802×9
3. 455×8
4. 274×6
5. 772×7
6. 607×5
7. 910×9
8. $1{,}023 \times 5$
9. $5{,}605 \times 7$
10. $2{,}345 \times 3$
11. 6×555
12. 8×909
13. 6×354
14. 4×747
15. 5×398
16. $5 \times 8{,}033$
17. $9 \times 6{,}765$
18. $5 \times 9{,}003$
19. $4 \times 4{,}888$
20. $9 \times 8{,}439$

Solve.

21. **Make a decision** Machine 1 fills 3,905 pies in an hour. Machine 2 fills 800 in 2 hours. Which machine would you buy? Why?

22. Fruity Juices packs 907 boxes on a train for delivery. Every day, 8 trains leave the station. How many boxes can Fruity Juices deliver in 6 days?

Multiply with Money page 193

Multiply.

1. $\$0.47 \times 5$
2. $\$0.16 \times 8$
3. $\$0.22 \times 7$
4. $\$0.38 \times 5$
5. $\$0.73 \times 4$
6. $\$1.53 \times 6$
7. $\$1.80 \times 9$
8. $\$2.77 \times 8$
9. $\$4.82 \times 3$
10. $\$3.33 \times 4$
11. $4 \times \$0.85$
12. $7 \times \$9.09$
13. $6 \times \$6.35$
14. $5 \times \$0.03$
15. $4 \times \$3.98$
16. $7 \times \$0.06$
17. $5 \times \$5.48$
18. $4 \times \$3.42$
19. $9 \times \$9.22$
20. $8 \times \$9.49$
21. $3 \times \$59.64$
22. $8 \times \$49.54$
23. $4 \times \$68.37$
24. $7 \times \$11.38$
25. $7 \times \$66.24$

Solve.

26. Tía buys 4 flower patches for her new denim jacket. Each patch costs $3.75. How much does Tía spend on patches?

27. Tía buys 3 yards of red ribbon and 2 yards of gold ribbon. Each yard of red ribbon costs $0.45. Each yard of gold ribbon costs $0.85. What is the total cost?

Problem Solvers at Work page 197

Solve. Explain your method.

1. Rashid wants to hike for at least 20 miles. The hiking paths he wants to use are 3 miles, 4 miles, 7 miles, 8 miles, and 9 miles long. Give three ways that Rashid can meet his goal.

2. Heather and her family are allowed 400 pounds of luggage. She and her two brothers each have luggage that weighs 70 pounds. Her parents' luggage weighs 160 pounds. If Heather carries an extra bag, what is the most it can weigh?

3. Karl bought 4 books. Three of the books cost $6.85, $24.99, and $29.50. He got $3.28 in change from $80. How much did the fourth book cost?

4. Rosa starts to list factors that give the product 192. She gives 8 and 24 as an answer. Give two more pairs of factors.

5. Marsha says she can solve 4×725 by finding 4×700 and 4×25 and adding both products. Is she correct? Why or why not?

6. **Write a problem** that can be solved by finding 7×44. Solve it. Did you use mental math, paper and pencil, or a calculator? Why?

Multiplication Patterns page 207

Multiply mentally.

1. 50 × 100
2. 30 × 40
3. 80 × 60
4. 60 × 90
5. 20 × 300
6. 400 × 90
7. 300 × 30
8. 80 × 400
9. 70 × 900
10. 40 × 500
11. 20 × 2,000
12. 60 × 7,000

ALGEBRA Find the missing number.

13. 30 × ■ = 27,000
14. ■ × 90 = 54,000
15. ■ × 4,000 = 280,000

Solve.

16. A recycling bin holds about 100 cans. About how many cans have been recycled when the bin has been filled 20 times?

17. Lenora receives 8¢ for each bottle she returns. She returns 10 bottles. Then she spends 35¢. How much money does she have left?

Estimate Products page 209

Estimate the product by rounding.

1. 38 × 42
2. 57 × 52
3. 405 × 24
4. 77 × 827
5. 29 × 541
6. 651 × 84
7. 11 × 5,602
8. 49 × 2,204
9. 18 × 74
10. 23 × 45
11. 304 × 26
12. 491 × 63
13. 3,217 × 53

Solve.

14. Zena's class spends 15 minutes picking up trash each week. Estimate how many minutes they pick up trash in 13 weeks.

15. Jay bikes 2 miles to work. Alex bikes twice as far as Jay. Frank bikes twice as far as Alex. How far is Frank from his place of work?

Multiply 2-Digit Numbers page 213

Find the product. Use any method.

1 19×40 **2** 20×14 **3** 11×46 **4** 34×27 **5** 23×12

6 26×21 **7** 34×16 **8** 41×32 **9** 29×19 **10** 44×14

11 14×19 **12** 21×30 **13** 32×32 **14** 26×15 **15** 19×44

Solve.

16 **Logical reasoning** The product of two numbers is 1,800. One number is twice the value of the other. What are the two numbers?

17 There are 12 inches in 1 foot. How many inches deep is a pond with a depth of 15 feet?

Multiply 2-Digit Numbers page 217

Multiply using any method. Remember to estimate.

1 21×50 **2** 35×13 **3** 24×15 **4** 13×37 **5** 41×23

6 72×31 **7** 18×44 **8** 29×53 **9** 66×33 **10** $\$0.12 \times 17$

11 38×20 **12** 19×52 **13** $\$0.24 \times 61$ **14** 12×32 **15** 74×11

Solve.

16 A farmer's stand sells 3 melons for $5.50. A market sells melons for $2 each. How much less do 6 melons cost at the stand than at the market?

17 Callan started picking strawberries at 8:30 A.M. She worked for 3 hours 20 minutes. When did she stop?

Problem-Solving Strategy: Use Alternate Methods page 223

Solve. Use the bar graph to answer problems 3–5. Explain your methods.

1 Simon, Greta, Malik, and Yoko work to save deserts, panthers, whales, or wetlands. Greta saves animals. Malik saves animals from boats. Yoko saves wetlands. Who works with which resource?

2 Copy the rectangle and show as many ways as you can to color one half.

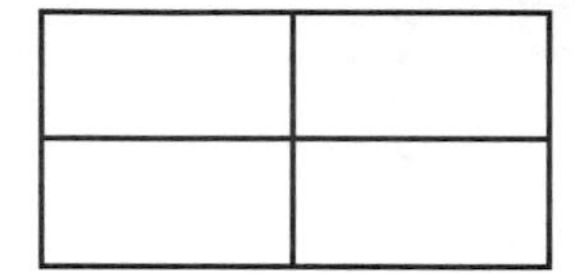

3 How many hours did students work during Week 1?

4 How many more hours were worked during the first two weeks than during the last two weeks?

5 During which weeks were the same number of hours worked?

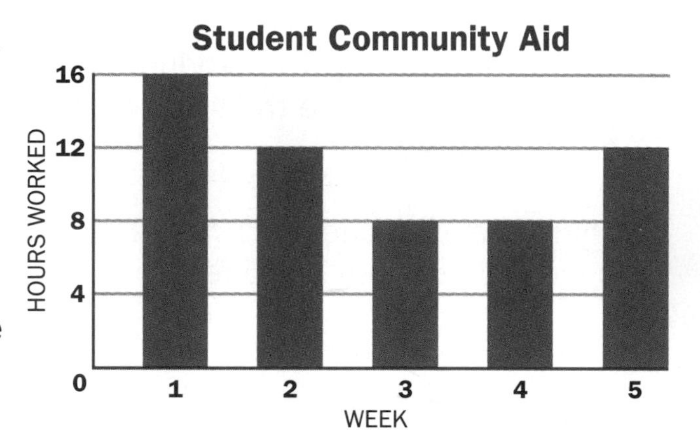

Multiply 3-Digit Numbers page 225

Multiply. Remember to estimate.

1 107×73 **2** 394×21 **3** 583×22 **4** 460×34 **5** 809×45

6 $\$6.41 \times 17$ **7** $\$0.29 \times 31$ **8** $\$1.98 \times 64$ **9** $\$5.03 \times 37$ **10** $\$2.40 \times 28$

11 42×118 **12** 22×372 **13** 36×511 **14** 13×309

Solve.

15 The Cho family conserves water and uses about 178 gallons less per month than they used to use. Estimate how much less water they will use over 2 years.

16 A regular light bulb lasts about 750 hours. An energy-saving bulb lasts about 10,000 hours. How many more hours does the energy-saving bulb last?

Multiply Greater Numbers page 227

Multiply mentally.

1. $3{,}000 \times 40$
2. $9{,}000 \times 70$
3. $11{,}000 \times 30$
4. $48{,}000 \times 20$
5. $31{,}000 \times 30$

Multiply using any method. Remember to estimate.

6. 879×72
7. $\$5.38 \times 21$
8. $5{,}309 \times 18$
9. $\$12.15 \times 83$
10. $4{,}006 \times 69$
11. $\$46.03 \times 24$
12. $5{,}131 \times 17$
13. $\$32.05 \times 36$
14. $8{,}060 \times 23$
15. $\$24.26 \times 21$
16. $27 \times 1{,}365$
17. $3{,}815 \times 34$
18. $\$28.16 \times 31$
19. $16 \times \$31.08$

Solve.

20. One club sold T-shirts to celebrate Earth Day. An adult's shirt sold for $15, and a child's shirt sold for $9. The club sold 1,230 adults' shirts and 462 children's shirts. How much money did the club make in sales?

21. At one time, it was estimated that 2 million, 100 thousand tons of used motor oil seeped into our waterways. Write this number in both standard and expanded forms.

Problem Solvers at Work page 231

Solve. Tell whether you can estimate to solve problems 1–2. Explain your methods.

1. The Morales family lowered the temperature in their house and saved from $30 to $50 each month from November through April. How much did they save over the 6 months?

2. The animal shelter had a race to raise money. Howie had pledges of $14.25 for each of the 15 miles he ran in the race. How much money did he earn for the shelter?

3. A grocery store gives customers a 3¢ discount each time they bring their own bag. How much will you save by using your own bag once a week for one year?

4. **Spatial reasoning** Copy the three shapes. Then try to fit them together to form a square.

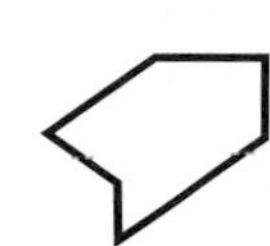

extra practice

Length in Customary Units page 243

Write the letter of the best estimate.

1. width of a door — **a.** 3 in. **b.** 3 ft **c.** 3 yd
2. length of a pencil — **a.** 7 in. **b.** 7 ft **c.** 7 yd
3. length of a baseball bat — **a.** 1 in. **b.** 1 ft **c.** 1 yd
4. width of a room — **a.** 12 in. **b.** 12 ft **c.** 12 yd
5. speed limit on the highway — **a.** 65 ft/h **b.** 65 yd/h **c.** 65 mi/h

Rename Customary Units of Length page 245

Complete.

1. 9 yd = ■ ft
2. 24 in. = ■ ft
3. 12 ft = ■ yd
4. 7 ft = ■ in.
5. 48 ft = ■ yd
6. 2 yd = ■ ft
7. 10 mi = ■ yd
8. 45 ft = ■ yd

Write >, <, or = .

9. 18 ft ● 5 yd
10. 96 ft ● 35 yd
11. 24 in. ● 2 ft
12. 2,000 yd ● 1 mi
13. 12 ft ● 132 in.
14. 3 yd ● 48 ft
15. 120 in. ● 12 ft
16. 3 mi ● 6,000 yd
17. 9 yd ● 36 ft
18. 360 in. ● 25 ft
19. 14 ft ● 168 in.
20. 7,000 yd ● 4 mi

Solve.

21. The length of a blue whale measures about 33 yards. The length of a finback whale measures about 84 feet. Which of these whales has a longer length? How much longer?
22. **Logical reasoning** The largest fish ever recorded was a whale shark over 41 ft long. It was about as long as:
    a. a closet.
    b. a schoolroom.
    c. an ocean liner.

Length in Metric Units page 249

Write the letter of the best estimate.

1 height of a woman **a.** 16 cm **b.** 16 dm **c.** 16 m

2 width of a teacher's desk **a.** 7 cm **b.** 7 dm **c.** 7 m

3 length of a minivan **a.** 5 cm **b.** 5 dm **c.** 5 m

4 height of a dog **a.** 60 cm **b.** 60 dm **c.** 60 m

5 length of a small boat **a.** 7 cm **b.** 7 dm **c.** 7 m

Solve.

6 **Write a problem** that has this answer: Michelle traveled 1,987 km altogether.

7 Ray buys 6 bags of lures packaged 12 in a bag. How many lures did he buy?

Perimeter page 251

Find the perimeter.

1

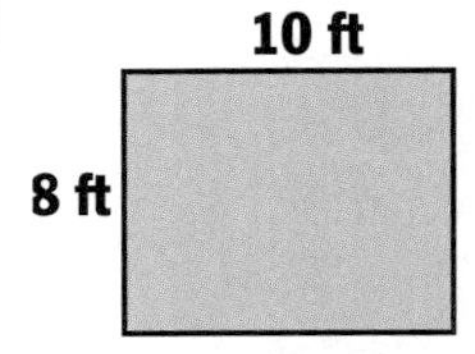

2

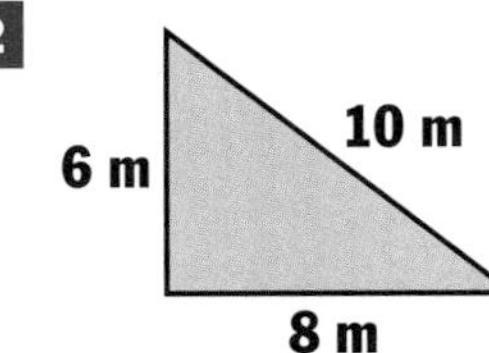

3

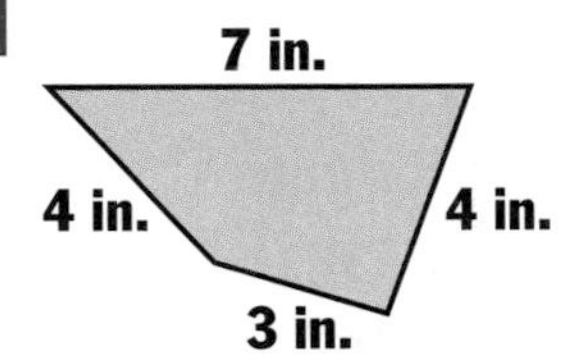

4

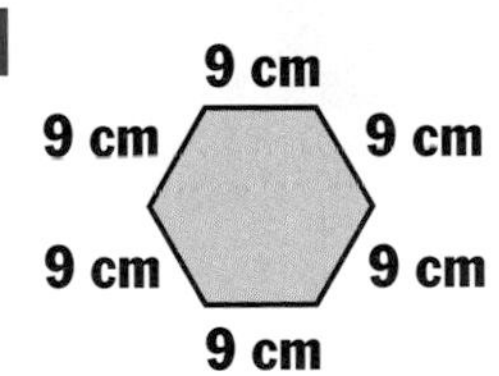

5

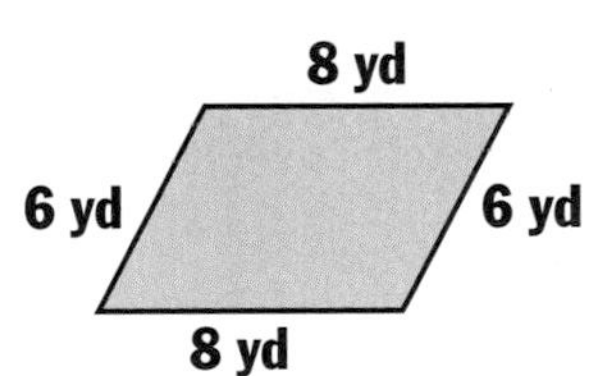

6

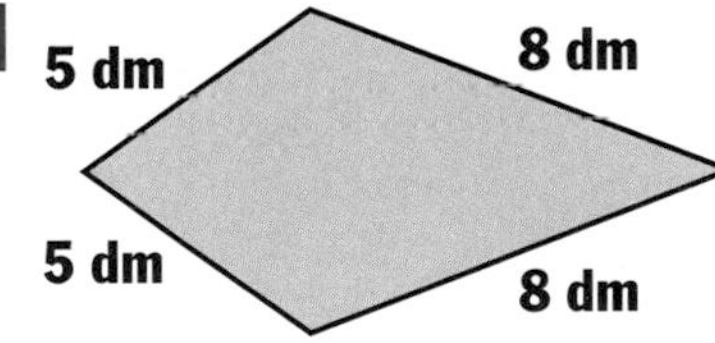

Solve.

7 The perimeter of Jack's sand castle is 3 ft. Each side of Este's square sand castle measures 8 in. Whose castle has a greater perimeter?

8 A ship's captain ordered 218 cases of soda for the ship's crew. There are 24 cans of soda in each case. How many cans did he order?

Capacity and Weight in Customary Units page 259

Choose the best unit to measure capacity. Write *cup, pint, quart,* or *gallon.*

1 bathtub
2 coffeepot
3 cocoa mug
4 auto gas tank
5 juice box
6 swimming pool
7 saucepan
8 water bucket

Choose the better unit to measure weight. Write *ounce* or *pound.*

9 pair of skates
10 paper clip
11 butter
12 thumbtack
13 pencil
14 bicycle
15 straight pin
16 big dog

Solve.

17 Kristy's thermos bottle holds 3 gallons. How many quarts of lemonade can Kristy put in the bottle?

18 Denny spends $1.50 each school day for lunch. How much will he spend for lunch in 4 weeks?

Problem-Solving Strategy: Use Logical Reasoning page 261

Solve. Explain your methods.

1 Copy the tangram at the right and try to arrange the pieces to make the pictures shown below. Then, try to make up your own picture using the pieces. Be sure to use all pieces and do not overlap them.

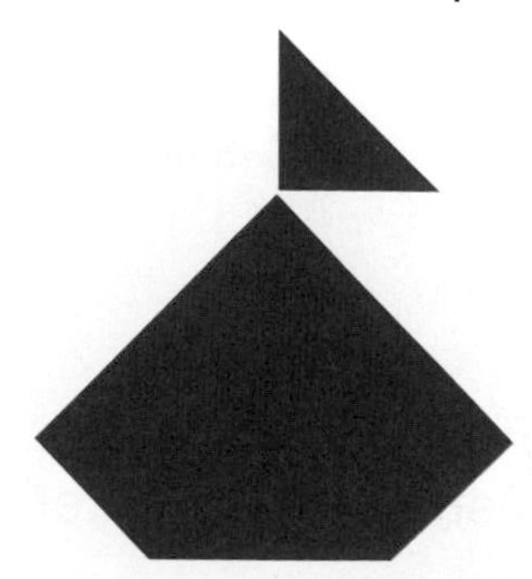

2 Jeffrey and his family took a boat ride from Medora to Sandy Point. They arrived at Sandy Point at 8:45 A.M. and left at 4:20 P.M. How much time did they spend at Sandy Point?

3 Dennis is saving money to make a trip to Plant City's aquarium. For every $3 he saves, his mother gives him $2. How much money will his mother have given him when he has a total of $30?

Capacity and Mass in Metric Units page 265

Choose the best unit to measure. Write *milliliter, liter, gram,* or *kilogram.*

1 How much liquid does a teaspoon hold?

2 How much do you weigh?

3 How heavy is a paper clip?

4 How much water does a bathtub hold?

Complete.

5 6,000 mL = ■ L **6** 9 L = ■ mL **7** 8 kg = ■ g **8** 3,000 g = ■ kg

Solve.

9 A pitcher contains 1 liter of juice. Rex uses the pitcher to fill three 100-mL containers and two 250-mL containers. Can he fill yet another 250-mL container? Explain.

10 Luke uses 18 posters for his report on ocean life. Five posters show fish, and seven posters show plant life. The rest show other marine animals. How many posters show other marine animals?

Problem Solvers at Work page 269

Solve. Explain your methods.

1 Caroline said, "If the temperature is 95 degrees tomorrow, we will swim in the lake." Did she mean 95° Celsius or 95° Fahrenheit? Explain.

2 While scuba diving, Enrique saw 8 fish, each with 5 red stripes. He also saw 9 fish, each with 4 blue spots. How many fish did he see in all?

3 **Logical reasoning** A symbol in a pictograph stands for 2 students' responses. If 8 students respond they like to swim and 4 students respond they like to water ski, how many symbols will you use to record these responses?

4 **Logical reasoning** Mei, Al, and Kay are wearing fishing hats. One is red, one is yellow, and one is orange. Neither girl is wearing a red hat. Mei wishes she had an orange hat. What color is each child wearing?

Division Patterns page 279

Divide mentally.

1 50 ÷ 5 **2** 90 ÷ 3 **3** 40 ÷ 4 **4** 120 ÷ 2

5 280 ÷ 4 **6** 560 ÷ 8 **7** 3,500 ÷ 7 **8** 2,400 ÷ 6

Solve.

9 Josie shares 480 pennies equally among 7 friends and herself. How many pennies will each have?

10 **ALGEBRA: PATTERNS** Write the next two numbers: 0, 3, 1, 4, 2, 5, ■, ■. Explain how you know.

Estimate Quotients page 281

Estimate by using compatible numbers.

1 62 ÷ 6 **2** 57 ÷ 3 **3** 178 ÷ 9

4 275 ÷ 4 **5** 556 ÷ 8 **6** 368 ÷ 6

7 432 ÷ 7 **8** 215 ÷ 5 **9** 3,545 ÷ 4

10 4,193 ÷ 6 **11** 3,125 ÷ 4 **12** 6,441 ÷ 9

Divide by 1-Digit Numbers page 285

Divide. You may use place-value models.

1 45 ÷ 7 **2** 76 ÷ 8 **3** 34 ÷ 5 **4** 53 ÷ 7

5 29 ÷ 4 **6** 47 ÷ 8 **7** 68 ÷ 9 **8** 36 ÷ 7

9 35 ÷ 2 **10** 85 ÷ 4 **11** 98 ÷ 3 **12** 58 ÷ 5

Solve.

13 Joe has 89 cubes. He groups them in piles of 5. How many piles of 5 can he make? How many cubes are left over?

14 **Logical reasoning** Al is standing in line between Joe and Mary. Minnie is directly in front of Joe. Two students are between Joe and Jeff. How are they lined up?

Divide by 1-Digit Numbers page 289

Divide. Remember to estimate.

1 $4\overline{)92}$ **2** $3\overline{)57}$ **3** $5\overline{)84}$ **4** $2\overline{)71}$ **5** $4\overline{)\$72}$

6 $7\overline{)89}$ **7** $6\overline{)95}$ **8** $9\overline{)358}$ **9** $8\overline{)579}$ **10** $2\overline{)195}$

11 $8\overline{)253}$ **12** $9\overline{)\$387}$ **13** $6\overline{)513}$ **14** $7\overline{)535}$ **15** $4\overline{)513}$

16 $5\overline{)128}$ **17** $3\overline{)196}$ **18** $7\overline{)501}$ **19** $9\overline{)318}$ **20** $4\overline{)145}$

Solve.

21 Jorge has 4 quarters, 3 dimes, and 2 nickels. How many 5-cent decals can he buy at the souvenir stand?

22 A roller coaster seats 36 people. Each seat holds 3 people. How many seats are on the ride?

Zeros in the Quotient page 291

Divide.

1 $8\overline{)854}$ **2** $2\overline{)419}$ **3** $9\overline{)960}$ **4** $3\overline{)306}$ **5** $7\overline{)914}$

6 $5\overline{)653}$ **7** $4\overline{)682}$ **8** $4\overline{)439}$ **9** $8\overline{)879}$ **10** $6\overline{)918}$

11 $91 \div 3$ **12** $83 \div 4$ **13** $\$303 \div 3$ **14** $\$847 \div 7$

Solve.

15 Craig has 618 compact discs. He stores them in small boxes with 6 discs in each box. How many boxes does he fill?

16 Jo leaves home and travels 3 blocks north, 2 blocks east, 5 blocks south, and 2 blocks west. How far is she from home?

Divide Greater Numbers page 293

Divide.

1 $5\overline{)816}$ **2** $6\overline{)\$8,526}$ **3** $4\overline{)9,741}$ **4** $9\overline{)4,613}$ **5** $6\overline{)\$5,124}$

6 $6\overline{)4,394}$ **7** $3\overline{)1,152}$ **8** $9\overline{)4,559}$ **9** $2\overline{)3,051}$ **10** $8\overline{)40,445}$

11 $2,945 \div 4$ **12** $4,995 \div 6$ **13** $1,672 \div 5$ **14** $\$29,344 \div 7$

Problem-Solving Strategy: Guess, Test, and Revise page 295

Solve. Explain your methods.

1 The sum of two dog-tag numbers is 57. Their difference is 33. What are the numbers?

2 A ticket to the dog show costs $8. Can Jake buy 14 tickets for $100? Explain.

3 Jean's collie weighs 25 pounds more than Dan's chihuahua and 62 pounds less than Joe's 97-pound rottweiler. How much does each dog weigh?

4 The Kennedy Dog Association is going to spend $72 for dog collars. It can buy either $6 or $9 collars. How many more $6 collars can it buy than $9 collars?

5 Mindy spent $52 for 2 doggie sweaters. One sweater cost $6 more than the other. How much did each sweater cost?

6 Jay has twice as much money as Ruth. Ruth has twice as much money as Irene. Irene has 45¢. How much money does Jay have?

Average page 301

Find the average.

1 Number of miles to and from school: 7, 13, 21, 19

2 Test scores: 88, 76, 91, 89

3 Number of minutes each student spends on homework: 35, 65, 80, 40, 30

4 Number of tickets sold for the school carnival: 235, 198, 206

Use the pictograph to find the answers to ex. 5–7.

Number of Rebounds	
Lee	● ◐
Abby	● ● ● ◐
Bo	● ● ● ● ●
Pat	● ●
Key: ● represents 2 rebounds. ◐ represents 1 rebound.	

5 How many rebounds were made in all?

6 Find the average number of rebounds.

7 How many more rebounds did Abby have than Lee?

Divide by Multiples of Ten page 303

Divide mentally.

1. 720 ÷ 80
2. 360 ÷ 60
3. $560 ÷ 70
4. 210 ÷ 30
5. $3,000 ÷ 50
6. 1,000 ÷ 20
7. 8,100 ÷ 90
8. 2,800 ÷ 70
9. 6,300 ÷ 90
10. 4,000 ÷ 80
11. $9,000 ÷ 30
12. 8,000 ÷ 40

Solve.

13. The Rodriguez family spent $264 for food on their 8-day vacation. Find the average spent per day.
14. The Rodriguez family brought 8 boxes of cereal on the trip. One box makes 8 servings. How many boxes will be left after 48 servings?

Divide by Tens page 307

Divide using any method.

1. $10\overline{)91}$
2. $20\overline{)75}$
3. $20\overline{)159}$
4. $20\overline{)113}$
5. $30\overline{)256}$
6. 128 ÷ 20
7. 157 ÷ 30
8. $120 ÷ 30
9. 132 ÷ 40
10. 194 ÷ 40
11. 161 ÷ 50
12. $420 ÷ 70
13. 123 ÷ 60
14. 187 ÷ 20
15. 125 ÷ 90

Solve.

16. Jim has 150 marbles, 10 times as many as Ken. How many does Ken have?
17. Alan drives 90 kilometers per hour. How many kilometers will he drive in 9 hours?

Problem Solvers at Work page 311

Solve. Explain your methods.

1. Betty uses one ribbon to make 3 hair bows. She wants to make 115 bows in all. How many ribbons does she need?
2. The cows and ducks in the barnyard have a total of 7 heads and 20 feet. How many cows and ducks are there in all?
3. **Logical reasoning** Tom's ride home is less than Jane's. Tami's ride is more than Harry's, but less than Tom's. Who has the longest ride?
4. Ryan has a test score of 88. What does he need to score on his next test to have an average of 90?

3-Dimensional Figures page 323

Describe each figure. Tell how many edges, vertices, and faces it has.

1 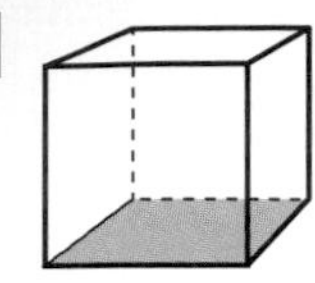

2 

3 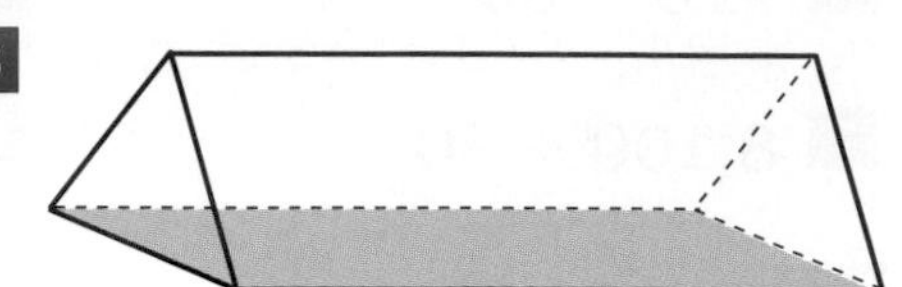

Solve.

4 How many more flat faces does a square pyramid have than a triangular pyramid?

5 Margarita has 38¢ in a piggy bank. There are 7 coins in the bank. What are the coins?

2-Dimensional Figures and Polygons page 327

Tell if the figure is open or closed.

1 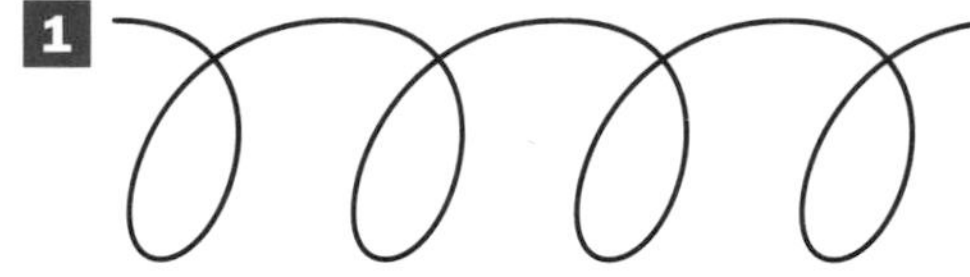

2 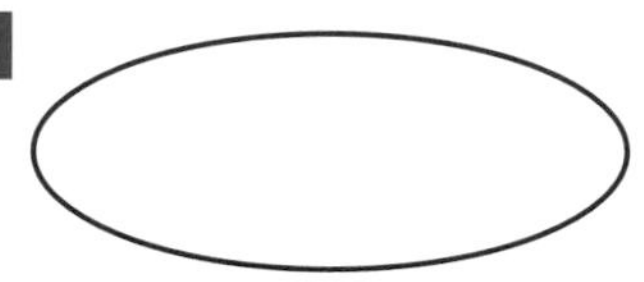

3 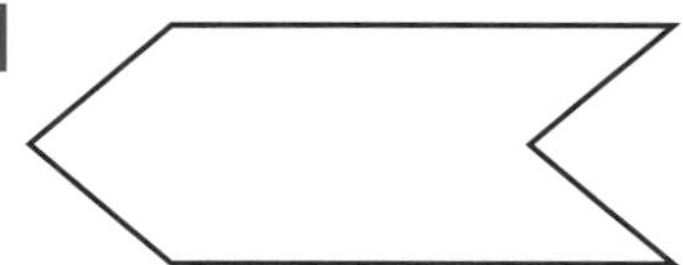

Name the polygon.

4 a polygon with 8 sides

5 a quadrilateral with 4 equal corners

Line Segments, Lines, and Rays page 329

Describe the figure.

1 G H

J K

2 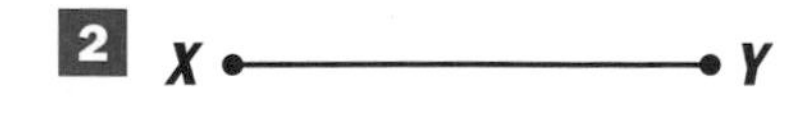

3 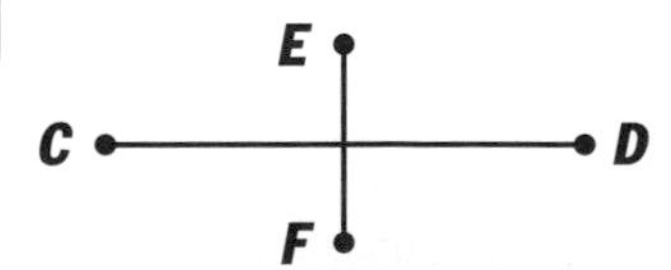

Solve. Use the figure at the right for problems 4–6.

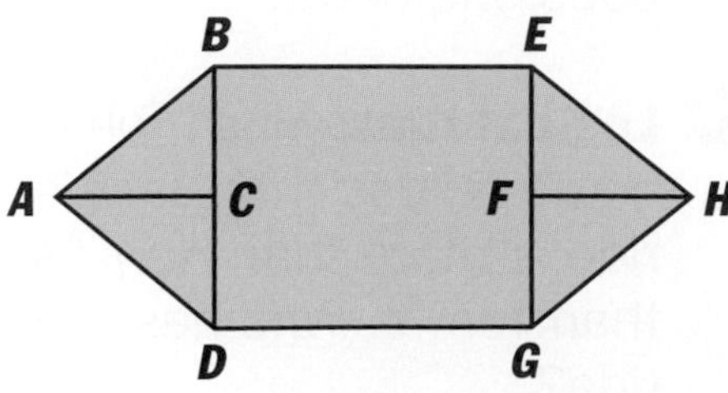

4 List two pairs of parallel line segments.

5 List two pairs of perpendicular line segments.

6 List two intersecting line segments that are *not* perpendicular.

Problem-Solving Strategy: Make an Organized List page 331

Solve. Explain your methods.

1 **Spatial reasoning** Study the diagram at the right. Then predict how many triangles there will be in the 10-sided figure that extends this pattern.

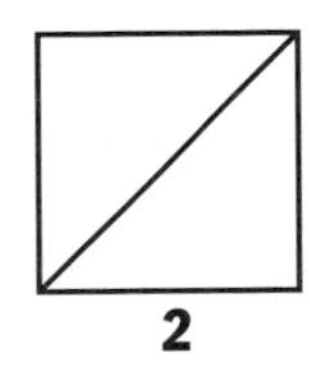

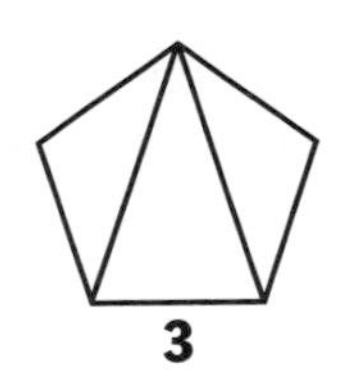

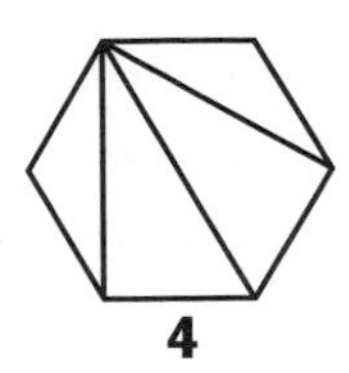

Angles page 335

Write *acute, obtuse,* or *right* for the angle.

1

2

3

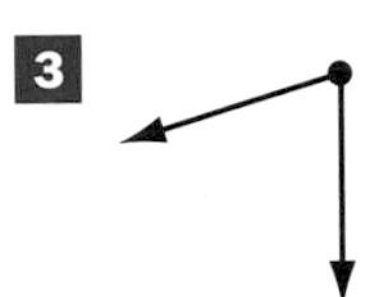

Solve.

4 Dan buys 1 gal of water; Mel buys 3 pints. How many pints do they have in all?

5 How many more right angles are in a square than in a right triangle?

Congruency and Similarity page 343

Write *congruent* or *similar* for the pair of figures.

1

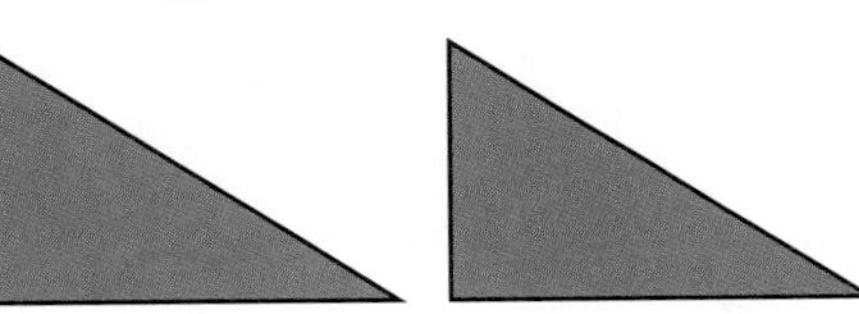

2

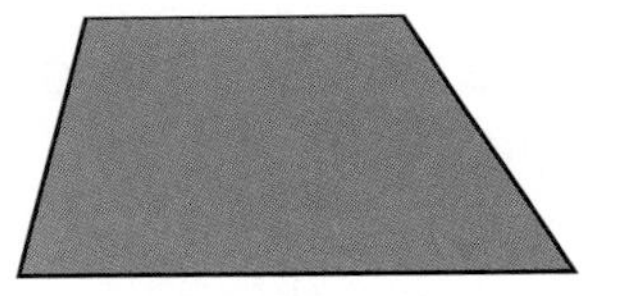

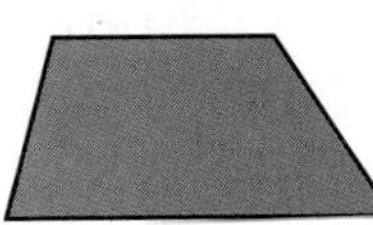

Solve.

3 Matthew hikes 2 kilometers. Jeremy hikes three times as far as Matthew. Zachary hikes half as far as Jeremy. How far does Zachary hike?

4 Jill uses 5 tubes of paint to paint 3 pictures. How many tubes will she use to paint 6 pictures?

5 **Spatial reasoning** Put 3 points (not in a straight line) on dot paper. Connect the points to make a polygon and give its name. Write instructions for another student to draw a congruent polygon.

extra practice

Symmetry page 345

Copy and complete the figure to make it symmetrical. Use dot paper.

1 

2 

3 

Solve.

4 How many lines of symmetry can you find in a square?

5 **ALGEBRA: PATTERNS** Complete.
1, 3, 6, 10, ■, ■, ■.

Slides, Flips, and Turns page 347

Tell how the figure on the left was moved to get the figure on the right. Tell how you can check that the two figures are congruent.

1 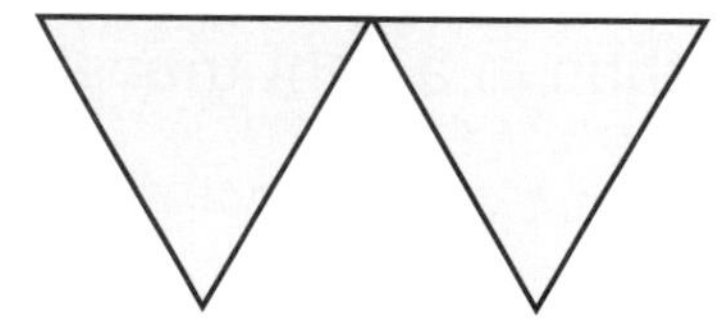

2 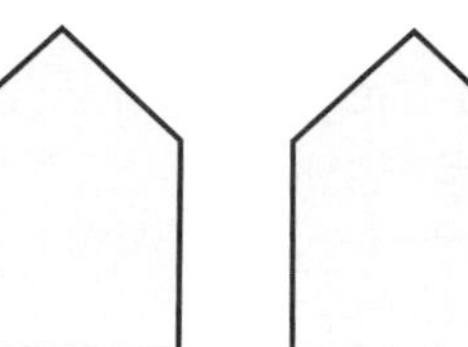

3 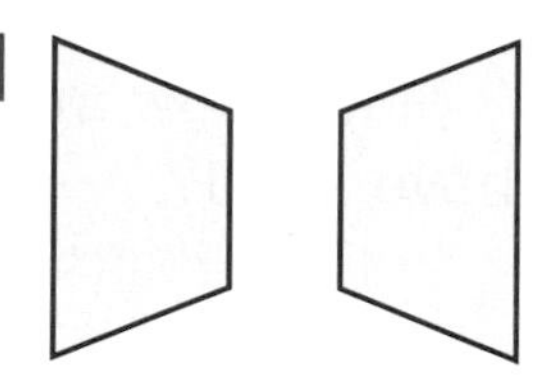

Solve.

4 You can move the left triangle onto the other triangle using a turn. What other two moves would accomplish the same thing?

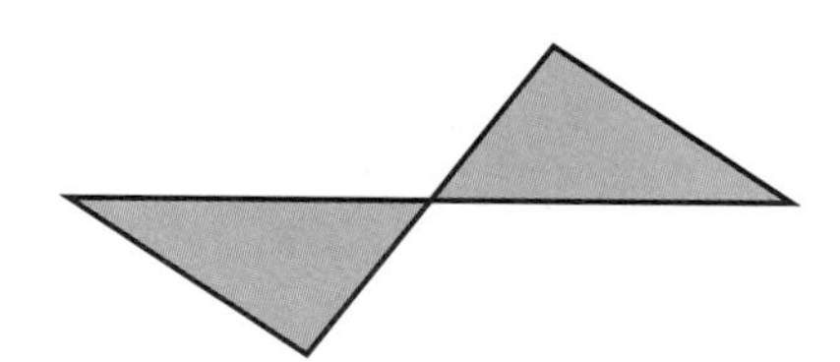

5 Choose any figure. Then draw a pattern using turns and flips.

Area page 351

Use graph paper to draw rectangles with the following areas. How many rectangles are possible for each?

1 10 square units

2 12 square units

3 16 square units

4 48 square units

Volume page 353

Find the volume for each rectangular prism.

1. length—5 in.
   width—4 in.
   height—3 in.

2. length—3 in.
   width—2 in.
   height—4 in.

3. length—9 in.
   width—3 in.
   height—5 in.

4. length—7 in.
   width—4 in.
   height—6 in.

5. length—12 in.
   width—8 in.
   height—4 in.

6. length—11 in.
   width—5 in.
   height—4 in.

Solve.

7. List the least and greatest volumes in ex. 1–6. Find the difference.

8. Amy's test scores in science were 85, 74, 96, and 89. Find the average.

Problem Solvers at Work page 357

Solve. Explain your methods.

1. Domingo can choose between a Monet art print and a Rubens portrait. He can also choose between a lion statue and an elephant statue. How many choices does he have? Tell what they are.

2. **Logical reasoning** At camp, there are teams only in baseball, volleyball, soccer, and swimming. Eli, Jon, Bob, and Lee each play on a different team. Eli's sport does not use a ball. Jon often kicks the ball. Bob does not play baseball. Who plays baseball?

3. Middleton Grade School is going to take a nature hike. They will use 9-passenger vans to transport 38 students and 3 adults to the trails. How many vans are needed?

4. Brooke buys a sandwich for $3.29, a soda for $1.19, and french fries for $0.89. She hands the clerk $10. How much change will she receive?

5. Dorita gave 3 cents to each of her 7 friends. Her mother then gave Dorita 6 pennies, so she had 9 pennies in all. How many pennies did she have to begin with?

6. There are 32 students in 2 classes studying rocks. There are 6 more students in Ms. Motsinger's class than in Mr. Brown's class. How many are in each class?

Part of a Whole page 371

Write the fraction for the part that is shaded.

1 

2 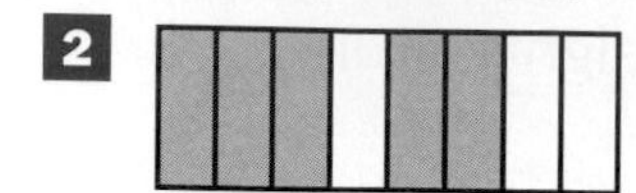

3 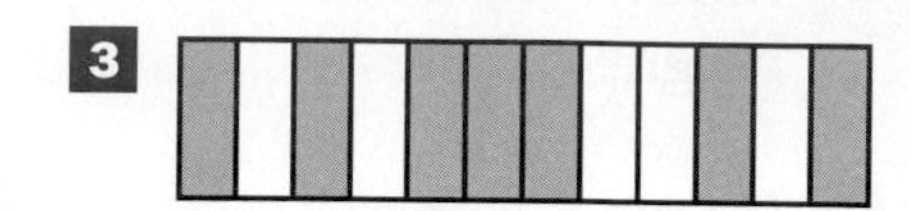

Solve.

4 A game board is separated into 12 equal sections. What fraction of the board are 5 sections?

5 Grace buys items for \$3.98, \$2.04, \$3.50, and \$0.75. Can she pay for them with a ten-dollar bill? Explain.

Part of a Group page 373

Choose the fraction that tells which part is blue.

1 

a. $\frac{7}{12}$ b. $\frac{4}{7}$ c. $\frac{1}{4}$

2 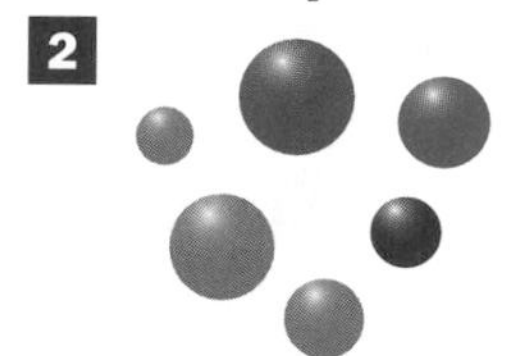

a. $\frac{3}{6}$ b. $\frac{1}{3}$ c. $\frac{1}{6}$

3 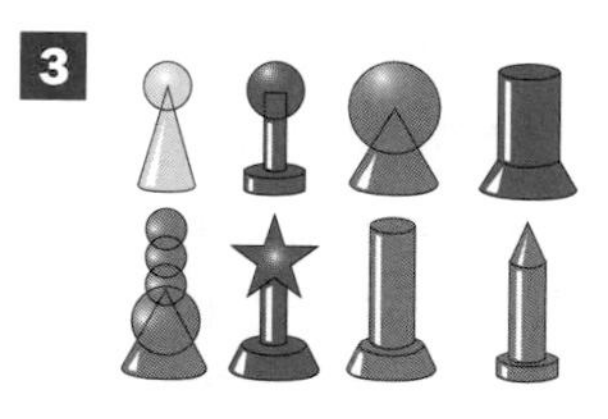

a. $\frac{3}{4}$ b. $\frac{3}{8}$ c. $\frac{1}{4}$

Solve.

4 Three out of four games use number cubes. What fraction of the games use number cubes? What fraction do *not* use number cubes?

5 Tiwa wants 20 new hair bows. They are sold in packages of 6. How many packages must she buy to have at least 20 bows?

Find a Fraction of a Number page 375

Find the answer. Use any method.

1 $\frac{1}{2}$ of 22 **2** $\frac{1}{5}$ of 20 **3** $\frac{3}{4}$ of 20 **4** $\frac{2}{3}$ of 15

5 $\frac{1}{6}$ of 24 **6** $\frac{3}{8}$ of 16 **7** $\frac{1}{3}$ of 18 **8** $\frac{5}{6}$ of 30

9 $\frac{3}{10}$ of 20 **10** $\frac{1}{4}$ of 24 **11** $\frac{3}{5}$ of 25 **12** $\frac{4}{4}$ of 9

Solve.

13 Three eighths of the 24 balls at Jay Elementary School are kickballs. How many of the balls are *not* kickballs?

Equivalent Fractions page 379

Complete the equivalent fraction.

1 $\frac{1}{2} = \frac{■}{12}$ **2** $\frac{9}{15} = \frac{3}{■}$ **3** $\frac{4}{24} = \frac{■}{6}$ **4** $\frac{5}{8} = \frac{10}{■}$

5 $\frac{2}{3} = \frac{8}{■}$ **6** $\frac{6}{8} = \frac{■}{4}$ **7** $\frac{4}{5} = \frac{8}{■}$ **8** $\frac{10}{25} = \frac{2}{■}$

Name three equivalent fractions for the fraction. Use any method.

9 $\frac{1}{8}$ **10** $\frac{5}{6}$ **11** $\frac{9}{27}$ **12** $\frac{6}{14}$ **13** $\frac{3}{8}$ **14** $\frac{16}{32}$

Simplify Fractions page 381

Is the fraction in simplest form? Write *yes* or *no*.

1 $\frac{6}{8}$ **2** $\frac{3}{5}$ **3** $\frac{35}{42}$ **4** $\frac{3}{12}$ **5** $\frac{25}{26}$ **6** $\frac{6}{15}$

Write the fraction in simplest form. Show your method.

7 $\frac{2}{8}$ **8** $\frac{6}{21}$ **9** $\frac{8}{10}$ **10** $\frac{9}{12}$ **11** $\frac{6}{16}$ **12** $\frac{15}{20}$

13 $\frac{8}{12}$ **14** $\frac{16}{32}$ **15** $\frac{24}{30}$ **16** $\frac{14}{20}$ **17** $\frac{9}{21}$ **18** $\frac{12}{48}$

Solve.

19 Lily has 20 marbles. 8 are red. What fraction of the marbles are red? Write in simplest form.

20 Write a number that rounds up to 41,200 and down to 41,170.

Compare Fractions page 383

Write >, <, or =. Use mental math when you can.

1 $\frac{6}{12}$ ● $\frac{2}{3}$ **2** $\frac{8}{10}$ ● $\frac{4}{5}$ **3** $\frac{3}{8}$ ● $\frac{1}{4}$ **4** $\frac{12}{16}$ ● $\frac{7}{8}$ **5** $\frac{2}{3}$ ● $\frac{9}{15}$

Write in order from greatest to least.

6 $\frac{2}{3}, \frac{1}{2}, \frac{5}{6}$ **7** $\frac{3}{8}, \frac{3}{4}, \frac{1}{2}$ **8** $\frac{3}{4}, \frac{5}{6}, \frac{5}{12}$

Solve.

9 A player in Wow! wins if she reaches the finish line first. Judy is $\frac{3}{4}$ of the way to the finish line, and Alice is $\frac{5}{8}$ of the way. Who is farther from the finish line?

10 Megan scores 10 points in Socko. Derek scores $\frac{3}{5}$ as many points as Megan, and Kirk scores twice as many as Derek. How many points does each student score?

Mixed Numbers page 385

Rename as a whole number or as a mixed number in simplest form.

1 $\frac{5}{3}$ **2** $\frac{7}{4}$ **3** $\frac{18}{5}$ **4** $\frac{13}{4}$ **5** $\frac{25}{6}$ **6** $\frac{24}{10}$

7 $\frac{10}{8}$ **8** $\frac{40}{6}$ **9** $\frac{16}{16}$ **10** $\frac{20}{5}$ **11** $\frac{52}{8}$ **12** $\frac{33}{9}$

Solve.

13 Each of two spinners for a game is divided into 8 equal parts. 9 parts are blue. Write a mixed number to describe the blue parts.

14 Demi brought 6 pints of water on a nature hike. Ashley brought 2 quarts. How many more pints did Demi bring than Ashley?

Probability page 391

Use the words *more likely, less likely, equally likely, certain,* or *impossible* to describe the probability.

1 picking a red marble

2 picking a yellow marble

3 picking a blue marble

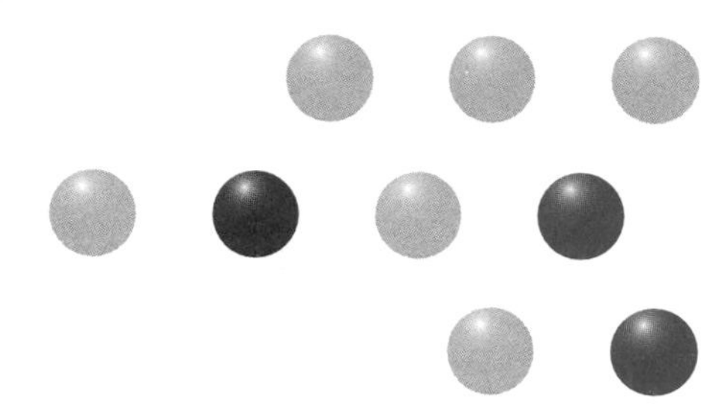

Solve.

4 **Logical reasoning** How can you arrange a marble display to make the probability of picking a red marble certain?

5 A clerk sold 25 games each month for 6 months. At this rate, how many will he sell in a year?

Fractions and Probability page 393

Find the probability of picking a letter from the cards below.

1 the letter *S* **2** the letter *I* **3** the letter *M* **4** the letter *P*

Problem-Solving Strategy: Conduct an Experiment page 395

Solve. Explain your methods.

1 Toss two pennies 30 times. Record the results as: 2 heads, 2 tails, or 1 head and 1 tail. Based on your experiment, which result is more likely?

2 Joy has 13 pairs of shoes in one closet and 8 pairs in another. If she puts them all in one closet with 7 pairs on each shelf, how many shelves will she need?

3 Garth wants to buy several booklets priced at $3 at the art museum. He can buy one booklet and get the next for half price. How many booklets can he buy if he has $25? How much money will he have left?

4 Four girls ran a 2-kilometer race. Uma was ahead of Mahala and Brianna. Lulu was directly after Uma. Two girls placed between Uma and Mahala. In what order did each girl place?

Predict and Experiment page 397

Use the spinner for problems 1–2. Explain your reasoning.

1 Is it reasonable to predict that the spinner will land on green 10 out of 30 times?

2 Can you ever expect the spinner to land on blue?

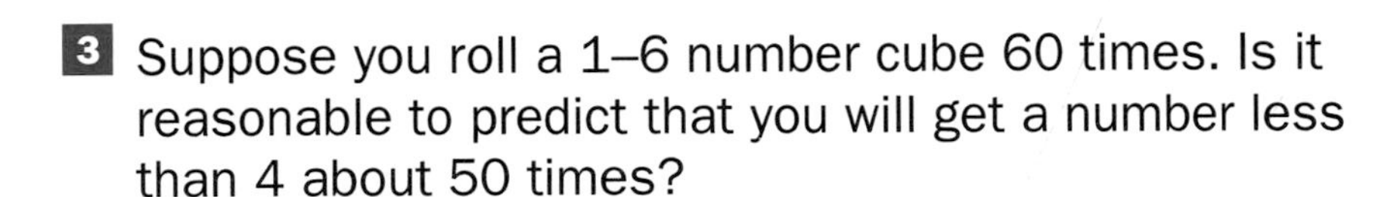

3 Suppose you roll a 1–6 number cube 60 times. Is it reasonable to predict that you will get a number less than 4 about 50 times?

4 Jay bought $\frac{1}{3}$ of the etchings that were on sale at the art gallery. He bought 6. How many are still on sale?

Problem Solvers at Work page 401

Solve. Explain your methods.

1 Delbert works in a video store and earns $18 for 3 hours of work. Irene works in a competing store and earns $42 for 7 hours work. Do Delbert and Irene receive the same amount of money per hour?

2 Alexandra has a $5-off coupon to use at the video store. She rents 3 videotapes at $1.98 each and buys a videotape for $19.98. How much money does she spend at the store?

extra practice

Add Fractions page 413

Add. You may use models if you wish.

1 $\frac{1}{3} + \frac{1}{3}$ **2** $\frac{5}{12} + \frac{7}{12}$ **3** $\frac{4}{5} + \frac{3}{5}$ **4** $\frac{1}{6} + \frac{1}{6}$ **5** $\frac{5}{8} + \frac{5}{8}$

6 $\frac{3}{10} + \frac{4}{10}$ **7** $\frac{2}{3} + \frac{2}{3}$ **8** $\frac{3}{4} + \frac{1}{4}$ **9** $\frac{5}{8} + \frac{7}{8}$ **10** $\frac{5}{6} + \frac{5}{6}$

Solve.

11 Ben ate $\frac{1}{4}$ of a pizza. Then, he ate another $\frac{1}{4}$. How much pizza did he eat in all?

12 **ALGEBRA: PATTERNS** Copy and complete the pattern. Then write the rule.
5, 20, 80, 320, ■

Add Fractions page 415

Add. Write the sum in simplest form.

1 $\frac{1}{5} + \frac{3}{5}$ **2** $\frac{5}{8} + \frac{3}{8}$ **3** $\frac{1}{4} + \frac{1}{4}$ **4** $\frac{7}{10} + \frac{1}{10}$ **5** $\frac{3}{8} + \frac{7}{8}$ **6** $\frac{5}{6} + \frac{1}{6}$

7 $\frac{1}{12} + \frac{7}{12}$ **8** $\frac{2}{10} + \frac{9}{10}$ **9** $\frac{3}{6} + \frac{5}{6}$ **10** $\frac{3}{4} + \frac{3}{4}$

Subtract Fractions page 419

Subtract. Write the difference in simplest form.

1 $\frac{3}{4} - \frac{1}{4}$ **2** $\frac{7}{8} - \frac{3}{8}$ **3** $\frac{4}{5} - \frac{2}{5}$ **4** $\frac{9}{10} - \frac{3}{10}$

5 $\frac{7}{12} - \frac{3}{12}$ **6** $\frac{4}{6} - \frac{1}{6}$ **7** $\frac{5}{6} - \frac{2}{6}$ **8** $\frac{7}{10} - \frac{5}{10}$ **9** $\frac{11}{12} - \frac{3}{12}$ **10** $\frac{9}{12} - \frac{5}{12}$

Solve.

11 Seven eighths of the students like ice cream. Four eighths prefer chocolate to vanilla. What fraction of the students prefer vanilla ice cream?

12 Ramona brought 5 gallons of orangeade to the picnic. Ramona brought 2 more quarts than Jane. How many quarts did they bring altogether?

Subtract Fractions page 421

Subtract. Write the difference in simplest form.

1 $\frac{5}{8} - \frac{4}{8}$

2 $\frac{3}{10} - \frac{2}{10}$

3 $\frac{7}{9} - \frac{1}{9}$

4 $\frac{3}{4} - \frac{1}{4}$

5 $\frac{5}{6} - \frac{3}{6}$

6 $\frac{11}{12} - \frac{1}{12}$

7 $\frac{4}{5} - \frac{1}{5}$

8 $\frac{5}{8} - \frac{1}{8}$

9 $\frac{10}{10} - \frac{2}{10}$

10 $\frac{5}{6} - \frac{1}{6}$

Write >, <, or =.

11 $\frac{5}{8} - \frac{2}{8}$ ● $\frac{7}{8} - \frac{3}{8}$

12 $\frac{3}{6} - \frac{1}{6}$ ● $\frac{5}{6} - \frac{4}{6}$

Solve.

13 Eileen cut a loaf of nut bread into eighths. She gave 3 pieces to Fred. What fraction of the loaf did she have left?

14 Dee buys 5 cans of corn for $2 at Mott's Grocery. One can of corn at Johnson's Mart costs 39¢. Which grocery charges more for 5 cans of corn? How much more?

Find a Common Denominator page 427

Name the common denominator.

1 $\frac{1}{2}$ and $\frac{3}{4}$

2 $\frac{3}{10}$ and $\frac{2}{5}$

3 $\frac{1}{2}$ and $\frac{2}{3}$

4 $\frac{2}{3}$ and $\frac{3}{4}$

5 $\frac{1}{4}$ and $\frac{5}{6}$

Write as equivalent fractions with common denominators.

6 $\frac{3}{4}$ and $\frac{5}{8}$

7 $\frac{5}{6}$ and $\frac{2}{3}$

8 $\frac{3}{4}$ and $\frac{5}{12}$

9 $\frac{2}{3}$ and $\frac{7}{9}$

10 $\frac{1}{2}$ and $\frac{3}{8}$

11 $\frac{1}{2}$ and $\frac{2}{3}$

12 $\frac{3}{4}$ and $\frac{1}{3}$

13 $\frac{2}{3}$ and $\frac{3}{5}$

14 $\frac{4}{5}$ and $\frac{1}{2}$

15 $\frac{1}{4}$ and $\frac{1}{6}$

Solve.

16 Partygoers ate $\frac{1}{4}$ of Nita's birthday cake. Her family ate $\frac{1}{3}$ of the same cake that night. Who ate more cake? Explain how you know.

17 Jerry places 18 marbles in a sack. 3 are white, 4 are green, 6 are blue, and 5 are red. If he picks one marble without peeking, what is the probability of getting a red marble?

18 A ship has 549 gallons of drinking water. The sailors drink 10 gallons each day. How many full days can they stay at sea?

19 Geraldine has $35.34 in her wallet. She cashes a check and then has $82.84. What is the amount of her check?

Add and Subtract Fractions with Unlike Denominators page 431

Find the equivalent fraction. Then add or subtract.

1 $\begin{array}{r} \frac{1}{4} = \frac{\blacksquare}{8} \\ + \frac{5}{8} = \frac{\blacksquare}{8} \\ \hline \frac{\blacksquare}{8} \end{array}$

2 $\begin{array}{r} \frac{1}{4} = \frac{\blacksquare}{4} \\ + \frac{1}{2} = \frac{\blacksquare}{4} \\ \hline \frac{\blacksquare}{4} \end{array}$

3 $\begin{array}{r} \frac{3}{10} = \frac{\blacksquare}{10} \\ - \frac{1}{5} = \frac{\blacksquare}{10} \\ \hline \frac{\blacksquare}{10} \end{array}$

4 $\begin{array}{r} \frac{5}{6} = \frac{\blacksquare}{6} \\ - \frac{1}{3} = \frac{\blacksquare}{6} \\ \hline \frac{\blacksquare}{6} \end{array}$

Add or subtract using any method. Write the sum or difference in simplest form.

5 $\frac{1}{6} + \frac{2}{3}$

6 $\frac{1}{4} + \frac{3}{8}$

7 $\frac{3}{5} + \frac{1}{10}$

8 $\frac{1}{12} + \frac{3}{4}$

9 $\frac{1}{4} - \frac{1}{8}$

10 $\frac{2}{3} - \frac{2}{9}$

11 $\frac{7}{8} - \frac{1}{2}$

12 $\frac{3}{5} - \frac{1}{10}$

Solve.

13 Shantelle and Donovan buy $\frac{1}{4}$ pound of pecans. They use $\frac{1}{8}$ pound to bake cookies. Then, they buy another $\frac{1}{4}$ pound. How many pounds of pecans do they then have?

14 There are 12 students in the marching band. How many different ways can they line up with the same number of people in each row? List each way.

Problem-Solving Strategy: Draw a Picture page 433

Solve. Explain your methods.

1 Use graph paper. Find two shapes with the same area but with different perimeters.

2 Jan is 3 times as old as Bob. When they add their ages, the sum is 16. What are their ages?

3 **ALGEBRA: PATTERNS** Draw the next three figures.

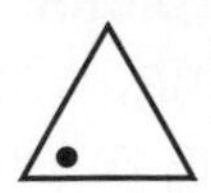 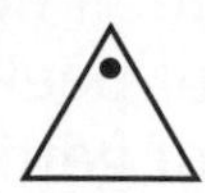 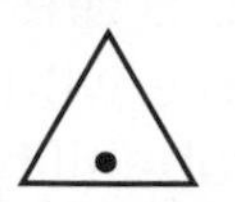  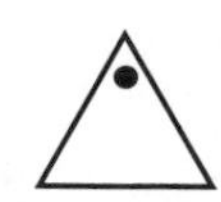  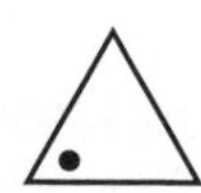

4 Brandon, Ashley, and Brooke each like a different kind of nut—pecans, peanuts, or almonds. Brandon does not like pecans. Ashley likes peanuts. Which is the favorite nut of each student?

Add and Subtract Mixed Numbers page 437

Add or subtract. Use fraction strips if you want. Rename and simplify when necessary.

1 $2\frac{1}{4} + 4\frac{1}{4}$
2 $1\frac{3}{8} + 2\frac{3}{8}$
3 $5\frac{2}{3} + 4\frac{2}{3}$
4 $4\frac{1}{2} + 2\frac{1}{2}$

5 $5\frac{5}{8} - 3\frac{3}{8}$
6 $7\frac{5}{6} - 3\frac{1}{6}$
7 $6\frac{3}{4} - 2\frac{1}{4}$
8 $7\frac{9}{12} - 4\frac{1}{12}$

9 $3\frac{2}{4} + 1\frac{3}{4}$
10 $9\frac{3}{8} - 7\frac{3}{8}$
11 $4\frac{5}{6} - 1\frac{3}{6}$
12 $4\frac{3}{10} + 2\frac{6}{10}$

Solve. Rename and simplify when necessary.

13 Henry mixed $2\frac{3}{8}$ pounds of hamburger and $1\frac{7}{8}$ pounds of ground veal to make meat loaf. How many pounds of meat did he mix in all?

14 Cynthia is riding to her cooking class. The distance is $3\frac{3}{10}$ miles. She has ridden her bike $1\frac{1}{10}$ miles. How many miles does she have yet to ride?

15 Naomi put decorative tape around the edge of the top of a recipe box. The top of the box measures 9 cm by 15 cm. How many centimeters of tape did she use?

16 How many days will be left in the year after September 21? Will your answer be different for a leap year? Explain how you know.

Problem Solvers at Work page 441

Solve. Explain your methods.

1 Forty-five students and 5 adults will travel in vans holding 8 passengers on a field trip to Mitzi's Cooking College. How many vans will be needed to take everyone on the field trip?

2 Ellie received these scores on some tests in her nutrition class: 85, 98, and 97. She needs an average of 95 points to get an A. What does she need to score on her next test to achieve an A average?

3 Jason earned $7 for mowing a lawn. He took the money to the bank and exchanged it for quarters. How many quarters did the bank give Jason?

4 Donna's train departs at 10:45 A.M. A car ride to the train station takes 55 minutes. How long will she have to wait at the train station if she leaves home at 8:55 A.M.?

Decimals Less Than 1 page 453

Write a decimal.

1 $\frac{4}{10}$ **2** $\frac{16}{100}$ **3** $\frac{39}{100}$ **4** $\frac{97}{100}$ **5** $\frac{6}{10}$ **6** $\frac{25}{100}$

7 $\frac{54}{100}$ **8** $\frac{8}{100}$ **9** $\frac{83}{100}$ **10** $\frac{72}{100}$ **11** $\frac{9}{10}$ **12** $\frac{6}{100}$

Solve.

13 Twenty-one out of 100 students have blond hair. Write a decimal for the part of the group that has blond hair.

14 Gina buys 60 computer disks. She pays $7.38 for each dozen. How much does Gina pay for all the disks?

Decimals Greater Than 1 page 455

Write a decimal.

1 $3\frac{7}{10}$ **2** $8\frac{35}{100}$ **3** $1\frac{12}{100}$ **4** $4\frac{3}{100}$ **5** $6\frac{82}{100}$ **6** $2\frac{8}{10}$

7 $5\frac{7}{100}$ **8** $7\frac{3}{10}$ **9** $8\frac{91}{100}$ **10** $1\frac{3}{10}$ **11** $2\frac{5}{100}$ **12** $3\frac{4}{10}$

Solve.

13 Write a decimal that tells how many sheets of stamps there are.

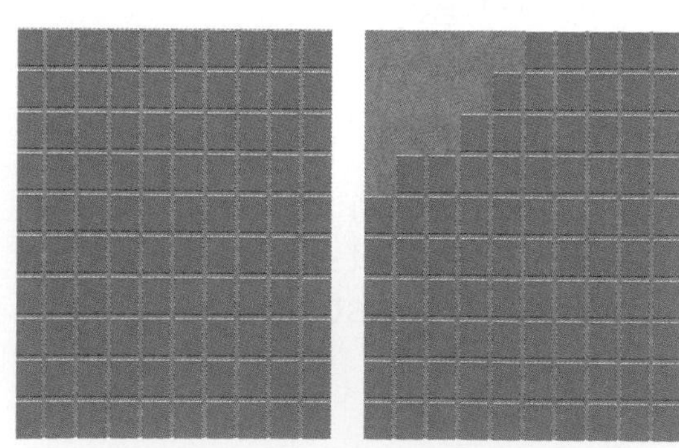

14 Renee made 8 calls to students' parents to tell them about the school fair. Jack gives her a list of 20 more names to call. She now has 45 calls to make. How many names did she have when she started calling?

Compare and Order Decimals page 457

Compare. Write >, <, or =.

1 0.4 ● 0.23 **2** 0.3 ● 0.30 **3** 4.17 ● 4.71 **4** 4.3 ● 4.30

Write in order from greatest to least.

5 0.05, 0.4, 0.34 **6** 1.5, 1.08, 2.1 **7** 3.5, 2.9, 3.08

Problem-Solving Strategy: Solve a Simpler Problem page 463

Solve. Explain your methods.

1 On Saturday, Dixie's Newsstand sold 328 newspapers at $0.50 each. On Sunday, 435 papers were sold at $1.50 each. How much more money did the stand receive for the Sunday papers than for the Saturday papers?

2 Paula works as a telephone operator from 8:45 P.M. until 5:15 A.M. She has a 30-minute break for a midnight snack and two additional ten-minute breaks. How long does Paula work?

3 **Spatial reasoning** How many triangles can you find in the figure below?

4 Jani saves a penny on Monday. Each day after that, she plans to save twice as much as she saves the day before. How much will Jani save in one week? How much in two weeks?

5 Bessie has 24 friends on the Internet. Of these, there are twice as many boys as girls. How many are girls? How many are boys?

6 The area of John's rectangular garden is 80 ft^2. The shorter sides each measure 8 ft. What is the length of each longer side?

Estimate Sums and Differences page 465

Estimate. Round to the nearest whole number.

1 3.9 + 4.3
2 4.7 + 9.2
3 5.2 − 3.9
4 21.8 − 4.3

5 13.62 − 8.51
6 7.49 + 8.64
7 9.75 − 4.38
8 10.9 − 5.4

9 6.2 + 1.9 + 3.5
10 3.87 + 2.35 + 1.09
11 8.94 + 6.71 + 9.06

Estimate. Round to the nearest dollar.

12 $5.48 + $7.65
13 $10.34 − $4.98
14 $5.39 − $2.13

15 $18.67 + $12.38
16 $16.75 − $8.98
17 $13.37 + $5.89

Solve.

18 Casandra buys an answering machine for $39.98, an extra cord for $10.19, and two tapes for $4.99 each. Can she pay for this purchase with 1 ten-dollar bill and 1 fifty-dollar bill? Explain.

19 **Logical reasoning** Abraham's CD-ROM sits between the computer and the modem. The printer sits farthest to the left. The computer sits next to the printer. Tell how they are lined up from left to right.

Add and Subtract Decimals page 467

Use decimal squares to add or subtract.

1 0.5 + 0.3
2 0.4 + 0.9
3 1.45 + 1.78
4 2.46 + 1.49
5 2.15 + 2.87

6 0.9 − 0.4
7 3.1 − 0.2
8 4.38 − 2.09
9 4.09 − 2.28
10 3.03 − 2.34

Solve.

11 Make the problem on the right and its answer reasonable by replacing each ■ with one of these numbers: 4.8, 5.1, or 0.3. Write a number sentence to represent the data.

> Al runs the mile in ■ minutes.
> Jim runs the mile in ■ minutes.
> Jim runs the mile in ■ fewer minutes than Al.

12 Some glaciers move about 210 inches in 70 years. About how many inches do they move each year?

13 Corky the hamster weighs $3\frac{3}{4}$ ounces. Cora weighs $3\frac{5}{8}$ ounces. Who weighs more?

Add Decimals page 469

Add using any method. Remember to estimate.

1 0.5 + 0.7
2 3.8 + 0.9
3 7.8 + 3.5
4 8.3 + 9.8
5 2.4 + 3.9

6 0.34 + 0.38
7 $3.49 + 1.98
8 5.36 + 4.25
9 4.39 + 8.27
10 7.16 + 3.04

11 1.23 + 3.45 + 2.89
12 34.15 + 14.28 + 9.01
13 $12.56 + 13.29 + 10.16
14 16.40 + 2.51 + 33.72
15 26.40 + 3.45 + 9.87

Solve.

16 Brian watches a 2.5-hour sports show, two half-hour sitcoms, and a 2-hour movie. How much time does he spend watching TV?

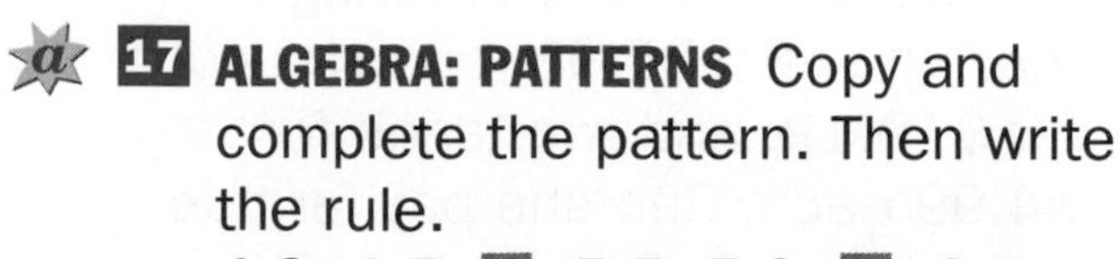

17 **ALGEBRA: PATTERNS** Copy and complete the pattern. Then write the rule.
4.3, 4.7, ■, 5.5, 5.9, ■, 6.7

Subtract Decimals page 471

Subtract using any method. Remember to estimate.

1 0.4 − 0.3

2 3.2 − 0.5

3 5.4 − 2.7

4 6.8 − 3.2

5 9.7 − 4.8

6 0.58 − 0.35

7 7.32 − 4.91

8 4.59 − 3.87

9 8.54 − 3.07

10 10.38 − 6.49

11 $24.38 − 15.19

12 $32.45 − 21.36

13 12.55 − 9.30

14 38.90 − 15.92

15 23.50 − 14.75

Solve.

16 A rain barrel holds 35 gallons of water. There are 42.5 quarts already in the barrel. How many more quarts will the barrel hold?

17 Suppose you toss a number cube whose sides are labeled 1, 2, 3, 4, 5, and 6. What is the probability you will toss an even number?

18 How many characters can a computer print on a page if there are 52 lines per page and 60 characters per line?

19 Carly had $38.75. She earned some money and then had $57.35. How much money did she earn?

Problem Solvers at Work page 475

Solve. Write a number sentence to record your work for problems 1–3. Explain your methods.

1 There are 2,143 jazz CDs, 4,154 country-and-western CDs, and 4,958 rock-and-roll CDs available for loan in the local library. How many music CDs does the library have to loan in all?

2 The average teen spends 6.23 hours watching TV each week from 8 P.M. to 11 P.M., while the average adult spends 8.61 hours. Who spends more time watching TV? How much more?

3 Eugene O'Neill won the Nobel Prize for Literature in 1936. He was born in 1888. How old was he when he won the prize? How many years have elapsed since he won the prize?

4 Mark can buy a Macintosh or an IBM computer. He can choose 8, 16, or 32 megabytes of RAM. How many different choices does he have? List the choices.

Databank

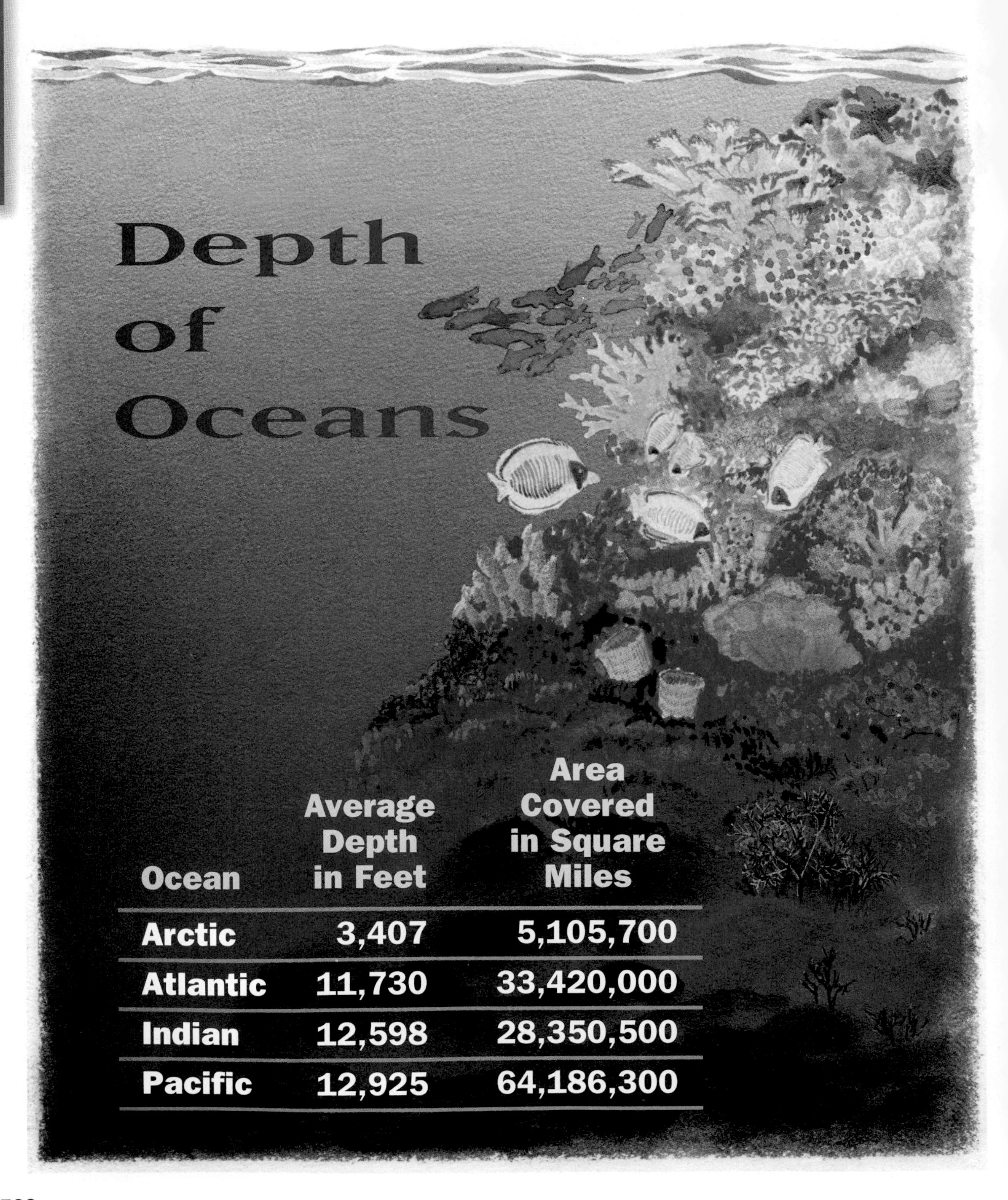

Depth of Oceans

Ocean	Average Depth in Feet	Area Covered in Square Miles
Arctic	3,407	5,105,700
Atlantic	11,730	33,420,000
Indian	12,598	28,350,500
Pacific	12,925	64,186,300

Endangered Animal Populations

Animal Species	Location	Estimated Population in 1970s	Estimated Population in 1990s
Arabian oryx	Asia	35	1,000
African elephant	Africa	1,300,000	600,000
Black rhinoceros	Africa	15,000	4,500
Florida cougar	North America	20	50
Hawaiian monk seal	Hawaiian Islands	2,000	1,000
Indian tiger	Asia	2,500	4,000
Japanese crane	Hokkaido, Japan	220	450
Mauritius parakeet	Africa	50	10
Nene *(similar to a goose)*	Oceania	750	500
Orangutan	Asia	150,000	100,000
Polar bear	Arctic region	5,000	40,000
Red wolf	Texas and Louisiana	100	300
Siberian tiger	Asia	130	400
Woolly spider monkey	South America	3,000	500

Heights of Mountains

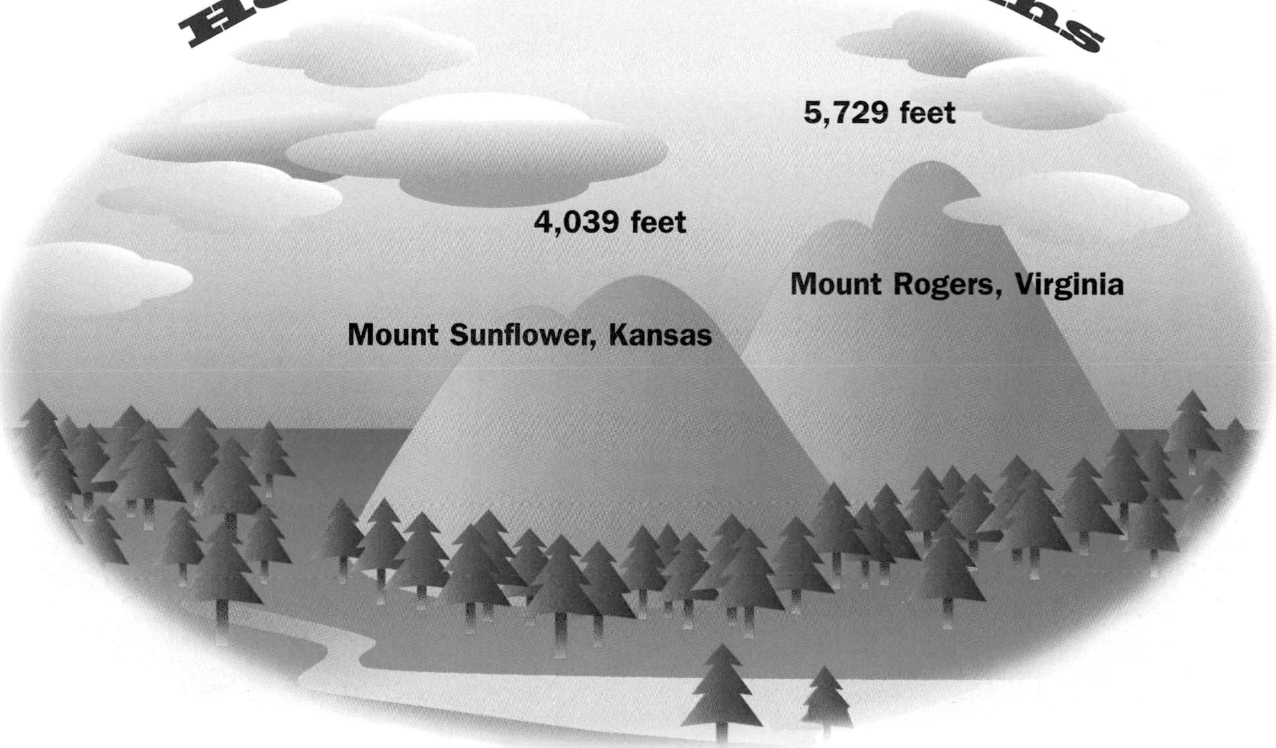

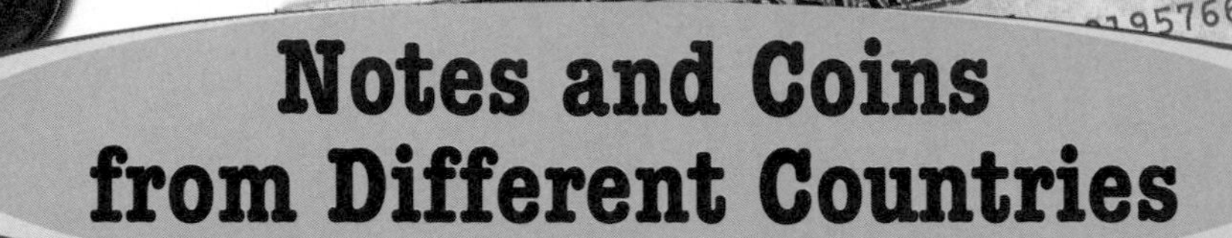

Notes and Coins from Different Countries

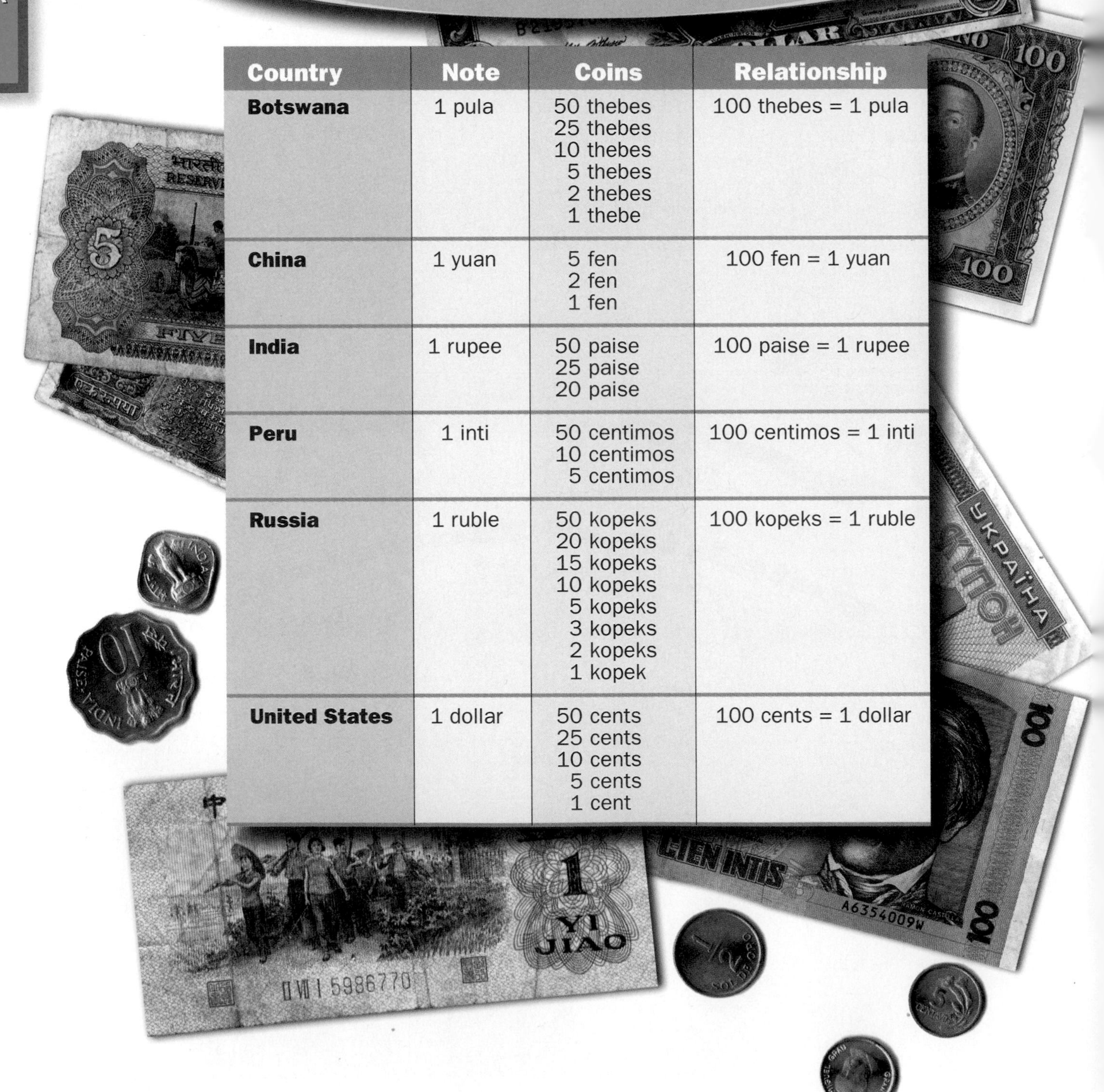

Country	Note	Coins	Relationship
Botswana	1 pula	50 thebes 25 thebes 10 thebes 5 thebes 2 thebes 1 thebe	100 thebes = 1 pula
China	1 yuan	5 fen 2 fen 1 fen	100 fen = 1 yuan
India	1 rupee	50 paise 25 paise 20 paise	100 paise = 1 rupee
Peru	1 inti	50 centimos 10 centimos 5 centimos	100 centimos = 1 inti
Russia	1 ruble	50 kopeks 20 kopeks 15 kopeks 10 kopeks 5 kopeks 3 kopeks 2 kopeks 1 kopek	100 kopeks = 1 ruble
United States	1 dollar	50 cents 25 cents 10 cents 5 cents 1 cent	100 cents = 1 dollar

DATABANK

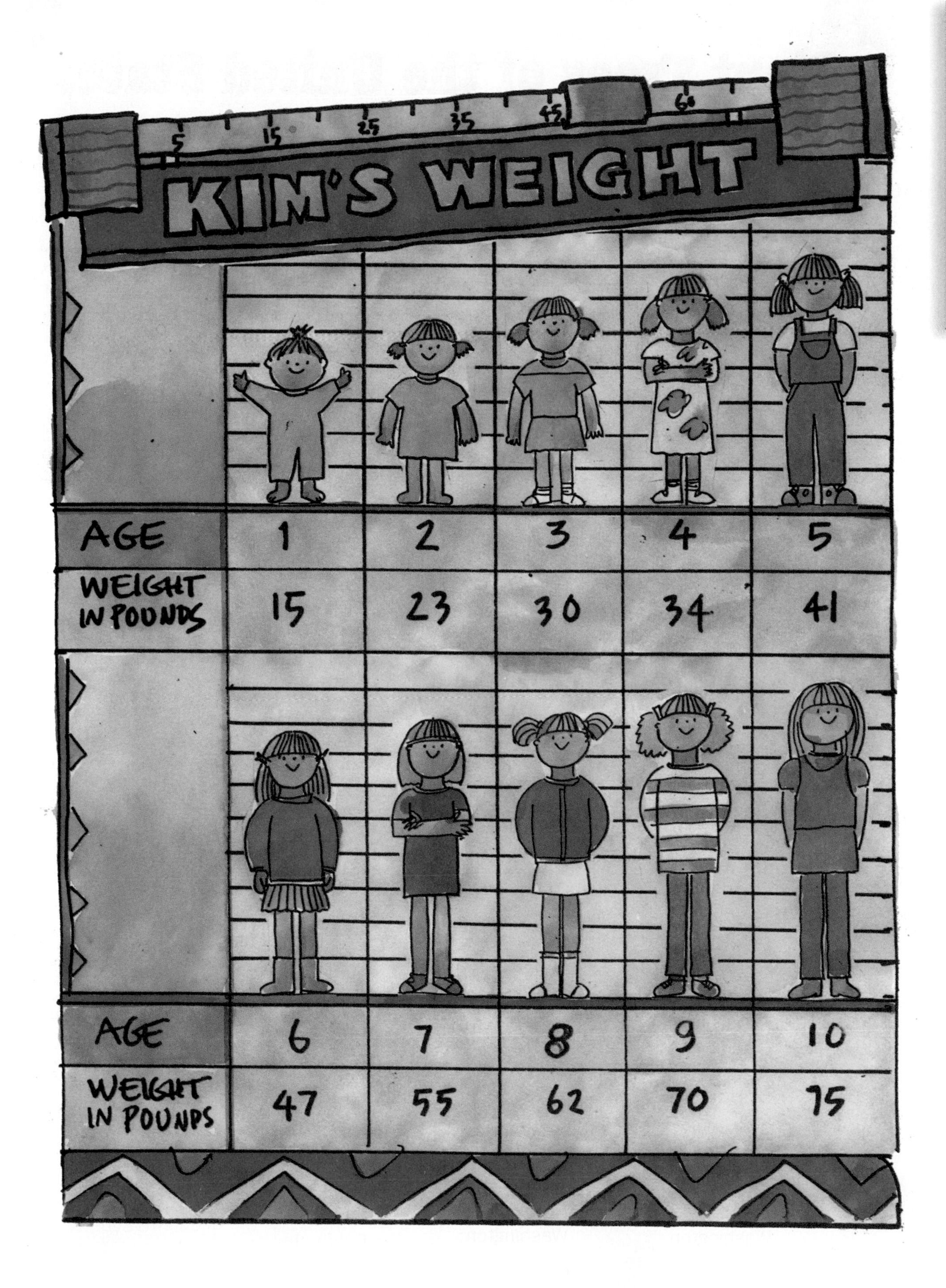
KIM'S WEIGHT

AGE	1	2	3	4	5
WEIGHT IN POUNDS	15	23	30	34	41

AGE	6	7	8	9	10
WEIGHT IN POUNDS	47	55	62	70	75

DATABANK

Giant Trees of the United States

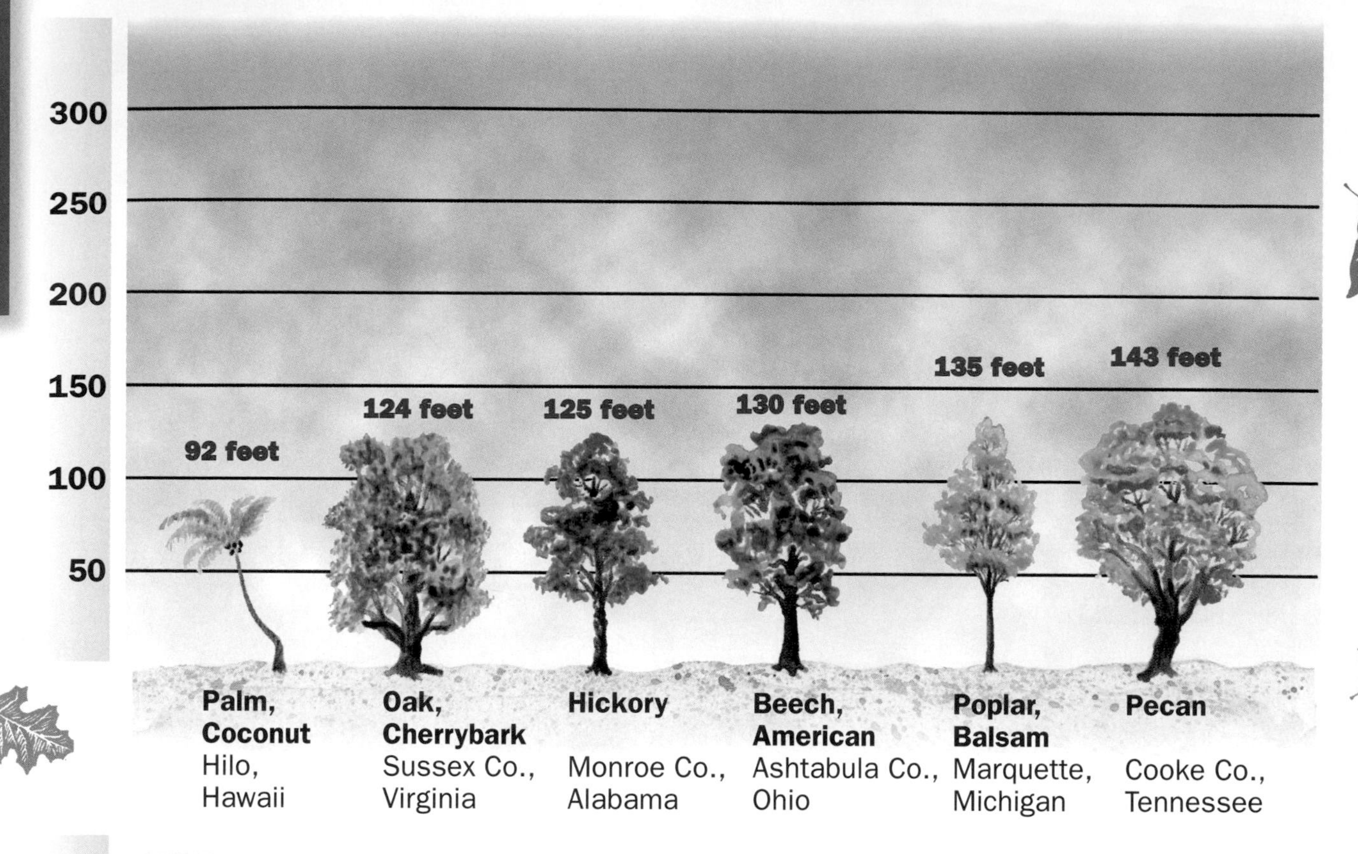

Orange Computer Company

Item	Price
X610 laptop computer	$3,899.00
Z14 external modem	$109.00
3.8Y CD-ROM drive	$299.00
76B laser printer	$549.00

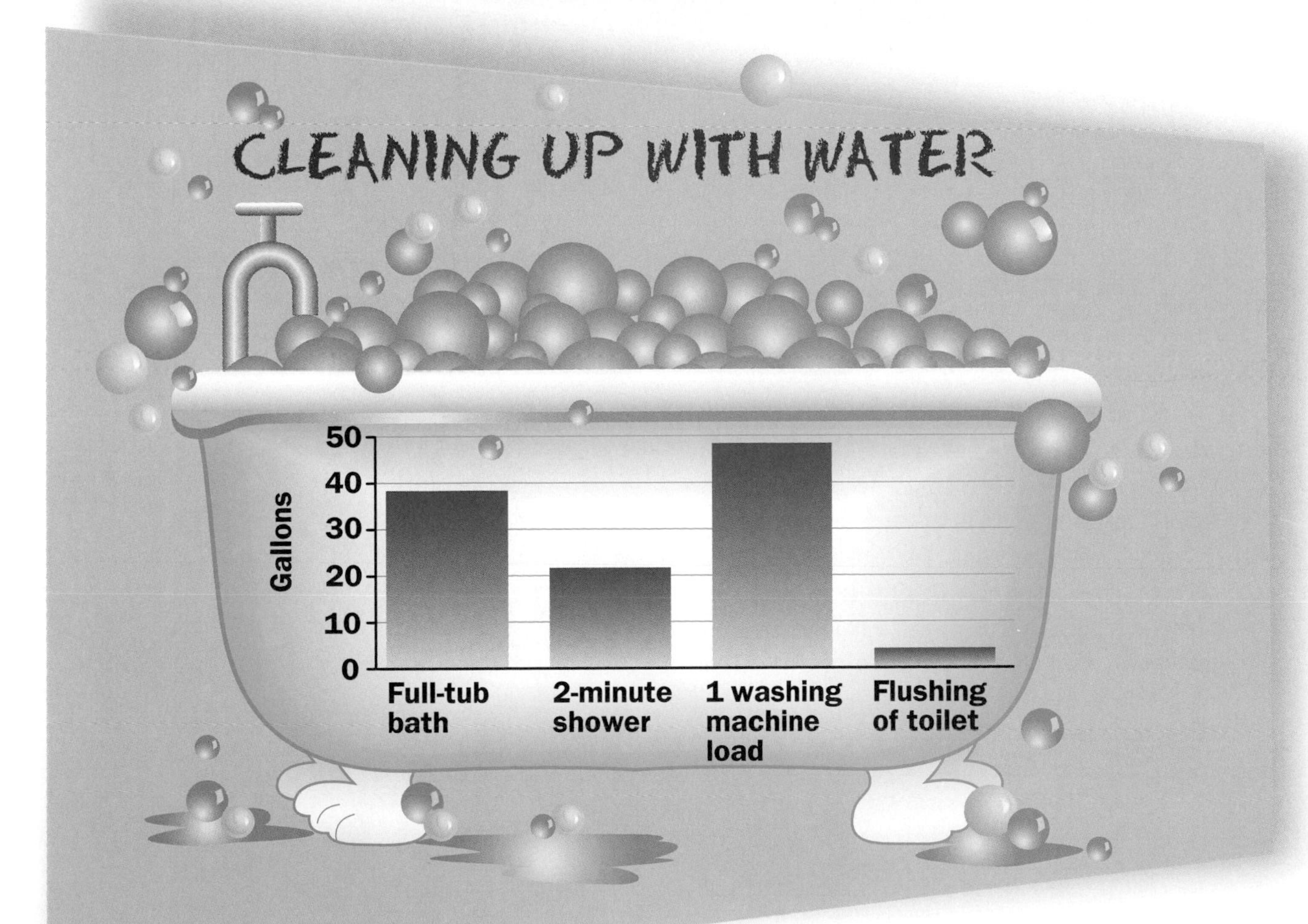

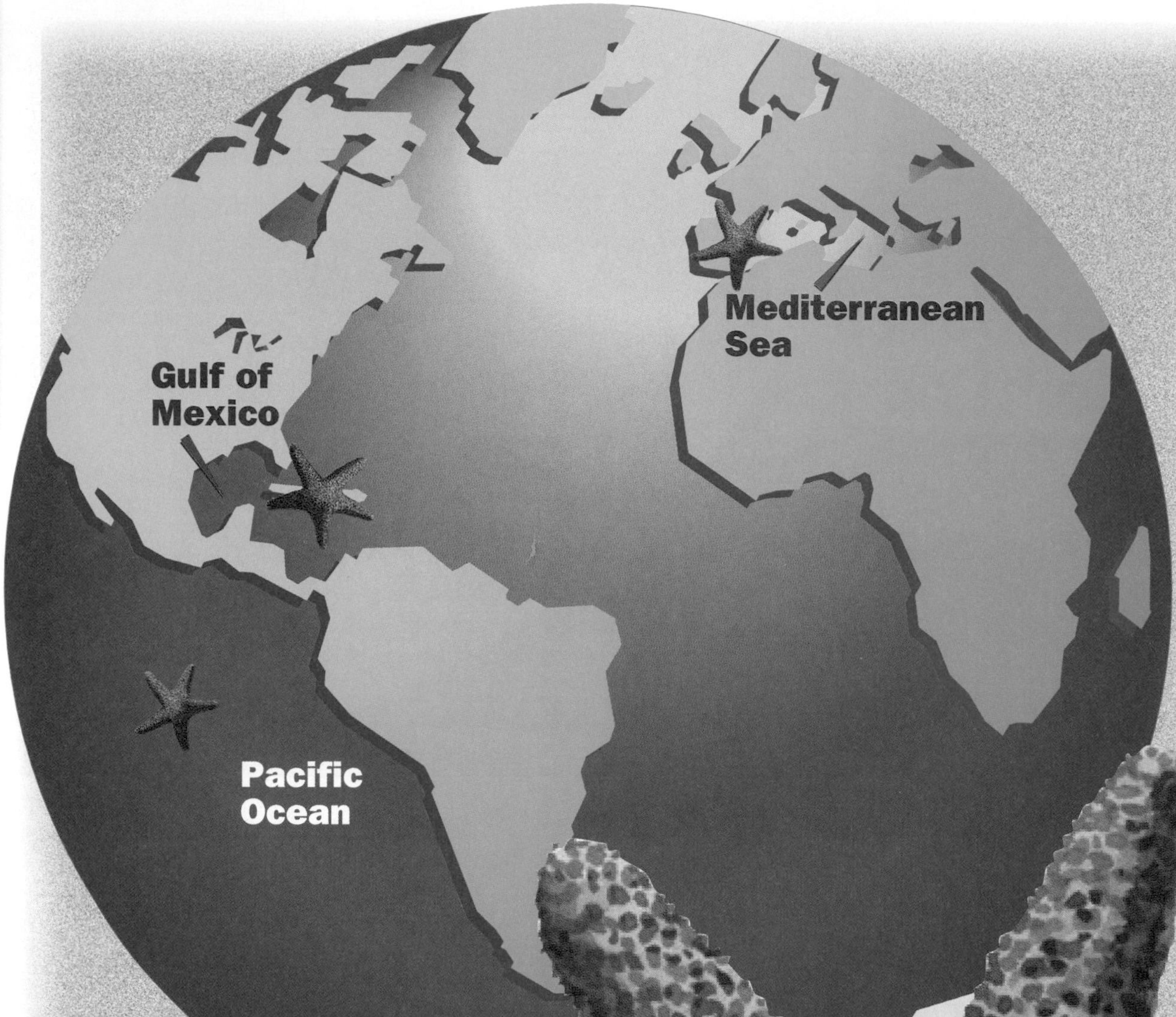

Gulf of Mexico

Mediterranean Sea

Pacific Ocean

Starfish

There are 1,600 known species of starfish. The largest starfish, found in the Gulf of Mexico in 1968, had an arm span measuring 54 in. but weighed only $2\frac{1}{2}$ oz. The heaviest, found in the Pacific Ocean, weighed 13 lb and had an arm span of 25 in. The smallest known starfish, which has an arm span of $\frac{1}{2}$ in., is found in the Mediterranean Sea.

POPULATION OF CAPITAL CITIES

State	Capital of State	Population of Capital
Arizona	Phoenix	983,392
Iowa	Des Moines	193,187
Hawaii	Honolulu	365,272
Idaho	Boise	125,738
Louisiana	Baton Rouge	219,513

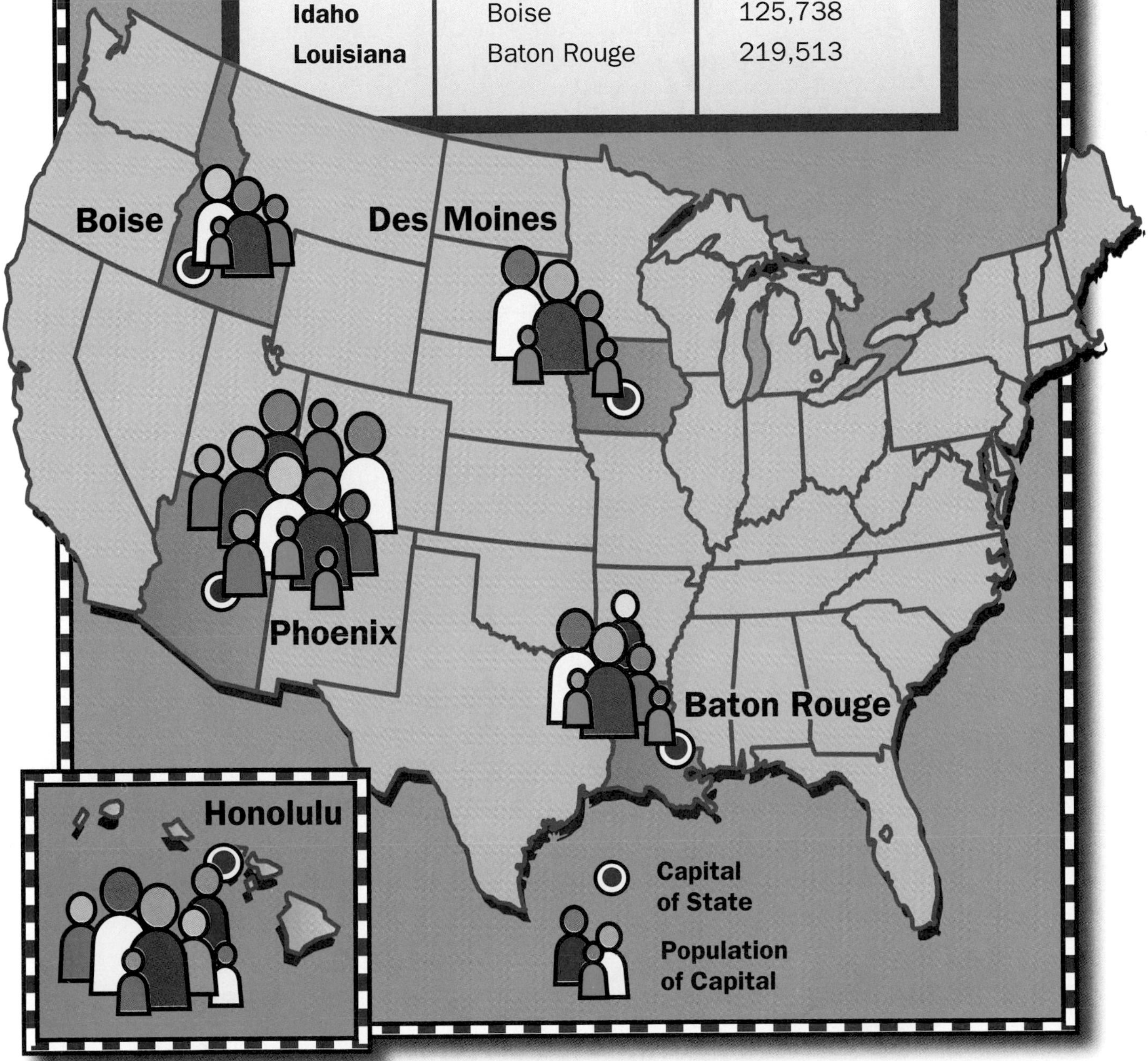

SHAPES IN REAL LIFE

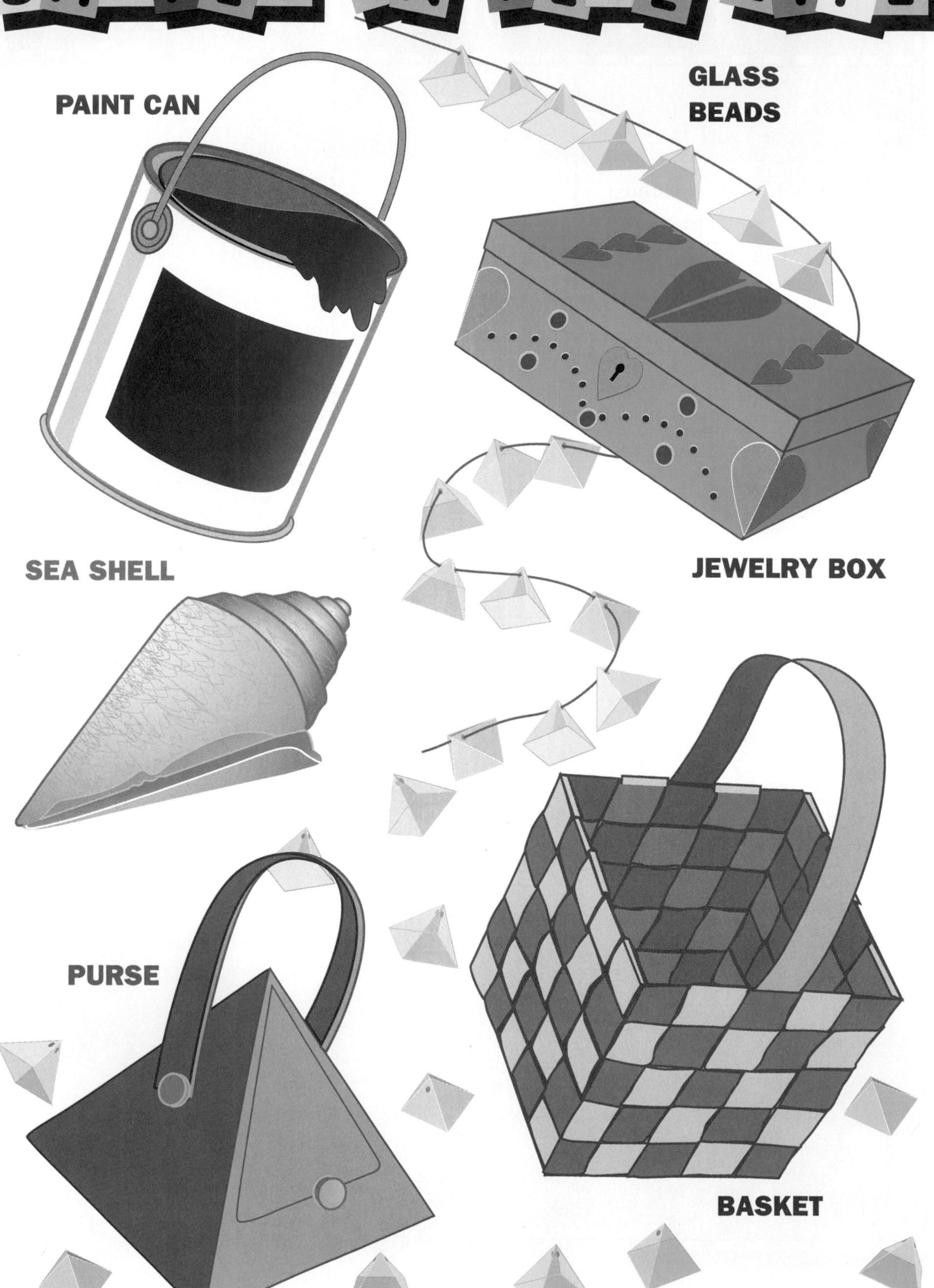

THINGS SOME FOURTH GRADERS IN COLORADO LIKE TO DO

Activity	Number of Students	
Color pictures	𝍸𝍸 1	11
Take photographs	𝍸𝍸𝍸	15
Watch it rain	𝍸 111	8
Collect foreign coins	1111	4
Collect old jewelry	111	3
Collect sports posters	𝍸 111	8
Swim	𝍸𝍸𝍸 111	18
Play roller hockey	𝍸𝍸 1111	14
Go fishing	𝍸𝍸 1	11
Garden	𝍸 11	7
Make electrical things	𝍸𝍸𝍸 1111	19

Apples Picked by Families

Family	Number of Pounds Picked
The Smiths	38
The Garcias	$25\frac{1}{2}$
The Lohs	$40\frac{3}{4}$
The Crowleys	51

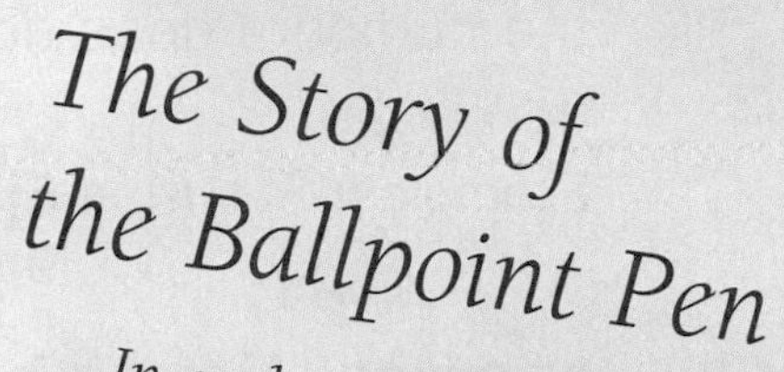

The Story of the Ballpoint Pen

In early times, people wrote with pens made of long reeds. Pens made of goose or swan quills were introduced in Europe around A.D. 600. Workable fountain pens were not invented until 1884. Fountain pens were different from quills in that they had metal points and they could hold their own supply of ink. The first fountain pen was developed by an American named Lewis Waterman.

The idea for a ballpoint pen was patented as early as 1888, but it took a long time to get the pen to work properly. Early ballpoints would only write on rough surfaces. They also tended to leak. The ink in a ballpoint pen is quite thick, but it still must be able to flow smoothly onto the ball so that it can be applied to the paper without skipping and dry almost immediately. It was not until 1944 that two Hungarian inventors named Laszlo and George Biro found the right consistency for the ink and produced a successful ballpoint pen.

Glossary

(*Italicized terms* are defined elsewhere in this glossary.)

A

abacus A counting board used to solve number problems by sliding beads along rods or wires.

acute angle An angle with a measure less than a *right angle.*

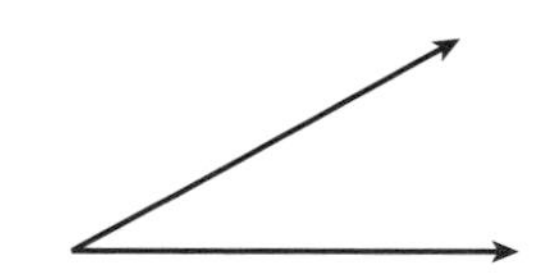

addend A number to be added.

Example: $5 + 4 = 9$
The addends are 5 and 4.

addition An operation on two or more numbers that tells *how many in all.*

Example: $9 + 3 = 12$ ← sum
↑ ↑
addends

A.M. A name for time between 12:00 midnight and 12:00 noon.

angle A figure formed by two *rays* with the same *endpoint.*

area The number of *square units* needed to cover a surface.

array Objects or symbols displayed in rows and columns.

o o o o
o o o o
o o o o
o o o o

Associative Property When adding or multiplying, the grouping of the numbers does not affect the result.

Examples: $3 + (4 + 5) = (3 + 4) + 5$
$2 \times (3 \times 9) = (2 \times 3) \times 9$

average A *statistic* found by adding two or more numbers and dividing their *sum* by the total number of *addends.*

Example: $92 + 84 + 73 = 249$
$249 \div 3 = 83$ ← average

B

bar graph A graph that displays *data* using bars of different heights.

C

capacity The amount a container can hold.

centimeter (cm) A *metric unit* of *length.* (*See* Table of Measures.)

circle A closed, curved *2-dimensional figure.* All the points on the circle are the same distance from the center.

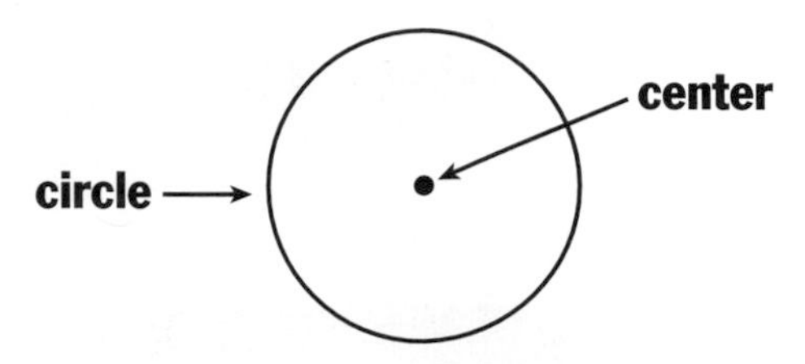

closed figure A figure that starts and ends at the same point.

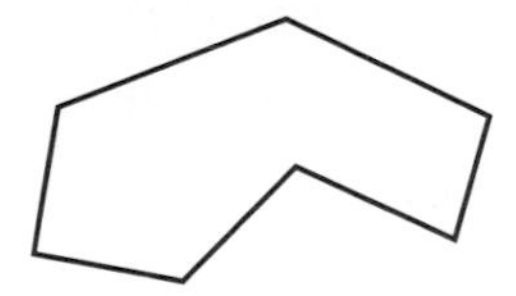

clustering *Estimating* a *sum* by changing the *addends* that are close in value to one common number and then multiplying that number by the number of *addends.*

Example: $47 + 55 + 59$
Estimate: $3 \times 50 = 150$

common denominator A *denominator* that is a *multiple* of the denominators of two or more fractions.

Example: 48 is a common denominator of $\frac{1}{12}$ and $\frac{1}{8}$.

Commutative Property When adding or multiplying, the order of the numbers does not affect the result.

Examples: $5 + 8 = 8 + 5 = 13$
$9 \times 3 = 3 \times 9 = 27$

compatible numbers Changing numbers to other numbers that form a basic fact to *estimate* an answer.

Example:
$133 \div 4$ becomes $120 \div 4 = 30$

composite number A whole number that has *factors* other than itself and 1.

Example: 8 is a composite number. Its factors are 1, 2, 4, and 8.

cone A *3-dimensional figure* whose base is a *circle.*

congruent figures Figures that have the same shape and size.

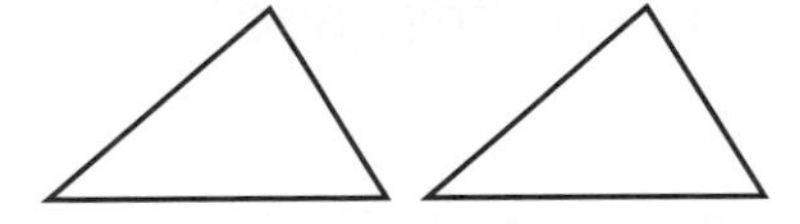

cube A *3-dimensional figure* with six square sides of equal *length.*

cubic unit A unit for measuring *volume.*

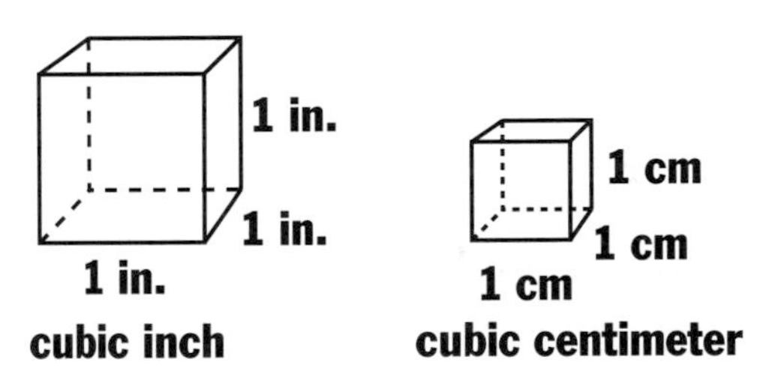

cup (c) A *customary unit* of *capacity.* (*See* Table of Measures.)

customary system A system of measurement whose basic units include *inch, ounce,* and *pound.* (*See* Table of Measures.)

customary unit A unit of measurement in the *customary system.*

cylinder A *3-dimensional figure* with two *faces* that are *circles.*

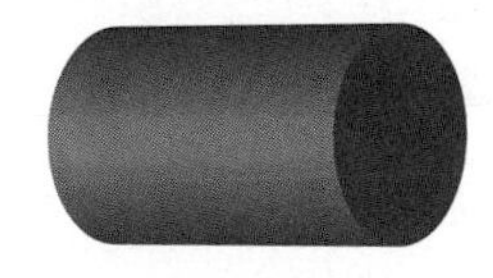

data Information.

decagon A *polygon* with ten sides and ten *angles.*

decimal A number that uses *place value* and a *decimal point* to show tenths, hundredths, and thousandths.

decimal point A period separating the ones and the tenths in a decimal number.

Examples: 0.3, 4.9, 24.752
↑ ↑ ↑
decimal points

decimeter (dm) A *metric unit* of *length.* (*See* Table of Measures.)

GLOSSARY

degree Celsius (°C) A *unit* for measuring *temperature.*

degree Fahrenheit (°F) A *customary unit* for measuring *temperature.*

denominator The number below the bar in a *fraction.*

Example: $\frac{3}{4}$ ← denominator

diagonal A *line segment* that connects two *vertices* but is not a side.

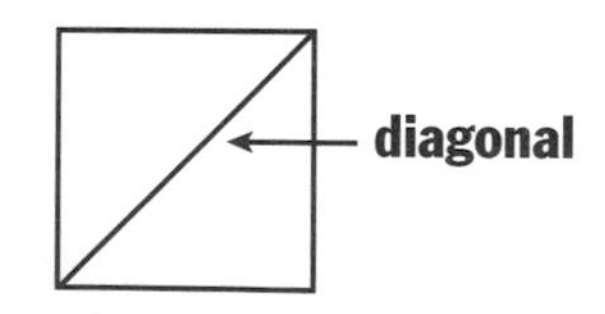

difference The number obtained by subtracting one number from another.

Example: 10 − 6 = 4 ← difference

dividend A number to be divided.

divisible by One number is divisible by another if the *remainder* is 0 after dividing.

division An operation on two numbers that tells *how many groups* or *how many in each group.* Division can also tell *how many are left over.*

Example : quotient
↓
9 R1 ← remainder
divisor → 5) 46 ← dividend

divisor The number by which the *dividend* is divided.

edge A *line segment* where two *faces* of a *3-dimensional figure* meet.

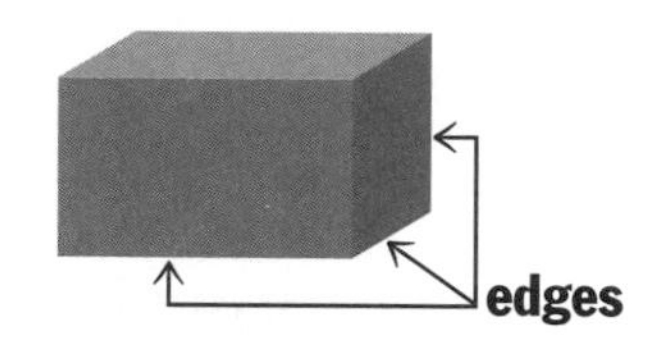

elapsed time The amount of time taken to go from start to finish.

endpoint A point at either end of a *line segment.* The beginning point of a *ray.*

equivalent fractions Two or more fractions that name the same number.

Examples: $\frac{1}{3}$, $\frac{2}{6}$, and $\frac{3}{9}$

estimate To find an answer that is close to the exact answer.

even number A number that ends in 0, 2, 4, 6, or 8.

expanded form A way of writing a number as the *sum* of the values of its digits.

Example: 1,489 can be written as 1,000 + 400 + 80 + 9.

exponent A number that tells how many times a given number is used as a *factor.*

Example: $10^3 = 10 \times 10 \times 10 = 1{,}000$
↑
exponent

face A side of a *3-dimensional figure.*

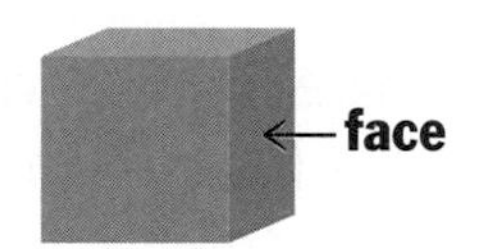

fact family A group of related facts using the same numbers.

Examples: $3 + 2 = 5$ $2 + 3 = 5$
$5 - 2 = 3$ $5 - 3 = 2$

$2 \times 4 = 8$ $4 \times 2 = 8$
$8 \div 2 = 4$ $8 \div 4 = 2$

factors Numbers that are multiplied to give a *product.*

Example: $7 \times 8 = 56$
The factors are 7 and 8.

favorable outcomes Winning results in a *probability* experiment.

flip To move a figure over a *line*; reflection.

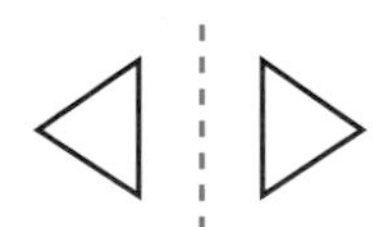

fluid ounce (fl oz) A *customary unit* of *capacity.* (*See* Table of Measures.)

foot (ft) A *customary unit* of *length.* (*See* Table of Measures.)

fraction A number that names part of a whole or part of a group.

Examples: $\frac{2}{3}$, $\frac{7}{10}$, $\frac{1}{100}$

gallon (gal) A *customary unit* of *capacity.* (*See* Table of Measures.)

gram (g) A *metric unit* of *mass.* (*See* Table of Measures.)

Grouping Property When adding or multiplying, the grouping of the numbers does not affect the result.

Examples:
$(3 + 5) + 7 = 15$ $(2 \times 4) \times 5 = 40$
$3 + (5 + 7) = 15$ $2 \times (4 \times 5) = 40$

height The distance from the base to the top of a figure.

heptagon A *polygon* with seven sides and seven *angles.*

hexagon A *polygon* with six sides and six *angles.*

improper fraction A fraction with a *numerator* that is greater than or equal to the *denominator.*

inch (in.) A *customary unit* of *length.* (*See* Table of Measures.)

intersecting lines Lines that meet or cross at a common point.

GLOSSARY

is greater than (>) Symbol to show that the first number is greater than the second.

Example: 439 > 436

is less than (<) Symbol to show that the first number is less than the second.

Example: 852 < 872

key The part of a graph that tells how many items each picture symbol stands for. (*See* pictograph.)

kilogram (kg) A *metric unit* of *mass. (See* Table of Measures.)

kilometer (km) A *metric unit* of *length.* (*See* Table of Measures.)

kite A *quadrilateral* with two pairs of touching *congruent* sides.

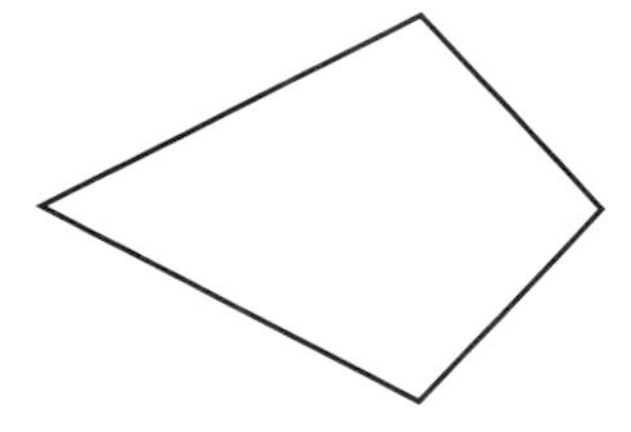

length The measurement of distance between two *endpoints.* (*See also* 2-dimensional figure, 3-dimensional figure.)

line A straight path that goes in two directions without end.

line graph A graph that uses lines to show changes in *data.*

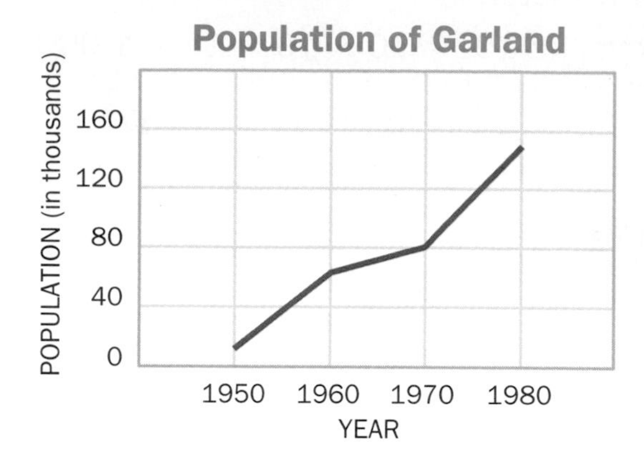

line of symmetry A line on which a figure can be folded so that its two halves match exactly.

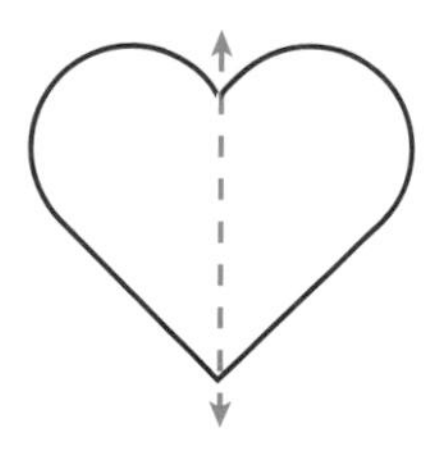

line plot A vertical *graph* that uses Xs above a line to show *data.*

Example:

What is Your Favorite Pet?

Dog	Cat	Fish	Snake
	x		
	x		
x	x		
x	x	x	
x	x	x	x

line segment A straight path that has two *endpoints.*

liter (L) A *metric unit* of *capacity.* (*See* Table of Measures.)

mass A measurement that indicates how much of something there is. It is measured by *kilograms* and *grams.*

median The middle number in a group of numbers ordered from the least to the greatest.

Example: The median of 4, 5, and 7 is 5.

meter (m) A *metric unit* of *length.* (*See* Table of Measures.)

metric system A decimal system of measurement whose basic units include *meter, liter,* and *gram.* (*See* Table of Measures.)

metric unit A unit of measurement in the *metric system.*

mile (mi) A *customary unit* of *length.* (*See* Table of Measures.)

milliliter (mL) A *metric unit* of *capacity.* (*See* Table of Measures.)

millimeter (mm) A *metric unit* of *length.* (*See* Table of Measures.)

mixed number A number that has a whole number and a *fraction.*

Example: $8\frac{5}{6}$

mode The number or numbers that occur most often in a collection of *data.*

Example: 2, 6, 1, 6, 4, 8, 6
The mode is 6.

multiple The *product* of a number and any whole number.

Example: 8 is a multiple of 4 because $2 \times 4 = 8$.

multiplication An operation that tells *how many in all* when equal groups are combined.

Example: $5 \times 8 = 40$
↑ ↑ factors; ↑ product

numerator The number above the bar in a *fraction.*

Example: $\frac{3}{5}$ ← numerator

obtuse angle An angle with a measure that is greater than a *right angle.*

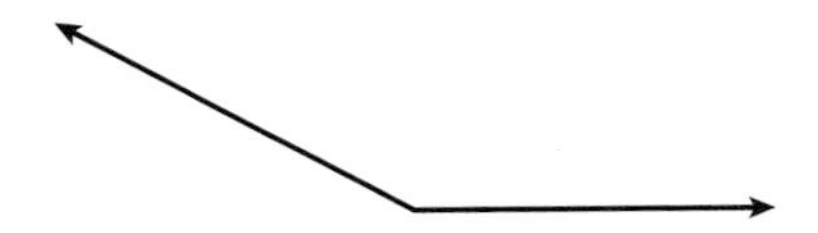

octagon A *polygon* with eight sides and eight *angles.*

odd number A number that ends in 1, 3, 5, 7, or 9.

open figure A figure that does not start and end at the same point.

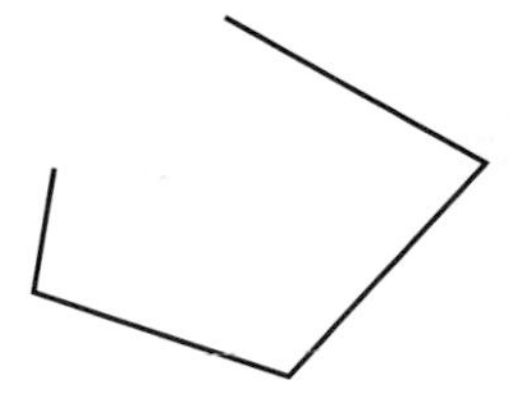

ordered pair A pair of numbers that gives the location of a point on a graph, map, or grid.

GLOSSARY

Order Property When adding or multiplying, the order of the numbers does not affect the result.

Examples: $8 + 9 = 17$ $\quad 4 \times 5 = 20$
$9 + 8 = 17$ $\quad 5 \times 4 = 20$

ordinal number A number used to tell order or position.

Example: second

ounce (oz) A *customary unit* of *weight.* (*See* Table of Measures.)

parallel lines Lines that never intersect.

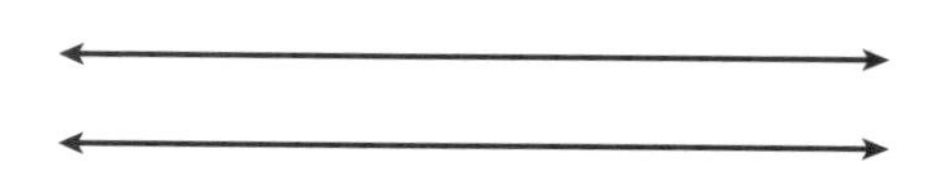

parallelogram A *quadrilateral* with both pairs of opposite sides parallel.

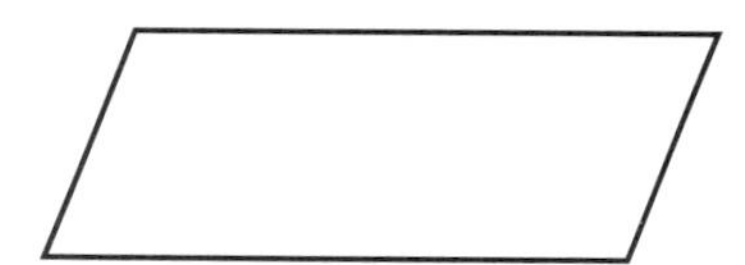

pattern A series of numbers or figures that follows a rule.

Examples: 1, 3, 5, 7, 9, 11, . . .

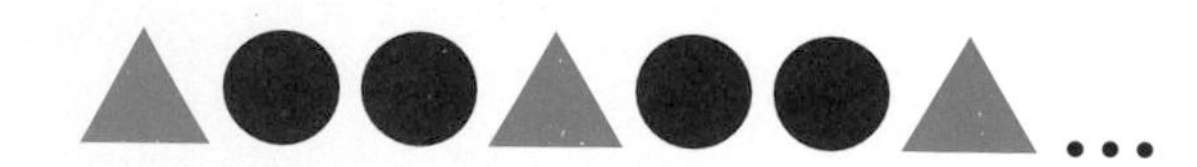

pentagon A *polygon* with five sides and five *angles.*

perimeter The distance around a *closed figure.*

period Each group of three digits in a *place-value* chart.

Example: 527,000

Thousands Period			Ones Period		
Hundred Thousands	Ten Thousands	Thousands	Hundreds	Tens	Ones
5	2	7	0	0	0

perpendicular lines Lines that intersect to form square corners.

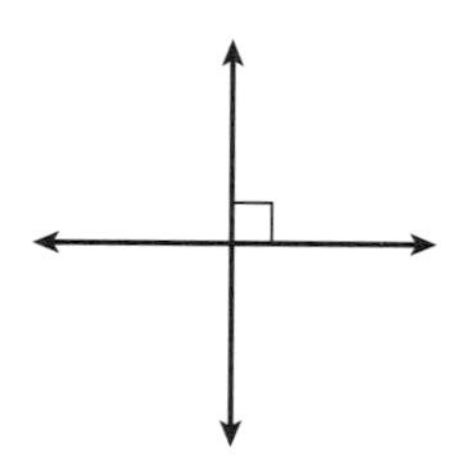

pictograph A graph that shows *data* by using picture symbols. A *key* tells how many items each picture symbol stands for.

pint (pt) A *customary unit* of *capacity.* (*See* Table of Measures.)

place value The value of a digit based on its position in a number.

P.M. A name for time between 12 noon and 12 midnight.

polygon A closed *2-dimensional figure* formed by *line segments.* The sides do not cross each other.

possible outcome Any of the results that could occur in a *probability* experiment.

pound (lb) A *customary unit* of *weight.* (*See* Table of Measures.)

prime number A *whole number* greater than 1 with only itself and 1 as *factors.*

Examples: 7 is a prime number. 2 is the only even number that is prime.

prism A *3-dimensional figure* with two parallel *congruent* bases and *rectangles* or *parallelograms* for *faces.*

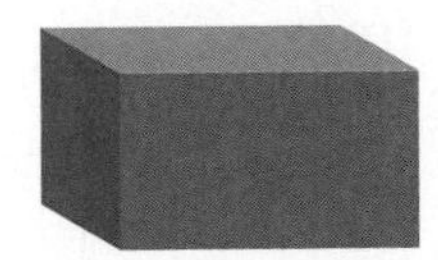

probability A number from 0 to 1 that measures the likelihood of an event happening.

product The result of *multiplication.*

Example: $6 \times 8 = 48$ ← product

pyramid A *3-dimensional figure* that is shaped by *triangles* on a base.

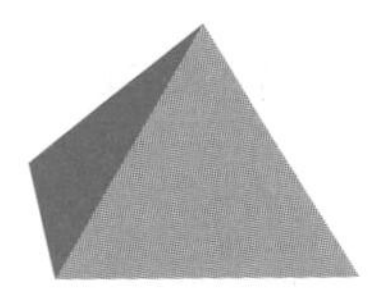

quadrilateral A *polygon* with four sides.

quart (qt) A *customary unit* of *capacity.* (*See* Table of Measures.)

quotient The result of *division.*

Example: $35 \div 7 = 5$ ← quotient

range The *difference* between the greatest and the least numbers in a group of numbers.

ray A *2-dimensional figure* that has one *endpoint* and goes on forever in one direction.

rectangle A polygon with four sides and four square corners.

rectangular prism A *3-dimensional figure* with six rectangular *faces.*

regroup To name a number in a different way.

Example: 23 can be regrouped as 2 tens 3 ones or as 1 ten 13 ones.

remainder The number left over after dividing.

Example: $43 \div 7 = 6$ R1← remainder

rhombus A *parallelogram* with all four sides the same *length.*

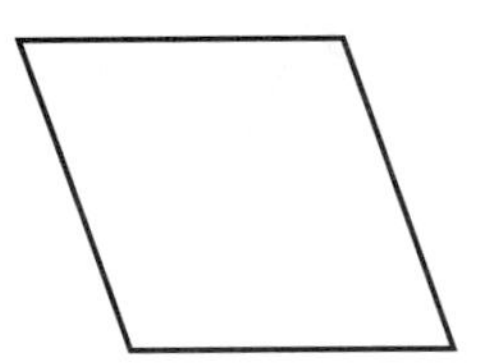

right angle An *angle,* or square corner, formed by *perpendicular lines.*

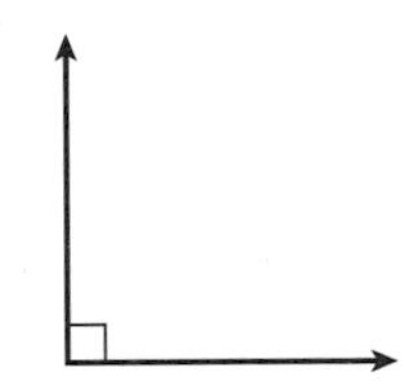

rounding (round) Finding the nearest ten, hundred, thousand, and so on.

Example: 868 rounded to the nearest hundred is 900.

scale Marks that are equally spaced along a line and are used to measure.

similar shapes Figures that are the same shape but are different sizes.

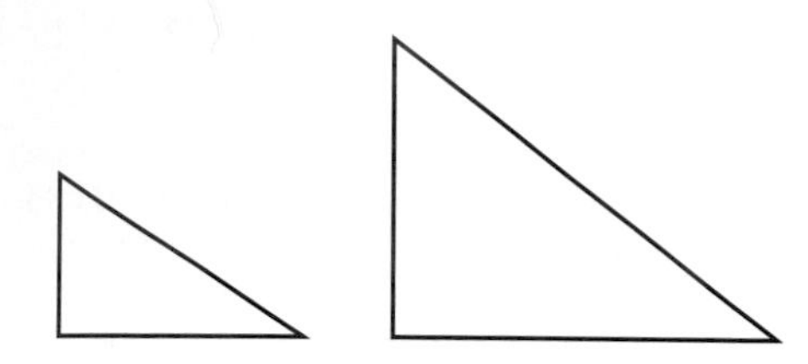

simplest form A *fraction* in which the *numerator* and *denominator* have no common *factor* greater than 1.

skip-count To count by twos, threes, fours, and so on.

Examples:
2, 4, 6, 8, 10, . . . 3, 6, 9, 12, . . .

slide To move a figure along a *line*; translation.

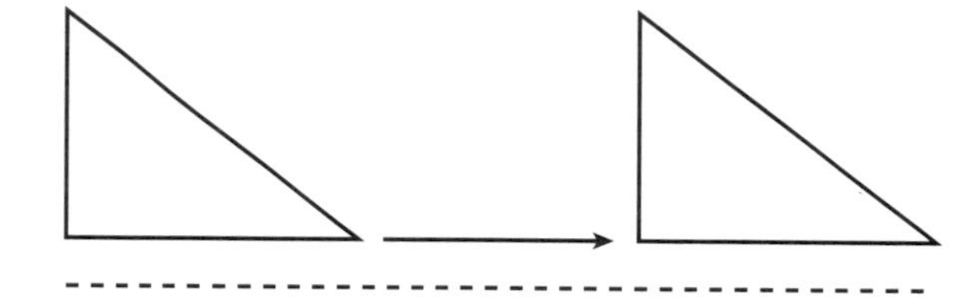

sphere A *3-dimensional figure* that has the shape of a round ball.

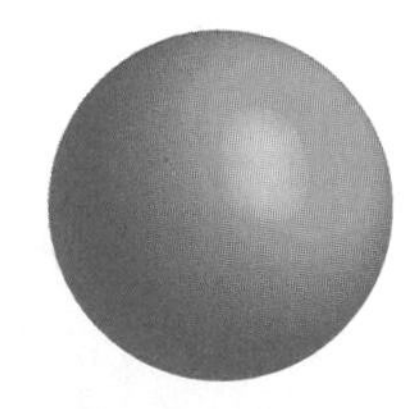

spreadsheet A computer program that arranges *data* and formulas in table form.

square A *2-dimensional figure* that has four equal sides and four square corners.

square pyramid A *pyramid* whose base is a *square.*

square unit A unit for measuring *area.*

Examples: square inch (in.^{2}), square foot (ft^2), square centimeter (cm^2), and square meter (m^2)

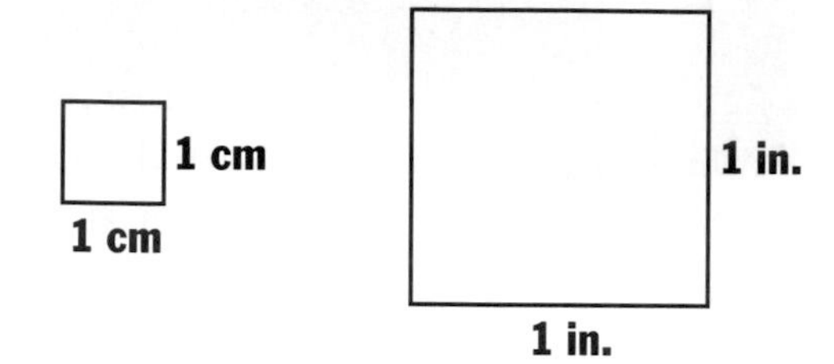

standard form The usual or common way to write a number.

statistics The collecting and arranging of *data* on a particular subject.

subtraction An operation on two numbers that tells *how many are left* when some are taken away. Subtraction is also used to compare two numbers.

Example: $14 - 8 = 6$ ← difference

sum The result of *addition.*

Example: $9 + 6 = 15$ ← sum

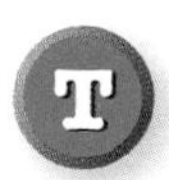

temperature A measurement that tells how hot or cold something is.

tessellation Related shapes that cover a flat surface without leaving any gaps; for example, the design on a checkerboard.

3-dimensional figure A figure that has *length, width,* and *height.*

time line A *line* that shows times or dates as points along a *line segment.*

trapezoid A *quadrilateral* with exactly one pair of *parallel* sides.

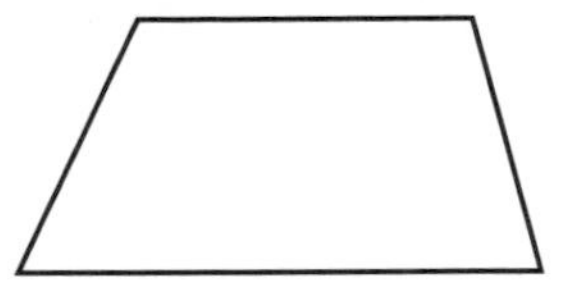

triangle A *polygon* with three sides and three *angles.*

triangular prism A *prism* whose opposite *faces* are *triangles.*

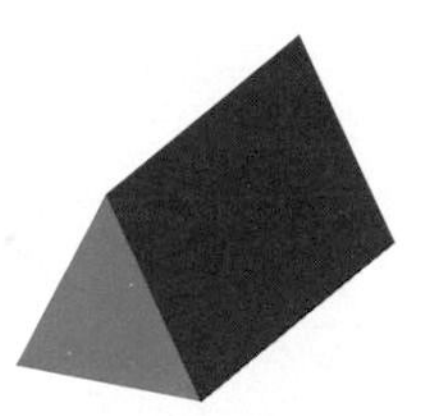

triangular pyramid A *pyramid* whose base is a *triangle.*

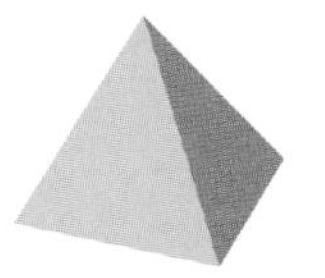

turn To rotate a figure around a point; rotation.

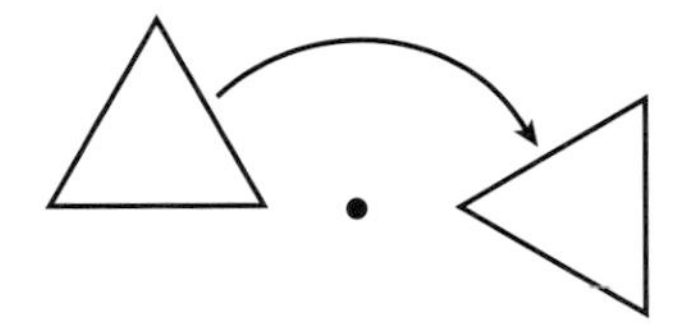

2-dimensional figure A figure that has only *length* and *width.*

variable A symbol used to represent a number or group of numbers.

vertex The common point of the two *rays* of an *angle,* two sides of a *polygon,* or three or more *edges of a 3-dimensional figure.*

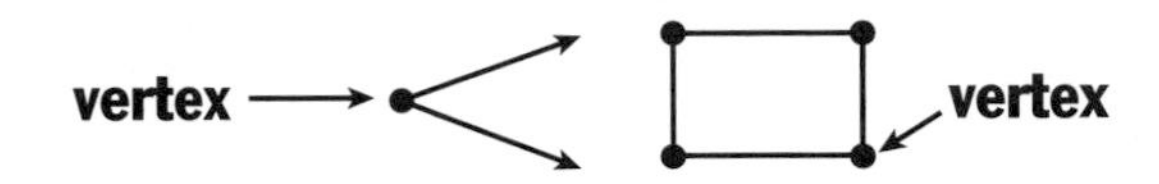

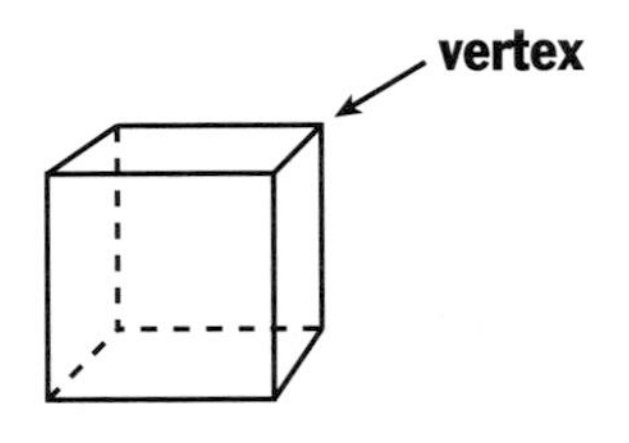

volume The number of *cubic units* that fit inside a *3-dimensional figure.*

weight A measure of how heavy something is.

width The measurement of the shorter of the sides of a *rectangle* that is not a *square.* (*See also* 2-dimensional figure, 3-dimensional figure.)

yard (yd) A *customary unit* of *length.* (*See* Table of Measures.)

GLOSSARY

Table of Measures

Time

60 seconds (s) = 1 minute (min)
60 minutes (min) = 1 hour (h)
24 hours = 1 day (d)
7 days = 1 week (wk)
12 months (mo) = 1 year (y)
about 52 weeks = 1 year
365 days = 1 year
366 days = 1 leap year

Metric Units

LENGTH	1 centimeter (cm) = 10 millimeters (mm)
	10 centimeters = 1 decimeter (dm)
	10 decimeters = 1 meter (m)
	1,000 meters = 1 kilometer (km)
MASS	1 kilogram (kg) = 1,000 grams (g)
CAPACITY	1 liter (L) = 1,000 milliliters (mL)

Customary Units

	1 foot (ft) = 12 inches (in.)
LENGTH	1 yard (yd) = 36 inches
	1 yard = 3 feet
	1 mile (mi) = 5,280 feet
	1 mile = 1,760 yards
	1 pound (lb) = 16 ounces (oz)
WEIGHT	1 cup (c) = 8 fluid ounces
CAPACITY	1 pint (pt) = 2 cups
	1 quart (qt) = 2 pints
	1 gallon (gal) = 4 quarts

Symbols

$<$	is less than	$^{\circ}$	degree	$\overrightarrow{AB}$	ray *AB*
$>$	is greater than	$\overleftrightarrow{AB}$	line *AB*	$\angle ABC$	angle *ABC*
$=$	is equal to	$\overline{AB}$	line segment *AB*	(5, 3)	ordered pair 5, 3

Index

INDEX

INDEX

INDEX